Ingo Klöckl

Handbook of Colorants Chemistry

Also of Interest

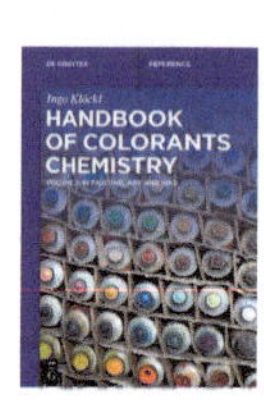

Handbook of Colorants Chemistry
Volume 2: in Painting, Art and Inks
Klöckl, 2023
ISBN 978-3-11-077700-0, e-ISBN 978-3-11-077712-3

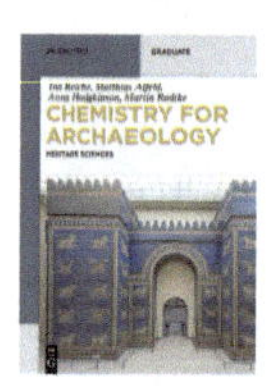

Chemistry for Archaeology
Heritage Sciences
Reiche, Alfeld, Radtke, Hodgkinson, 2020
ISBN 978-3-11-044214-4, e-ISBN 978-3-11-044216-8

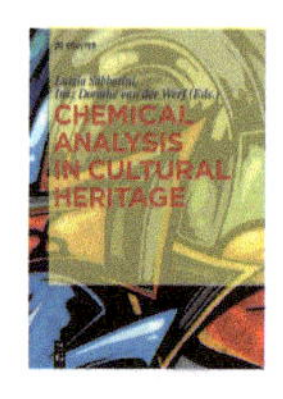

Chemical Analysis in Cultural Heritage
Sabbatini, van der Werf (Eds.), 2020
ISBN 978-3-11-045641-7, e-ISBN 978-3-11-045753-7

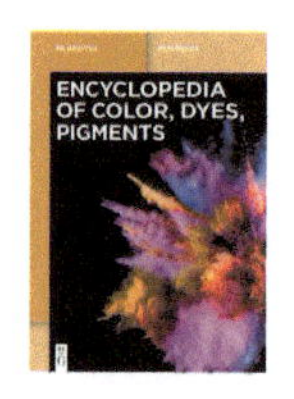

Encyclopedia of Color, Dyes, Pigments
Pfaff, 2022
Volume 1. Antraquinonoid Pigments – Color Fundamentals
ISBN 978-3-11-058588-9, e-ISBN 978-3-11-058807-1
Volume 2. Color Measurement – Metal Effect Pigments
ISBN 978-3-11-058684-8, e-ISBN 978-3-11-058710-4
Volume 3. Mixed metal Oxide Pigments – Zinc Sulfide Pigments
ISBN 978-3-11-058686-2, e-ISBN 978-3-11-058712-8

Ingo Klöckl

Handbook of Colorants Chemistry

Volume 1: Dyes and Pigments Fundamentals

DE GRUYTER

Author
Dr. rer. nat. Ingo Klöckl
St. Leoner Strasse 16
68809 Neulußheim
Germany
i.kloeckl@2k-software.de

ISBN 978-3-11-077699-7
e-ISBN (PDF) 978-3-11-077711-6
e-ISBN (EPUB) 978-3-11-077728-4

Library of Congress Control Number: 2022942567

Bibliographic information published by the Deutsche Nationalbibliothek
The Deutsche Nationalbibliothek lists this publication in the Deutsche Nationalbibliografie; detailed bibliographic data are available on the Internet at http://dnb.dnb.de.

Cover image: aga7ta / iStock / Getty Images Plus
Typesetting: VTeX UAB, Lithuania
Printing and binding: CPI books GmbH, Leck

www.degruyter.com

For all those who, through curiosity, have discovered or may discover their interest in the fascinating field of chemistry

Foreword

Nothing shows us the beauty of our world as vividly as its colors. For our distant biological ancestors, there were predominantly smells. However, at some point in our long evolution into modern humans, we dared to swap the dull magic realm of scents for the bright precision of our eyes. And yet colors are also a magical realm that holds many a secret. Unlike shape, density, or surface texture, color is not an inherent property of an object but only our perception of how the object reflects or absorbs visible light. Moreover, our eyes show us only a tiny fraction of the immense spectrum of electromagnetic radiation that fills our universe. The wavelengths, and thus the frequencies, of this spectrum span 16 orders of magnitude—from the 10 to 20 km long radio waves of some military transmitters to the gamma rays of imploding galaxies, which are only a thousandth of a nanometer short. Life on our planet mainly registers wavelengths between 300 and 1000 nm. This range includes the ultraviolet, with wavelengths below 400 nm, which unlike many insects, we cannot see; the range from blue to green to red, that is, from 400 to about 750 nm, which means light to us; and finally, infrared rays, with wavelengths above 800 nm, which some animals perceive as light but we perceive only as heat.

Ultraviolet was probably the first color that life on our planet saw. This spectral part of sunlight meant danger, as it destroyed many biological building blocks. Cells developed a sensor for ultraviolet and blue light that controlled the direction of rotation of their flagella to avoid these dangerous rays. Since these flagella act as propulsion propellers, the cells could now not only see the harmful short-wave light but also avoid it. A descendant of this blue light sensor is still found today in many primitive protozoa.

This ingenious blue light sensor probably also served the cells as a construction manual for a solar collector, thanks to which they could feed on the energy of sunlight. Cells shifted the blue sensor's absorption to yellow-orange to capture as much of the sunlight's energy as possible. The cells coupled this solar collector to a system that converted the captured light into chemical energy. With its help, the cells could now power energy-hungry processes such as growth, cell division, and movement, or synthesizing fat, sugar, and proteins. This primitive photosynthesis is still found today in some single-celled organisms that thrive in the salt-rich margins of the Dead Sea or spoiled cured fish. Ultimately, however, this form of photosynthesis proved to be a dead end because it did not efficiently convert light from the sun into chemical energy. When later cells used chlorophyll as a solar collector, ushering in modern photosynthesis, photosynthesis that evolved from the blue light sensor remained limited to a few primitive single-celled organisms.

So, life learned very early to see the world in two colors—blue and yellow-orange. And now that it had seen color, it no longer wanted to do without it. With their extensive and information-rich genetic material, the complex modern cells created three, four, or even five different variants from the primitive blue light sensor, which opened

https://doi.org/10.1515/9783110777116-201

up a vast and differentiated color spectrum for them. Even more, these modern cells were able to couple the signals of these different color sensors separately to increasingly complex nervous systems.

Our eyes are equipped with five different light sensors, all closely related chemically and probably descended from the primordial blue light sensor mentioned earlier. One of these sensors is not used for vision but for the daily calibration of our "circadian" body clock. Another sensor is found in the rod cells of our retina. This sensor is susceptible to light and, therefore, we use it in dim light. However, this high light sensitivity comes at a price because our retinal rods do not detect color or fine detail. In bright light, we use three color sensors in the cone cells of our retina—one for blue, one for green, and one for red. These sensors are not very sensitive to light, but they show us fine detail—and color. Since each of these three color sensors can detect about a hundred different intensities of color, and our brain compares the signals from the three sensors, we can see not just three but one to two million colors. Older animals, such as insects and birds, have up to five different color sensors and cannot only distinguish many more colors than we humans can, but some can also see ultraviolet or infrared light to which we are blind. When the first mammals evolved, they mainly hunted at night, leaving some of their color sensors to atrophy, leaving only two of them. Almost all mammals—such as dogs, horses, cats, and cows—therefore see only about 10000 different colors—about the same as "color blind" humans. Only when intelligent apes wanted to distinguish ripe from unripe fruit against the background of multicolored leaves did they again develop a third color sensor, which allowed them to see the world in a new blaze of color. So, humans and our close relatives, the great apes, are the only mammals that can see millions of colors.

In this impressive book, Ingo Klöckl describes the magic realm of colors from a chemist's perspective. The synthesis of modern dyes with intoxicating color depth and impressive stability was one of the great triumphs of nineteenth and twentieth century chemistry, and the development of rewritable digital data carriers or catalysts for light-driven water splitting suggests that the time of color chemistry is far from over. Ingo Klöckl describes the bewildering variety of dyes available today and gives us detailed information on how they can be produced, categorized, and compared with each other. This book is a masterpiece, a true magnum opus that reveals to us in each chapter a new wonder from the world of colors. The wealth of information it imparts to us is almost mind-boggling, yet it is an exciting read for anyone who is no stranger to chemistry. The book also builds a most welcome bridge between science and art, which have become increasingly distant from each other in recent centuries, forgetting their common roots. May not only natural scientists but also painters and art scholars pick up this book and lose themselves in the magical realm of colors.

Basel

Dr. Gottfried Schatz†

Foreword to the English edition

Dear readers, painting scientists, and researching painters, when the German edition appeared, I never expected it would appeal to so many fellow human beings since the balancing act between painterly observations and chemical-physical theories presupposes a profound understanding in many areas or at least interest. However, I was proven wrong, so this English edition will hopefully accompany and support your work in the vast, intricately interwoven field of art and technology.

Unfortunately, the unhelpful division into natural science and humanities also divides the view of our world as the gods of the Greeks divided the spherical people. Bucklow's work [10] showed me how a holistic, Platonically oriented understanding of the science of painting looked in the Middle Ages and embraced (al-)chemistry and painting equally.

My deepest hope is that you will also see this bridging of art and science as a contribution to an overall humanistic understanding of the world, as was familiar to many great minds of science.

https://doi.org/10.1515/9783110777116-202

Acknowledgment

My first and most important thanks go to my enchanting wife, Claudia, who once again had to spend a large part of her time with a fanatically lecturing author and who, through constant gentle urging, got me to finish the incubation of the book. Only this made the publication possible at all. Moreover, most important, she dedicated vast amounts of her time to translating the text into readable English, urging me to improve and clarify numerous statements. She introduced me to the secrets of translation-oriented writing and the benefits of a concise language.

Sadly, the author of the foreword, Dr. Schatz, passed away, to whom I will forever be grateful for the insightful discussions and his foreword, which so well expressed the magic of color. Furthermore, I would like to thank Mr. D. Widmer for important book recommendations and information on writing inks, Dr. S. Hunklinger for an engaging discussion on the color of semiconductors, Dr. T. Vilgis, Dr. B. Schneppe, and G. Bosse for a discussion of the topic of egg white binders and clarea, Dr. W. Müller for his feedback on the composition of acrylic paints, Dr. B. Born and Mrs. R. Ardal-Altun for a long, informative and entertaining telephone conversation on the production of artists' papers, Dr. G. Kremer for many valuable suggestions, improvements and information from the practice of a paint manufacturer, Dr. K.-O. Schäfer and Dr. W. Thiessen for information on ink production, and Dr. G. Schatz for sending additional material. I would especially like to thank Dr. Kremer, and again Dr. Schatz, for their positive evaluation of the manuscript. They encouraged me, once as a practical color chemist and once as a versatile natural scientist, to pursue the book's aim. They were right, as the friendly and positive letters from colleagues and experts prove, and I thank them for their interest and constructive comments.

I would also like to thank the artists Mrs. S. Steinbacher and Mr. H. Karlhuber, with whom I learned old-master and contemporary painting practice and who were thus "question suppliers."

Finally, I would like to thank Dr. R. Sengbusch, Mrs. Bentkuvienė, and the production team of De Gruyter-Verlag as competent publication partners.

https://doi.org/10.1515/9783110777116-203

Table of contents for volume 1

Table of contents for volume 2

1 Introduction

Why write a book on such a particular topic as the chemistry of painting? Since I paint myself, the question of the nature of the materials used was evident to me, and it was with enthusiasm that I began my research. However, I quickly discovered that most of the known books dealt more with the technique of painting and less with the nature of the materials, mentioned only in passing. The question “What is in the tube of yellow oil paint?” was replaced by “What causes yellow ocher or chrome yellow to appear yellow?” And more questions followed: “Why does the linseed oil in the tube form a clear film? And why is the chrome yellow purer and more intensive than the yellow ocher? What happens when paintings are damaged by age?”

What I imagined was a book not about painting but the *chemistry* of painting; to write it, I had not only to work through a whole physical-chemical library but also unexpected sections such as dairy farming, adhesives, professional cake baking, and the art of restoration.

It was fascinating not only to gain deeper insight into the nature of my painting materials but also to experience the great wealth of subject areas that are necessary to explain the basic materials and principles of painting, encompassing quantum mechanics, solid-state and semiconductor physics, inorganics and organics, biochemistry, and the chemistry of natural products and colloids. I had to read books on the technology of dairy products as well as technical articles on frying oil used for fast food or the problem of discoloration of olives and apples when cut with iron knives. I became knowledgeable about the Kamares style of Minoan vases and the pressure dependence of the *s* orbitals of lead in certain minerals in the depths of the earth. Some books took me back to 1831 or even into the Middle Ages. Some research results were only obtained in papers as late as 2010; many questions are still open or waiting for a conclusive clarification. Several research projects aim to answer them in the future. Some results were only possible with the most modern methods of analysis. Using the latest high-tech analytics in museum laboratories has significantly expanded our knowledge in recent years by gaining deeper insights; so far, seemingly proven facts may even need to be adapted.

There were some surprising discoveries. After reviewing some medieval paint recipes, I learned how I could obtain basilisks if I were in need of them—astonishing was also the high number of articles revolving around modern methods to analyze the composition of contemporary artists’ materials whose origins lie not precisely in the dark past. For example, in the case of acrylic paints or modern inks, the accompanying information contains surprisingly vague statements such as “an anionic acrylate” or “linseed oil-based.” These are products guarded by company or production secrets. Even conservators then have to go to an analytical laboratory to find out how they best treat their treasures.

https://doi.org/10.1515/9783110777116-001

My intention with this book is to summarize the widely scattered knowledge to provide comprehensive information about the nature of painting materials to natural scientists interested in art and art lovers curious about the nature of materials.

Dear reader, whatever your interests are, let me show you the path leading along the boundary between painting and science. Accompany me on tour through both fields using ►Figure 1.1 as our map that illustrates the sights along our way. Let us divide our route into several stages:

- First, we will become acquainted with the basic physical process of light-matter interaction (from ►Section 1.5 on), followed by the chemical principles of color (►Chapter 2). To do so, we will look into four fundamental mechanisms of chemically induced color impressions.
- Next, we look at the colorants as the actual means of artistic expression. Based on the knowledge gained so far, we discuss inorganic (►Chapter 3) and organic (►Chapter 4) pigments relevant to art and their causes of color. Likewise, we give an overview of dyes suitable for graphics and drawing (►Chapter 5).
- Painters do not use only colorants but, in fact, whole paint systems. For colorants to be applicable as expected by a particular technique and to form durable layers of paint, they must integrate into complex systems, e. g., oil or acrylic paints. We will first learn about the components of such systems, their functions, and modes of action (►Chapter 6) and then proceed to examine actual paint systems in detail, including painting such as oil, acrylic, and watercolor, as well as drawing, e. g., pastels or pencils (►Chapter 7).
- Finally, we conclude with an examination of materials and paint systems of ink-based graphics, such as writing, drawing and printing, and modern media such as inkjet printers or color laser printers (►Chapter 8).

What makes the field of painting chemistry very exciting is the interaction of chemical and physical factors to induce the final visible color of a pigment.

- Chemical causes. The molecular structure is primarily responsible for the color of a pigment or a dye. Type, oxidation state, and linkage of atoms determine if, at which wavelengths, and how intensively an atom or molecule absorbs visible light, and thus appears colored.
- In pigments, several physical phenomena join this primary cause. Large, spatially extended objects are subject to scattering, favoring specific wavelengths and emission geometries. Other processes such as developing surface plasmons can add new absorption bands. These phenomena usually depend on the pigment particles' size, morphology, and shape.
- In addition, supramolecular phenomena participate in creating a final color impression. Individual molecules or ions can mutually interact in solid or crystallized pigment particles, changing or amplifying absorption bands, or creating new ones. Uniform crystalline structures facilitate such interactions, while disturbances of a regular structure, in contrast, suppress them. Also, broad distribu-

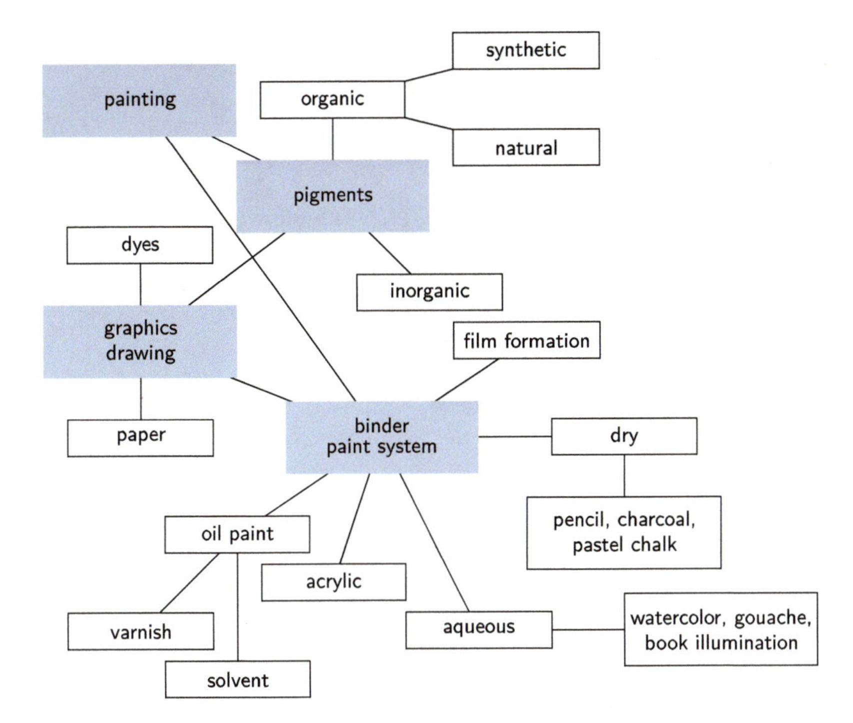

Figure 1.1: Tour overview: The starting point is painting or graphics and drawing. To understand color in these applications, we will travel to the main sights along our route: pigments, dyes, binders, and their characteristics.

tions of values for a given property, such as particle diameter, have adverse effects on the color quality of pigments, usually yielding dull colors.

►Figure 1.2 illustrates various factors using the example of red ocher:

- The primary cause of color from a chemical point of view is a ligand field (LF) transition within an Fe^{III} ion, located in the blue and green-yellow spectral range, and an intensive charge transfer (CT) transition in the near-UV. Both transitions cause blue to green light absorption, resulting in a yellow to red color impression. However, compared with yellow ocher, which has almost the same chemical structure, why is the one red and the other yellow? What we see is the influence of other factors.
- Sintered red ocher acquires a distinct cool purple tint (colcothar or Caput Mortuum), while raw red ocher shows an orange-red to red color. Since sintering does (at least in this case) not change the chemical composition, something different must go on here. As a pure physical phenomenon, electron oscillations can occur throughout the ocher particle, so-called plasmons, giving rise to intensive absorption bands (SP). The location and fine structure of this band depend on the size and shape of the particle.

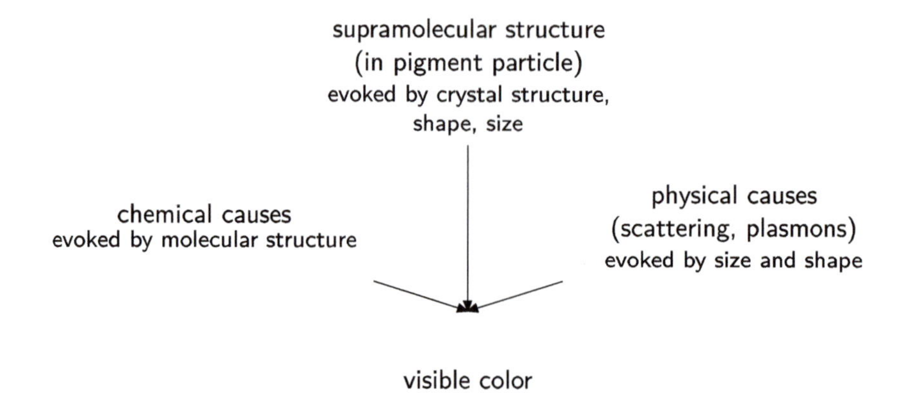

(a) Interaction of chemical, physical, and crystallographic causes to induce the visible color of pigments.

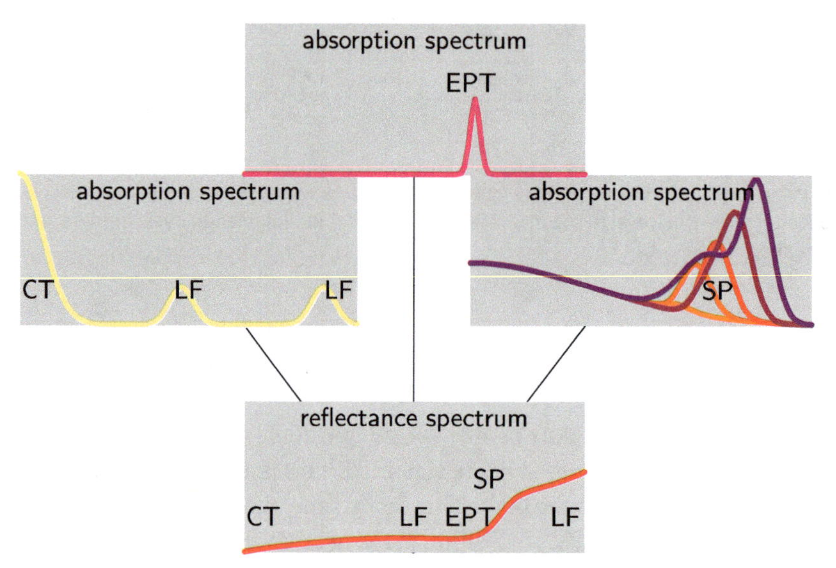

(b) Individual effects. 1. The primary color is determined by bound electrons in the molecular structure, i. e., CT and LF transitions in the Fe^{III} cation. 2. Pigment grains of a specific diameter develop absorption bands from surface plasmons (SP), distinguishing between coarse-grained purple and fine-grained orange-red ocher. 3.The supramolecular interaction of many iron ions in a pigment particle gives rise to an electron pair transition (EPT) between neighboring Fe^{III} cations, which is decisive for distinguishing between red and yellow ocher.

Figure 1.2: Interaction of individual chemical and physical phenomena when inducing a color expression, exemplified by red ocher.

- The spatial proximity of iron ions in the red ocher's crystal lattice, determined by its crystallographic structure, results in a magnetic coupling, which manifests itself as a new absorption band (EPT) in the green-yellow range. The EPT transition effectively causes absorption of yellow light and intensifies the red color impression considerably. Due to the slightly different crystal lattice of yellow ocher, this EPT transition occurs in the blue-green spectral range so that a pure yellow color results. We see that a supramolecular phenomenon, the interaction of independent iron ions, causes the distinction between yellow and red ocher.

1.1 Further (and underlying) literature

i

My goal cannot be to write textbooks on all of the fields we have to enter. I will present the basic statements of these subjects as far as necessary to understand aspects of painting. For interested readers, I recommend the following excellent works for further study, although this selection cannot be exhaustive and has to omit many good publications:

- A comprehensive introduction to all chemical and physical aspects of color is available [311]. Focusing on physics are [311, 312, 314, 321] and [322]. The chemistry of color is covered by [2–7], [206, 207, 211], presenting the chemical theoretical and spectroscopic fundamentals. Group theory relating to physics and chemistry is covered by [221] and [222].
- On the chemistry of modern colorants, there are some standard works available [11–16], for (textile) dyeing [18–21, 23].
  Pigments throughout art history are summarized by [47–54, 70], and medieval book illumination is the focus of [108]. Historical considerations of prehistoric and early times are presented in [109]. Specializing in printing and ink colorants are [148–156, 164] and [785].
- The chemistry and composition of artists' inks and methods of investigation are the subject of [74, 75, 90, 95–97, 99, 102–107, 116–121] and [186, 188]. Specializing in binders are [182–185, 190–192] and the standard work [181] as well as [188]. Valuable information on colloid and polymer chemistry is provided by [212–220].
- Representatives of the wealth of literature on natural products are [269–272, 281, 292].
- In the field of food, [277] and [278] are worth to be mentioned, and specialized works are [274–276, 279, 280].
- For minerals (ligand field theory), [332] as well as [7, 223, 224, 226–228, 233] are of interest.
- Chemical-technical fields are covered by [199–202].

The bibliography is organized thematically and provides additional resources and original works beyond those mentioned above.

On our route, I will sometimes fall back on time arrows or other graphical representations of the chronological development of the discoveries. However, I do not claim absolute art historical accuracy. You can refer to the principal and collective works mentioned above and the references given in the tables. Instead, I would like to depict an overview of the chemistry behind the inventions and discoveries in art and represent the primary traces of its change over time. We shall not be concerned with 20 or 50 years more or less. Otherwise, we would have to define the precision of our date, which means to answer the following questions: when was it first mentioned and how (oral? or written?), when occurred the first find of a painting, a discovery, or manufacture? What if a pigment initially served only as a color glaze for over a thousand years before a painter found it useful for himself? What if, unbeknownst to us, it was initially sold only by a small Parisian paint merchant for a long time before the public took notice of it? You see, this is where diversions arise, trying to lure us away to other exciting stories. If you feel tempted, follow the interactive time arrow caringly prepared by the professional historians of the Metropolitan Museum of Art [477].

1.2 Pigments as the basis of painting

The essential element of any artistic expression is color, striking and flamboyant in painting or extremely subdued in monochrome drawings and prints. Therefore, let us look at the beginning at some artists' palettes used at different times by artists to bring color into their work. The aim is to present an overview of the essential pigments and their changes over time; the primary sources for the references given in the tables are [11, 12, 14–16, 41, 47–55, 57, 63, 64, 67–70, 74, 75, 96, 97, 99, 105, 106, 113–121, 127–132, 137]. If you are interested in how scientists obtain such information, [103] provides a valuable source.

1.2.1 Pigments of the ancient and early medieval world

Since early times, humankind desired to change their living place with the help of colors, and to express spiritual ideas, as the Neolithic cave paintings show in overwhelming beauty. The sedentary groups of the Neolithic and the peoples of the Orient decorated their houses and palaces with colorful illustrations. From Minoan times, 2000 years before Christ, we find on Crete sensual portraits of the palace ladies and maritime landscapes that the brush of a contemporary artist could have just as well created. Likewise, in Egypt, ancient Greece, and Italy, the peoples of antiquity created world-class works of art full of color and perspective. The colorants used for this purpose (►Tables 1.1 and 1.2) retain their significance into the early Middle Ages, representing the four fundamental chemical causes of color, ►Chapter 2.

The most intensively colored, pure white, yellow, and red pigments are *semiconductors* of the lead group: lead white (white) $PbCO_3 \cdot Pb(OH)_2$, massicot (yellow) PbO, red lead (orange-red) Pb_3O_4, and litharge (red) PbO. The sulfides, cinnabar or synthetic vermilion (red) HgS, and orpiment (yellow) As_2S_3 are also semiconductors. They show, in principle, intensive, pure colors, but only the hues white, yellow, and red, ►Section 2.2. They occur naturally as sulfides or can be readily produced from natural mineral sources by the metallurgical means of antiquity, representing early achievements in quasi-industrial chemical synthesis.

d block metals (transition metals) can exhibit coloring that spans the entire range by electron transitions between *partially occupied d orbitals*, ►Section 2.3. However, this mechanism is generally "forbidden" and of low intensity; the pigments are therefore weak in color. Nevertheless, it is crucial for the existence of green- and blue-colored compounds, such as some copper minerals, extracted from copper ores since the Bronze Age.

Table 1.1: Overview of commonly used pigments in the prehistoric and early medieval world. The arrows indicate the appearance ↑ and the disappearance ↓ of pigments. Italicized terms indicate less commonly used pigments. For early antiquity, such a classification is not possible throughout.

White	Yellow/orange/brown	Red/purple	Green	Blue	Black
Paleolithic, 400000 BC–10000 BC [110–115, 476, 478, 479, 487]					
–	*Yellow ocher*, *umber*	Red ocher, hematite	–	–	Iron manganese oxides, manganese dioxide, soot, charcoal
Early Orient, 11000 BC–2000 BC [120, 122, 407]					
Calcite, gypsum	*Yellow ocher*, *umber*	Red ocher, hematite, *cinnabar*, *manganese purple*	*Malachite*	*Egyptian blue*, *azurite*, *lapis lazuli*	Charcoal, soot, *bitumen*, *galena*
Egypt, 3000 BC–300 BC [23, 120, 124, 402, 407, 480, 481, 487, 488, 491, 495, 496, 771, 772]					
Kaolin, lime, gypsum, anhydrite, huntite	Yellow ocher, *umber*, ↑ orpiment, *jarosite*	Red ocher, hematite, *burnt umber*	↑ Egyptian green ↓, ↑ copper chlorides ↓, ↑ *malachite*, *copper green*	↑ Egyptian blue, *azurite*	Plant black, soot, bone black
Minos/Crete, Mycenae, Aegean, 2000 BC–1000 BC [120, 125, 126, 407, 482–486]					
Calcite	Yellow ocher, *umber*	Red ocher, hematite	*Green earth*, *malachite*	Egyptian blue, *blue amphiboles*	Plant black, soot, *hematite*, *pyrolusite*, *manganite*
Ancient Orient, 11000 BC–500 BC [120, 122, 407]					
Calcite, lime, gypsum	*Yellow ocher*, *umber*	Red ocher, hematite, *cinnabar*, *manganese purple*	*Malachite*	*Egyptian blue*, *azurite*, *lapis lazuli*	Soot, charcoal, *bitumen*, *galena*

Table 1.2: Overview of commonly used pigments in the ancient and medieval world. The arrows indicate the appearance ↑ and the disappearance ↓ of pigments. Italicized terms indicate less commonly used pigments.

White	Yellow/orange/brown	Red/purple	Green	Blue	Black
Greece, 1000 BC–100 BC [120, 127, 487, 488, 491–494], [53, ch. 9], [129, 386, 409, 495]					
↑Lead white, kaolin, chalk, lime, gypsum	Yellow ocher, orpiment, umber, terra di Sienna	Red ocher, ↑cinnabar, hematite, ↑red lead, ↑madder, *realgar*, *purple*	Malachite, paratacamite, ↑verdigris, green earth	Egyptian blue, azurite, ↑*ultramarine*, *indigo*	Plant black, bone black, soot
Rome, 500 BC–500 [120, 487, 488, 491, 495, 497, 498], [53, ch. 9], [49, 132]					
Lead white, chalk, calcite, gypsum, kaolin	Yellow ocher, ↑*massicot*↓, orpiment, ↑yellow lake, umber	Red ocher, cinnabar, madder, realgar, red lead, purple↓	Green earth, verdigris, malachite, (par)atacamit↓	Egyptian blue, azurite, ultramarine, *indigo*	Plant black, bone black, soot
Late Antiquity, 200–900 [120, 499–501, 514]					
Lime	Yellow ocher, *umber*	Red ocher, hematite, *red lead*	Green earth, *verdigris*, *copper green*	*Egyptian blue*, *false blue*	Charcoal
(Pre-)Carolingian, 700–1000 [120, 133, 134, 136, 501, 502, 511, 513–515]					
Lime, *lead white*, *St. John's white*	Yellow ocher, *massicot*, *lead-tin yellow*, *umber*, *terra di Sienna*	Red ocher, hematite, *cinnabar*, *red lead*, *red lake*	Green earth, *verdigris*, *copper green*	*Egyptian blue*, *ultramarine blue*, *azurite*, *false blue*	Plant black, *bone black*
Romanesque, 1000–1300 [120, 133–135, 501, 507, 509, 510, 512–514]					
Lime, St. John's white, *lead white*	Yellow ocher	Red ocher, hematite, *cinnabar*, *red lead*	Green earth, *verdigris*, *copper green*	Ultramarine, *azurite*, *false blue*	Plant black

Azurite	Blue	$Cu(OH)_2 \cdot 2\,CuCO_3$
Chrysocolla	Blue-green	$(Cu, Al)_2H_2Si_2O_5(OH)_4 \cdot x\,H_2O$
Malachite	Green	$Cu(OH)_2 \cdot CuCO_3$
Atacamite, Paratacamite	Green	$Cu_2(OH)_3Cl$
Basic copper hydroxy chloride	Green	$Cu(OH)Cl \cdot Cu(OH)_2$

The few copper-free bluish and greenish pigments owe their color to the presence of mixed-valence iron compounds in which electron transitions occur between the oxidation states Fe^{II} and Fe^{III} (*intervalence charge transfer, IVCT*, ►Section 2.4):

Vivianite	Blue	$Fe^{II}_3(PO_4)_2 \cdot 8\,H_2O$ with Fe^{III}
Blue amphiboles (glaucophane, riebeckite)	Blue	$Na_2(Mg, Fe^{II})_3(Al, Fe^{III})_2[(OH)_2\vert Si_8O_{22}]$
Green earth, celadonite/glauconite	Green	$(K, Na)(Al, Fe^{III}, Mg)_2(Si, Al)_4O_{10}(OH)_2$

In iron oxides, charge transfer can also occur from the oxide ligands to the metal (*oxygen-metal charge transfer, OMCT*). Examples from ancient palettes are ocher earths, which contain colored oxides and hydroxides of iron and manganese in addition to clay minerals and red hematite.

Goethite	Yellow	$FeOOH$
Jarosite	Yellow	$(K, Na)Fe_3(SO_4)_2(OH)_6$
Hematite	Red	Fe_2O_3

The charge transfer mechanism induces intensive colors. However, in natural colorants, especially in colored earths, the color-inducing elements are usually present only in low concentrations, so blue amphiboles or green earths are not intensive pigments despite the intensity of CT transitions.

The fourth important mechanism is coloring numerous organic compounds. Intensive electron transitions occur between complex *molecular orbitals* of natural colorants, ►Section 2.5.

Carmine	Magenta	Carminic acid Al lake
Madder lake	Red	Mixture of anthraquinone Al lakes
Purple	Red-purple	Dibromine indigo

White materials that are not semiconductors do not have an inherent “white” color but solely appear white (►Section 1.6.6) due to grinding, and consequently high light scattering: gypsum and anhydrite, chalk, lime, and huntite, a calcium magnesium carbonate.

1.2.2 Pigments from the late Middle Ages onward

The significant works exhibited in our art museums, from Trecento painters onwards to those of the Rococo, essentially rely on the Greco-Roman palette. It contains numerous natural colored minerals and only a few artificial pigments, mostly lead compounds, ►Table 1.3. Lead-tin yellow and later Naples yellow extend the palette of semiconducting lead pigments by pure and attractive yellow shades. Semiconductors also include *aureum musivum* or mosaic gold SnS_2, a glittering, golden pigment for gilding and lettering. New additions are *lakes*, comprising colored plant or animal extracts and a metal cation. Textile dyeing, which already, for many centuries, used plant extracts and alum (potassium aluminum sulfate) for fixing, was the inspiration for lakes.

Madder lake	Red	Anthraquinone Al complex (anthraquinone Al lake)
Carmine	Red	Anthraquinone Al complex (carminic acid Al lake)
Stil de grain	Yellow	Flavone Al complex (rhamnetine Al lake)
Brazilwood lake	Red	Neoflavone-metal complex (brasilein metal lake)

These pigments are metal-organoligand complexes, exhibiting subtle shades through electronic transitions between molecular orbitals of metal and ligand character. However, due to the limited range of ligands naturally occurring (for anthraquinones and flavonoids, there are only hydroxyl groups causing alternation), their color is limited to the red to purple and yellow to orange range, ►Chapter 4.

In the Bronze Age, chemistry evolved from a metallurgical craft based on experience; then it took up oriental and cabbalistic elements in its further development [268]. It fascinated secret lodges and kings alike with alchemy until it developed into a less glamorous but more systematic science in the eighteenth century. Then, with Prussian blue, chemistry starts to influence the palette seriously, and thus, the possibilities of artistic expression.

In book illumination, for which lightfastness of colorants is of lesser importance, a series of complexly composed plant and fruit extracts containing flavones, anthocyanins, and anthraquinones were used, extending the color range to purple, blue, and green [108]. Copper pigments of different compositions (one assumes mixtures of acetates, chlorides, and hydroxides) also contributed to the blues and greens.

Table 1.3: Overview of pigments in use since early Italian painting. The arrows indicate the appearance ↑ and the disappearance ↓ of pigments. Italicized terms indicate less commonly used pigments.

White	Yellow/Orange/Brown	Red/Purple	Green	Blue	Black
Early Italy (Duecento–Trecento), 1270–1370 [49, 145, 518]					
Lead white	↑ Lead-tin yellow, yellow ocher, orpiment, yellow lake	Red ocher, vermilion, ↑ brazilwood lake, red lead, carmine, ↑ lac dye	Green earth, verdigris, malachite, ↑ copper resinat	Azurite, ultramarine, *vivianite*	Lamp black
Renaissance, Mannerism, 1400–1550 [141, 403, 404, 418, 419, 519–553]					
Lead white, chalk, gypsum	↑ Lead-tin yellow, yellow ocher, orpiment, yellow lake, sienna, umber	Red ocher, vermilion, madder, carmine, red lake, realgar, red lead ↓	Verdigris, green earth, malachite, copper resinat	Azurite, ultramarine, smalt, *indigo*	Plant black, bone black, soot, manganese black
Baroque, 1550–1750 [139, 140, 554–564]					
Lead white, chalk, gypsum	Lead-tin yellow ↓, yellow/orange/brown ocher, orpiment ↓, yellow lake, sienna, umber, ↑ Vandyke brown	Red ocher, vermilion, madder, kermes, *realgar*, carmine	Verdigris, green earth, copper resinat, malachite ↓	Azurite, ultramarine, smalt, indigo	Plant black, bone black, soot
Rococo, 1750–1790 [565–572]					
Lead white	↑ Naples yellow, yellow ocher, orpiment, orange ocher, umber, Vandyke brown	Red ocher, vermilion, madder ↓, carmine, *realgar* ↓	Verdigris ↓, green earth	Azurite, ultramarine, ↑ Prussian blue, smalt ↓, indigo	Plant black, bone black, soot

1.2.3 Pigments from Romanticism and Classicism to Classical Modernism

The rapid development of scientific chemistry from the nineteenth century onwards, especially with the discovery of numerous *d* block elements, led to a wealth of new intensive colors based on *d* electron transitions, semiconductors, and charge transfer mechanisms. These systematically opened up the long sought-for color ranges of blue, green, orange, and purple shades.

Alizarin crimson	Red	Alizarin aluminum complex
Cobalt violet	Purple	Cobalt arsenate or phosphate
Emerald green	Green	Copper arsenate
Viridian	Green	Hydrated chromium oxide
Cadmium yellow	Yellow	Cadmium sulfide
Cadmium orange	Orange	Cadmium sulfide
Chrome yellow	Yellow	Lead chromate
Manganese blue	Blue	Barium manganate

►Table 1.4 gives an overview of the newly available colorants. The formation of *mixed metal oxides* (MMO) results in high-quality, stable pigments that even withstand ceramic firing; for this purpose, they are used in the ceramic industry today, e. g., we have the following:

Cobalt blue	Blue	Cobalt aluminate
Cerulean blue	Blue	Cobalt stannate

These pigments are based on a natural mineral, e. g., spinel $MgAl_2O_4$. Replacement of Mg or Al ions with chromogenic metal ions such as Co provides the color. *d* electron transitions, i. e. ligand-field transitions, are responsible for coloring those chromogenic ions.

Lake pigments improved considerably in the nineteenth century since the emerging chemical industry could provide anthraquinones, the starting ingredient of the significant red lakes, in pure form and in large quantities. Also, the process of laking itself was systematically examined and developed for the manufactory scale. Independent of the ever-changing composition of natural raw materials, manufacturers could now provide colorants of high color purity and constancy, and even standardize them. An example is alizarin crimson, the synthetic form of madder lakes. Later on, chemists could even plan the synthesis of anthraquinones of a given structure. (In fact, progress in chemistry dedicated to coloring agents was so fast and economically so relevant that the "tar color industry" and later on the chemical industry, in general, gave wealth to whole nations, e. g., England and especially Germany. Many

Table 1.4: Overview of commonly used pigments since the Romantic period. The arrows indicate the appearance ↑ and the disappearance ↓ of pigments. Italicized terms indicate less commonly used pigments.

White	Yellow/orange/brown	Red/purple	Green	Blue	Black
Romanticism, Realism, Classicism, 1790–1850, 1850–1900 [69, 573–585, 587]					
Lead white, ↑ zinc white	Naples yellow ↓, yellow ocher, yellow lake, sienna, ↑ chrome yellow, ↑ cadmium yellow, ↑ *strontium yellow* ↓, ↑ *cadmium orange*, umber, Cassel earth	Red ocher, vermilion, ↑ alizarin crimson, carmine	Green earth, ↑ emerald green, ↑ viridian	Ultramarine, Prussian blue, ↑ cobalt blue, ↑ coelin blue, *azurite* ↓ *smalte* ↓	Bone black, plant black, iron oxide black, carbon black
Impressionism, Classic Modernism, 1850–1940 [144, 146, 588–597]					
Lead white, zinc white	*Yellow ocher*, chrome yellow ↓, *barium yellow*, cadmium yellow, *Naples yellow*, *sienna*, *cadmium orange*, *chrome orange*	*Red ocher*, vermilion ↓, alizarin crimson, carmine, *red lead*, ↑ *cobalt violet*, ↑ *manganese violet*	*Green earth*, emerald green ↓, viridian	Coelin blue, ultramarine, Prussian blue ↓, cobalt blue	Bone black, plant black

manufactories founded in that time grew to corporations of national significance up to World War II, e. g., Agfa, Hoechst, Bayer, or BASF. Today, they are partly reorganized in still well-known corporations. However, the history of the advent and rise of industrial chemistry is a fascinating subject in its own right, as is the fact that the tarry materials, which could deliver anthraquinones and aromatic amines, were itself results of previous industrial-chemical revolutions: the introduction of coke instead of coals for iron smelting and reprocessing for steelmaking in England, and the introduction of coal gas for urban illumination. Both left behind vast amounts of tarry compounds deposited outside the cities until the chemical industry discovered their potential.)

1.2.4 Modern pigments

The twentieth century has brought a new wealth of pigments to avant-garde and contemporary art. The impressive data sheets of well-known paint manufacturers (e. g., [984–986]) reflect the many innovations in inorganic and organic pigments, ►Tables 1.5 to 1.7. Other sources of information are the national and international standards dedicated to artist's colors: ASTM D4302-05, ASTM D5098-05a, and ASTM D5067-05 [981], and the "Ullmann" [199, keyword "Artist's colors"].

We can identify the composition of modern artists' paints directly on the label of tubes and packaging or look them up in the manufacturers' catalogs. Similar to a car's license plate, the label on the product indicates details such as

Naples yellow light
PBr24/PY53

to determine the hue and chemical composition of the colorant using the *Colour Index* designations [1]. The Colour Index is a collective work that records countless colorants. Moreover, just as a license plate contains information about the location where the car was registered, we can tell from the first letters which colorant type and which hue is present: the "P" stands for "Pigment," the "Br" for "Brown," and the "Y" for "Yellow." The numbers represent a sequential numbering of the colorants. The above "license plate" thus provides us with the following information:

- The Naples yellow in question contains a brown pigment with the number 24, namely "Pigment Brown 24." The Colour Index tells us that it is a chromium antimony titanate $(Ti, Cr, Sb)O_2$.
- It also contains a yellow pigment with the number 53, the "Pigment Yellow 53," chemically a nickel antimony titanate: $(Ti, Ni, Sb)O_2$.

The Colour Index also covers dyes, distinguished by further identification letters, ►Table 1.8.

Table 1.5: Inorganic pigments in artists' paints by D. Smith (∘) [983], by Winsor and Newton (•) [984], by Schmincke (⋄) [985], and by Charbonnel (⋆) [986].

Pigment	Oil	Watercolor	Acrylic	Linol print	Copper print	Chemical composition
PW4	•⋄∘	•⋄∘	⋄		⋆	ZnO
PW6	•⋄∘	•⋄∘	•⋄	⋄	⋆	TiO_2
PY35	•⋄	•⋄	•⋄			(Cd, Zn)S
PY40		•∘				$K_3[Co(NO_2)_6]$
PY42	•⋄∘	•⋄∘	•⋄	⋄	⋆	FeOOH
PY43	•⋄∘	•⋄∘	•⋄			FeOOH
PY53	•⋄∘	•⋄∘	⋄			$(Ti, Ni, Sb)O_2$
PY119		•⋄	⋄			$(Zn, Fe)Fe_2O_4$
PY159		•⋄				$(Zr, pR^{IV})SiO_4$
PY184	•⋄	•⋄∘	•⋄			$4\,BiVO_4 \cdot 3\,Bi_2MoO_6$
PY216		•⋄				$(Ti, Zn, Sn)O_2$
PO20	•⋄	•⋄	•⋄			Cd(S, Se)
PR101	•⋄∘	•⋄∘	•⋄	⋄⋆	⋆	α-Fe_2O_3
PR102		•∘	•		⋆	α-Fe_2O_3
PR108	•⋄	•⋄	•⋄			Cd(S, Se)
PR233	∘	•⋄∘	•			Cr_2O_3 in $CaO \cdot SnO_2 \cdot SiO_2$
PV14	•	•∘				$Co_3(PO_4)_2$
PV15	•⋄∘	•⋄∘	•			$Na_{6-8}Al_6Si_6O_24S_{2-4}$
PV16	•⋄∘	•⋄				$(NH_4)MnP_2O_7$
PB27	•⋄∘	•⋄∘		⋆	⋆	$(Na, K\,NH_4)Fe^{III}[Fe^{II}(CN)_6]$
PB28	•⋄∘	•⋄∘	•⋄			$CoAl_2O_4$
PB29	•⋄∘	•⋄∘	•⋄	⋄⋆	⋆	$Na_{6,9}Al_{5,6}Si_{6,4}O_{24}S_{4,2}$
PB35	•⋄	•⋄∘	•			$CoO \cdot nSnO_2$
PB36	•⋄∘	•⋄∘	•			$Co(Al, Cr)_2O_4$
PB74	•⋄	•⋄				$(Co, Zn)_2SiO_4$
PG17	•⋄∘	•⋄∘	•⋄			Cr_2O_3
PG18	•⋄∘	•⋄∘	⋄			$2\,Cr_2O_3 \cdot 3\,H_2O$
PG19	⋄	⋄∘	⋄			$ZnCo_2O_4$
PG23	•	•				FeO-silicates
PG26	•⋄	•⋄	•⋄			$Co(Cr, Al)_2O_4$
PG50	•⋄∘	•⋄∘	•⋄			$(Co, Ni, Zn)_2TiO_4$
PBr6		⋄∘	⋄			$FeOOH/Fe_2O_3/Fe_3O_4$
PBr7	•⋄,∘	•⋄∘	•⋄	⋄⋆	⋆	$x\,Fe_2O_3 \cdot y\,FeOOH \cdot z\,MnO_2$
PBr24	•⋄	•⋄	•⋄			$(Ti, Cr, Sb)O_2$
PBr33	⋄	⋄				$(Zn, Fe)(Fe, Cr)_2O_4$
PBk6, 7, 8	•⋄∘	•⋄∘	⋄	⋆	⋆	C
PBk9	•⋄∘	•⋄∘	•⋄	⋄⋆	⋆	$Ca_3(PO_4)_2/C$
PBk10	⋄	⋄∘	•⋄			C
PBk11	•⋄∘	•⋄∘	•⋄	⋄	⋆	Fe_3O_4
PBk19	•	•	•			C
PBk28	⋄					$Cu(Cr, Fe)_2O_4$

Table 1.6: Organic pigments in artists' paints by D. Smith (∘) [983], by Winsor and Newton (•) [984], by Schmincke (⋄) [985], and by Charbonnel (⋆) [986].

Pigment	Oil	Water-color	Acrylic	Linol printing	Copper printing	Chemical class
PY3	•⋄∘	⋄∘	•⋄	⋄	⋆	Monohydrazone yellow ►Section 4.11.4.1
PY65	•∘	•⋄∘	•			Monohydrazone yellow ►Section 4.11.4.1
PY74	•⋄∘		•⋄	⋆	⋆	Monohydrazone yellow ►Section 4.11.4.1
PY83	∘	∘	•	⋆	⋆	Diarylide yellow ►Section 4.11.4.2
PY97	∘	•∘				Monohydrazone yellow ►Section 4.11.4.1
PY110	•⋄	•⋄∘	•⋄			Isoindolinone ►Section 4.15
PY128	•⋄					Diarylide yellow ►Section 4.11.4.2
PY139	•	•∘	•			Methine ►Section 4.15
PY151	∘	⋄∘				Benzimidazolone ►Section 4.11.4.5
PY154		•⋄	⋄	⋄		Benzimidazolone ►Section 4.11.4.5
PY155	⋄	⋄	⋄			Diarylide yellow ►Section 4.11.4.2
PY175	∘	•⋄∘				Benzimidazolone ►Section 4.11.4.5
PO43		∘				Perinone ►Section 4.13
PO48	∘	∘				Quinacridone ►Section 4.12
PO49			•			Quinacridone ►Section 4.12
PO62	•∘	•⋄∘		⋄		Benzimidazolone ►Section 4.11.4.5
PO67	•⋄		⋄			Pyrazolo-quinazolone
PO71	⋄	⋄∘	⋄			DPP ►Section 4.14
PO73	•∘	•∘	•			DPP ►Section 4.14
PR83:1	•⋄∘	•⋄∘			⋆	Alizarin lake ►Section 4.6.4.1
PR112			•			Naphthol AS ►Section 4.11.4.4
PR122	•⋄∘	•⋄∘	•⋄	⋄	⋆	Quinacridone ►Section 4.12
PR144		⋄				Diarylide yellow ►Section 4.11.4.2
PR149	•∘	∘	•			Perylene ►Section 4.13
PR168		∘				Anthanthrone ►Section 4.6.4.3
PR170	∘	∘	•			Naphthol AS ►Section 4.11.4.4
PR171		∘				Benzimidazolone ►Section 4.11.4.5
PR175		∘				Benzimidazolone ►Section 4.11.4.5
PR176		∘				Benzimidazolone ►Section 4.11.4.5
PR177	•∘	⋄∘	•			Anthraquinone ►Section 4.6.4.2
PR178		⋄∘				Perylene ►Section 4.13
PR179	⋄	•⋄∘	•⋄			Perylene ►Section 4.13
PR187	•	⋄				Naphthol AS ►Section 4.11.4.4
PR188	•∘	•⋄∘		⋆	⋆	Naphthol AS ►Section 4.11.4.4
PR202		⋄∘				Quinacridone ►Section 4.12
PR206	⋄∘	•⋄∘	•⋄			Quinacridone ►Section 4.12
PR207	⋄	⋄	⋄			Quinacridone ►Section 4.12
PR209	•∘	•∘	•			Quinacridone ►Section 4.12
PR242	⋄	⋄				Dihydrazone ►Section 4.11.4.2
PR254	•∘	•⋄∘	•⋄	⋄		DPP ►Section 4.14
PR255	•⋄∘	⋄∘	•⋄	⋄		DPP ►Section 4.14
PR264	⋄	•⋄∘	⋄	⋄		DPP ►Section 4.14

Table 1.7: Organic pigments in artists' paints by D. Smith (∘) [983], by Winsor and Newton (•) [984], by Schmincke (⋄) [985], and by Charbonnel (⋆) [986].

Pigment	Oil	Water-color	Acrylic	Linol printing	Copper printing	Chemical class
PV19	• ⋄ ∘	• ⋄ ∘	• ⋄		⋆	Quinacridone ►Section 4.12
PV23	• ⋄ ∘	• ⋄ ∘	• ⋄	⋆	⋆	Dioxazine ►Section 4.9
PV29		• ⋄ ∘	•			Perylene ►Section 4.13
PV32		∘				Benzimidazolone ►Section 4.11.4.5
PV42	⋄	⋄ ∘				Quinacridone ►Section 4.12
PV55	∘	• ⋄ ∘				Quinacridone ►Section 4.12
PB15	• ⋄ ∘	• ⋄ ∘	• ⋄	⋄ ⋆	⋆	Phthalocyanine ►Section 4.10
PB16	⋄	• ⋄ ∘	⋄			Phthalocyanine ►Section 4.10
PB60	• ⋄ ∘	• ⋄ ∘	• ⋄			Indanthrone ►Section 4.6.4.3
PB82		∘				Indigoide ►Section 4.7.2
PG7	• ⋄ ∘	• ⋄ ∘	• ⋄	⋄ ⋆	⋆	Phthalocyanine ►Section 4.10
PG36	• ⋄ ∘	• ⋄ ∘	• ⋄			Phthalocyanine ►Section 4.10
PBr25	•	∘				Benzimidazolone ►Section 4.11.4.5
PBr41	⋄	⋄				Dihydrazone ►Section 4.11.4.2
PBk31	• ⋄	• ⋄ ∘	• ⋄			Perylene ►Section 4.13

Table 1.8: Labeling of pigments and dyes according to the Colour Index system.

Type	Color (Yellow, orange, red, purple (violet), blue, green, black, white, brown)
Pigment	PY, PO, PR, PV, PB, PG, PBk, PW, PBr
Acid dye	AY, AO, AR, AV, AB, AG, ABk, ABr
Reactive dye	RY, RO, RR, RV, RB, RG, RBk, RBr
Direct dye	DY, DO, DR, DV, DB, DG, DBk, DBr
Vat dye	VY, VO, VR, VV, VB, VG, VBk, VBr
Disperse dye	DSY, DSO, DSR, DSV, DSB, DSG, DSBk, DSBr

If we chemically classify the variety of new pigments, mixed metal oxides' success becomes apparent. The principle of replacing atoms of a mineral with those providing color was applied to minerals other than spinels, resulting in modern, temperature-resistant pigments.

Cobalt turquoise	Blue-green	Cobalt chromium aluminate
Nickel titanium yellow	Yellow	Nickel titanate
Titanium white	White	Titanium dioxide

In most of them, *d* block elements are incorporated into the crystal lattice, colored by electron ligand field transitions. Some, such as titanium white, act as semiconductors.

A deeper understanding of semiconductor physics allows specifying a desired band gap width (the desired color) and determining the chemistry supporting this, e. g., by suitable alloying elements.

The vast number of today's organic pigments, colored by MO transitions and delivering precisely defined shades, express the success of organic structures for pigment design.

1.3 Overview of pigments

In the chronological overview, we have touched on the essential chemical mechanisms of color formation and given a few examples of pigments. Let us now consider the chemistry of significant artists' pigments, grouped by color. ►Tables 1.9–1.13 state the mechanisms responsible for each pigment's color. The tables only list those pigments for which statements about the history of their use are possible. The abbreviations SC, LF, CT, and MO denote the four chemical mechanisms described in ►Chapter 2.

The tables schematically indicate the period and extent of its use for each pigment. For a detailed description of time and usage, refer to the relevant technical literature, as we focus only on the relationship between chemistry and period of application.

1.3.1 Color range white, black

White pigments, ►Table 1.9, constitute a problem since an ideal, pure white pigment possesses a 100 % uniform reflection over the complete visual spectral range and a high refractive index. Compounds with absorption bands in the visual range cannot achieve this, so we use semiconductors almost exclusively today. Fortunately, semiconducting lead compounds are easy to produce, and already in Greek times, a white, lead-based pigment was widely available. It was not until the nineteenth century that new white semiconductors replaced the lead-based ones, namely zinc white (zinc-based) and later the well-known titanium white (titanium-based).

The other possibility to produce white pigments is the fine grinding of colorless compounds. A high scattering of white incident light physically creates a white color impression, ►Section 1.6.6. However, the cheap compound most widely available in nature, calcium carbonate in all its forms, has only a low refractive index, so it does not appear white but transparent in highly refractive media such as oil. Therefore, we can use it only in watercolors and to prepare grounds.

In contrast, an ideal black pigment must possess a total absorption over the entire optical spectral range to induce the color impression of "black." Since the earliest times, the black range has been represented by an excellent pigment: carbon.

Table 1.9: History of use of essential white and black painting pigments. The thickness of the bars indicates the extent of their application from prehistoric times to the present day [16, 48, 49, 51, 52, 97, 99, 116–121, 495] as well as sources from ►Tables 1.2–1.7.

Name	Composition		Period of application
White			
Gypsum PW25	$CaSO_4$	Scattering	
Chalk PW18	$CaCO_3$	Scattering	
Lime w., St. John's w. PW18	$CaCO_3$	Scattering	
Huntite PW 18:1	$CaMg_3(CO_3)_4$	Scattering	
Lead white PW1	$2PbCO_3 \cdot Pb(OH)_2$	SC	
Barite, barytes PW22	$BaSO_4$	Scattering	
Blanc fixe, permanent white PW21	$BaSO_4$	Scattering	
Lithopone PW5	$ZnS/BaSO_4$	SC	
Zinc white PW4	ZnO	SC	
Precipitated calcium carbonate (PCC) PW18	$CaCO_3$	Scattering	
Titanium white PW6	TiO_2	SC	
Black			
Lamp black, Chinese ink, soot PBk6	C, small particles	MO	
Charcoal bk., vegetable bk., vine bk., plant bk. PBk8	C, morphology acc. to precursor	MO	
Magnetite, iron oxide black PBk11	Fe_3O_4	CT	
Bone black, ivory bk. PBk9	$C \cdot Ca_3(PO_4)_2 \cdot CaCO_3$	MO	
Carbon black PBk7	C, small particles	MO	
Black spinel PBk20–PBk30	$(Cu, Co, Fe, Ni)(Cr, Fe, Mn)_2O_4$	CT/LF	
Organic pigments	Miscellaneous	MO	

In a solid, carbon's atomic orbitals fuse to so many molecular orbitals that their mutual spacing practically disappears, leading to a quasi-continuous, uniform absorption (MO type of color formation). In black iron and manganese oxides, several oxidation states of the elements are present, e. g., magnetite Fe_3O_4 or $Fe^{II}Fe^{III}_2O_4$. They exhibit such intensive IVCT transitions in the UV that the broad edge of their absorption bands reaches into the entire optical spectral range and absorbs all visible light; thus, the oxides appear brown to black. Most other colorants are out of the question since they possess distinct absorption bands, and thus an uneven absorption.

1.3.2 Color range yellow-orange-brown

Yellow is a rewarding color for the artist. Each of the four chemical mechanisms for color formation can produce a yellow color impression, so procurement of yellow colorants at any time was relatively simple. Therefore, yellow colorants are available in many hues, ►Table 1.10.

Table 1.10: History of use of essential yellow, orange, and brown painting pigments [16, 48, 49, 51, 52, 97, 99, 116–121, 495] and sources from ►Tables 1.2–1.7.

Name	Composition		Period of application
Yellow			
Yellow ocher, yellow earth PY43	α-FeOOH	CT/EPT	
Mars y., iron oxide y. PY42	α-FeOOH	CT/EPT	
Orpiment, king's y. PY39	As_2S_3	SC	
Realgar PY39	As_4S_4	SC	
Massicot, lead yellow PY46	PbO	SC	
Lead-tin yellow	Pb_2SnO_4 (Type I), $PbSn_2SiO_7$ (Type II)	SC	
Indian yellow NY20	Mg salt of euxanthic acid	MO	
Naples yellow PY41	theor. $Pb_2Sb_2O_7$	SC	
Barium y., lemon y. PY31	$BaCrO_4$	CT	
Chrome yellow PY34	$Pb(Cr, S)O_4$	CT	
Cadmium yellow light PY37	(Cd, Zn)S	SC	
Cadmium yellow PY35	CdS	SC	
Zinc yellow PY36	$Ka_2O \cdot 4\,ZnCrO_4 \cdot 3\,H_2O = 3\,ZnCrO_4 \cdot Zn(OH)_2 \cdot Ka_2CrO_4 \cdot 2\,H_2O$	CT	
Cobalt y., aureolin PY40	$Ka_3[Co(NO_2)_6]$	CT	
Chrome titanium y. PBr24	$(Ti, Cr, Sb)O_2$	SC	
Nickel titanium yellow PY53	$(Ti, Ni, Sb)O_2$	SC	
Zinc iron brown PY119	$(Zn, Fe)Fe_2O_4$	CT	
Bismuth vanadate and molybdate PY184	$4\,BiVO_4 \cdot 3\,Bi_2MoO_6$ and $BiVO_4$	CT	
Organic pigments	Miscellaneous	MO	
Orange			
Chrome orange PO21	$PbO \cdot PbCrO_4$	CT	
Cadmium orange PO20	Cd(S, Se)	SC	
Cerium sulfide orange light PO78	$Ce_2S_3 \cdot La_2S_3$	SC	
Cerium sulfide or. PO75	Ce_2S_3	SC	
Cadmium cinnabar PO23	(Cd, Hg)S	SC	
Organic pigments	Miscellaneous	MO	
Brown			
Umber PBr6, PBr7, PBr8	$xFe_2O_3 \cdot yFeOOH \cdot zMnO_2$ (little Fe_2O_3)	CT/EPT	
Terra di Sienna PBr6, PBr7, PY43	$xFe_2O_3 \cdot yFeOOH \cdot zMnO_2$ (little Mn)	CT/EPT	
Chrome iron brown PBr29	$(Fe, Cr)_2O_3$	LF+CT/EPT	
Mars brown PBr6	$FeOOH/Fe_2O_3/Fe_3O_4$	CT/EPT	
Manganese titanium brown PY164	$(Ti, Mn, Sb)O_2$	SC	
Organic pigments	Miscellaneous	MO	

Organic compounds, in particular, show absorptions in the near-UV range, which can often easily be shifted into the adjacent blue spectral range, inducing a yellow color impression. Therefore, we know numerous organic yellow colorants; natural sources include plant and flower extracts. These extracts predominantly provide dyes for dyeing purposes and have only limited value to painting because of their low resistance to light and chemicals. Only in modern times have organic chemists developed stable yellow pigments in all shades of yellow, including the formerly unknown, pure shades of orange and brown.

Fortunately, stable pigments that are colored due to ligand field transitions in Fe^{III} were widely available as natural colorants. As ocher earths and iron oxides, yellow to brown earth pigments have been applied in the arts for millennia until today.

A pure yellow can also result from charge transfer mechanisms, e. g., in the chromate anion $CrO_4^{2\ominus}$. Due to the lack of natural minerals containing the chromate group, this mechanism was not represented on earlier palettes and only added in the nineteenth century by chrome yellow.

Finally, yellow is one of the colors produced by semiconductors. Oxide and sulfide minerals deliver suitable semiconductors, which can also be transformed into new, artificial pigments. Since antiquity, natural orpiment and artificial massicot, lead-tin yellow, and Naples yellow have been popular additions to the earth color palette as pure, bright yellows, admittedly at the cost of high toxicity. Cadmium yellow, followed by cadmium orange, are successes of early inorganic chemistry, cadmium orange being the first true orange pigment.

1.3.3 Color range red, purple

The situation for red pigments has been similarly favorable as for yellow pigments, their production being easily possible already in antiquity, ►Table 1.11. All four chemical mechanisms can also induce red color impressions; however, the purer the color, the more toxic the pigment.

Organic extracts from plants and colored woods were of particular interest not to painters, but first to dyers and provided the classic red lakes: madder lake, carmine, and cochineal red. Later on, modern chemistry supplies countless synthetic red pigments in all hues. Due to advances in inorganic chemistry in the nineteenth century, chemists also provided pure, purple pigments, at the beginning mainly colored by LF transitions, later on by MO transitions.

As hematite widely occurs both pure and in ocher earths, it is the classic example of a natural red pigment on a charge transfer basis. Since the beginning of art, red ocher has prevailed on palettes; however, pure, synthetic iron oxides replaced hematite over time.

Since the earliest times, semiconductors were the unchallenged predominant source of pure reds, cinnabar or its artifical form vermilion being the classic example.

Table 1.11: History of use of important red and purple painting pigments [16, 48, 49, 51, 52, 97, 99, 109, 116–121, 495] as well as sources from ►Tables 1.2–1.7.

Name	Composition		Period of Use
Red			
Red ocher, red earth, sinoper PR102	α-Fe_2O_3	CT/EPT	
Mars red, iron oxide red PR101	α-Fe_2O_3	CT/EPT	
Madder lake PR83	Anthraquinone Al lake	MO	
Alizarin crimson	Alizarin Al lake	MO	
Kermes lake	Kermesic acid Al lake	MO	
Carmine	Carminic acid Al lake	MO	
Vermilion PR106	HgS	SC	
Red lead PR105	Pb_3O_4	CT	
Burnt sienna PR101, PR102, PBr7	Fe_2O_3	CT/EPT	
Burnt umber PR101, PR102, PBr7, PBr8	$Fe_2O_3 \cdot zMnO_2$	CT/EPT	
Chrome red PR103	$PbO \cdot PbCrO_4$	CT	
Cadmium red PR108	Cd(S, Se)	SC	
Ultramarine pink PR259	$Na_{3,6}(NH_4)_{0,25}(H_3O)_{1,94} \cdot [Al_{4,8}Si_{7,2}O_{24}] \cdot S_{3,01}$	MO	
Molybdate red PR104	$Pb(Cr, S, Mo)O_4$	CT	
Cerium sulfide red PR265, PR275	Ce_2S_3	SC	
Cadmium cinnabar PR113	(Cd, Hg)S	SC	
Organic pigments	Miscellaneous	MO	
Purple			
Purple	$C_{17}H_8Br_2N_2O_2$	MO	
Han purple	$BaCu[Si_2O_6]$	LF	
Ultramarine violet PV15	$Na_{6,08}(NH_4)_{0,17}(H_3O)_{1,28} \cdot [Al_{5,36}Si_{6,64}O_{24}] \cdot S_{3,83}$	MO	
Cobalt violet PV14	$Co_3(PO_4)_2$	LF	
Manganese violet PV16	$(NH_4)MnP_2O_7$	LF	
Ammonium cobalt phosphate PV49	$NH_4CoPO_4 \cdot H_2O$	LF	
Lithium cobalt phosphate PV47	$CoLiPO_4$	LF	
Organic pigments	Miscellaneous	MO	

Yellow, orange, and dark red cadmium pigments, together with the cerium sulfides, are creations of modern times, only made possible by advances in inorganic chemistry in the nineteenth century, especially by discovering elements such as cadmium.

1.3.4 Color range blue

In the blue range of the spectrum, difficulties in pigment production occurred at all times before the eighteenth century [429], ►Table 1.12. Unfortunately, semiconductors

Table 1.12: History of use of essential blue painting pigments [16, 48, 49, 51, 52, 97, 99, 109, 116–121, 495] as well as sources from ►Tables 1.2–1.7.

Name	Composition		Period of use
Blue amphiboles, e. g., glaucophane or riebeckite	$Na_2(Mg,Fe^{II})_3(Al,Fe^{III})_2[(OH)_2\|Si_8O_{22}]$	CT	
Egyptian blue PB31	$CaCu[Si_4O_{10}]$	LF	
Han blue	$BaCu[Si_4O_{10}]$	LF	
Azurite PB30	$Cu(OH)_2 \cdot 2\,CuCO_3$	LF	
Blue verditer, artificial azurite PB30	$Cu(OH)_2 \cdot 2\,CuCO_3$	LF	
Vivianite	$Fe_3(PO_4)_2 \cdot 8\,H_2O$ with Fe^{III}	CT	
Ultramarine PB29	$Na_{6,3}[Al_{4.79)}Si_{7,21}O_{24}]S_{3,74}$	MO	
French ultramarine (artificial ultramarine) PB29	$Na_{6,3}[Al_{4.79)}Si_{7,21}O_{24}]S_{3,74}$	MO	
Indigo	$C_{16}H_{10}N_2O_2$	MO	
Artificial indigo PB66	$C_{16}H_{10}N_2O_2$	MO	
Smalt PB32	Cobalt silicate glass	LF	
Prussian blue, Berlin bl. PB27	$(Na,K\,NH_4)Fe^{III}[Fe^{II}(CN)_6] \cdot nH_2O$	CT	
Cerulean blue PB35	$CoSnO_3$	LF	
Cobalt bl., Thénard's bl. PB28	$CoAl_2O_4$	LF	
Manganese blue PB33	"$BaMnO_4 \cdot BaSO_4$"	LF	
Cobalt blue turquoise PB36	$Co(Al,Cr)_2O_4$	LF	
Phthalocyanine blue PB15	Copper phthalocyanine	MO	
Organic pigments	Miscellaneous	MO	

cannot produce blue colors, thus eliminating natural minerals as the most important source of pure and consistent hues.

MO-based organic colorants with pure, blue hues exist, but manufacturers require a modern chemistry tool set to design and produce them. So, it is easy to shift absorption bands from the UV to the near blue spectral range to obtain yellow pigments, and it is so difficult to move them further toward the yellow spectral range to achieve pure blue pigments. The only classical, bright blue pigment is ultramarine, an inorganic MO colorant that has always been extremely valuable because of its far-away deposits in Afghanistan. Furthermore, extracting and purifying the pigment from its natural environment as a precious stone was a skillful craft.

The only affordable blue compounds are based on LF transitions in Cu^{II}. Artists in antiquity used various copper minerals, but over two millennia, azurite gained dominance. However, copper compounds are unstable and interact with many other pigments; on top of that, the LF transitions of copper are low in intensity. The only alternative for copper is cobalt. Co^{II} often occurs in its compounds in tetrahedral coordination so that the spectroscopic selection rules do not apply. Therefore, Co^{II} compounds are intensively colored, usually blue. For painting purposes, cobalt has been used only since the fifteenth century as a blue glass called smalt. However, cobalt is present only in small quantities in smalt, so smalt is a weak colorant. True cobalt pigments could not be used before the recognition of cobalt as an element and its systematic chemistry in the nineteenth century.

In the middle of the seventeenth century, we observe an exuberant abundance of blue color. This abundance, however, should not hide the disadvantages that the painter of this lavish period faced [427]:

- The primary sources of pure blue were still the well-known blue pigments ultramarine, greenish azurite, and smalt.
- Ultramarine was still very expensive.
- High-quality azurite became rarer and more expensive. As a result, it disappeared entirely from the repertoire in the eighteenth century.
- Artificial azurite (blue verditer) was used only by decorative painters, as it was gritty in texture and had a greenish tint.
- Smalt was weak in color and faded in oil.
- Indigo, a well-known vat dye, was also known for fading.

At the beginning of the eighteenth century, there was still only a minimal set of stable blue pigments available. Bluish rocks offered no alternative. In those, the essential IVCT transition between Fe^{II} and Fe^{III} ions appears. However, the second oxidation state occurs only as an impurity, so the color strength is insufficient when blueish rocks are ground to a pigment. Therefore, discovering a mixed-valence iron compound came just at the right time. Prussian blue, found around 1704, is the classic example of a charge transfer transition that induces bright blue hues. It was, therefore, quickly adopted. Barely 30 years after its discovery, however, reports of fading, especially in white mixtures, increased so that painters continued to wait for persistent blue pigments.

Finally, systematic chemistry provided these long-sought permanent blues. Like smalt, cobalt blue and cobalt turquoise pigments widely used today rely on LF transitions in tetrahedrally coordinated cobalt. However, their production became possible only when cobalt could be isolated as metal and methodically converted into its compounds. Only the synthesis of ultramarine in the nineteenth century and the discovery of phthalocyanines and other organic blue pigments in the twentieth century created a solid basis for today's blue and green pigments.

1.3.5 Color range green

It seems paradoxical: The dominant color green of forests and meadows all over the world surrounded the painters, and yet green was, for the longest time, the most problematic color on the palette, ►Table 1.13. One can divide the known green pigments before the nineteenth century into

Table 1.13: History of use of essential green painting pigments [16, 48, 49, 51, 52, 97, 99, 109, 116–121, 495] as well as sources from ►Tables 1.2–1.7.

Name	Composition		Period of application
Chrysocolla	$(Cu, Al)_2H_2Si_2O_5(OH)_4 \cdot x\,H_2O$	LF	
Atacamite, Paratacamite	$Cu_2(OH)_3Cl$	LF	
Egyptian green	Green copper silicate glass	LF	
Malachite PG39	$Cu(OH)_2 \cdot CuCO_3$	LF	
Green verditer, artificial malachite PG39	$Cu(OH)_2 \cdot CuCO_3$	LF	
Green earth PG23	FeO-silicates	CT/LF	
Verdigris, Spanish gr. PG20	$Cu(CH_3COO)_2 \cdot Cu_3(OH)_2$	LF	
Rinmann's green, Cobalt green PG19	$ZnCo_2O_4$	LF	
Scheele's Green PG22	$n\,CuO \cdot As_2O_3 \cdot m\,H_2O$	LF	
Emerald green, Schweinfurt green, Mitis green PG21	$Cu(CH_3COO)_2 \cdot 3Cu(AsO_2)_2$	LF	
Chromium oxide green PG17	Cr_2O_3	LF	
Viridian PG18	$Cr_2O_3 \cdot 2\,H_2O$	LF	
Chrome gr., cinnabar gr. PG15	Chrome yellow+Prussian blue	CT+MO	
Cadmium green PG14	Cadmium yellow+ultramarine	SC+MO	
Fast chrome green PG48	Chrome yellow+phthalocyanine blue	CT+MO	
Cobalt turquoise PG50	$(Co, Ni, Zn)_2TiO_4$	LF	
Phthalocyanine green PG7	Copper phthalocyanine chlorinated	MO	
Phthalocyanine green PG36	Copper phthalocyanine halogenated	MO	
Organic pigments	Miscellaneous	MO	

- minerals (malachite, green earth)
- metal salts, mostly of copper (verdigris/Spanish green, artificial malachite, basic copper chlorides)
- plant extracts (sap green for watercolors)

However, they were unsatisfactory since they either showed a noticeable blue tint, as all copper pigments do or, in the case of green earth, displayed only a dull shade of green. In any case, there was no pure medium green of intensive color [396]. Consequently, practices for mixing green from other pigments had developed early on. Typical combinations were the following.

Green earth, verdigris, or malachite	with lead-tin yellow or yellow varnish	and black where appropriate
Azurite, ultramarine, indigo, Prussian blue	with yellow ocher, lead-tin yellow, Naples yellow, yellow varnish, orpiment	

The components could be mixed beforehand on the palette or directly on and in the painting. The actual composition reflects the pigments commonly used in each epoch. A change is noticeable across the seventeenth and eighteenth centuries. Painters replaced the expensive blue pigments azurite and ultramarine with new, more affordable ones such as Prussian blue and indigo. Similarly, lead-tin yellow recedes in favor of Naples yellow. As we will see, nature follows this path of color mixing in green earths: a yellow Fe^{II} compound is mixed with a blue one, comprising Fe^{II} and Fe^{III} and colored blue by IVCT transitions.

With increasing mastery of oil painting techniques and the emergence of the Old Master glazing technique, artists could achieve green hues by *optical mixing*, i. e., by layering colored glazes, e. g., a blue and a yellow one. For this purpose, opaque colors such as lead-tin yellow or Naples yellow were unsuitable, so yellow lakes or transparent ochers were used from the seventeenth century onward.

Though mixed greens provided quite lovely hues, still there remained problems:

- Copper pigments often darkened or affected other pigments by reacting with the medium.
- Many of the pigments used were known to lose color more or less rapidly, predominantly yellow and red lakes. Nevertheless, indigo, smalt, Prussian blue, and even orpiment were also considered unsafe. Varnishes could provide protection.

The underlying difficulty is that pure green cannot be obtained by absorption of a single complementary wavelength. The color closest to a pure green that we can achieve by a single absorption band in the red spectral range is blueish green. Charge transfer bands do not appear in this range for materials commonly used as colorants, while semiconductors are principally not capable of inducing green color impressions. The development of organic blue and green pigments requires a considerable shift of an absorption band from the UV region into the yellow or red spectral range, which only modern organic chemistry has achieved, tailoring electron donors and acceptors.

An alternative to color mixing is the creation of *two* absorption bands around the green spectral range, a task for a modern research chemist. Nature realizes this in two green copper compounds, namely malachite and porphyrins (chlorophyll). Both possess absorption bands in the blue and the yellow spectral range. However, green copper compounds have the same disadvantages as blue ones in that they are unstable, of low intensity, and moderately toxic. Brilliant green copper pigments from the industrial era were even highly toxic (though on behalf of their arsenic content). The discovery of chrome in modern times made it possible to develop stable, brilliant green chrome oxide hydrates.

However, even today, mixing green is the best way to achieve pure or intensive neutral green colors.

1.4 Paint systems, definitions

Before we turn to details, some terms need to be clarified to facilitate understanding of the following chapters. A *paint system* comprises all colored substances, binders, solvents, auxiliaries, ground, and supports required for a particular painting technique, e. g.:

- The paint system of oil painting includes colorants (pigments), binders (linseed, poppy, or safflower oil), possibly resins, and a solvent (turpentine oil). Ground is, e. g., a chalk ground or gesso, supports are, e. g., canvas or wooden panels.
- The paint system of writing or drawing ink comprises colorants (pigments or dyes), a solvent (water), auxiliaries such as humectants, surfactants, and pH buffers. A common support is a paper or board.

According to DIN, the german industry standard, the term *colorant* applies to all substances with inherent coloring properties, whether pigments or dyes.

Pigments

Colorants that are insoluble in the binders and solvents of a paint system are called *pigments*. The painter fixes them to the ground or support by employing the binder. The actual artists' color or *paint* is a dispersion of pigments in the binder and solvent. It can be fluid (e. g., an ink), semi-viscous (e. g., watercolor pigments in a gum solution) or viscous and pasty (e. g., oil paint). The binder can also act as a solvent, such as oil in oil paints. All artists' pigments are insoluble in linseed oil and water and are suitable for producing oil, acrylic, watercolor paints, and printing inks. (When discussing industrial printing inks, we must take care since some industrial printing techniques use strong or aromatic solvents capable of dissolving some classes of organic pigments.)

Pigmented paintings tend to be opaque. Artists can apply bodily paints (in more or less viscous masses such as oil paint) using an impasto technique, lending visible texture to the painting. Due to their thickness, impasto layers are opaque. If the artist decides to apply paint thinner or dilute it with a solvent, the paint layer may be thin but still opaque. The transparency of a pigment depends on its refractive index and its particle size. Some pigments such as titanium white, vermilion or ocher are highly opaque. An artist must select pigments with a low refractive index or particular grades to obtain glazes. Manufacturers can grind some pigments to a grade at which their scattering power is minimal so that they appear transparent, ►Section 1.6.6. Usually, manufacturers indicate these grades in the trade name, e. g., "transparent yellow ocher" in contrast to (opaque) "yellow ocher." Today, it is often possible to adjust the opacity of a pigment smoothly from glazing or transparent to opaque: yellow ocher, e. g., is a superior opaque pigment but can be transparent in specific particle sizes. Applications such as pigmented inkjet inks also require pigments with small particle sizes.

Dyes

Colored substances molecularly dissolved in binder and solvent are called *dyes*. Since there are no colorant particles, they do not scatter light, and the solution is clear and transparent. Due to their immateriality and lack of body, painters do not generally use dyes for painting but employ them occasionally to color varnishes. Besides, their persistent solubility in common solvents such as water renders most of them unsuitable for storage and exhibition.

Their main application is graphic work, drawing (inks), printing, writing (inks), and dyeing. Dyes are adsorbed by the support (paper, textiles) and fixed mainly by secondary interactions such as van der Waals forces, ►Section 6.1.4. These interactions are weak on a large scale but significant in the molecular dimensions of the dye.

By *laking* (►Section 2.6), an artist can give body to some dyes for painting. Adding a metal salt to a suitable dye solution precipitates the dye to a carrier as a sparingly soluble substance ("lakes"), and thus imparts materiality. If the carrier is clay or chalk, the artist obtains glazing pigments since the white carrier is transparent in most binders due to its low refractive index. Significant examples of lakes are carmine red (Al lake of water-soluble carminic acid from cochineal lice) and madder lake (Al lake of alizarin from madder roots).

1.5 Basic physical processes, spectra

To understand what the actual task of the chemist is in producing color, we have to know what makes color come into being in the first place. For this purpose, we will now have a closer look at the elementary absorption process of (visible) radiation by a (colored) body before we turn to its physics in ►Section 1.6.

Painting owes its existence to the process of electronic excitation shown in ►Figure 1.3. By irradiating with the light of a suitable wavelength λ, frequency ν or angular frequency ω, electrons of a compound can be transferred from one energy level E_{GS} (the *ground state*) to a higher energy level E_{ES} (the *excited state*), ►Figure 1.3(a). The energy $\Delta E = h\nu$ consumed by the transfer is missing in the reflectance spectrum and appears as an absorption band, ►Figure 1.3(b). If the wavelength is within the visible range of the electromagnetic spectrum, we perceive a seemingly colored compound. Energy and wavelength are related to each other as follows:

$$\lambda = \frac{c}{\nu} = \frac{2\pi c}{\omega}, \qquad \Delta E = h\nu = \frac{h}{2\pi}\omega = E_{ES} - E_{GS} \tag{1.1}$$

In the reflectance spectrum, we specify the reflectance $R(\lambda)$, i. e., the fraction or percentage of irradiated light of wavelength λ. $R(\lambda)$ ranges from 0 % (total absorption, ideal black) to 100 % (total reflection, i. e., no absorption at all, ideal white).

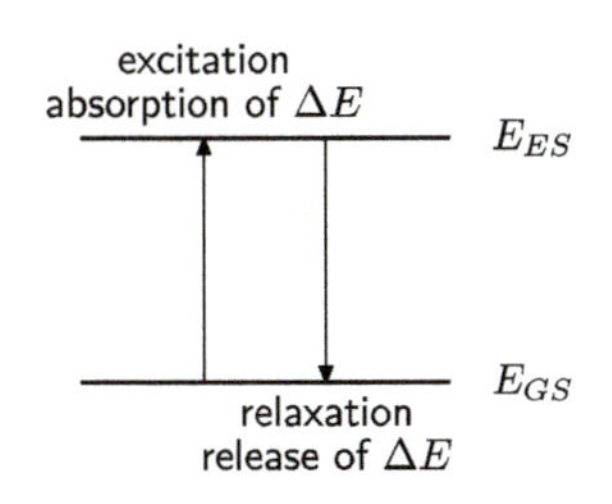

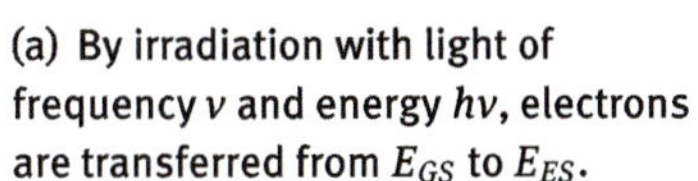
(a) By irradiation with light of frequency v and energy hv, electrons are transferred from E_{GS} to E_{ES}.

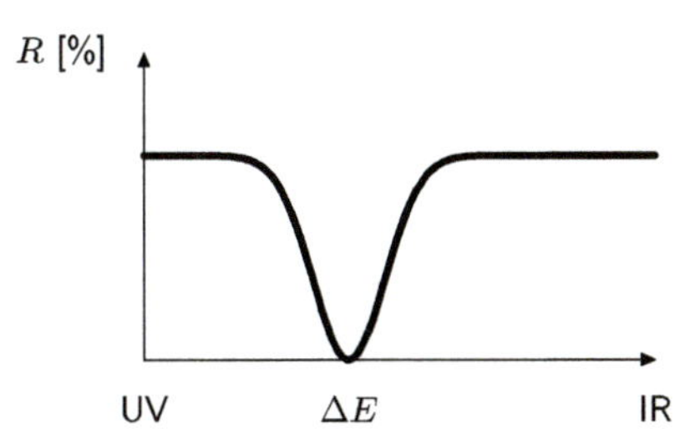

(b) The absorbed energy hv is missing in the reflectance spectrum and appears as an absorption band at frequency v or energy ΔE, respectively.

Figure 1.3: Basic principle of absorption. As an example, a material (possessing two energy levels E_{GS} and E_{ES}) is irradiated with white light. E_{GS} is called ground state and E_{ES} is called excited state. Here, $\Delta E = E_{ES} - E_{GS} = hv$ applies. In spectroscopy, ground and excited states are often called initial and final states, $\Delta E = E_{if} = E_f - E_i$.

1.5.1 Emission colors

The first primary type of a spectrum comprises a single, more or less sharp peak, ►Figure 1.4. Such a spectrum is created by a self-luminous body emitting light of a specific wavelength, which is not the case in painting, but very much so in a video installation. The color perceived corresponds to the wavelength of the peak's maximum and its "width." The peak's maxima would be located in the blue and yellow spectral range in the example. The width of a peak can be defined as its full width at half-maximum, called FWHM or bandwidth, if a normal distribution describes the peak. The sharper the peak is, the purer is the color, visualized by the distinct "narrow" yellow peak in contrast to the "broad" blue one in ►Figure 1.4. If the peak broadens in the visible range, it resembles more and more the regular white light, and the perceived color is pale, not very distinct, and vanishes more and more.

A compound showing such a spectrum can exhibit all rainbow colors, as the figure shows, especially the distinct shades purple (420 nm), blue (460 nm), green (525 nm), yellow (575 nm), orange (600 nm), and red (650 nm). The specified wavelengths are to be taken as typical values since the exact color impression depends strongly on the precise shape of the spectrum.

1.5.2 Absorption color

The counterpart to the spectrum just discussed is one in which only light of a particular wavelength is absorbed. When we irradiate a body with white light, and the body is capable of absorbing the energy of a specific wavelength through an electronic transition, as shown in ►Figure 1.3, we get an idealized reflectance spectrum containing a

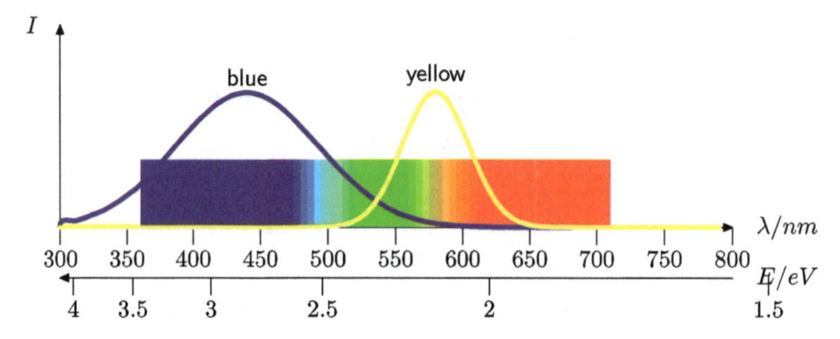

Figure 1.4: The ideal emission spectrum of a self-luminous compound. Depicted is the intensity I. The perceived color corresponds to the wavelength λ of the peak's maximum and its "width," e. g., the full width at half-maximum (FWHM) or bandwidth. The smaller the bandwidth of the peak, the purer the hue (yellow in this case). The reciprocal relationship between energy $h\nu$ of the radiation and wavelength λ can also be seen. Color perceptions and wavelength ranges are typical values.

"hole" or "absorption band" at this wavelength, ►Figure 1.5(a). This scenario is typical for artists' pigments. Also typical is that absorption bands, in reality, are by no means sharp, ►Figure 1.5(b).

The mixture of the residual light determines the color impression of such a spectrum. As a rule of thumb, we perceive the complementary color to the color corresponding to the wavelength of the peak. If a body absorbs light of 590 nm, this corresponds to yellow light, so the complementary color and the color impression of this body is blue, ►Table 1.14. Note that this situation is the reverse of that in ►Figure 1.4: The color of light emitted at 580 nm is yellow, but the color of a body absorbing at 580 nm is blue.

Table 1.14: Correlation between absorbed wavelength λ_{max}, the color of the corresponding emission, and the perceived color [205, Chapter 34]. A pure green color impression can hardly be created.

Color of λ_{max}	λ_{max}/nm	Color perception	Energy/eV
Purple	380–430	Yellow-green	3.26
Blue	430–480	Yellow	2.75
Cyan blue	480–490	Orange	2.57
Blue-green	490–500	Red	2.52
Green	500–560	Magenta	2.33
Yellow-green	560–580	Purple	2.17
Yellow	580–595	Blue	2.13
Orange	595–605	Cyan blue	2.11
Red	605–750	Blue-green	2.04
Magenta	750–770	Green	1.65

Compounds exhibiting an absorption peak can take on all colors from yellow-green to purple to blue-green except for pure medium green. ►Figure 1.5(b) illustrates re-

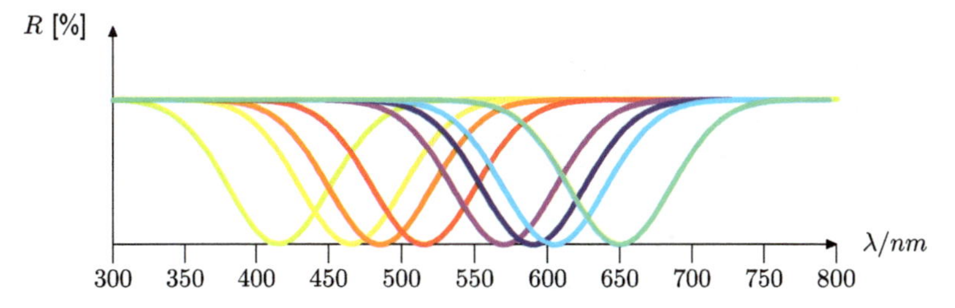

(a) The ideal reflectance spectrum of an absorbing colorant. The color of each line corresponds to the perceived color of the light reflected by the body (pigment particle, stained surface) for absorption wavelengths from around 400 to 650 nm.

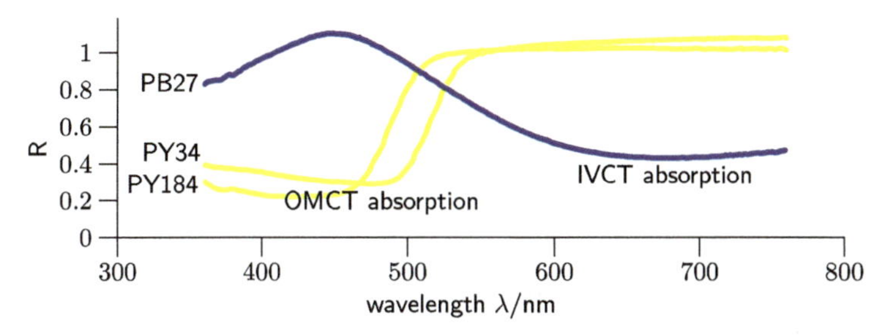

(b) Measured reflectance spectrum normalized to an arbitrary unit. In this example, the IVCT transition in Prussian blue (PB27) leads to intense and broad absorption in the yellow-red spectral region. It induces the complementary, i. e., blue, color of the pigment. Likewise, OMCT transitions in vanadium yellow (PY184) and chrome yellow (PY34) lead to intense and broad absorptions in the blue spectral region, thus inducing the pigments' complementary, yellow color.

Figure 1.5: Reflectance spectrum of a colorant with a single absorption band.

flectance spectra for chrome yellow and Prussian blue, [9, ch. 7, app. D] contains more reflectance spectra for important pigments.

Multiple absorption bands

In practice, colorants possess several absorption bands in the visible and especially in the adjacent UV spectral range. The resulting color is then no longer easy to predict. However, we can consider an essential, particular case that produces the green hue described.

To generate a pure green color impression, we combine two absorption bands, one in the yellow and the other in the blue spectral range. An "emission peak" of visible green light forms in the intervening green spectral range. This peak of green light is what the well-known mixture of blue and yellow colorants aims at and one goal of colorant design for scientists. As an idealization, ►Figure 1.6(a) shows how two absorption bands around λ_1 and λ_2 (which need not be located completely within the visible spectral range) give rise to a "peak" around λ_3.

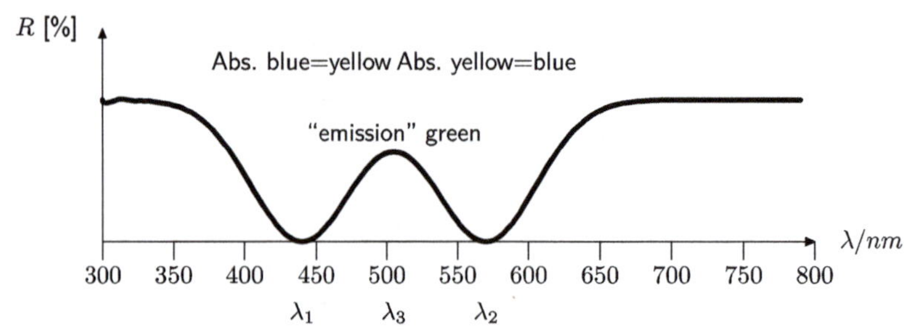

(a) The ideal reflectance spectrum of a green colorant caused by two absorption bands; one absorbs the red, the other the blue spectral part. Green light passes.

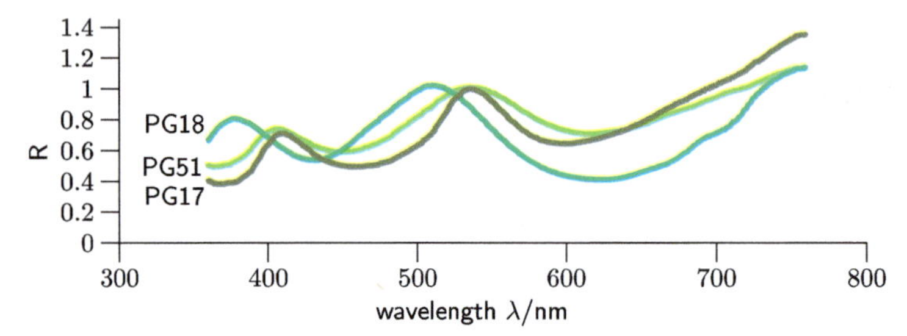

(b) Measured reflectance spectrum normalized to any unit: Victoria green (PG51), viridian (PG18), and chromium oxide green (PG17). ►Section 2.3.5 discusses a mechanism for producing multiple absorptions for the chromium oxide pigments.

Figure 1.6: Reflectance spectrum of a pigment comprising two absorption bands, producing a green color impression.

Influence of bandwidth

The color impression depends not only on the absorption wavelength but also on the bandwidth of the absorption band. A narrow or sharp absorption band is more purely colored (higher chroma) than a broad band. However, it is also weaker in color because only a tiny amount of white light is absorbed. As example, didymium glass (neodymium-praseodymium glass used for safety glasses) possess an absorption band about 30 nm in width in the yellow range and appears weakly bluish. In contrast, the band in the intensely blue cobalt glass is about 200 nm in width, absorbing a considerable amount of light.

If we increase the bandwidth to such an extent that light is absorbed in large parts of the visible spectral range, the color impression is lost, and dark colors are perceived. Broad absorption bands in the yellow range appear blueish-black, those in the blue range brown-gray.

Influence of intensity

The intensity of an absorption band influences the color impression. At low intensity (e. g., due to inadequate concentration of the colorant), the wavelength of maximal absorption determines the color impression. The absorption at neighboring wavelengths is below the threshold of the eye's perception. However, this is no longer the case at higher intensities; we begin to perceive the absorption of the neighboring wavelengths, which shifts the hue.

Solutions of a yellow dye provide good examples. At low concentrations, only the absorption of blue light is perceptible; we see a yellow solution. At higher concentrations (extreme case: the colorant as powder), the absorption peak gains intensity and broadens at its base, so that purple, green, and yellow components begin to be perceptibly absorbed. The color impression of a concentrated solution or the powder is orange.

When the absorption intensity is very high, the peak becomes so broad that noticeable absorption occurs throughout the complete spectral range, resulting in a brown, dark blue, or even black color impression. The most intensive organic colorants have an almost blackish appearance as solids or when undiluted. Many oxide minerals are also colored dark brown or nearly black because of intensive absorption in the near-UV region (CT transition from oxygen to metal). One edge of the absorption band extends from UV into the visible spectral range and absorbs purple, blue, green, and to some extent also, yellow light. Faint yellow and red light pass. (Another effect of deeply colored pigments is the appearance of a metallic luster called bronzing, ►Section 1.6.7 at p. 65.)

1.5.3 Color by absorption at a band edge

Semiconductors exhibit a particular absorption behavior. In a simplified model, they completely absorb all light up to a specific wavelength and reflect all light above that wavelength, ►Figure 1.7. The steeper the resulting edge in the reflectance spectrum, the purer the color impression. The position of the edge corresponds to an energy value called "band gap" since it is the energy difference between the valence and conduction band of the semiconductor, ►Section 2.2. In reality, thermal energy at room temperature feed some transitions. Consequently, realistic band edges are slightly smeared.

The sum of all reflected light determines the color and ranges from black (the edge is located in the IR, all light is absorbed) to brown (some red light is reflected), red (all red light is reflected), and yellow to white (the edge is in the UV, all light is reflected). ►Figure 1.7 depicts white, yellow, orange, and red semiconducting pigments, their reflectance spectra with varying positions of the band edge, and the resulting colors.

1.5.4 RGB and CMY primaries, tristimulus theory, metamerism

The emissive colors in ►Section 1.5.1 allow us to mix "all" colors using only the three primary colors red, green, and blue. The process is called *additive color mixing* and

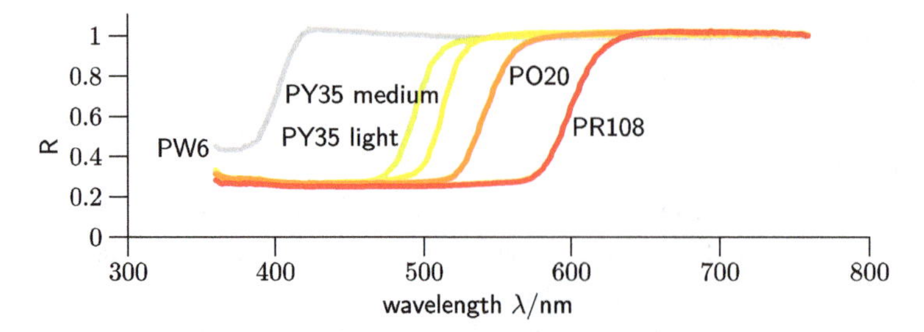

Figure 1.7: Measured reflectance spectra of semiconductors with different band gaps: PW6 (titanium white, TiO_2, white, large band gap), PY35 (cadmium yellow, CdS, yellow, smaller band gap), PO20 (cadmium orange, Cd(S, Se), orange, even smaller band gap), and PR108 (cadmium red, Cd(S, Se), red, smallest band gap) normalized to arbitrary unit. Absorbance changes abruptly at band gap energies.

has been used for decades in color television sets and tube monitors, in which three colored phosphors produce the primary colors. Today, red, green, and blue LEDs or organic light-emitting diodes (OLEDs) take over this role, but the principle of additive color mixing remains. It extends to applications such as "ambient lighting," i. e., residential lighting with variable colors. When discussing printing inks, drawing inks, and writing inks, we will learn about three other primary colors (or process colors) in the CMYK process (four-color printing): cyan, magenta, and yellow. These are primary colors in *subtractive color mixing*, used in printing, painting, and most other artistic techniques.

Today, the four-color printing process CMYK or 4C is standard for commercial and personal color printing. Manufacturers add more colors for more strict requirements regarding color reproduction; a well-known system was the six-color or CMYKOG system, adding orange and green. Photo inkjet printers still use 6C or CcMmYK, a six-color printing process that additionally employs lighter cyan and magenta ink. However, these are not primary colors in the sense of subtractive color mixing but deal with imperfections of colorants by replacing mixtures of primary colors with dedicated colorants. An equivalent amount of an actual orange colorant with better coloristic qualities is employed instead of mixtures of yellow and magenta. Similarly, green colorants replace mixtures of yellow and cyan, and purple colorants replace cyan and magenta.

In additive and subtractive color mixing, three primary colors achieve "all other" colors. Behind this is the realization that the foundation of our color perception is a light receptor sensitive to red, green, and blue light. ►Figure 1.8(b) shows the spectral sensitivity of the color receptor as so-called *tristimulus values* or *color matching functions*. These functions represent the intensity of the receptor's excitation as a function of wavelength, i. e., blue light around 420 nm leads to strong excitation of the blue-sensitive receptor but not of the green-sensitive receptor. The functions illustrate that the choice of red, green, and blue light as primary light colors is extremely reasonable since they show maximum adoption to the tristimulus functions.

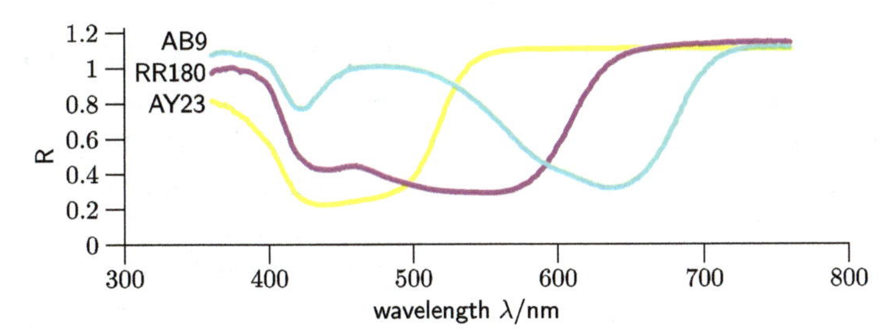

(a) Reflectance spectrum of writing inks suitable as primary colors for four-color printing, e. g., inkjet color mixing, normalized to an arbitrary unit: yellow (AY23, FY4, E102, tartrazine), magenta (RR180) and cyan (AB9, FB2, E133), all from the RIMIK™ Rainbox Ink Mixing Kit, Octopus GmbH.

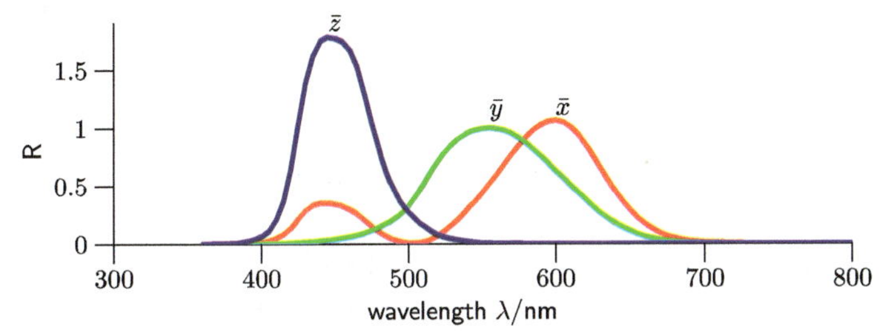

(b) Tristimulus values (2° color matching functions, CIE 1931, http://cvrl.ioo.ucl.ac.uk).

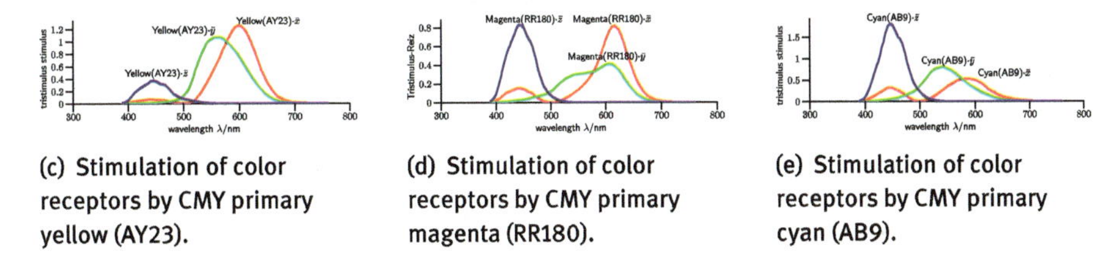

(c) Stimulation of color receptors by CMY primary yellow (AY23).

(d) Stimulation of color receptors by CMY primary magenta (RR180).

(e) Stimulation of color receptors by CMY primary cyan (AB9).

Figure 1.8: Excitation of color receptors by cyan, magenta, and yellow primary colors (CMY) for four-color printing, i. e., convolution integrals of reflectance spectra of primary colorants with tristimulus values of red, green, and blue receptors.

The overlap, or convolution, integral of the tristimulus curves with the colorant's reflectance spectrum is how our eye converts a spectral sensory stimulus into a triplet of values (R, G, B) to encode a color stimulus. The values are calculated according to

$$R = \int \bar{x}(\lambda) \cdot R(\lambda)\, d\lambda \tag{1.2}$$

$$G = \int \bar{y}(\lambda) \cdot R(\lambda)\, d\lambda \tag{1.3}$$

$$B = \int \bar{z}(\lambda) \cdot R(\lambda)\, d\lambda \tag{1.4}$$

where $\bar{x}(\lambda)$, $\bar{y}(\lambda)$, and $\bar{z}(\lambda)$ are the tristimulus functions for red-, green-, and blue-sensitive receptors, respectively, and $R(\lambda)$ is the reflectivity of the colorant.

In this process, *different* reflection spectra $R(\lambda)$ and $R'(\lambda)$ may lead to *the same* values (R, G, B), thus giving the same color impression. This phenomenon is called *metamerism*. Metamerism reveals itself only when one observes objects that apparently have the same color with a different light source. The inherent spectral characteristics of the colorants (reflectance or absorbance spectra) remain the same. However, the reflection depends on the external light source, so two new reflectance spectra are now present. It is unlikely that this pair of spectra will also map to the same color values, and we now perceive the different reflectance spectra as differently colored.

►Figure 1.8 gives an example of how the spectra of a set of CMY primary colors (writing or inkjet inks) are perceived. ►Figure 1.8(a) shows the reflectance spectra of three writing inks suitable as primary yellow, magenta, and cyan inks. ►Figures 1.8(c) to (d) illustrate the stimulus intensity for each receptor and each primary color. Calculating the integrals, we obtain $(R, G, B) = (0.45; 0.44; 0.11)$ for the yellow ink, $(R, G, B) = (0.40; 0.30; 0.31)$ for the magenta ink, and $(R, G, B) = (0.26; 0.31; 0.42)$ for the cyan ink. We do not focus on absolute values and are only interested in the relations in each triple. We see that a yellow color stimulates the red- and green-sensitive receptors, a magenta color stimulates the red- and blue-sensitive receptors and a cyan color stimulates the blue- and green-sensitive receptors. Web designers may recognize the familiar formulas for colors on the monitor: yellow = red + green, magenta = red + blue, cyan = blue + green. Due to the strong overlap of the green and red tristimulus functions, red- and green-sensitive receptors are similarly stimulated, in contrast to the web color formula.

The R, G, and B values directly define process colors such as CMYK yellow, magenta, cyan, and black. These are not defined by a specific spectral reflectance but according to ISO 2846 by color perception, quantified by their CIELAB color values L* (lightness), a* (green-red value), and b* (blue-yellow value). The mentioned standard and some technical guidelines clarify this for process colors employed in sheet-fed offset printing on LWC-I paper, ►Table 1.15. Within the stated tolerances, printing inks for four-color printing may thus have different chemistry. ►Figure 1.9 shows reflectance spectra of watercolors, which are also suitable as CMY primary colors but are different from the ink colorants.

Table 1.15: Example definition of primary colors of the CMYK process according to ISO 2846-1 using the example of printing inks for sheet-fed offset printing on LWC-I paper [825].

Primary color	L* (lightness)	a* (green-red)	b* (blue-yellow)	ΔE^*_{ab}
Cyan	57	−37	−46	5
Magenta	48	73	−6	5
Yellow	86	−2	89	5
Black	20	1	2	5

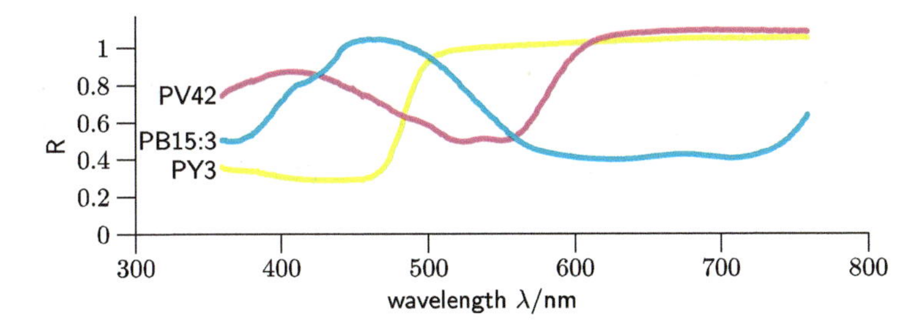

Figure 1.9: The reflectance spectrum of watercolors suitable as primary colors for four-color printing, normalized to an arbitrary unit: PY3 (Lemon yellow, Schmincke Horadam no. 215), PV42 (Magenta, Schmincke Horadam no. 352, primary magenta), and PB15:3 (Helio cerulean, Schmincke Horadam no. 479).

1.6 The interaction of light and matter

We dwell a bit on physics and consider a model describing light-matter interaction at an atomic level to better understand macroscopic pigment properties, such as transparency, reflection, and scattering. A detailed treatment of all phenomena is given in physics and optics textbooks and in [311, Chapters 8, 11, 13], [312, 314, 327, 328], [329, Chapter 3]. We will follow [312] in deriving some significant relations.

1.6.1 Basics of dielectric materials

If an electric field **E**, e. g., the field of light, meets matter, then **E**

- aligns existing electric dipoles in the material to itself, or
- induces new electric dipoles in the material and aligns them with itself

Both processes together are expressed by the *polarization* **P** of the material. A dielectric material responds to an external field **E** with a polarization **P**. The magnitude of the response **P** to **E** is called *dielectric susceptibility* χ. χ is a tensor describing how the material responds to an electric field:

$$\mathbf{P} = \epsilon_0 \chi \mathbf{E} \tag{1.5}$$

$$P_{i,i=x,y,z} = \epsilon_0 \sum_j \chi_{ij} E_j + \epsilon_0 \sum_j \sum_k \chi_{ijk} E_j E_k + \cdots \tag{1.6}$$

In ►equation (1.6), we have developed χ into a power series, which we can break off after its linear term. This operation is possible because the quadratic and all higher terms are only necessary for high radiation intensities, leading to nonlinear optics, which, e. g., is essential for laser technology. The first term represents the linear dependence at low radiation intensities, which is common when viewing paintings.

P is essential for us because it describes the interaction of light with matter, and we can derive it from models for simple cases. We can thus gain deeper insights into the nature of optical phenomena. These manifest as color, scattering, and reflection and are significant for pigments, painting materials, and varnishes.

Association with D and ϵ_r

When developing the basic electrodynamic equations, Maxwell used the quantities **D** (electric displacement) and ϵ_r (relative dielectric constant) to describe electric phenomena within solids instead of the polarization and susceptibility used today. **D** and ϵ_r are not direct experimental quantities like **P** and χ; we can establish their relationship using the plate capacitor:

$$\mathbf{D} = \mathbf{D}_0 + \mathbf{P} = \epsilon_0\mathbf{E} + \epsilon_0\chi\mathbf{E} = \epsilon_0(1+\chi)\mathbf{E} = \epsilon_0\epsilon_r\mathbf{E} \tag{1.7}$$

$$\epsilon_r = 1 + \chi \tag{1.8}$$

Types of polarization

Now that we know what polarization means—the question of the elementary mechanisms arises so that we can work out the tensor χ theoretically. One distinguishes four types of polarization:

- *Space charge polarization* or *surface polarization* occurs in the smallest pigment particles, surfaces of all kinds, at grain boundaries, and in all interfaces between objects. It bases on existing dipoles in the volume or surface. Under the influence of **E**, they orient themselves according to the electric field.
- *Ionic polarization* occurs in ionic crystals and ionic compounds. It bases on existing dipoles, usually compensating each other. Under the influence of **E**, their atoms are slightly shifted from their equilibrium position so that the dipole moment changes. An example of such a material is common salt, NaCl.
- *Orientation polarization* occurs in liquids and gases. It bases on existing dipoles. Under the influence of **E**, they orient themselves according to the electric field. An example material is water, H_2O.
- *Electronic or atomic polarization* occurs in all materials. It bases on negatively charged electron clouds and positively charged atomic cores. Under the influence of **E**, a displacement of the electron clouds relative to the atomic cores takes place and creates a dipole moment. This polarization occurs in all atoms; it is ubiquitous.

All polarization types occur side by side and result in a complex dielectric susceptibility. Thus, e. g., in SiO_2, the atomic polarization of a nonspherical sp^3-hybridized Si atom overlaps with ionic and covalent polarization contributions. The space charge polarization cannot generally be described, and discussions often omit it.

1.6.2 Microscopic view: the oscillator model

We can group the four polarization types for further microscopic analysis:
- In electronic and ionic polarization, the distances of electrons, protons, or ions change in response to **E**, resulting in (electrostatic) restoring forces. A suitable model is the dipole oscillator, and a prominent phenomenon is the *dipole resonance*.
- In the case of orientation polarization, no restoring forces occur; the dipoles aligned in response to **E** reorient statistically. Here, slow *dipole relaxation* occurs after **E** decays.

To show how the discussion of **P** and χ can help us, we focus only on electronic polarization in a solid. We assume that monochromatic light of angular frequency ω illuminates the solid. The oscillator model of Lorentz describes the dipoles caused by **E** and restoring forces. According to this model, the solid contains innumerable dipole oscillators of these types:
- Atomic oscillators of bound charges (electrons and protons or atomic nuclei) separated from the field **E** and returned by an electrostatic restoring force. In quantum mechanics, the oscillation corresponds to a transition from one energy state to another, excited by **E**.
- Vibrational oscillators of polar molecules with mechanical vibrational modes can be excited.

We will examine only atomic oscillators with bound electrons and exclude metallic substances with free electrons (described by the Drude model) and semiconductors.

Starting from an atomic oscillator, we can calculate the macroscopic polarization **P**. We assume an isolated dipole consisting of two electric charges $+q$ and $-q$ at positions $\mathbf{r}_+$ and $\mathbf{r}_-$ that has a dipole moment

$$\vec{\mu} = q \cdot (\mathbf{r}_+ - \mathbf{r}_-)$$

A solid comprises countless of these dipoles. If we expose the solid to an electric field **E**, the dipoles orient themselves according to the field, and we can consider **P** as the sum of all dipole moments per unit volume:

$$\mathbf{P} = \frac{1}{V} \sum \vec{\mu} = \langle \vec{\mu} \rangle \cdot N_V$$

where $\langle \vec{\mu} \rangle$ is the average dipole moment, and N_V is the dipole density. In atomic dipoles, **E** displaces electrons and the nucleus relative to each other. Since the change of the nuclear position is (by its mass) considerably smaller than the change of the electron position, we set $\mathbf{r}_+ = 0$ and $\mathbf{r}_- = \mathbf{x}$ so that the dipole moment is given by

$$\vec{\mu} = -q \cdot \mathbf{x} \tag{1.9}$$

Furthermore, we assume that $\mathbf{x} \parallel \mathbf{E}$ and that the solid is a cuboid, oriented perpendicular to $\mathbf{E}$. Then the dipole moments inside the solid will mutually compensate in pairs. However, net charges will remain on the two side surfaces A perpendicular to $\mathbf{E}$, in the form of a single layer of dipoles, having no direct neighbors to compensate for their moment. It is then $\mathbf{x} \perp A$. We can calculate $\mathbf{P}$ by summing up the net dipole moments in a thin layer $V = A \cdot x$ of thickness x:

$$P = \frac{1}{V_A}\sum_A -qx = -\frac{x\sum_A q}{V_A} = -\frac{x\sum_A q}{xA} = -\frac{\sum_A q}{A}$$

We have dropped the vector character of all quantities parallel to $\mathbf{E}$ and reduced $\mathbf{x}$ to a one-dimensional x in this direction. P is then the magnitude of the polarization vector for which $\mathbf{P} \perp A$ also holds. $\sum_A$ and V_A are related to the thin volume layer of area A and thickness x. The result shows that the polarization expresses a surface charge distributed in A. We have reduced the calculation of polarization to that of the deflection x or $\mathbf{x}$ of electrons in a field $\mathbf{E}$.

To calculate x, we describe in a highly simplified way the atomic oscillators driven by $\mathbf{E}$ as a harmonic oscillator with mass m_0, a damping coefficient γ, and charge q by the differential equation

$$m_0\frac{d^2}{dt^2}x + m_0\gamma\frac{d}{dt}x + m_0\omega_0^2 x = -qE(\omega, t) = -qE_0 e^{-i\omega t} \tag{1.10}$$

The equation has a general solution $x(\omega, t)$ of the form

$$x(\omega, t) = X_0(\omega) \cdot e^{-i\omega t} \tag{1.11}$$

$$X_0(\omega) = -\frac{qE_0}{m_0}\frac{1}{\omega_0^2 - \omega^2 - i\gamma\omega} \tag{1.12}$$

with a time-dependent and a frequency-dependent part. If E_0 and X_0 are complex numbers, we also have considered any phase shifts. ω_0 is the resonant frequency of the oscillator. The polarization of a solid of N dipole oscillators is then

$$P_{\text{dipole}} = N\vec{\mu} = -Nqx = \frac{Nq^2E}{m_0}\frac{1}{\omega_0^2 - \omega^2 - i\gamma\omega} \tag{1.13}$$

We use the result to introduce, via the dielectric shift D, the complex *dielectric function* $\epsilon(\omega) = \epsilon_1(\omega) + i\epsilon_2(\omega)$:

$$\begin{aligned} D = \epsilon_0\epsilon_r E &= \epsilon_0 E + P = \epsilon_0 E + P_{\text{other}} + P_{\text{dipole}} \\ &= \epsilon_0 E + \epsilon_0\chi E + \frac{Nq^2E}{m_0}\frac{1}{\omega_0^2 - \omega^2 - i\gamma\omega} \end{aligned}$$

$$\epsilon_r = 1 + \chi_{\text{other}} + \underbrace{\frac{Nq^2E}{\epsilon_0 m_0} \frac{1}{\omega_0^2 - \omega^2 - i\gamma\omega}}_{\chi_{\text{dipole}}} \tag{1.14}$$

$$\epsilon_1(\omega) = 1 + \chi_{\text{other}} + \frac{Ne^2}{\epsilon_0 m_0} \frac{\omega_0^2 - \omega^2}{(\omega_0^2 - \omega^2)^2 + \gamma^2\omega^2} \tag{1.15}$$

$$\epsilon_2(\omega) = \frac{Ne^2}{\epsilon_0 m_0} \frac{\gamma\omega}{(\omega_0^2 - \omega^2)^2 + \gamma^2\omega^2} \tag{1.16}$$

We separated the polarization into the part induced by atomic dipole oscillators, which we were able to calculate approximately, and a part caused by other polarization types, generally described by a constant χ_{other}. There exist no other polarization types in a solid except the electronic polarization. Given that in the solid, there are i different dipoles with different resonant frequencies ω_i, the result for ϵ_r would be

$$\begin{aligned} D = \epsilon_0\epsilon_r E = \epsilon_0 E + P &= \epsilon_0 E + P_{\text{dipole}} \\ &= \epsilon_0 E + \frac{Nq^2E}{m_0} \sum_i \frac{1}{\omega_i^2 - \omega^2 - i\gamma\omega} \\ \epsilon_r &= 1 + \frac{Nq^2E}{\epsilon_0 m_0} \sum_i \frac{1}{\omega_i^2 - \omega^2 - i\gamma\omega} \end{aligned} \tag{1.17}$$

We derived this result from inspection of a specially shaped solid under particular circumstances, but regarding our subject, it is also valid for bodies in general and also fluids. The dipole resonance of the oscillating dipoles gives rise to typical optical phenomena at $\omega \approx \omega_0$ due to their resonant frequency ω_0:

- Absorptions in the UV and VIS spectral range by atomic oscillators having high resonant frequencies.
- Absorptions in the IR range by vibrational oscillators having low resonant frequencies.

We must consider two cases for ω.

$\omega = \omega_0$, resonance and absorption

If the incident light is in resonance with the oscillator, energy is efficiently extracted from the radiation field, transferred to the oscillator, and amplifies its oscillations considerably, ►Figure 1.10(a). The energy of the light must be precisely equal to the difference between two energy levels of the oscillator so that the light can raise an electron to a higher level. The electron can return to the ground state under reemission of radiation of the same frequency. In contrast to the directed field **E** of the incident light, reemission occurs uniformly in all directions, called *elastic scattering*, ►Figure 1.10(b). In a many-particle system, the energy of the excited electron can also be transferred to lattice vibrations, i. e., thermalized.

(a) Excitation of a single oscillator to oscillate by an incident electromagnetic field.

(b) Undirectional reemission of the absorbed energy.

Figure 1.10: Excitation and relaxation of an individual oscillator after irradiation with light.

At the macroscopic level, we observe an *absorption*, contributing to both the *color* of the body and its *heating*.

$\omega \neq \omega_0$, off-resonance and transparency

The incident light drives the oscillator, but no efficient energy transfer ensues. Again, the oscillator reemits radiation of the same frequency undirected in all spatial directions (elastic scattering), ►Figure 1.10(b). However, we now observe a phase shift due to the forced oscillations: for $\omega < \omega_0$, the oscillator follows the external field in phase but with a phase shift. For $\omega > \omega_0$, the oscillator follows the fast external field only inversely phased.

Macroscopically, these processes lead to *transparent* bodies since no energy is extracted from the incident light. The light passes in the solid from one oscillator to another in a relay race. In the process, all phase shifts add to each other. From the outside, it seems as if the light propagates within the body at a slower speed. The visible effect of this slowing down is the *refractive index n* of the body. Since the oscillators can no longer follow the fast changes of the incident field, the described phenomena are especially observable at $\omega < \omega_0$.

Transition to macroscopic description variables

The presented oscillator model starts from a harmonic oscillator with resonance frequency ω_0 and allows for the calculation of the polarization, and thus the response of a body to incident light. The most important result is the complex dielectric function ϵ_r, from which we can derive the complex refractive index $n+ik$ and, subsequently, the macroscopically observable quantities refractive index n, absorption coefficient k, and extinction α:

$$n + ik = \sqrt{\epsilon_r(\omega)} \tag{1.18}$$

$$n(\omega) = \sqrt{\frac{\epsilon_1 + \sqrt{\epsilon_1^2 + \epsilon_2^2}}{2}} \tag{1.19}$$

$$k(\omega) = \sqrt{\frac{-\epsilon_1 + \sqrt{\epsilon_1^2 + \epsilon_2^2}}{2}} \tag{1.20}$$

$$\alpha(\omega) = \frac{2\omega k}{c} \tag{1.21}$$

The four quantities describe the optical properties of a body in general terms, but what happens to the radiation each oscillator reemitted? In the transition from a single microscopic oscillator to a macroscopic body, the reemitted radiation from all individual oscillators, ►Figure 1.11, is superimposed, and a macroscopic observable phenomenon emerges. In the particle model of light, we observe that some light rays are reflected at the same angle as the incident light (reflection). In addition, we observe that other light rays travel into the body at a larger angle (transmission and light refraction). In the wave model of light, we also notice a reflected and a transmitted wavefront. The transmitted wavefront, passing through the body, has a lower propagation velocity v than the incoming wave c, or a longer wavelength $\lambda_2 > \lambda_1$, due to all phase shifts summed up, and it is, therefore, inclined at an angle $\theta' < \theta$ against the surface of the body.

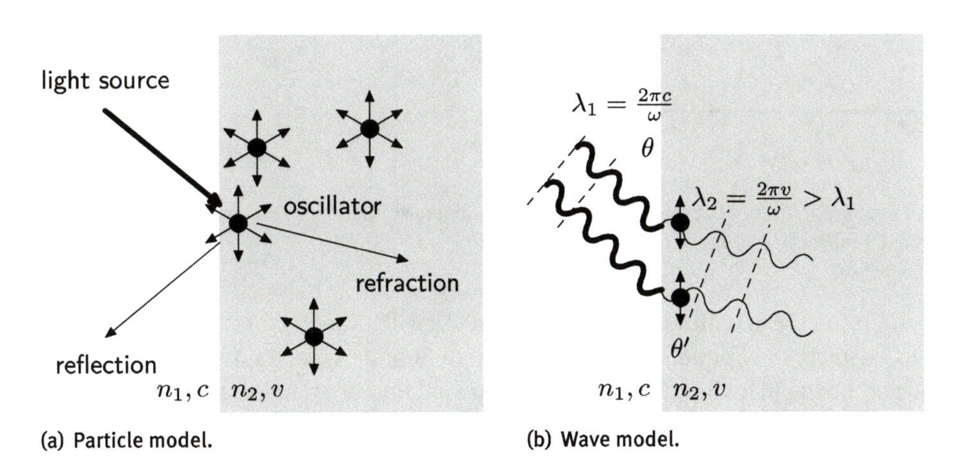

(a) Particle model. (b) Wave model.

Figure 1.11: Irradiation of a body with light. The reemitted radiation of all microscopic oscillators superimposes to a total wave comprising several components: transmitted light, refracted light, and reflected light. The intensity difference between incident and outgoing radiation can be ascribed to absorption. θ is the entrance angle, θ' is the exit angle. n_1 and c are the refractive index and propagation velocity in the environment, n_2 and v in the body.

The microscopic examination of an illuminated body helps us to understand three vital, macroscopically observable phenomena, which are directly responsible for optical effects related to paintings:
- absorption of light of a specific wavelength in the body, ►Section 1.6.3
- refraction and transmission of light through the body, ►Section 1.6.5
- scattering and reflection by the body, ►Section 1.6.6

Subsequently, instead of using cumbersome frequencies ω, we will switch to wavelengths that are more familiar when speaking about color phenomena in painting, i. e., instead of a resonance frequency ω_0, we speak of the resonance wavelength λ_0, etc. The relation is $\lambda\nu = \frac{\lambda\omega}{2\pi} = c_{\text{medium}}$, c being the speed of light of the body or medium in question.

1.6.3 Macroscopic view: absorption

We quantify the absorption strength by the *absorption coefficient* $k(\lambda)$, which depends on the wavelength. It becomes maximum at the resonant wavelength λ_0 of the oscillator (►Equation (1.20) and ►Figure 1.12).

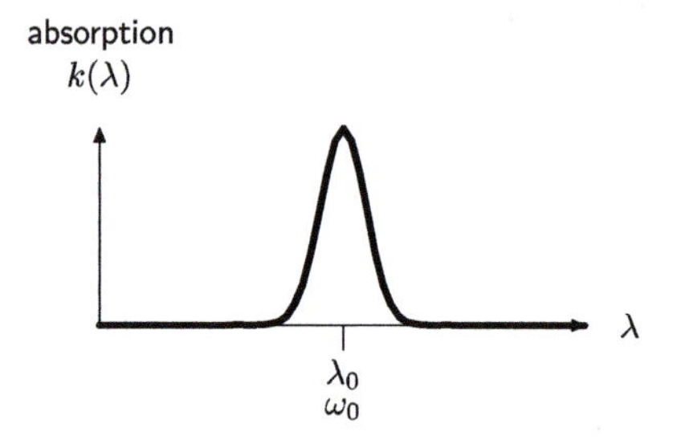

Figure 1.12: Graph of the absorption coefficient according to ►equation (1.20) for an oscillator with resonant frequency ω_0 or resonant wavelength λ_0.

Depending on the magnitude of λ_0, we are dealing with different forms of energy. ►Figure 1.13 depicts the frequency range from UV to IR and marks the typical absorption bands occurring in bodies. For this text, we are not interested in infrared absorptions, which excite thermal lattice vibrations and are the subject of IR spectroscopy, a helpful analytical technique for art examination. Instead, we focus on excitations of electronic transitions in the UV and visual spectral range since they induce the impression of the *color* of a body (►Section 1.5 for the relation between λ_0 and color perception).

The spectrum drawn in black in ►Figure 1.13 belongs to a colorless (transparent) body since no visible light is absorbed. We try to achieve absorption in the visible spectral range when designing a pigment by *chromophores*, e. g., shown in yellow in the spectrum. Mostly, however, absorptions are in the UV range, so we have to shift them into the visual spectral range by *auxochromes* (a so-called *bathochromic shift*). The discussion of resonant wavelength and methods to maximize the intensity of k provides the central part of the following chapters of this book.

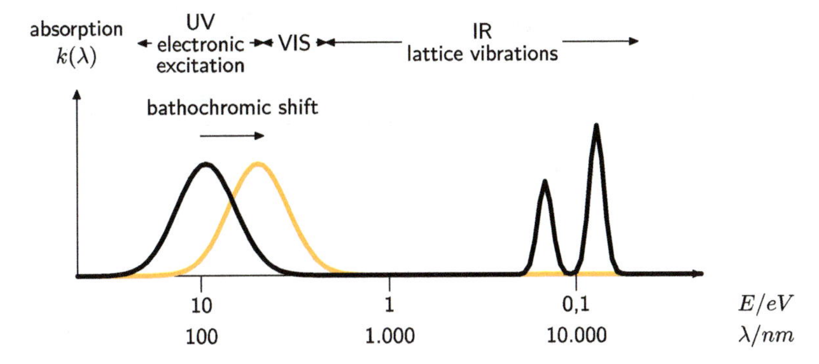

Figure 1.13: A typical graph of the absorption coefficient k over a broad spectrum range. Shown are several resonance frequencies and their typical excitation modes. Most compounds absorb in the UV range and are colorless (black graph). Therefore, they can only be used as colorants if their absorption bands are shifted into the visual spectral range by chemical modifications (bathochromic shift, yellow graph). Colorants possess absorption bands in the visual spectral range and exhibit chromaticity (yellow graph).

1.6.4 Macroscopic view: Absorption by size-dependent collective excitations, surface plasmons

The descriptive quantities $\epsilon(\omega)$, $n(\omega)$, and $k(\omega)$ or $\epsilon(\lambda)$, $n(\lambda)$, and $k(\lambda)$ determine the optical properties of an atom, ion, or molecule. They depend on the resonance frequencies ω_0 or the resonance wavelengths λ_0 of the electronic transitions, e. g., between molecular orbitals. They are, therefore, primarily determined by the electronic structure of the compound.

In the transition from the isolated oscillator to clusters of a few oscillators and then to a macroscopic solid, phenomena occur which do not originate in the oscillator or chromophore itself. Instead, they are induced only through the interaction of many chromophores. They cause a dependence on the size and shape of the chromophore particle [328], [329, Chapter 4]. The phenomena are studied in the context of the physics of (metal) clusters and can be attributed to two effects.

- Intrinsic effects are size- and shape-dependent changes in volume and surface properties such as ionization potential, binding energy, crystal structure, chemical reactivity, and the location of energy levels in compounds. Since the location of energy levels determines λ_0, a dependence $\epsilon = \epsilon(\lambda, \lambda_0, r, g)$ on size r and shape g of pigment particles results due to the intrinsic effects.
- Extrinsic effects are size- and shape-dependent optical responses to external (electromagnetic) fields or forces. These include the development of collective electronic oscillations (Mie resonances or surface plasmons).

►Table 1.16 shows a schematic division into suitable size classes. We are interested in clusters from a few nanometers to micrometers, represented by pigment particles.

Table 1.16: Classification of materials in terms of size to describe size-dependent optical appearances [328].

	Very small clusters	Small clusters	Large clusters	Bulk material
Number of atoms	< 20	< 500	$< 10^7$	$> 10^7$
Particle radius r	< 1 nm	< 4 nm	< 100 nm	> 100 nm
Mathematical treatment	Molecular orbital theory	Jellium theories	Solid-state theories (electrodynamics)	
Phenomena	Single/multielectron excitation Quantum size effect	Collective electron excitation (Plasmons)		Solid state spectra
Influence on ϵ	λ_0, r	λ_0, r	$\lambda_0(r)$	$\lambda_0 \approx$ const (material constant)
	Intrinsic size effect		Extrinsic size effect	

Absorption phenomena occur in them that cannot be accounted for by electronic transitions alone, as we have already indicated in the Introduction in ►Figure 1.2.

For *small* particles, intrinsic effects become potent, strongly changing the optical quantity $\epsilon = \epsilon(\lambda, \lambda_0, r, g)$ on a minor scale r. For example, LF transitions only arise when a free ion is introduced into an environment of ligands. Semiconductors provide another example: the band gap depends on particle size (quantum size effect) since atomic energy levels evolve into a crystalline band structure comprising many molecular orbitals, ►section 2.2, specifically ►p. 88.

Large pigment particles lie between small clusters and bulk material and are described using electrodynamic theories. These consider atomic or molecular oscillators, delays, and reflections of incident light waves in extended bodies and free electrons' excitations in the particle's volume. Their optical spectra can be characterized by both atomic or molecularly determined absorptions and collective resonances (Mie resonances or *surface plasmons*), which depend strongly on the size and shape of the particle, the incident wavelength, and the angle of observation.

These phenomena are based on the interaction of the electromagnetic field of light with the particle. Mie calculated in 1901 how an incident field behaves in spherical particles, which contain some hundreds to millions of electric oscillators and whose size is some multiples of the irradiated wavelength. The results are expressions for the field in the spheres' inner and outer regions, i. e., for the transmitted, reflected, and scattered waves. Those allow us the calculation of macroscopically observable quantities such as absorption and reflectivity. Corresponding calculations for nonspherical bodies are mathematically challenging to perform; we obtain their solutions numerically.

The expressions for scattered waves allow deriving conditions for resonances that lead to maxima in the extinction. These conditions can be satisfied by the electronic structure ϵ of the material. Suitable geometrical dimensions of the body, as in the case of rainbows or droplets, provide them as well.

Contributions of bound electrons

We will inspect almost exclusively ways to create energy levels for *bound* electrons by chromophores, between which electronic transitions can occur in the visible region of the spectrum. Such shifts determine the course of ϵ as shown approximately as harmonic oscillators with a resonance at λ_0. If λ_0 is in or near the visible spectral region, the pigment particles appear colored.

Contributions of free electrons, (surface) plasmons

Some materials used for pigments have *free* electrons in addition to bounded ones and exhibit additional resonances. These can explain otherwise incomprehensible phenomena, e. g., the dependence of the color of iron oxides on grain size and shape. Such materials include metals, semiconductors, alkali, and noble metals, and in the field of pigments, in particular semiconducting metal oxides and sulfides. Furthermore, free electrons helps us to explain the metal's properties such as reflection.

Electrons in the conduction band (plasma) are excited by the irradiated field to collective oscillations, so-called *surface plasmons*, which lead to oscillating net charges at the particle surface. The field sustains them if its wavelength is resonant with the oscillatory motion, ►Figure 1.14. The energy corresponding to the absorption is perceived as color at a resonant frequency in the visible range.

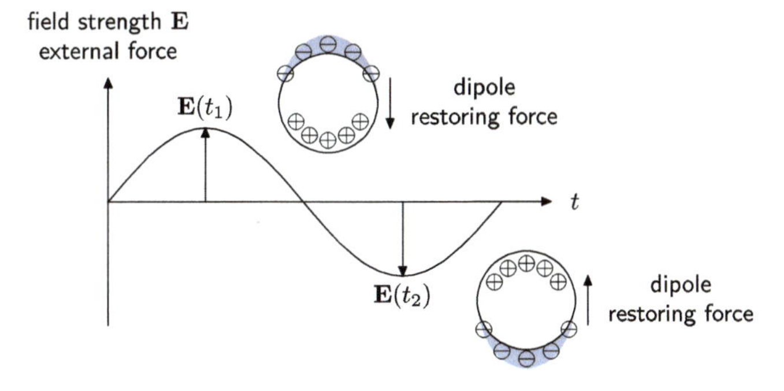

Figure 1.14: Excitation of a surface plasmon in a colloid particle. The light incoming from left to right generates an electric field depending on polarization, shown here in the up-down direction. The free-conduction electrons follow the field collectively. An oscillating dipole of free electrons (blue ⊖) and positively charged metal cations ⊕ is formed on the metal surface, causing a repulsive force and electron oscillations. Surface plasmons correspond to the natural frequencies of this oscillation; absorption of light occurs when the incident light is resonant to the plasmon frequency [661].

Textbooks of solid-state physics explain the physics of plasmons in more detail [322, 330, 331]. Burda et al. [659] and Riss and Diwald [354] address all questions around the size and shape dependence of materials, particularly the color of colloidal metals, as they are present in the gold ruby glass. Economou and Ngai [660] is a theoretical introduction to a *solid-state plasma* formed by atomic hulls and conduction electrons in the solid state. The cause of the deep color of metal and semiconductor colloids is explained in [33, 34, 330, 331, 661–663] and in great detail in [328, 329].

Basic surface plasmon dependencies

Surface plasmons show a pronounced dependence on the size and shape of the particle and appear only at specific spatial ratios: Gold colloids of 99 nm diameter show absorption at 575 nm, those of 22 nm, one at 525 nm [659]. Water waves provide a familiar example in containers or cups; their resonant frequencies and vibration modes depend strongly on the size and shape of the respective container. Since plasmon resonances are not symmetry or Laporte-forbidden like ligand field transitions, their extinction coefficients are very high ($\epsilon \approx 1 \cdot 10^9\ \mathrm{mol^{-1}\ cm^{-1}}$), and the colors are intensive.

►Figure 1.15 shows some plasmon resonances' dependencies in metal colloids on important particle parameters schematically [329, p. 20]. These dependencies apply in the same way to semiconductor particles such as iron oxide or cadmium sulfide. Therefore, they give us vital clues to the parameters that must be monitored during the preparation of pigment particles to achieve consistently high pigment quality.

A crucial factor for the purity of the resulting color is the size distribution of the particles. The more consistent the particle size, the sharper the absorption peaks of the resonances and the purer the observable colors, ►Figures 1.15(a) and (b). This observation explains a manufacturing method for natural ocher in times passed: ocher soils are slurried in water several times, separating the ocher particles into batches of different but homogenous sizes. The more homogenous the sizes, the purer and brilliant the color.

In producing glasses containing colloidally distributed colorants (gold, red, and yellow semiconductor particles), the formation of aggregations may alter the colors obtained. While isolated metal particles give rise to sharp resonances and pure colors, the peaks broaden as the colloidal particles aggregate since the aggregates have increasingly irregular shapes and varying sizes, ►Figure 1.15(c).

In ►Figure 1.15(d), we see the predominant influence of the particle shape: spherical and cylindrical particles differ clearly in the form of the plasmon resonance. We understand this influence if we consider that the charge that accumulates in the surface regions of the particle determines the restoring forces of the electron oscillation. The shape of the particle directly determines the volume available for this charge. This dependence, e. g., becomes vital in yellow ocher particles: The color of acicular yellow ocher varies between greenish-yellow and warm yellow [65, Chapter 6.4] depending on the length of the needles. Gold nanotubes whose length-to-diameter ratio is varied

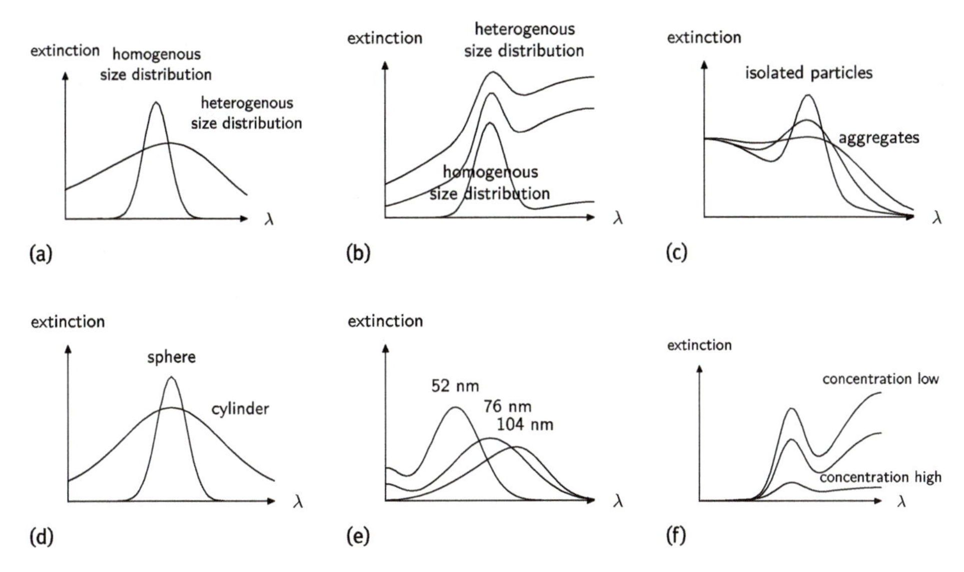

Figure 1.15: Dependence of the plasmon resonance on various parameters (schematically drawn after [329, Chapter 2]). (a) and (b): Broadening of resonance with heterogeneous size distribution in Fe_2O_3 (a) and silver (b), particles of uniform size show narrow band resonances. (c): Influence of aggregation of gold particles on resonance/color. Isolated particles show sharp resonances, while the accumulations of different sizes and heterogeneous shapes lead to broad absorptions. (d): Influence of the shape of gold particles on resonance. Spherical particles yield sharp resonances; deviations lead to broadened peaks. (e): Broadening the resonance during the growth of a gold colloid due to aggregation of the particles. As the degree of aggregation increases, the particles become larger and more irregular, the effective particle size increases, and the peak broadens. (f): Transition of a sharp plasmon resonance from silver particles to the nonspecific solid spectrum as the particle concentration increases to the bulk material.

illustrate the influence of shape. The generated plasmon resonance can extend over almost the entire visual spectrum [663].

►Figure 1.15(e) displays a color change from red to blue during the growth of gold colloids; here, the effective particle size grows due to particle aggregation. The example of iron oxide shows that parallel to this shift, multipole resonances emerge above a specific particle size.

Finally, ►Figure 1.15(f) shows how resonance gets lost as a phenomenon of tiny particles; as the concentration of the particle increases, the solid-state spectrum emerges that no longer exhibits plasmon resonance.

Surface plasmons with dipole character: example gold ruby glass

Beautiful representatives of surface plasmons, in the truest sense, appear in the form of the bright red-gold ruby glass or the yellow-colored silver glass, ►Section 3.9.1.2. In these, gold or silver particles are colloidally distributed, fulfilling all the conditions for the emergence of surface plasmons. ►Figure 1.16 shows the observed absorption spectra for order 2–10 nm particles. We detect a sharp absorption peak around 520–530 nm in the green in the gold ruby glass (resulting in a red color) and about 405 nm in the blue in the silver glass (yielding a yellow color). The figure shows how the peak evolves and gains intensity as the particle size increases. However, the absorption frequency remains approximately constant in the narrow size range where the resonance occurs. If the particles grow to about 300 nm, the peaks broaden due to the addition of higher multipole vibrations, shift to higher wavelengths, and lose intensity. If the particles are too large, the peak is wholly lost, and the glass becomes grayish-cloudy due to big embedded metal particles.

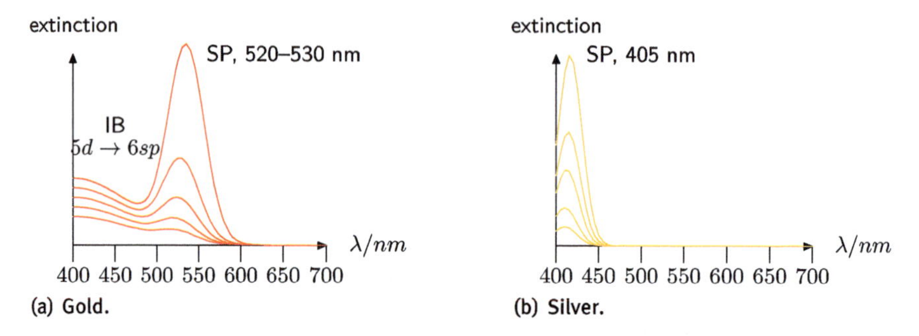

Figure 1.16: Dependence of plasmon resonance (SP) on the particle size of gold (a) and silver (b) colloids in gold ruby and silver glasses, respectively, in a size range of about 2–10 nm (schematically drawn after [329, Chapter 6.2]). The SP peaks are not yet broadened by higher multipole vibrations or shifted to higher wavelengths for such small particles. For gold, in addition to the SP in the small wavelength region, we see the interband transitions $5d \rightarrow 6sp$ (IB) of the noble metals, which are in the SP region for gold and copper.

The noble metals gold, silver, and copper used for annealing coloring show particularly intensive colors: in addition to the actual conduction electrons, interband transitions (gold: $5d \rightarrow 6sp$, silver: $4d \rightarrow 5sp$, copper: $3d \rightarrow 4sp$) transfer *d* electrons into the conduction band available for plasmon excitation [329, Chapter 6.2]. These interband transitions are also visible in gold particles at low spectrum wavelengths.

Surface plasmons with multipole character: example iron(III) oxide

The intense color of the gold ruby glass originates from a single sharp absorption peak. However, surface plasmons induce several peaks in the absorption spectrum of the particle, having different *multipole orders*. We see only the intense peak of dipole oscillation for small particles, but for larger particles or materials with other electronic structures, we can observe higher multipole oscillations. The higher vibrational modes broaden the absorption peak of the dipole vibration; they develop more clearly as the particle size increases. ►Figure 1.17 schematically shows sharp dipole peaks for

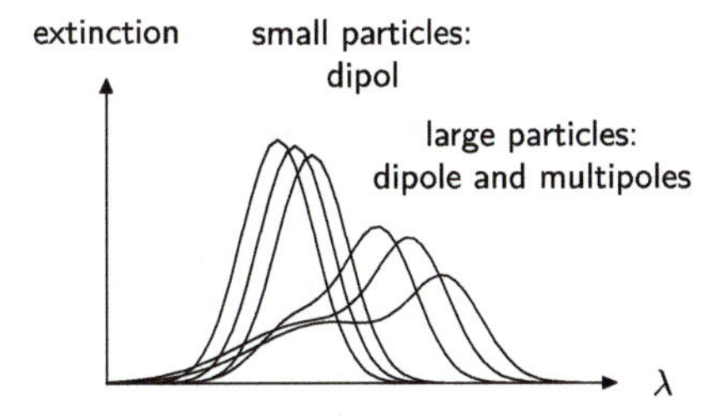

Figure 1.17: Dependence of plasmon resonance on the particle size of aluminum particles in size range 10–90 nm (schematically drawn after [329, Chapter 6.1]). Small particles show large size-independent and sharp dipole resonances at low wavelengths, while larger particles show broad superpositions of dipole and multipole resonances at higher wavelengths.

small aluminum particles, which broaden with increasing particle size due to the formation of multipole oscillations at higher wavelengths.

Iron oxides also have the character of semiconductors and can create surface plasmons. The required free electrons come from the conduction band of iron oxide, which forms from empty iron orbitals. The valence band is formed from the partially filled *4sp(Fe)* and filled *2p(O)* MOs. In ►Figure 2.34, the conduction band corresponds to MOs starting from $3a^*_{1g}$, the valence band to the nonbonding oxygen MOs $1t^*_{1g}$, $1t^*_{2u}$, the oxygen metal MOs $2t_{1u}$, $3t_{1u}$, and the underlying MOs with oxygen character. The split *3d(Fe)* orbitals lie between the bands [223, Chapter 4.2.2.2].

In painting practice, multipole resonances are significantly involved in the color impression of red and yellow ocher pigments and explain why bluish-purple hues appear when sintering red ocher, ►Figures 1.18 and 3.14, and [634], [329, Chapter 6.11]. Small particles of red ocher are dominated by LMCT absorption in the blue spectral region and an electron pair transition in the green one, leading to the orange-red color. With increasing particle size, a plasmon resonance with dipole character develops from about 100 nm in the green spectral region, joined by higher multipole resonances in the yellow one and then the red one. Large particles shift the color impression to the bluish-purple hue of the sintered red ocher (Caput Mortuum). Analogous resonances also change the color impression of the particles of yellow ocher.

1.6.5 Macroscopic view: transmission, refraction, dispersion

Let us now turn to transmitted light. In a macroscopic body, all oscillators, taken together produce a consistent wavefront passing through the body.

Refractive index

Being outside of a body, the velocity of propagation *v* of the light waves inside the body appears smaller than its velocity in air or vacuum *c*.

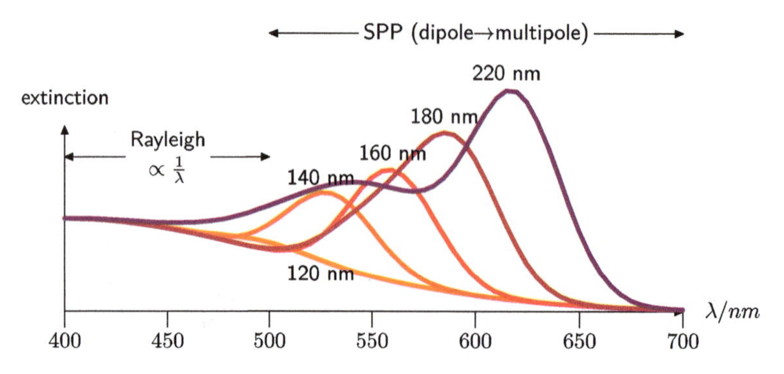

Figure 1.18: Dependence of plasmon resonance on particle size (100–220 nm) of iron(III) oxide (schematically drawn after [329, Chapter 6.11]). For tiny particles, only the CT and LF absorptions in the UV and VIS/IR are visible (orange-red color). As particle size increases, plasmon resonance develops, shifting from the low wavelengths of dipole oscillation around 500 nm to the high wavelengths of higher multipole oscillations. As a result, small particles are orange-red, while the resonance development in large particles produces a bluish-purple color impression. ►Figure 3.14 shows measured spectra of red iron oxide pigments.

Since the wavelength is proportional to the propagation speed ($\lambda \propto v$), it is smaller inside the body than outside. At the boundary surface of the body, however, the incoming and outgoing waves must match so that all wave crests and troughs of both sides coincide. Due to the different wavelengths, this is only possible if the direction vector of the outgoing wave is inclined against the incoming one. In the particle model, we can see this as well; here, the beam is refracted toward the perpendicular. Snellius has formulated a simple law relating entrance angle θ and exit angle θ' (or, in the wave image, the propagation velocities):

$$\frac{\sin\theta'}{\sin\theta} = \frac{n_2}{n_1} = \frac{c}{v} = n \tag{1.22}$$

The material constant n is specific to air and body and is called the *refractive index*. It was already derived from the microscopic model with ►equation (1.19). n_1 is the refractive index of air with the value of 1.000272; ►Table 1.17 shows some numerical values for substances from the artists' environment.

The theoretical treatment (►equations (1.19) and (1.20)) shows that n and the absorption coefficient k are coupled and often combined into the complex refractive index $n + \mathbf{i}k$. It is strongly wavelength-dependent and undergoes significant changes at the resonant frequencies at which a body absorbs light, ►Figure 1.19. This leads to dispersion and refraction of light in prisms.

High refractive index glass, lead crystal

According to ►equations (1.15) and (1.16), we can increase the dielectric constant, hence n and k, if we increase the oscillator density N as much as possible.

Table 1.17: Refractive index n_D at 589 nm (sodium D-line) and dispersion $d = n_B - n_G$ (B-line: 686.7 nm, G-line: 430.8 nm) for some selected substances.

Substance	n_D	d	Substance	n_D	d
Air	1.000		Linseed oil aged	1.60	
Water	1.333		Flint glass, crystal glass	1.6	
Ethanol	1.361		Dense flint	1.7	0.040
Toluene	1.496		Sapphire	1.77	0.018
Linseed oil	1.48		Zinc oxide	1.99	
Polymethyl methacrylate	1.50		Zircon	2.15	0.039
Window glass	1.5	0.010	Diamond	2.418	0.044
Borosilicate crown glass	1.5	0.010	Rutile	2.76	
Polyester, alkyd resins	1.55				
Calcium carbonate	1.57				

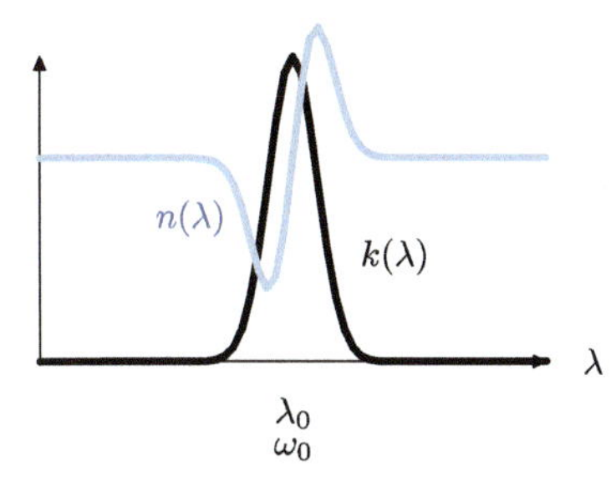

Figure 1.19: Wavelength dependence of refractive index n (blue) and absorption coefficient k (black) according to ►equations (1.19) and (1.20).

Scholze [31, Chapter 3] discusses the refraction of light from glasses. Based on the equation for the refractive index,

$$n^2 = \frac{1+2Y}{1-Y}, \quad Y = \frac{4\pi N_L}{3}\frac{\alpha\rho}{M_r}, \tag{1.23}$$

which Lorentz and Lorenz derived from Maxwell's equation, it becomes clear how high oscillator density can be achieved and what other influencing factors exist. N_L is the Loschmidt number, α is the polarizability, ρ is the density, and M_r is the average molecular weight of the compound that forms glass. We see from the equation that to achieve high refractive indices, high polarizability α and low molar volume $\frac{M_r}{\rho}$ are necessary.

The anions of the glass mass have high polarizability and contribute significantly to the refractive index, especially the oxygen anions, which are abundant in number. Depending on the role of the oxygen atoms in the glass, their contribution varies. In pure SiO_2 glasses, the oxide anions appear as bridging members and have little polarizability, so these glasses are low refractive. In alkali silicate glasses $SiO_2 \cdot R_2O$, we find many oxide anions as separation-point oxygens. They are strongly polarizable due to their partial charge, so these glasses are higher refractive:

structure with bridging oxygen structure with separation oxygen

In the series Li → Na → K, the polarizability increases. However, potassium silicate glasses have a lower refractive index than lithium silicate glasses because the molar volume also plays a role, which is lower in the case of lithium glasses. The cross-linked SiO_4 tetrahedra have many cavities in which the small lithium ions find space, while the large potassium ions require a structural expansion so that the molar volume of the lithium glasses is smaller while their density and refractive index are higher in total.

Y is additive concerning the glass components, so we can compare the contributions of different cations to the refractive index, ►Table 1.18. We note the increase in polarizability for larger or more highly charged cations and the low value for the silicon cation with a closed electron shell. The high value for Pb^{II} cations is striking; it explains why lead crystal glass is particularly highly refractive and shows a sparkling brilliance ►equation (1.25). The refractive index is based on the high polarizability of the two 6*s* electrons, which are not localized in bonds and can be readily excited to oscillate by external radiation fields [30, Chapter 7.5].

Table 1.18: Molar refraction $4\pi N_L \cdot \frac{1}{3} \cdot \alpha$ of different cations in glasses [31].

Cation	Molar fraction	Cation	Molar fraction	Cation	Molar fraction
$Na^{\oplus}$	0.44	$Ba^{2\oplus}$	4.02	$Pb^{2\oplus}$	9.13
$K^{\oplus}$	2.07	$Ca^{2\oplus}$	1.18	$Si^{4\oplus}$	0.084

Dispersion

►Figure 1.20 shows the progress of *n* over an extensive frequency range when absorptions occur in both the IR and UV ranges. We see that *n* changes significantly in the visible spectral range when these absorptions are strong. We also see both the long-wavelength slope of UV absorptions and the short-wavelength slope of IR absorptions, *n* becoming smaller toward longer wavelengths. Blue light is therefore refracted more strongly than red light. The difference of *n* at both ends of the visible spectral range is called *dispersion*; ►Table 1.17 lists typical values.

Dispersion is the cause of many optical phenomena: both the splitting of white light into prism and rainbow and the fire of gemstones, especially diamond with its

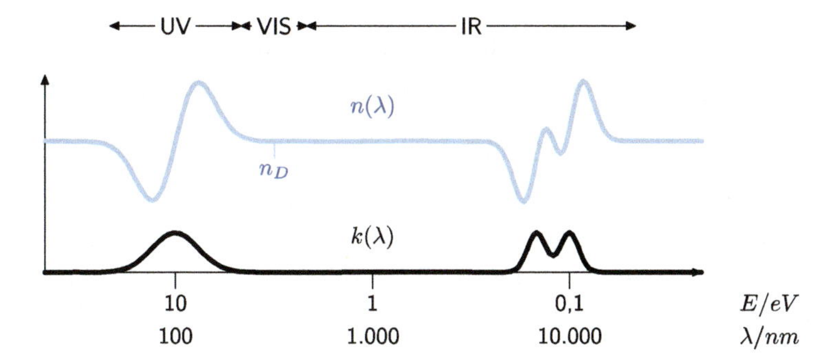

Figure 1.20: Typical progress of the refractive index (blue curve) with three resonant frequencies over a broad spectrum range. Dispersion occurs as the refractive index changes in the visible spectral range due to absorption peaks in the UV and IR range (black curve). In the example, n falls with increasing wavelength (normal dispersion). The frequently quoted value n_D is the refractive index at the wavelength of the sodium D-line (589 nm).

brilliant luster. While high dispersion is essential for gems (especially the just mentioned diamond), it must be low in eyeglasses or photographic lenses to obtain as few optical artifacts as possible. Accordingly, glasses are divided, among other things, into high dispersion flint glasses and low dispersion crown glasses, among others, ►Section 3.9.

1.6.6 Macroscopic observation: scattering, reflection, brilliance

We have seen that a macroscopic body absorbs a part of the incident radiation and thus appears colored (►Section 1.6.3), refracts another part when passing from the direction of incidence and, therefore, seems transparent, ►Section 1.6.5. It also reflects a third part on which we want to focus in the following. Depending on the magnitude of the body, the reflection appears in different ways [313, Chapter 4]:

- A macroscopic body with a perfectly smooth surface forms a reflected wavefront due to the superposition of many reemitted radiation components concentrated in a preferred direction. It is a reflection, or at high intensity, brilliance on the surface, ►Figure 1.21(a). Since the wavelength remains unchanged during reemission, the reflection leads to a mirroring of the incident light.
- On close inspection, the surface of even a polished body is not ideally flat but rough. It consists of many small specular surfaces pointing only approximately in the same direction. Consequently, a part of the light is not reflected in emission direction but diffusely, i. e., in all directions, ►Figure 1.21(b). The proportion of diffuse reflection increases with the roughness of the surface.

  However, diffusely reflected light is a macroscopic phenomenon. Even these tiny surfaces still consist of so many oscillators that their scattered light is superimposed to form a macroscopic wave that traces an image of the light source. It is not related to the scattered light emitted in many directions by a single or a few oscillators.

- If the size of the body approaches the wavelength of the light, only a few oscillators are involved. In addition to geometrically small bodies, a very rough surface with a strongly fluctuating refractive index can induce many tiny scattering centers. Microscopic effects, namely elastic scattering, determine the oscillators' emergence and propagation of *scattering light*. We are no longer dealing with reflection but with *Mie scattering*, ►Figure 1.21(c). Scattered light emitted under the Mie regime has a complex directional dependence without dominant directions and is easily perceived as "white." The fraction of diffusely scattered light increases as the body size decreases, eventually reaching a maximum of half the wavelength.
- Finally, if the size of the body falls below the wavelength of the light and we approach a system of a single oscillator, we observe *Rayleigh scattering*, which dominates in the forward-backward direction and is bluish to the side, ►Figure 1.21(d).

The reflected or scattered radiation wavelength does not change because they are forced oscillations following the exciting wave with the same frequency. Johnston-Feller [9, Chapter 6] explains the microscopic causes in detail.

Scattering

Microscopic excitation and reemission processes appear at the transition from a single oscillator to a solid, depending on the scale. Mie started to study the interaction of a light wave with matter in 1908. He analyzed the behavior of a set of oscillators in a particle in response to an external electromagnetic field. The mathematical treatment is complex and can be found in textbooks of optics [313] or in [327, 328], [329, Chapter 4], [337].

Some results describing the dependence of absorption or scattering on the incident light's wavelength and the pigment particle's size are essential for us [17, Chapter 5]. ►Table 1.19 summarizes some results, which essentially divide the optical behavior of particles into two classes. The distinction is due to the wavelength λ of the incident radiation:

- Tiny particles below the order of λ can be described by Rayleigh scattering, which assumes that the scattering centers oscillate as dipoles like microscopic Lorentz oscillators.
- Particles on the order of λ and above are described by Mie scattering, which depends on particle dimension and shape. The equation of van de Hulst shown in the table approximates spherical particles. The complexity of Mie scattering arises from occurring multipole oscillations and phase shifts within the large particles.

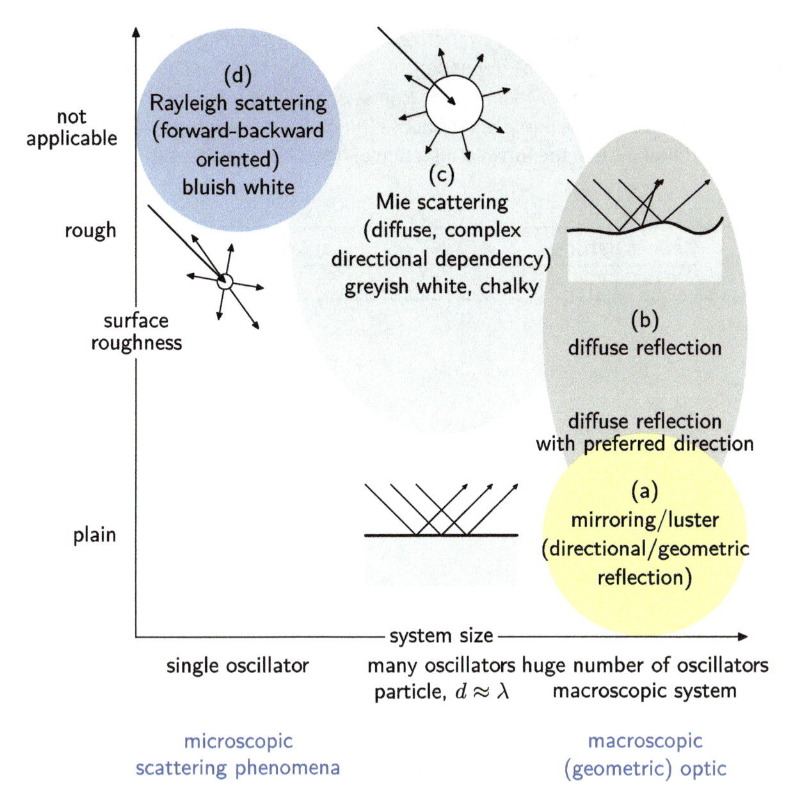

Figure 1.21: The transition from reflection into Rayleigh scattering when reducing the scale [313, Chapter 4]. (a): A macroscopic body with a smooth surface. The scattered light from its countless oscillators superimposes to form a reflected wave concentrated in the outgoing direction (geometric optics). (b): The actual surface irregularities are still many-particle systems that form reflected waves but reflect in different directions depending on the local surface slope. Diffuse reflection adds to the reflected main beam that increases with the surface roughness. (c): In the size range of multiple wavelengths of light (among others in pigment colloids or very rough surfaces with strongly changing refractive index, thus many scattering centers), diffuse *scattering* (Mie scattering) occurs, which shows a complicated angular dependence, but is undirected. (d): Single oscillators, having molecular dimensions, exhibit Rayleigh scattering that favors the forward and backward directions.

Scattering centers, i. e., small particles at which scattering takes place, are found in abundance in a paint system, always provided they are sufficiently small: above all, of course, the pigments' particles, added filler particles, opacifiers, and matting agents, emulsions, dispersions or air inclusions formed unintentionally during aging of the binder, dust, bubbles, fibers, and roughness on the surface or in deeper layers. Scattering affects the appearance of artworks to a great extent in all places where scattering centers occur: the primers or in imprimatura layers, in pigmented paint layers, intermediate or final varnish layers, on the painting surface, and at all interfaces between

Table 1.19: Schematic dependence of absorption strength and scattering intensity of particles of different size classes [17, Chapter 5], [337]. The dependencies on particle size r and incident wavelength λ are expressed by the parameter $\alpha = \frac{r}{\lambda}\pi n_0$, $\beta = 2\alpha|n-1|$. n_0 is the refractive index of the medium surrounding the particle, n is the particle's refractive index. θ is the angle at which the scattered radiation is observed in relation to the forward direction of light. k is the absorption coefficient of the particle.

Small particles	**Medium particles**	**Large particles**
Rayleigh regime	**Mie regime**	
Absorption Q_a		
$\propto \frac{r}{\lambda}k$	$\propto \frac{1}{r}\frac{r}{\lambda}$	
Scattering Q_s		
$\propto \frac{8}{3}\alpha^4(\frac{n^2-1}{n^2+2})^2$	$\propto 2 - \frac{4}{\beta}\sin\beta + \frac{4}{\beta^2}(1-\cos\beta)$	$\propto \frac{3}{2r}(\frac{n-1}{n+1})^2$
$\propto \frac{r^4}{\lambda^4}$	(van de Hulst)	(Fresnel)
$\propto (1+\cos^2\theta)$	$\propto f(\theta)$ complex	

these layers. It changes the perceived color and color quality and determines opacity, transparency, glazing ability, surface, and depth light.

Rayleigh scattering

Rayleigh scattering occurs at tiny scattering centers, typically molecules and atoms, e. g., in air. Excitation of the dipoles and the radiation occurs so that the scattered radiation is maximum in the direction of light propagation ($\theta = 0\,°$) and its opposite ($\theta = 180\,°$), ►Figure 1.22(a).

Due to the strong dependence of the scattered radiation on λ^{-4}, short-wave purple or blue light is scattered considerably more strongly than long-wave red light. We see the result when looking at acrylic dispersions, emulsions such as milk and latex, or smoke: in transmitted light, these dispersions appear reddish, but in lateral light, they appear bluish. In the same way, Rayleigh scattering is responsible for the blue daytime sky; the sun seems yellowish in the viewing direction because blue components are scattered away laterally; we perceive these as sky blue in the direction of the sky. Evening red occurs because the low sun's rays travel a long way through the scattering atmosphere and lose their blue components almost entirely.

Rayleigh scattering is significant for painting precisely because of this bluish color impression. From the early medieval wall painting of the Carolingian period, but still also in pictures by Rubens and Van Dyck, we find mixtures of finely ground charcoal with lead white or lime white, as well as glazes of charcoal on white, the so-called "false blue" [54, 501, 502, 509, 510, 513, 514], ►p. 215. Charcoal absorbs evenly and provides, mixed with white, a grayish hue. Fine charcoal particles can impart a bluish color impression due to the preferential scattering of blue light. For a distinct color impression, specific particle morphologies are necessary that do not occur with other

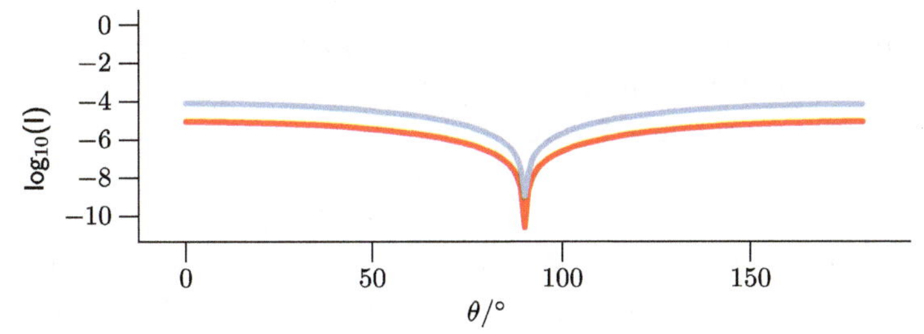

(a) Rayleigh scattering (particle size 25 nm).

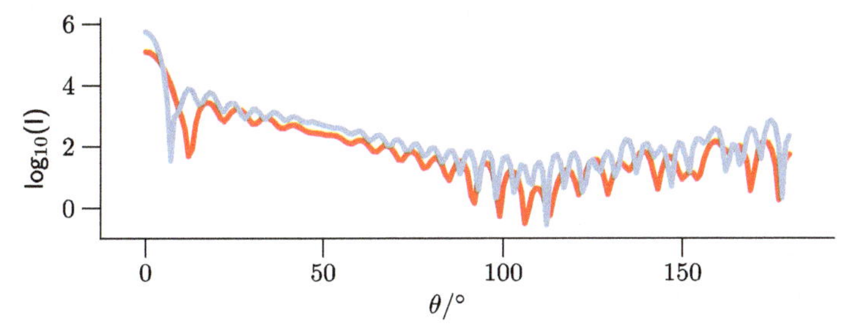

(b) Mie scattering (particle size 2.5 μm).

Figure 1.22: Dependence of scattering intensity I on observation angle θ for Rayleigh scattering and Mie scattering (calculated with Mieplot [333]). We recognize the order of magnitude higher scattering intensity of blue versus red light of Rayleigh scattering and the preference of scattering in forward and backward directions (0 ° and 180 °). Blue curve λ = 450 nm, red curve λ = 650 nm. The angular dependence of Mie scattering is complicated.

black pigments such as lamp black. Painters, therefore, specifically used the Rayleigh effect of charcoal particles.

Mie scattering

This type of scattering occurs at scattering centers on the order of λ and above. For painting physics, Mie scattering is the dominant scattering mode because pigment particles, in general, are too large for Rayleigh scattering. The re-emission of dipoles and multipoles combined with the phase shift yields a complex spatial distribution of intensity; the high fractions of scattering in all directions explain the diffuse light impression. However, the scattering in the direction of the light always dominates. ►Figure 1.22(b) shows the dependence of scattering intensity on angle θ. Due to the complicated dependence of scattered light on r, θ, and λ, scattered light in inhomogeneous particle systems such as clouds, fogs, and latex dispersions mixes to produce a white (gray) color impression. Diffuse scattering from painting surfaces results in a chalky, cloudy, and white appearance; scattering from fine pigment powders brightens as particle size decreases.

An essential result of the Mie calculations is that the scattering spectrum, and thus the perceived hue depends not only on the inherent color of the particle but also on particle size and wavelength. For example, ►Figure 1.23 depicts a Mie simulation for particles of a white pigment such as titanium white with a refractive index of 2.7. It shows that at a radius of 75 nm, the maximum of the scattering is in the blue at 400 nm and drops rapidly to almost zero toward the red so that the particle has a blue tint (light blue curve). If we enlarge the particle to 100 nm, the maximum scattering shifts to yellow at 550 nm, so mainly yellow and red components are scattered. Consequently, we perceive a tint of warm red (black curve). At 125 nm radius, the scattering maxima have shifted even further to higher wavelengths (yellow curve). Therefore, small titanium white particles are bluish-white, while larger particles appear warm-white.

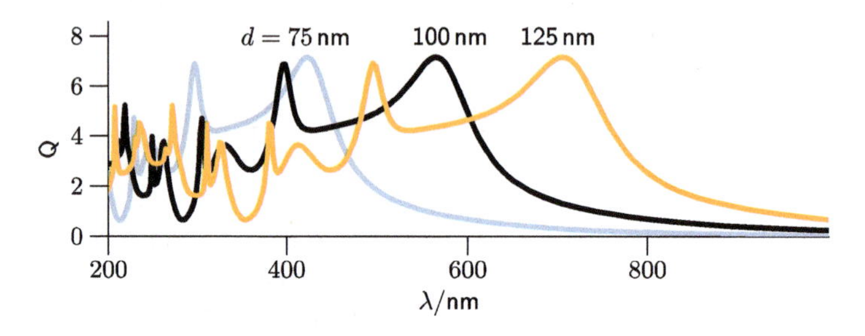

Figure 1.23: Dependence of Mie scattering Q on wavelength λ for particles with particle radius 75 nm (light blue), 100 nm (black), and 125 nm (yellow), refractive index 2.7 (e. g., titanium white) calculated with Mieplot [333]. Small particles show a blue tint.

►Figure 1.24 shows the dependence of scattering on particle diameter for red light (650 nm). For each refractive index, we see an optimal particle diameter for which maximum scattering occurs, e. g., 0.2 µm for $n = 2.7$ (titanium white) and 0.7 µm for $n = 1.6$ (chalk) in air. The same calculation for blue light shows that the maxima shift toward smaller particle diameters (0.15 µm for $n = 2.7$ and 0.45 µm for $n = 1.6$).

Thus, it becomes evident that the scattering functions have a very complex structure. We can also see these relationships simplified using the approximation of van de Hulst [17, Chapter 5]. If we plot Q against β, we see that Q has a global maximum for $\beta \approx 4$, which signifies that for this β, respectively, α, the scattering becomes maximal. The equation

$$\beta \approx 4 = \frac{2r\pi n_0(n-1)}{\lambda} \tag{1.24}$$

approximately links the particle diameter r, for which the scattering becomes maximal, with the wavelength λ. If we decrease r, the wavelength λ at which maximum scattering occurs must also decrease, i. e., shift into the blue range.

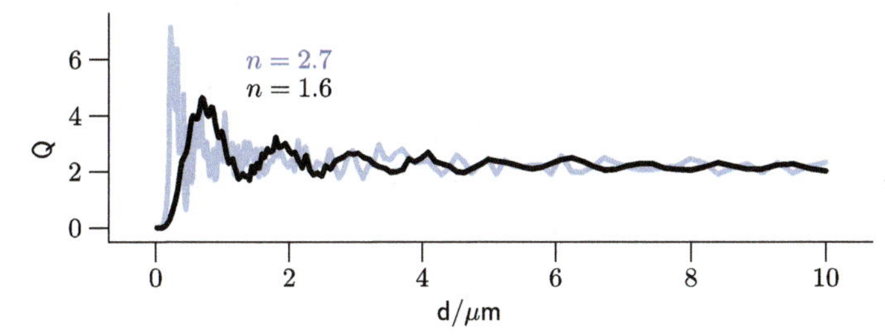

(a) Refractive index of medium n_M = 1 (e. g., air). Due to the large difference in the refractive indices of the pigments to n_M, strong scattering takes place on small particles of both pigments. The scattering optimum for high refractive titanium white is about 200–300 nm; for low refractive chalk, about 0.7 μm particle size. Both white pigments are distinctly visible and opaque in this medium.

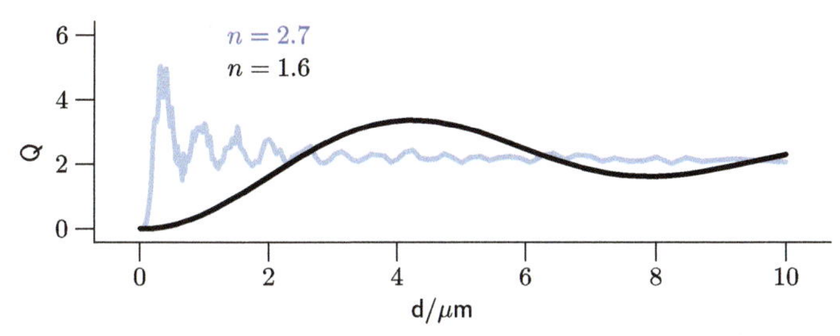

(b) Refractive index of medium n_M = 1.5 (e. g., linseed oil). Only the refractive index of titanium white still has a sufficiently large difference to n_M, and only at titanium white strong scattering occurs at small particles. Its scattering optimum is at about 200–350 nm. At small particles of low refractive chalk, practically no scattering occurs anymore. Only titanium white is distinctly visible and opaque in this medium.

Figure 1.24: Dependence of scattering Q on refractive index and particle diameter d. Light blue line: scattering of a particle with refractive index 2.7 (titanium white), black line: scattering of a particle with refractive index 1.6 (chalk), calculated with Mieplot [333] for the red light of wavelength 650 nm.

Reflectivity and luster

For large reflecting bodies, the intensity of the reflections is connected to refractive index n and absorption coefficient k with

$$R = 100\,\% \times \frac{(n-1)^2 + k^2}{(n+1)^2 + k^2} \tag{1.25}$$

We achieve high reflectivity or *luster* by a high refractive index or absorption coefficient. Colorless glass ($n = 1.5$, $k = 0$) shows a reflectivity of 4 %. Silver as a metal has a high absorption coefficient ($n = 0.18$, $k = 3.6$), its strong reflectivity expresses in $R \approx 95\,\%$. Intensively colored compounds such as potassium permanganate, iodine, graphite, and some organic pigments have such high absorptions that their crystals show metallic luster.

(Pearl) luster pigments

(Pearl) luster pigments exhibit a particular form of luster. We can obtain a lustrous pigment by incorporating tiny reflective metal platelets. However, we cannot imitate the exceptional iridescence of pearls in this way, since in their case, many reflections and refractions lead to the delicate luster. Pfaff [16, 473] offer details on this exciting group of pigments.

1.6.7 Consequences of absorption: metallic luster, metallic colors, bronzing

As a prominent optical property, electrically conductive compounds, especially metals, display typical metallic luster and characteristic colors. Gold overlays and metal backgrounds came to characterize early Italian and Byzantine painting and the art of amelioration and Verre eglomise.

Metallic luster

Drude and Lorentz gave a mathematical treatment of metals. According to the combined Drude–Lorentz model, a metal can be described as a gas of free-charge carriers (electrons), originating in the valence band [312, Chapter 7.1]. The complex dielectric function ϵ_r of that electron gas is approximated by

$$\epsilon_r(\omega) = 1 - \frac{\omega_p^2}{(\omega^2 + i\gamma\omega)}, \quad \omega_p = \sqrt{\frac{Ne^2}{\epsilon_0 m_0}} \tag{1.26}$$

In this equation, ω_p (or λ_p, when expressed as wavelength) is the so-called *plasma frequency* of the metal. It is the resonance frequency of the entire free-electron gas and depends on the charge density per volume N. Since N is huge, describing the number of free electrons in metals, ω_p (λ_p) is located in UV for metals, e. g., $\lambda_p = 115\,\text{nm}$ for copper and $\lambda_p = 138\,\text{nm}$ for silver and gold. In the visual spectral range, $\omega \gg \gamma$ so that ►equation (1.26) reduces to

$$\epsilon_r(\omega) = 1 - \frac{\omega_p^2}{\omega^2} \tag{1.27}$$

With ►equations (1.18) and (1.25), we link the complex dielectric function $\epsilon_r(\omega)$ to the complex refractive index $n + ik$ and to the reflectivity R. Three important cases arise:

- $\omega < \omega_p$ or $\lambda > \lambda_p$: For wavelengths larger than the plasma frequencies' wavelength, $\epsilon_r < 0$ and $n + ik$ is an imaginary number ik, k being the absorption coefficient. With increasing λ, ik increases, the metal is highly absorbing for low energies in the visible spectral range. ►Equation (1.25) yields $R = 100\,\%$ so that the metal is highly reflective in this spectral range, a typical feature of metals, matching our everyday experience with metals.

- $\omega = \omega_p$ or $\lambda = \lambda_p$: For wavelengths identical to the plasma frequencies' wavelength, $\epsilon_r = 0$, $n + ik = 0$, $R = 100\,\%$.
- $\omega > \omega_p$ or $\lambda < \lambda_p$: For wavelengths smaller than the plasma frequencies' wavelength (UV spectral range), $\epsilon_r > 0$ and $n + ik$ is a real number n. $R \in\,]100\,\%, 0\,\%[$ quickly approaches zero for small wavelengths. $R = 0\,\%$ means absorption, reflectivity vanishes; the metal becomes transparent in the UV.

►Figures 1.25(a) and (b) illustrate the graphs of ϵ_r and $n + ik$; ►Figure 1.25(c) depicts in black the typical maximum reflectivity $R \approx 100\,\%$ of a metal for $\lambda > \lambda_p$, modeled as free-carrier gas. Above the plasma frequency, free electrons can no longer follow an external field, energy intake and absorption are zero, and the metal rapidly becomes UV-transparent, the so-called *UV transparency of metals*.

What can we learn from this result? The model confirmed the metal's high absorption coefficient necessary for luster phenomena. Irradiating a metal with an electromagnetic field, i. e., light, readily excites the free-valence electrons, inducing a current. Due to their high mobility, the electrons follow the external field from low frequencies up to the plasma frequency, extracting energy from the field; the absorption is high up to this frequency, ►Figure 1.25(b) (blue graph). The resulting reflectance R is so significant (in theory 100 %) for the high values of k below the plasma frequency that we perceive a metallic luster. Suppose the reflections of many electrons overlap in a spatially extended conductor. In that case, they represent an image of the incident light source, i. e., a macroscopic *geometric reflection* instead of a diffuse one. Geometric reflection happens for gold and silver coatings and substrates. Since R is, in theory, unity in the visual spectral range, metals completely reflect white light and should appear whitish and highly reflective, ►Figure 1.25(c) (black graph).

In colloidally dissolved metals, the metal particles are so small that electric currents forced by external electromagnetic fields cannot form. As a result, no reflection occurs; the metal colloid shows no luster and appears black due to absorption. In some cases, e. g., for gold ruby glass, absorption phenomena such as plasmon resonances occur, yielding brilliant colors, ►Section 1.6.4.

White, gray and dark metals

Metals exhibit properties of a free-electron gas due to their unbound valence electrons, yielding, in theory, a highly reflective whitish appearance. In reality, in addition to these free-carrier absorptions, metals possess absorptions caused by interband (IB) transitions of bound electrons between various bands in the metal's band structure [312, Chapter 7.3.2]. We can express this symbolically as

$$R(\lambda) = R_{\text{free-carrier}}(\lambda) - \text{IB}(\lambda)$$

IB transitions reduce reflectivity since they occur in concurrence with free-carrier absorptions. Light penetrates a short distance into the metal, and photons can initiate

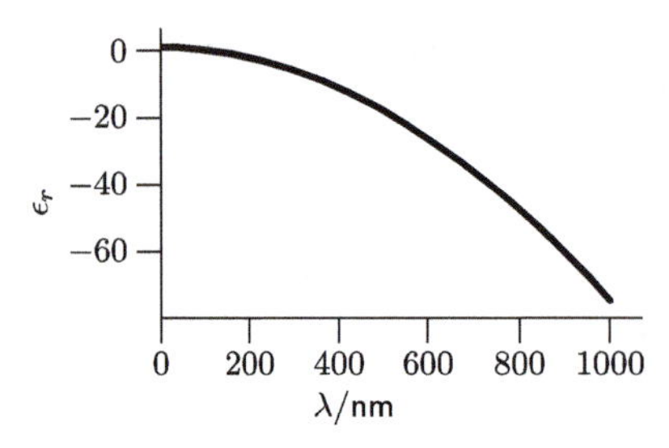

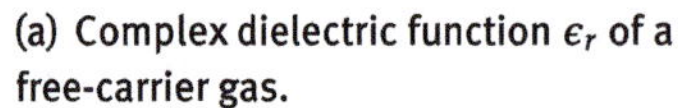
(a) Complex dielectric function ϵ_r of a free-carrier gas.

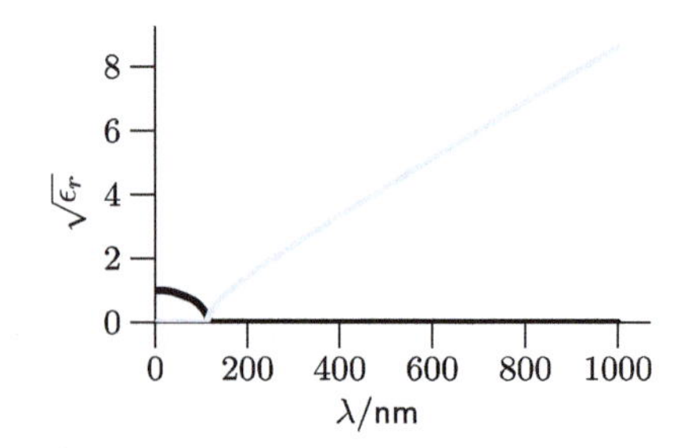

(b) Complex refractive index $n + ik$ of a free-carrier gas. Black: n. blue: k (absorption coefficient).

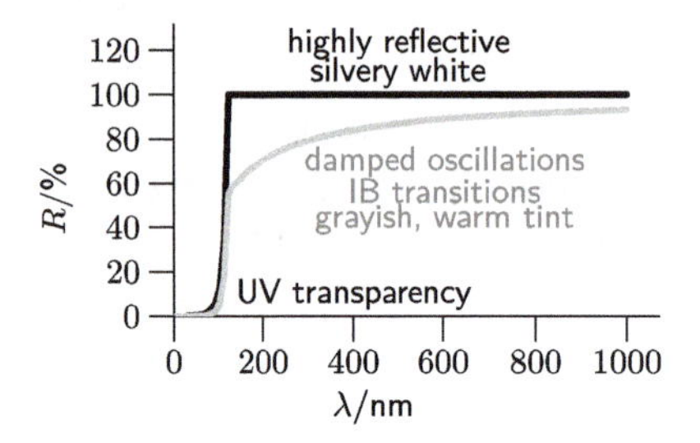

(c) Reflectivity R. Black: idealized for a free-carrier gas, $\gamma = 0$. Gray: real metal, damped oscillations, $\gamma > 0$, and IB transitions throughout complete energy range, reducing the free-carrier reflectivity.

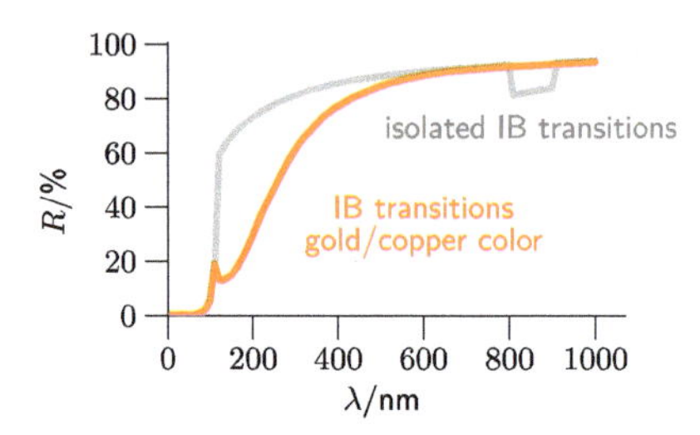

(d) Reflectivity R of metals with additional localized IB transitions. Gray: IB in IR does not change visual appearance. Yellow: IB in the blue to the green spectral range yields yellow, gold, or copper color.

Figure 1.25: Graph of essential optical properties of a metal according to the Drude–Lorentz model of a free-carrier gas [312, Chapter 7.1]. Carriers are the metal's valence electrons. Calculated are ϵ_r, $n + ik$ and R for λ_p=115 nm, e. g., copper [312, Chapter 7.1]. For $\lambda < \lambda_p$, the *UV transparency of metals* is distinctly visible ($R \approx 0$ %). For $\lambda > \lambda_p$, the metal is highly reflective ($R = 100$ %). Damping and particularly IB transition reduce the free-carrier reflectivity, causing grayish, copper-red, or golden colors or blueish tints instead of white, depending on the IB pattern, density of states and band structure [312, Chapter 7.3].

IB transitions or free-carrier effects. (For a more formal discussion, a single equation for ϵ_r would comprise terms for free-carrier and IB transitions, and this ϵ_r would yield more complex dependencies of $n + ik$, eliminating the artificial subtraction.)

The gray graph in ►Figure 1.25(c) depicts this more realistic scenario, respecting damped oscillations ($\gamma > 0$) and IB transitions, reducing R significantly. Damping reduces R over the entire spectral range, but only to a small extent. The decrease of R is more significant due to IB transitions from metal bands below the Fermi energy to unoccupied bands above the Fermi level. Since IB transitions can occur at a whole range of energies, they attenuate R over a broad spectral range, depending on IB intensity and, therefore, band structure.

A more or less equal distribution of IB transitions, i. e., a homogenous absorption of the metal in the visual spectral range, reduces R uniformly and induces the grayish or dark appearance typical for most metals. For example, in grayish aluminum, $R \approx$ 80–90 % due to broad IB transitions of $3s$ and $3p$ electrons into states above the Fermi level in the visual spectrum. In contrast, silver's intense IB transition $4d \rightarrow 5s$ occurs in UV, so silver's color is dominated by the nearly maximum reflectivity of the free-electron gas and appears shimmering white.

Color of metals

As in aluminum, some bands are parallel, separated by an energy equivalent to λ_{IB}. The density of states is high for $\lambda = \lambda_{IB}$ in such a case since many IB transitions of this energy are possible. In the model, such an intense IB transition would distinctively decrease R around λ_{IB}, ►Figure 1.25(d) (gray graph). An IB transition in IR does not influence aluminum's visual appearance.

In contrast, copper possesses intense IB transitions $3d \rightarrow 4s$ from tightly bound and narrow $3d$ band into a wide $4s$ band, starting from ≈ 560 nm (yellow) onwards. As a result, blue, green, and yellow light is absorbed rather than reflected, resulting in copper's red color, ►Figure 1.25(d) (yellow graph). In the case of gold, the absorption due to IB transitions $5d \rightarrow 6s$ occurs at slightly higher energies than copper in the greenish-yellow range, and bulk gold, therefore, exhibits its unique golden-yellow color in incident light by absorbing blue and some green light. (However, according to ►equation (1.25), the absolute absorption intensity is maximum in the yellow and red spectral range since the reflectivity is maximum. This statement can be confirmed by placing a very thin gold foil for a light source. The transmitted light is blue-green due to the intense absorption and reflection in the yellow and red spectral range.)

The causes of the other metal colors are similarly due to the IB transitions depending on the metal's band structure. A more in-depth discussion of the processes can be found in [311, 312, 314].

Bronzing

Deeply colored pigments give the appearance of colored bronze when applied at greater concentrations, the so-called *bronzing*. Like metals, highly absorbing pigments reflect the absorbed light as luster; the luster, therefore, shows a color complementary to the intended color of the pigment. An intense blue pigment strongly absorbs yellow and therefore has a yellow luster, while a cyan pigment absorbs orange and shows orange bronzing. As shown in [428], the effect occurs with all pigments but usually cannot be perceived.

Bronzing is especially noticeable on a black background since the incident light scattered back from a bright background is more intense and outshines the bronze luster. Furthermore, the scattered light, having passed through the color layer twice (after entering the color layer and after reflection from the substrate), is colored by the pigment. Consequently, on bright substrates, colored layers and reflected light almost exclusively show the expected color of the pigment.

Bronze luster is only visible on bright substrates if the pigment has a high absorption coefficient or is highly concentrated in the paint layer. Unfortunately, the latter case can occur in printing since, for many printing inks, the solvent and parts of the binder wash away into the paper. In that case, the pigment concentration on the paper rises sharply, and the usually undesirable bronze sheen appears. In particular, blue-purple pigments are critical when printing since they intensely absorb in the entire visible spectral range. This nearly complete absorption reduces the scattered light on white paper so that even weak bronzing is clearly visible. Examples of this effect are Prussian blue and alkali blue pigments used to tone black inks.

High absorption coefficients are caused either by inherently intense absorptions of a pigment or by interactions of pigment particles, yielding delocalization of electrons across the particles involved. Such intermolecular interactions occur particularly with indigo, alkali blue, Prussian blue, quinacridones, perylenes, diketo-pyrrolo-pyrrole pigments, i. e., either planar molecules that can align well with each other in parallel layers and staples. It also occurs with molecules that strongly interact via amino-carboxyl groups.

1.6.8 Consequences of scattering: opacity, white pigments, and depth light

The complicated mutual interactions between absorption, reflection, and scattering at the painting surface, in the depth of the paint layer, and at the boundary layers of varnish and paint layers, as well as pigment particles, leads to several phenomena that influence the perceived color quality [9, Chapter 6].

Opacity and transparency, opaque and glaze pigments

In the interactions of pigmented paint layers, imprimatura, and ground, the scattering is decisive for a pigment's opacity and *glazing properties*. Highly scattering pigment particles allow only a tiny amount of residual light to penetrate to underlying layers so that a paint layer of such particles is highly opaque. On the other hand, if only little light is lost by scattering, much residual light is available to penetrate deeper layers and return to the observer as *deep light*. Depending on the coloring power of the pigments in these layers, i. e., on the size of their absorption coefficients, deep light is colored more or less intensively. A paint layer consisting of particles that scatter little thus has a glazing or transparent effect. This detrimental effect occurs when paintings age, ►Figure 7.18 in ►Section 7.4.11.6.

Strong scattering: rough surfaces, lean binders, opaque pigments, and white pigments

The intensity of scattered light increases as

- pigment particles become smaller (up to a certain point, ►Figure 1.24),
- the difference Δn in the refractive index increases (►Figure 1.24) between different paint layers or between binder/medium and pigment, or
- the number of scattering centers increases.

Since scattered light has not penetrated the paint layers of the painting and has therefore not participated in any absorption processes, scattering does not change the frequency of the incident light. Therefore, scattered light corresponds to the incident light, mostly daylight or lamplight, and is usually whitish, resulting in a characteristic matt or chalky color quality.

To a certain extent, the *size of the pigment particles* can control the scattering intensity and opacity of the pigment. Opacity increases with decreasing particle size up to a point where the pigment has maximum opacity. Such an opaque pigment dominates the color effect in the painting, ►Figures 1.26(a) and (b). We have seen that scattering is maximum for pigment particles of the order of λ. In contrast, glazing pigments employ particles smaller than that dimension. Synthetic iron oxide pigments provide an example familiar to the painter: the glazing types ("transparent ocher") have particle sizes up to 0.05 µm, opaque ones from 0.05–1.0 µm [191]. ►Figure 3.11 or, more generally, ►Figure 1.24 illustrates how the limiting value of 0.05 µm is identified by employing the calculated plot of scattering Q versus particle size d. Q is practically zero only for $d < 0.05$ µm. Above, Q always assumes finite values, and for $0.05 < d < 1.0$ µm, Q is maximum.

Regardless of opacity, scattered light always lightens a finely ground pigment, providing *white* pigments; see below.

A significant *difference in refractive index* Δn is present when pigments directly contact low-refractive air, as with pencil and pastel crayon drawings. Here, only a minimum of binder fixes the pigments. The high scattering gives these colors their characteristic matt, chalky quality. (The fact that such drawings nevertheless show something of the background has to do with a very thin pigment application or a very rough paper surface that breaks up the color layer).

A high *number of scattering centers* plays a role in objects with dry, rough, or natural surfaces, such as unpolished, naturally rugged mosaic stones or rough, lean paint layers of acrylic, tempera, and watercolor. In all of these painting systems, the binder scarcely embeds the pigments, and a microscopically fissured surface forms during water evaporation or already exists. The scattering centers are formed by a multitude of microscopic irregularities, impurities of any kind in the untreated surface and pigment particles, and by strong local fluctuations of the refractive index, which constantly alternates between that of air, that of the pigment, that of the impurities, and

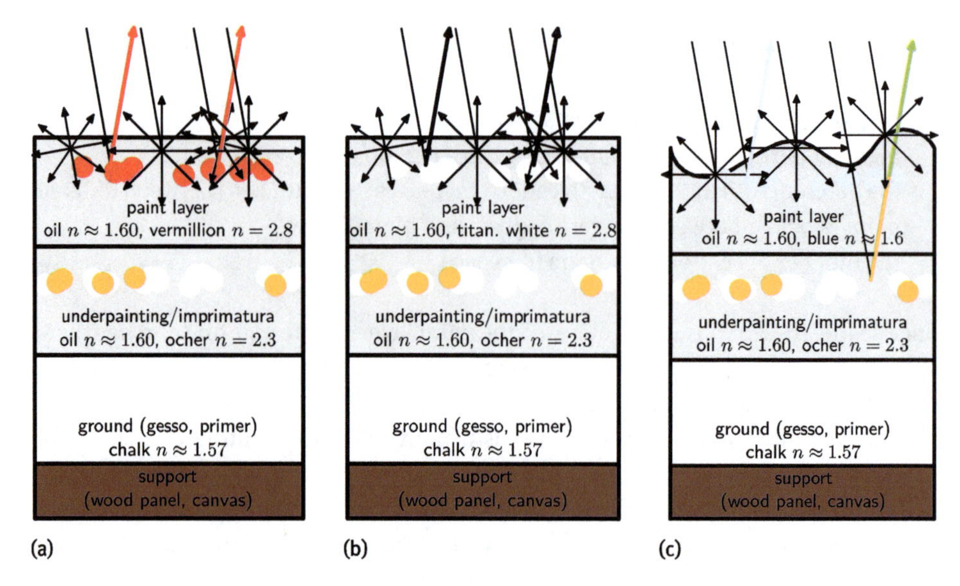

Figure 1.26: Development of surface light and matt, chalky color quality due to intensively scattered light [9, Chapter 6]. The scattered light corresponds to the spectrum of the incident light, being in general white. (a) and (b): Intensive scattering by highly refractive, opaque chromatic, and white pigments (n large, Δn large). The pigment and the scattered light determine the color effect. (c): Intensive scattering due to the rough surface of a drawing or a lean paint layer (tempera, acrylic, watercolor) in which evaporation of the water components has created a fissured surface with strong local fluctuation of n and Δn, resulting in numerous scattering centers on the surface. Adding a matting agent such as silica also leads to roughening the surface and new scattering centers. The pigment and the scattered light determine the color effect. In addition, diffuse reflection occurs.

that of the binder, ►Figure 1.26(c) [9, Chapter 6]. The high scattering yields these painting systems' typically matt, chalky color quality. Unintentionally, this occurs when most delicate air bubbles emerge within the binder due to detrimental effects, as sometimes happens with the blue pigment smalt, leading to a graying of the affected image areas, ►Section 7.4.11.

Suppose a pigment is to be opaque and colored at the same time. In that case, it must scatter strongly but also exhibit such intense absorption in the visible spectral range that as much as possible of its inherent color is absorbed in the scattered light, even though hardly any light can penetrate the paint layer. Thus, scattering offers the possibility of obtaining a particular hue: white. When grinding colorless crystals or particles of an intrinsically colorless substance, the fraction of scattered light increases the more finely it is ground until, with suitable particle size, it is present as a pure white powder with maximum scattering. The quality of a white pigment, i. e., its suitability for various paint systems, is increased by the highest possible refractive index.

Low scattering: deep light, fat binders and varnishes

We observe the opposite of the whitish-chalky color impression in intensive scattered light when we polish the pieces of mosaic or wet them with water, varnish paintings, or apply the fat paint system of oil painting. The treated surfaces and fat paint layers show intense and bright colors while their composition remains unchanged, ►Section 7.4.10. The difference between color-intensive fat oil painting layers and matt-chalky lean tempera or acrylic layers is distinct and contributed (among others) to the rise of oil painting in the Quattrocento and among the Old Netherlanders, at the expense of (egg) tempera painting.

Polishing, applying a water film, or varnishing evens out the surface irregularities, *reduce the scattering centers*, and achieve a macroscopically smooth surface. Also, fat binders that form bodily homogeneous oil films reduce scattering centers, *suppress refractive index fluctuations*, and offer a smooth surface instead. Now, geometric reflection instead of scattering occurs on the smooth surfaces, and it begins to shimmer. Part of the incident light is reflected, but most of it is refracted into the water, varnish, or paint layer. In the paint layer, the pigment particles absorb parts of it, and the residual emerges as colored depth light, ►Figure 1.27(b).

If the difference Δn of refractive indices disappears at an interface, e. g., between binder and pigment particle, the particle becomes transparent, ►Figure 1.27(c) [9, Chapter 6]. This phenomenon brings us to the role of the medium or binder. If we apply pigments with different binders, some pigments become practically transparent and hardly perceptible in specific binders. We impressively experience this with white pigments: Marble or chalk ($CaCO_3$, $n = 1.6$) appear opaque white in air ($n = 1.0$), while they are transparent in linseed oil ($n = 1.6$). Lead white ($n = 2.0$) or titanium white with a high refractive index ($n = 2.7$), in contrast, have an opaque appearance even in oil. Zinc white ($n = 2.0$) is a partly glazing pigment excellently suited for achieving atmospheric effects due to this medium value.

►Figure 1.24 shows simulations of the Mie scattering occurring in these cases: in air, both chalk and titanium white show strong scattering, and thus opacity at the optimum particle sizes of 200–700 nm, ►Figure 1.24(a). In contrast, if particles of the same size are bound in linseed oil, only titanium white shows scattering and opacity, ►Figure 1.24(b). The chalk particles no longer scatter light at all and become transparent. The predominance of titanium white in today's paints is based on the fact that it has by far the highest refractive index of all white pigments and is intensively white in all media. Both marble and chalk can only be used as fillers in oil-based paints, not as pigments.

►Tables 1.17 and 1.20 contain refractive indices of important painting media and pigments. Classical opaque pigments such as vermilion, Naples yellow, or ocher have high refractive index that causes a high opacity in all media. Low-refractive pigments such as ultramarine, a classic glazing pigment, must be mixed with white to achieve a

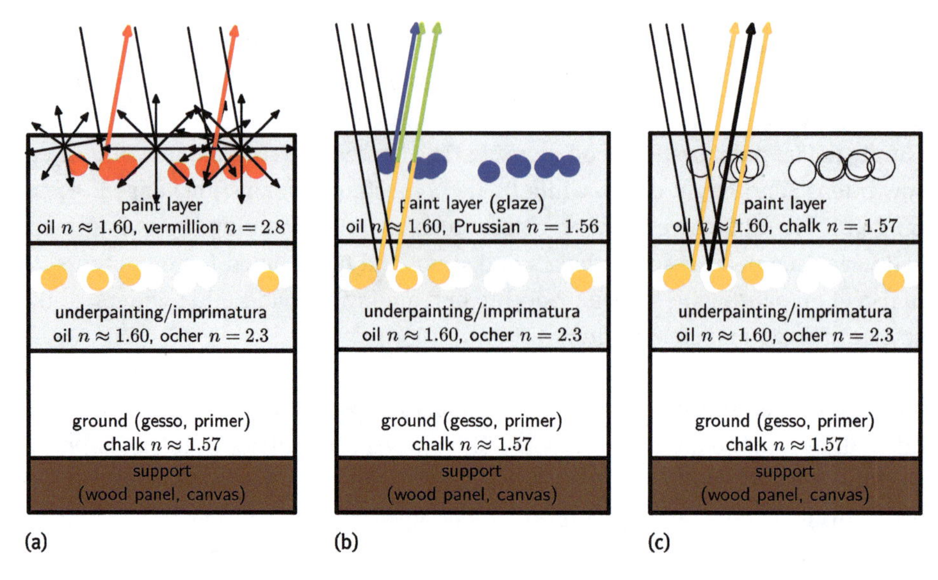

Figure 1.27: Opaque paint layers and the emergence of deep light and intensive, luminous color quality due to low scattered light in glazes [9, Chapter 6]. Incident light is refracted into the paint layers, participates in absorption processes there, and falls to a high degree into the eye of the observer as intensively colored, brilliant deep light. The medium has a refractive index n_M. (a): Opaque paint layer, due to the high refractive index of chromatic pigments (n large, Δn large), the imprimatura is covered and does not participate in the color impression. (b): Glazing effect due to the low refractive index of chromatic pigments (n moderately large, Δn small), the imprimatura shines through and participates in the color impression. (c): Unintended lack of opacity due to the low refractive index of the white pigment ($n \approx n_M$, $\Delta n \approx 0$). Such white pigments appear almost transparent and can only be used as fillers that shall not participate in the artistic vision. Therefore, the color impression is dominated by the first sufficiently opaque layer, the imprimatura.

certain degree of hiding power; otherwise, we perceive them as a transparent colored layer, a glaze.

1.7 Summary: physical factors influencing pigment properties

We are now familiar with the influence of the microscopic interaction between light and matter on perceptible painting properties of spatially extended pigment particles, [474], [9, Chapter 6] elaborates this topic further. Alone by varying the optical-physical context, a variety of pigments can be produced which, though having the same chemical composition, exhibit different hues (cool-bluish, warm-reddish), painting properties (glazing, opaque), and processing properties (aggregation tendency, viscosity influence).

Table 1.20: Refractive index n_D of some painting pigments at 589 nm (sodium D-line) [16, 48, 543].

Pigment	n_D	Pigment	n_D	Pigment	n_D
Chrysocolla	1.45–1.55	Smalt	1.46–1.55	Calcite	1.486–1.658
Ultramarine	1.50	Copper resinate	1.52	Verdigris	1.53–1.56
Scheele's green	1.550–1.749	Prussian blue	1.56	Calcium carbonate	1.57
Egyptian Blue	1.591–1.636	Red lake	≈ 1.6	Green earth	1.62
Viridian	1.62–2.12	Vandyke brown	1.62–1.69	Baryte white	1.638–1.648
Malachite	1.65–1.88	Indian yellow	1.67	Bone black	1.68
Emerald green	1.71–1.78	Cobalt yellow, aureolin	1.72–1.76	Azurite	1.73–1.84
Cobalt blue	1.74	Lead white	1.94–2.09	Zinc white	2.00–2.02
Naples yellow	2.01–2.28	Yellow ocher (goethite)	2.36	Lead-tin yellow	2.29–2.31
Chrome yellow	2.3–2.7	Molybdate orange, red	2.3–2.65	Cadmium yellow lemon	2.356–2.378
Orpiment	2.4–3.02	Chrome orange, chrome red	2.42–2.7	Red lead	2.42
Bi-V yellow	2.45	Chromium oxide green	2.5	Cadmium sulfide hexag.	2.506–2.529
Realgar	2.54–2.7	Cerium sulfide	2.7	Titanium white	2.72
Red ocher (hematite)	2.87	Vermilion	2.819–3.146		

1.7.1 Particle size

The size of the pigment particles has a decisive influence on the color, transparency or opacity, coloring power, and applicability (viscosity in the paint) of the pigment via the mechanisms mentioned in ►Table 1.21:

- As particle size decreases, the scattering and hiding powers of a pigment increase to a maximum. Opaque pigments have particle sizes near this maximum. As the particle size decreases further, the pigment becomes translucent. Example: glazing or opaque iron oxide pigments, ►Section 1.6.8.
- As scattering increases, the hue of the pigment becomes progressively lighter, and its coloring power decreases because white scattered light increases relative to deep light, which is colored by absorption. Example: azurite can only be used coarsely ground since fine grinds no longer show any discernible color [48, Volume 2].
- Small particles preferentially scatter short-wave blue light. Small particle sizes result in cool-blue color tints, larger ones in warm-red ones. Example: false blue caused by fine charcoal particles, ►Section 1.6.6.
- At suitable particle dimensions, surface plasmons can develop and contribute to the color by intensive, sharp absorptions. For small particles, the plasmon reso-

Table 1.21: Mechanisms by which particle size can affect the perceived pigment property.

Small particles	Medium particles	Large particles
Size quantization (semiconductor), white-yellow-red-black	Evolution of surface plasmons (in blue, in red)	
← Transparency increases	Maximum scattering Maximum opacity Maximum brightness	← Scattering increases ← Opacity increases ← Brightening of the hue increases
← Intensity of blue scattered light increases	→ Intensity of red scattered light increases	

nances occur in the blue region of the spectrum, and thus quenching it; for larger particles, they occur in the red region. Example: red iron oxide pigments that are red-orange when finely ground but blue-purple (colcothar or Caput Mortuum) if coarsely ground, ►Section 1.6.4 at p. 50.

- For tiny particles, the size quantization effect can change the hue. If the coloring is based on a band gap in a semiconductor (vermilion, cadmium yellow), decreasing the particle size leads to an apparent color change toward red → yellow → white; increasing it leads to a change toward red → brown → black, ►Section 2.2.2. Example: vermilion is orange to light red depending on the grind, ►Section 3.4.1 at p. 236.

Regardless of its chemical structure, the color purity of a pigment can be increased by carefully separating grain sizes. In the Japanese manufacturing tradition, pigment particles are hammered into 14 precisely defined grain size classes to improve their brilliance. The brightness of the natural ocher has also been considerably enhanced by frequent suspending into classes of grain size.

1.7.2 Crystal structure and particle shape

Crystal structure and particle shape determine the surface properties, and thus the pigment's color and application properties. The *crystal structure* determines how many and which side faces a microscopic pigment crystal has. Furthermore, it defines how pigment molecules are incorporated into the crystal's lattice. The microscopic pigment crystals generally assemble into macroscopic pigment particles (aggregation and agglomeration, ►Section 6.3) of various shapes (spherical, cylindrical, cubic, acicular). The *particle shape* determines which areas of the crystals and molecules are exposed. It thus imparts distinct chemical and electrical properties to the particle, which differ from those of an isolated molecule.

Pigments interact with their environment via these particle surfaces and exhibit collective properties. These relate, among other things, to scattering power (influencing brightness, color intensity, and opacity), excitation of electronic transitions of particular symmetry (influencing color and causing dichroism), and surface energy (influencing dispersibility and aggregation tendency).

Crystal structure

The crystal structure determines the placement of the pigment molecules. It therefore sets the chemical and electrical conditions for the color-bearing elements of the molecules (e. g., metal cations or molecular orbitals) compared to their isolated state. They react by adapting their electronic properties, and thus the number and type of possible transitions, wavelengths, and intensities. We recognize this in pigments of the same chemical composition but different crystal structures (allotropy):

- Example semiconductor pigment: yellow lead(II) oxide represented an early yellow pigment under the name massicot. Its second crystal modification is red and was traded as litharge.
- Example LF-based pigment: iron(III) oxide is, as hematite, a well-known red earth pigment (red ocher), but brown as maghemite. Iron oxide hydrate is yellow as goethite (yellow ocher) and orange as lepidocrocite, enabling the paint industry to produce pigments in the range yellow-orange-red-brown-purple from a base material "iron oxide."
- Example MO-based pigment: the β phase of copper phthalocyanine, crucial for the printing industry, is an ideal cold blue for four-color printing; its α modification is blue with a red tint.

The change from tetragonal red lead oxide to orthorhombic yellow oxide and vice versa also changes its band structure and the size of its band gap. The change from the cubic lattice of lepidocrocite or maghemite to the hexagonal one of goethite or hematite alters the position and distribution of electric charges and the strength of magnetic interactions. The altered ligand field strength changes the size of the orbital splitting of the iron ions. Organic molecules, especially flat-structured ones such as phthalocyanines, assemble into stacks in which electrons of many molecules interact so that the electron densities responsible for the color-bearing transitions depend on the crystal structure. As a result, pigments such as diketo-pyrrolo-pyrrole can acquire a black color, ►Section 4.14.

Interstitial compounds

Strong interaction between a color-bearing system and its surroundings can occur, especially with *interstitial pigments* such as Maya blue, ►Section 4.7.1. It consists of indigo sitting in the tubes of a clay mineral, changing its dark blue color to a bright turquoise blue. As we now know, the intensive blue color of an indigo crystal is due to a chemical structural element (a chromogen) in conjunction with the electronic interaction of many indigo molecules. In the interstitial compound, this interaction is reduced and replaced by interactions with the host lattice, which is noticeable as a color change [430, 431].

Crystal structure defects, lattice distortions, crystal growth rates

Under natural, i. e., mostly nonideal formation conditions, pigment crystals exhibit a high number of construction defects, which can lead to significant color deviations:

- For example, the emerald green production by uncontrolled or uncontrollable precipitation reactions yielded a product whose color varied with the conditions of manufacture [48, Volume 3].
- The synthetic azo pigment PY13 exists immediately after its synthesis by precipitation reaction as a fine amorphous powder of dull yellow [474]. If one heats the pigment carefully for a time ("annealing"), it assumes a pure yellow color.
- Iron and copper minerals (e. g., ocher earths, azurite), formed hydrothermally in natural deposits by precipitation from (aqueous) solutions or weathering, often exhibit dull brown to black-brown or blue to green color mixtures. Large amounts of water, hydroxide, and other atomic groups that are not part of the mineral's composition can be included during their formation. Severe color defects are caused by incorporating colored foreign atoms such as Cr, Co, Ni, V, or Mn [65, Chapter 6.4]. Textbooks of crystal chemistry and crystallography can provide further information on this [332]. The pure-colored ocher earths are obtained by crushing the nodules and earths, suspending, and sorting them by grain size and properties.

In the examples, pigment crystals originate from precipitation reactions. As a result, many crystals form abruptly or in an uncontrolled environment or by long-lasting, variable weathering. During such a rapid or uncontrolled crystal formation, it is likely that many ions or molecules will not find their ideal place and settle in a "wrong" position or an unfavorable orientation. Molecules and ions from impurities present in the solution will also be mistakenly coprecipitated or coincorporated. Colorless substances in the mixture can prevent, promote, or alter the dissolution and precipitation behavior of the interesting color-bearing structures.

Each perturbation distorts the ideal, pure crystal lattice of a substance and leads to local changes in the color-producing structures' chemical and electrical environment. As a result, instead of the ideal and typical color, slightly deviating local absorptions occur everywhere, which mix to form broad absorption bands and dull brownish overall color. In addition, all the physical color phenomena described in this chapter can happen in the irregular particle mixture and cause further variation of the colors.

With synthetic pigments, we find a solution by controlling crystal growth. If the crystals form slowly, their constituents have time to occupy the most thermodynamically favorable sites and grow an ideal, uniform lattice. Heating the raw PY13 serves the same purpose: prolonged heat application allows misaligned molecules to dissolve and gradually become better positioned.

Something similar happens during the preparation of natural ocher: frequent suspension separates the particles more or less according to size, shape, and chemical properties. Each of the resulting classes shows higher homogeneity of its composition. Although this does not eliminate the structural defects, at least similarly defective or similarly composed particles are better grouped to obtain bright yellow and red ocher soils.

Particle shape

The particle shape determines the application properties of clustering, dispersibility, and suitability for grinding via the surface of a pigment particle. It also determines which surface plasmons can be formed in suitable materials and directly influences the color of the pigment, ►Section 1.6.4 at p. 50.

- Example: A fine crystalline pigment that does not form aggregations has a large surface area due to many small particles, and thus high surface energy. The color powder will minimize this energy by forming clusters. Accordingly, a higher energy input will be necessary for dispersion, so one must ground the colorant more intensively or add a dispersant.
- Example: The color of the yellow ocher varies from lemon yellow to yellow-orange depending on the length of the acicular particles [65, Chapter 6.4].

2 The chemistry of color

The previous chapter has shown us how physics induces color by light absorption in the visible spectral range. (Strictly speaking, color arises only through our sensual perception of the resulting wavelength mixture as a *color impression*. Hereafter, we shall always implicitly assume this strict distinction.) In this chapter, we proceed to chemistry and see how chemists can implement absorption processes, and thus obtain colored substances for use in painting. We will learn about four mechanisms realized in distinctive chemical building blocks called *chromophores*. These chromophores represent classes of colorants by themselves or serve as a basis for a more complex classification of colorants. We will consider each of the four classes and their representatives in the form of actual colorants in more detail in this chapter. Unfortunately, we must disregard numerous attractive and significant physical absorption possibilities in everyday life, but [311] offers a comprehensive treatise on all the options of producing color, and [233] gives a detailed overview of mechanisms active in minerals and inorganic matter. Finally, ►Section 2.6 illustrates the process of laking, which allows us to obtain paintable pigments from dyes.

2.1 Chemical absorption mechanisms

From the chemist's point of view, we dispose of four basic mechanisms to induce light absorption, and thus color impressions. The mechanism and their designations we use for reference in this textbook are:

- electronic transitions from the valence band to the conduction band across the band gap in semiconductors (designation SC), ►Section 2.2
- electronic transitions between *d* or *f* orbitals, split in the electric field of crystals or ligands in *d* and *f* metal complexes (designation LF), ►Section 2.3
- charge-transfer transitions (designation CT) between ligands L and metals M (LMCT, MLCT) or between metals of different oxidation states (IVCT), ►Section 2.4
- electronic transitions between molecular orbitals (designation MO or inner ligand), ►Section 2.5

The typical features arising from these mechanisms in absorption spectra are indicated in ►Figures 2.1 and 2.2.

►Figure 2.3(a) locates these mechanisms in the periodic table of elements. We can see that each mechanism collects in specific groups or periods and, based on this, we can derive what chemistry and properties the colorants will possess:

- SC occurs in semiconductors, formed from *p* block elements and group 12 metals, with chalcogenides acting as anions. Oxides, sulfides, and selenides of groups 12, 14, and 15 metals also occur naturally as minerals, being especially important for painting in the past.

https://doi.org/10.1515/9783110777116-002

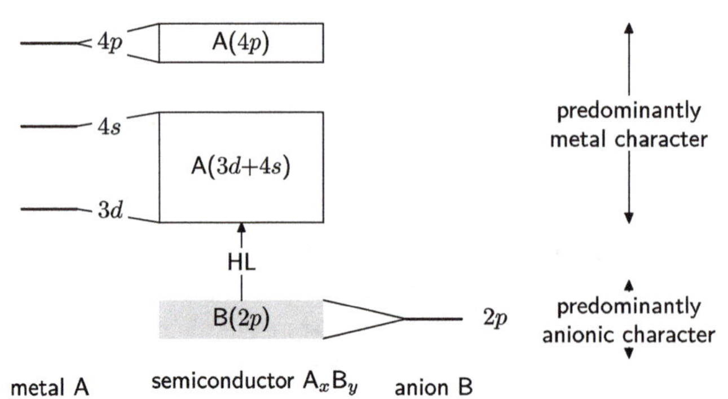

(a) Flat band structure and schematic SC transition in A_xB_y. In this example, $3d$ and $4s$ electrons form the conduction band, and the anion's $2p$ electrons form the valence band.

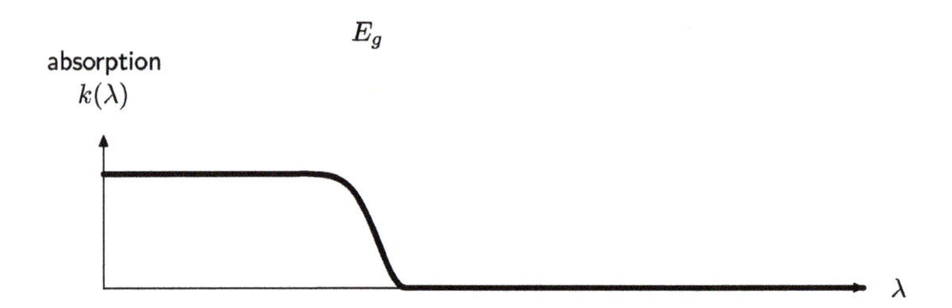

(b) Typical feature in the absorption spectrum due to the SC mechanism. Compared to ►figure 2.2, we do not see individual absorption peaks but a band edge. ►Figure 1.7 illustrates this by the example of actual pigments.

Figure 2.1: Flat band structure, schematic SC transition, and absorption spectrum (SC type) in a binary semiconductor A_xB_y.

Since SC causes no absorption peaks in the spectrum but a sharp band edge instead, SC transitions cannot induce arbitrary colors. The resulting colors are limited to the sequence of white–yellow–red–black with intermediate shades (►Figure 1.7 shows this using common pigments). The hue depends on the energy of the band edge and its purity on the steepness of the edge. Usually, colors induced by SC are among the purest. The transition strength is high, so colors are intensive ($\epsilon \approx 10^3$–10^6 [768, p. 21]).

– LF is a phenomenon of the *d* and *f* block elements. It can produce all colors since the transition between *d* and *f* orbitals can be of any energy, depending on the energy difference between the participating orbitals. The chemical environment can lift degeneracies of *d* or *f* orbitals in a complicated way, enabling multiple transitions. As a result, the absorption spectrum possesses multiple absorption bands, possibly resulting in duller colors.

  The colors depend strongly on the chemistry of the metal compound, and small changes may cause considerable variations. In practice, they are determined pri-

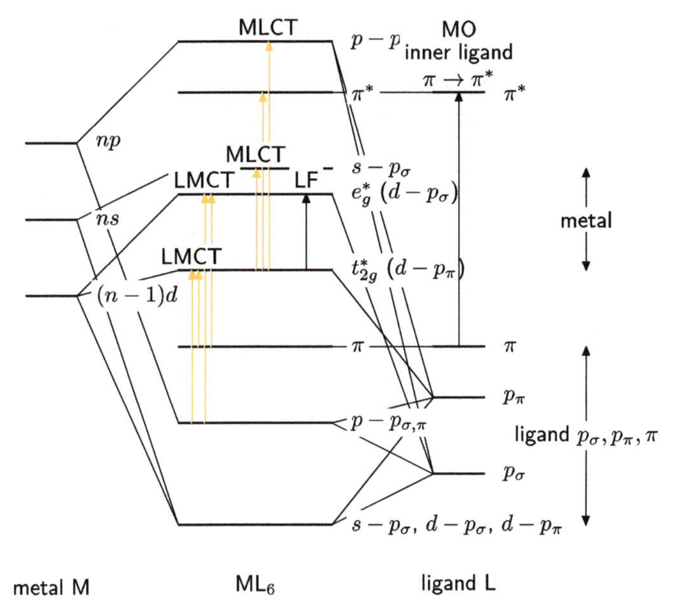

metal M ML6 ligand L

(a) Schematic classification of CT, LF, and MO transitions in the MO diagram of ML_6. CT transitions from ligand to metal typically appear in the high-energy (blue) spectral range. As they can lead to yellow pigments, they are marked in yellow.

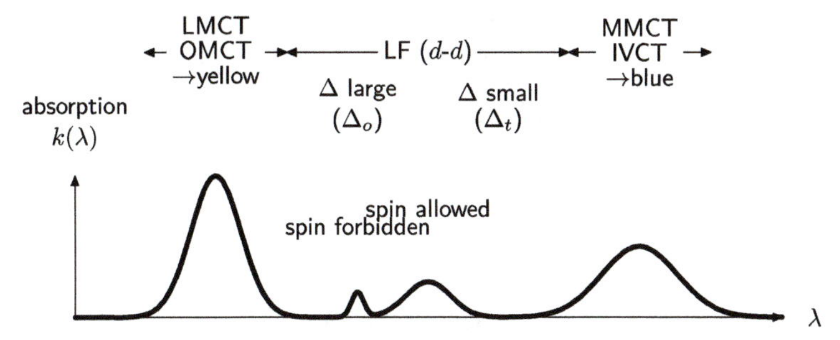

(b) Typical features in the absorption spectrum due to the different color-generating mechanisms: intense CT transitions between ligand and metal in the high-energy (blue) spectral range to the left (LMCT, often OMCT), weak LF transitions between metal d orbitals in the middle-energy spectral range, and intense CT transitions between metals of different valence state in the low-energy (yellow to red) spectral range to the right (MMCT or LMCT). Ligand field splitting Δ is lower in tetrahedral coordination (Δ_t) and tends to result in colors in the blue and green regions of the spectrum. ►Figure 1.5(b) shows CT spectra of measured pigments, and ►figure 1.6(b) shows measured LF spectra.

Figure 2.2: Different types of electronic transitions and absorption spectrum of an ML_6 complex (M = metal, L = ligand).

1	2	3	4	5	6	7	8	9	10	11	12	13	14	15	16	17	18
H																	He
Li	Be											B	C	N	O	F	Ne
Na	Mg	Sc	Ti	V	Cr	Mn	Fe	Co	Ni	Cu	Zn	Al	Si	P	S	Cl	Ar
K	Ca	Y	Zr	Nb	Mo	Tc	Ru	Rh	Pd	Ag	Cd	Ga	Ge	As	Se	Br	Kr
Rb	Sr	La	Hf	Ta	W	Re	Os	Ir	Pt	Au	Hg	In	Sn	Sb	Te	I	Xe
Cs	Ba	Ac	Rf	Db	Sg	Bh	Hs	Mt	Ds	Rg	Cn	Tl	Pb	Bi	Po	At	Rn
Fr	Ra											Nh	Fl	Mc	Lv	Ts	Og

(a) With consideration of all color active elements.

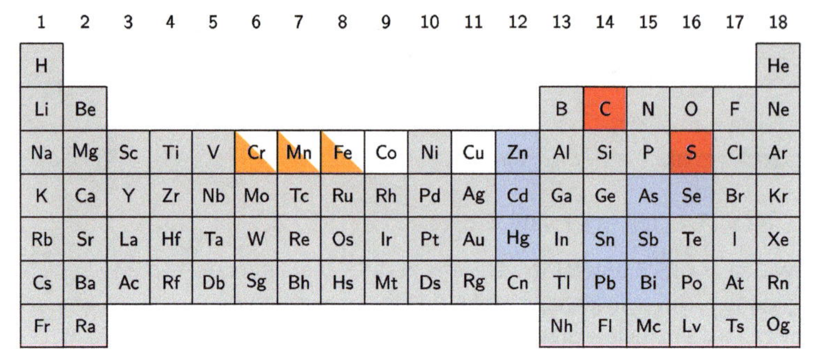

(b) With consideration of the elements actually used in pigments.

Figure 2.3: Localization of the dominant color-generating mechanisms in the PTE (group designation according to [194]). Red: transitions between molecular orbitals (MO); orange: charge transfer transitions (CT); white: ligand field (LF) transitions; light blue: band transitions in semiconductors (SC). The mechanisms occur predominantly in distinct regions of the PTE.

marily by the stability and availability of the complexes formed from metals and anions.

Depending on the composition of the complex and the symmetry around the metal cation, some electronic transitions are forbidden by so-called spectroscopic selection rules. The intensities of forbidden transitions are significantly lower than those of SC transitions (for *d* electron transitions $\epsilon \approx 10^2$, for *f* electron transitions $\epsilon \approx 10^1$ [768, p. 21]).

d block elements occur in crystalline minerals and amorphous materials such as glass. Practical examples are Mn, Cr, Fe, Co, Ni, and Cu, which occur naturally as oxide, hydroxide, or carbonate minerals and cover the color range brown, red, yellow, blue, and green.

– CT is mainly found in complex anions built with metals from groups 6 to 9 and in compounds of elements with multiple oxidation states. Therefore, colors resulting from this mechanism are not limited to a particular spectral range. Instead, the stable compounds of suitable metals and anions determine the achievable colors.

Since isolated transitions cause the colors, they exhibit high purity, and since no selection rules forbid the transition, they are very intensive ($\epsilon \approx 10^3$–10^4 [768, p. 21]).
Chromates, manganates, and molybdates represent the LMCT type (ligand-metal CT), occurring in complexes of highly charged metal cations with oxide ligands. Another example is hematite, an abundant iron(III)-based mineral of natural origin and known for its intense red color.
Metal pairs or multivalent metals lead to the MMCT or IVCT type (metal-metal or intervalence CT). Some intensely colored natural minerals and gemstones, such as the bluish rocks, olivines, and sapphire, contain Fe, Mn, and Pb in different valence states (IVCT) or metal pairs such as Fe/Ti (MMCT).

- MO is not limited to specific elements but occurs predominantly in carbon and sulfur compounds. MO transitions are intense (e. g., for transitions with $\pi \rightarrow \pi^*$ character $\epsilon \approx 10^3$–10^6 [768, p. 21]) and can occur in all spectral regions.
  Carbon-based MO transitions lead to numerous organic colorants. The plethora of colors achieved by MO transitions is impressively demonstrated by the colorful appearance of flowers, fruits, and animals, not regarding all modern organic colorants introduced in ►Chapter 4.
  Sulfur-based MO transitions cause polysulfide colorations, as in hot volcanic springs, sulfur melts, and the mineral lazurite (ultramarine).

Not all elements are employed to produce colorants, either because the metals are too rare or the colored compounds are unstable. For the actual components used for painting purposes, ►Figure 2.3(b) locates the mechanisms in the PTE. Regarding color, ►Figure 2.4 illustrates the correlation of colors versus mechanisms, also summarized in Table 2.1.

1	2	3	4	5	6	7	8	9	10	11	12	13	14	15	16	17	18
H																	He
Li	Be											B		N	O	F	Ne
Na	Mg	Sc	Ti	V	Cr	Mn	Fe	Co	Ni	Cu	Zn	Al	Si	P	S	Cl	Ar
K	Ca	Y	Zr	Nb	Mo	Tc	Ru	Rh	Pd	Ag	Cd	Ga	Ge	As	Se	Br	Kr
Rb	Sr	La	Hf	Ta	W	Re	Os	Ir	Pt	Au	Hg	In	Sn	Sb	Te	I	Xe
Cs	Ba	Ac	Rf	Db	Sg	Bh	Hs	Mt	Ds	Rg	Cn	Tl	Pb	Bi	Po	At	Rn
Fr	Ra											Nh	Fl	Mc	Lv	Ts	Og

Figure 2.4: Localization of typical colors obtainable by metal compounds in pigments in the PTE (group designation according to [194]). Similar colors occur mechanistically for elements in specific ranges.

Table 2.1: Possible chemical-mechanistical causes of hues.

Hue	SC	LF	CT	MO
White	√			
Yellow	√	√	√	√
Red	√	√	√	√
Green		√		√
Blue		√	√	√
Brown	√	√	√	√
Black	√		√	

2.2 SC: band gap transitions in semiconductors

Since semiconductors are of the most significant importance to any modern material scientist, we can limit ourselves to a simplified overview of the topic and suggest inorganic and physics textbooks for further details (see [237, 315–326]) on the topic of semiconductors in minerals [223, 233].

Solids consisting of many atoms or molecules can exhibit properties that differ significantly from individual, isolated constituents. For this text, peculiarities in the electronic structures that can cause color are of interest, as is the case when the solid is a *semiconductor*:

- The color is not caused by absorption peaks but by a continuous absorption cut off by a sharp edge, i. e., the absorption drops rapidly to zero above a particular wavelength. This wavelength corresponds to the *band gap* between valence and conduction band. (Eventually, we can observe a fine structure with weak peaks under certain circumstances, but this is beyond our discussion.)
- Due to the absorption edge, semiconductor colors are limited to the sequence of white-yellow-red-black, depending on the wavelength (energy) of the band edge.
- Quantum mechanical selection rules do not forbid absorptions. Therefore, semiconductors show intensive colors.
- The purity of the color depends on the steepness of the edge but is generally very high.
- Classical semiconductors used for painting are chalcogenides of group 12–15 elements, especially sulfides, oxides, and selenides of arsenic, antimony, lead, zinc, cadmium, and mercury. Newer research in inorganic chemistry shows that elements such as lanthanum and tantalum can replace the toxic heavy metals arsenic, mercury, cadmium, and lead without loss of color, ►Section 2.2.3. Due to alloy formation, anions and cations can be relatively freely combined, and semiconductor materials are not limited to binary compounds, gaining numerous shades of color.

The structure of semiconductors and the development of the collective properties becomes understandable if we start from single atoms, combine them into pairs, then clusters, and keep increasing their number. In this process, "molecular orbitals" (MO) form from atomic orbitals (AO). The MOs proceed to combine more and more MOs until finally, at vast atomic numbers N, we can view them as "crystal orbitals" encompassing the entire crystal. In effect, they form by the entirety of the atoms in the crystal; ►Figure 2.5 [237, 318, 350, 351, 353]. Albright et al. [210, Chapter 13] derives the formation of bands from AOs from the point of view of MO theory.

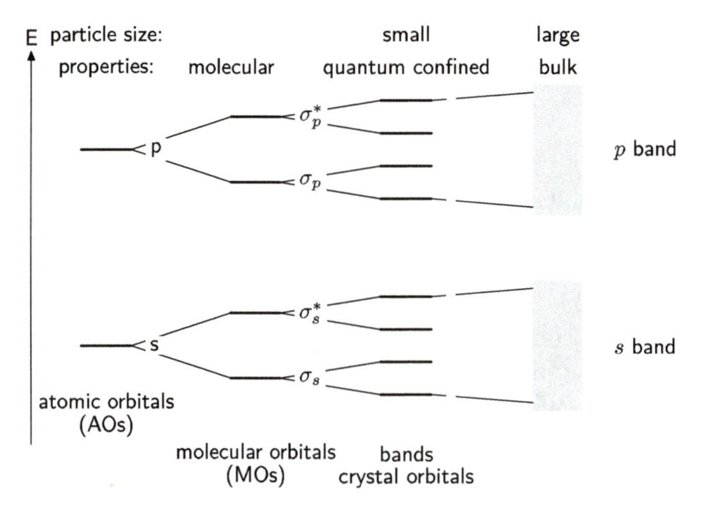

Figure 2.5: Formation of bands (crystal orbitals, crystal-wide molecular orbitals) by formation of molecular orbitals (MO) from an increasing number of atomic orbitals (AO) [210, Chapter 13]. AOs of *s* character combine to *s* bands, AOs of *p* character combine to *p* bands, etc. *s* and *p* MOs can overlap; bands then show both *s* and *p* characters proportionally. Small particles possess a larger band gap than large particles since the band gap decreases with the number of participating MOs.

For N atoms, N of these crystal orbitals emerge. Since each combination of MOs results in a pair of new MOs with higher and lower energy, the energy differences of the crystal orbitals decrease and eventually become infinitesimally small ($\propto 1/N$). The individual orbitals develop into *bands* comprising countless crystal orbitals. The number of MOs in an infinitesimal energy interval is called the *density of states*; the number of electrons occupying one of these crystal-wide MOs is called the *occupation number*. Both of these data are energy-dependent. In this simple "flat-band" model, we can speak of *s*, *p*, and *d* bands, depending on the original AOs.

The resulting bands may be "impure," i. e., comprising several types of AO, depending on the energy distribution of MOs. A *d* band, e. g., may also exhibit some *p* or *s* characteristics when *s*- and *p*-type MOs overlap energetically with *d*-type MOs during band formation.

Another model of semiconductors focuses on a delocalized electron gas in the potential of the positive atomic hulls. The energy distribution is calculated without relating the resulting bands to the principal or azimuthal quantum numbers.

In both models, bands are no longer characterized by individual energy but by their energy ranges, densities of states, and occupation numbers. The collective or bulk properties emerge when so many atoms or molecules join that the resulting particle is in the size range of around 10 nm; i. e., when the material is colloidally distributed. Clusters of size 2–50 nm are also called *quantum dots* or *Q-particles*. In the literature, the reverse process, the emergence of molecular (quantized) properties as a reaction to decreasing particle size, is called *quantum confinement* or *size quantization effect*. In this intermediate range between bulk and isolated constituent, properties depend on the particle's size.

2.2.1 Valence and conduction band

Two bands are particularly significant [210, p. 315]: the *valence band* and the *conduction band*. The valence band (VB) is the highest energy band completely or partially occupied by electrons. The conduction band (CB) is the lowest energy band unoccupied, possibly containing the delocalized electrons that cause thermal and electrical conductivity. Both bands are separated by the energy or *band gap* E_g. Solids can be classified according to the occupation number of their conduction band and the size of their band gap:

- The VB is partially occupied, or empty bands overlap with the filled VB so that valence electrons can freely occupy states of equal or higher energy, becoming delocalized or conduction electrons; the solid is a metal or a conductor, ►Figure 2.6(a). Since there is no band gap, even small amounts of external energy initiate transitions into slightly higher states. The delocalized electrons mediate thermal and electrical conductivity.
- The VB is filled, and the CB is empty. E_g is large (more than a few eV). The solid is an insulator since it requires too large an energy supply to transfer electrons from the VB to the CB, which cannot be provided thermally or optically, ►Figure 2.6(b).
- The VB is filled, and the CB is empty. E_g is moderately large (a few eV); the solid is a semiconductor because valence electrons can be transferred from the VB to the CB by supplying a small amount of energy (e. g., by heat at room temperature or by exposure to light), ►Figure 2.6(c). The insulating material thus becomes increasingly conductive.

The highest occurring electron energy in the solid is called *Fermi energy* E_F. Its value relative to the valence or conduction band's energies can also be used to classify solids as conductors or semiconductors/insulators.

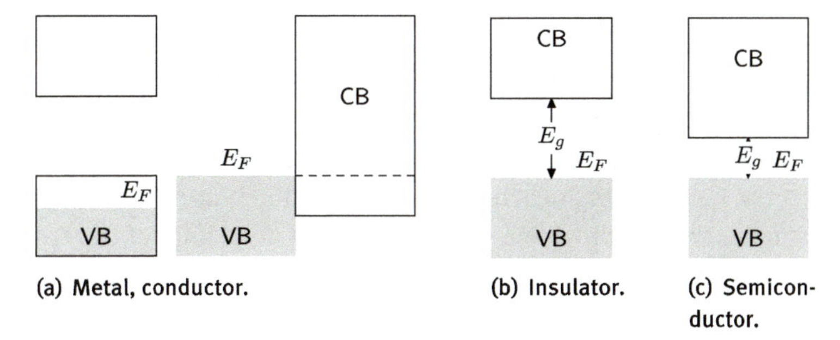

Figure 2.6: Flat-band models of three different relative arrangements of the highest occupied band (the valence band VB) and the lowest unoccupied bands (conduction bands CB) and the origin of the band gap E_g [210, p. 315]. States occupied by electrons are shaded in gray, and unoccupied states within the bands are depicted in white. The highest energy occupied by electrons is called *Fermi energy* E_F. (a): The VB is partially occupied, or empty bands overlap so that valence electrons can freely occupy higher energy states, becoming delocalized or conduction electrons in metals and conductors. (b): The VB is completely filled, the CB is empty and separated from VB by a large band gap in insulators. (c): As in (b), but the band gap is small, and valence electrons can be thermally or optically excited to the CB in semiconductors.

Complex band structure diagrams and dispersion curves

The "flat-band diagrams," as we use them in this book, are strong simplifications of actual band structures since we depict only the band edges, i. e., the smallest and largest potential energy of an electron in a band. This simplification is acceptable for understanding semiconductor colors, especially since realistic representations of electron energies in crystals and solids are quite complicated and depend on the spatial position. Aven and Prener [316] gives a detailed introduction to band theory; Müller [237, Chapter 10], Reinhold [267, Chapter 10] offers a description suitable for chemists.

We can reduce crystalline colorants to *elementary cells* due to their crystal symmetry. This cell, strung together any number of times, always yields crystals of the substance in question and contains all the information about the (periodic) electron structure. It can assume a cube or hexagonal-prismatic shape or many others.

The entirety of curves and lines of the energy diagram of a crystal represents the *dispersion relation* of an electron, i. e., the influence of the crystal lattice's electromagnetic potential on an electron. Each curve represents the dependence of the energy E of the electron along specific paths through the crystal, using as a "pathfinder" the *wave vector* k. In solid-state physics, we can think of k as the continuous counterpart of the discrete quantum numbers we use in chemistry to denote a particular orbital. Since many orbitals fuse into bands in the solid state, we need to move from discrete numbers to continuous designations.

Theoretically, we could follow an infinite number of arbitrary paths through the crystal and record an electron's energy. In practice, the paths along which we calculate or measure electron energies are called *high-symmetry lines*. They connect selected *high-symmetry points* (critical points) typical for a crystal lattice. High-symmetry points are denoted by capital Greek or Latin letters, e. g., Γ, X, K, or L. Examples are the midpoint of the unit cell, the midpoints of the bounding surfaces of the cell, or the vertices or midpoints of lines connecting such points. High-symmetry lines are denoted with Greek capital letters, e. g., the Λ-line indicating the path from Γ to L.

Although the electric field differs at each spatial point of the unit cell, the representations along symmetry lines provide a sufficiently accurate picture of the energetic conditions even for spatial regions outside the symmetry lines and points. Kobayashi et al. [640] shows an example of a cell and its associated cutting pattern; Bouckaert et al. [349] explains the symbolism. However, it is sufficient to combine all those curves into several flat blocks, which are distinctly separated from each other.

2.2.2 Color

The concept of a valence band filled with electrons and an empty conduction band allows the transfer of electrons from the valence band to the conduction band by absorbing electromagnetic energy (light). This transfer induces color if the transition energy occurs in the range of optical energies. The band gap provides the *minimum energy* necessary for the electronic transition and represents a decisive quantity for the perceived hue. In a realistic model, transitions can occur between the dispersion curves, resulting in a complex spectrum whose fine structure depends on the density of states, occupation numbers, and transition probabilities. Phillips [649] illustrates such a realistic scenario in more detail.

However, since we have ignored all details of dispersion curves in favor of simplicity in the flat band model, we further ignore any subtleties and pretend that there is only one major transition between the band edges. This transition corresponds to the band gap and its energy E_g, and it marks the steep band edge in the absorption spectrum, ►Figure 2.1. The band gap represents the minimum energy required to enable the transition. At higher energies (lower wavelengths), the absorption remains high; when incident light of higher energy is absorbed, the transitions end within the conduction band instead at its lower edge, ►Figure 2.7(a). Similarly, electrons of lower energy than E_F can also be excited, leading to transitions with higher energies. Thus, absorption in SC-type pigments is constantly high up to an energy corresponding to the band gap E_g. ►Figure 2.8 represents the color as a function of the band gap. (In addition, ►Figure 2.8 depicts the emission color corresponding to the band gap, i. e., the color emitted by an LED made of the semiconductor.)

Characteristics of colors

An important observation is that SC-based colorants can produce pure white and black. Semiconductors, possessing only a small band, gap exhibit an evenly high absorption over a wide range of wavelengths and, therefore, a deep black color. In

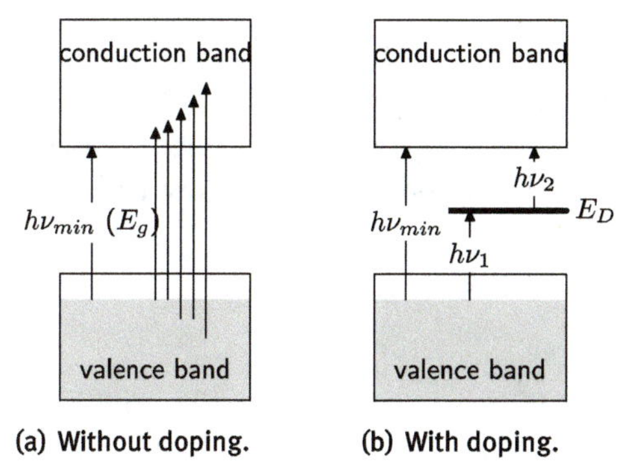

(a) **Without doping.** (b) **With doping.**

Figure 2.7: (a): Formation of the absorption edge in an absorption spectrum when incident light falls onto semiconductors. Any energy above E_g or $h\nu_{min}$ enables an electronic transition; the absorption remains high above this energy. The simplifying flat-band model is used to depict the band structure. (b): Reduction of band gap by an additional level E_D due to doping elements. The new level can be an electron donor (ν_2) or an electron acceptor (ν_1), i. e., only the energy $h\nu_1$ or $h\nu_2$ has to be supplied instead of $h\nu_{min}$ to trigger a transition. The transition induces color when $h\nu_1$ or $h\nu_2$ falls into the visible spectral range.

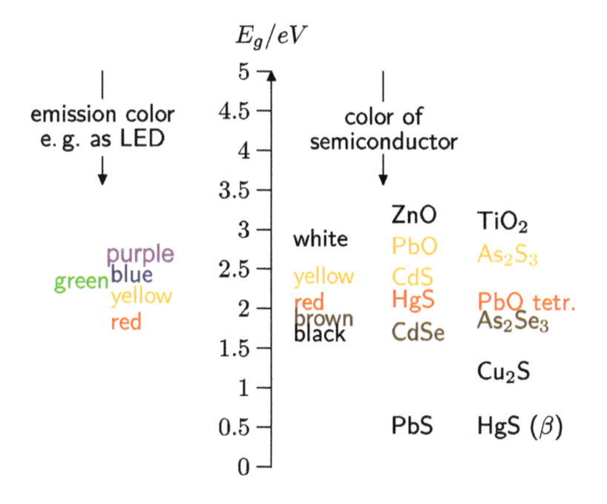

Figure 2.8: Correlation between band gap E_g and color. Left: The emission color when the semiconductor material works as a LED. Right: The perceived color of a semiconductor and some semiconductor pigments.

contrast, organic black pigments must simulate this absorption behavior by placing multiple overlapping absorption bands side-by-side. Pure white is achieved by grounding an insulator or a semiconductor with a large band gap to a fine powder. While the insulator or semiconductor is colorless (transparent) when solid, we perceive the powder as pure white due to its intensive white scattered light.

The steeper the absorption edge, the more purely colored the substance is. At temperatures above zero, some electrons are already excited by thermal energies. They do not reach the conduction band but higher energies within their original band so that the occupation number does not abruptly drop to zero at E_F. An intermediate region forms around E_F, which is partially (de)occupied by electrons, as ►Figure 2.1 suggests by the slope of the absorption edge. The edge is steep enough that SC-based pigments always show brilliant colors at room temperature. Compare chrome yellow, cadmium yellow, or cadmium red with yellow and red ocher earths.

Variation of band gap and color with crystal size

The band gap decreases with an increasing number of atoms involved in the pigment particle, i. e., it decreases with the crystal size, ►Figure 2.5. Therefore, in this type of chromophore, the size of the pigment crystals influences their color. Increasing the particle size leads to the marked color change red → brown → black; decreasing the particle size shifts the color to red → yellow → white. Murray et al. [352] shows the band gap variation between 400 and 700 nm when CdSe clusters are enlarged from 1 nm to 10 nm. Weller [353] demonstrates the perfect semiconductor color sequence white → yellow → orange → red → brown → black for Cd_3P_2 particles of different sizes. Cinnabar is significant for painting purposes since it is deep dark red for particle sizes $> 5\,\mu m$ and bright orange-red for particle sizes $< 5\,\mu m$ [58]. The practical consequence is that grinding influences the color of semiconductor pigments.

2.2.3 SC-based chromophores

Today we know numerous semiconducting solids, e. g., the metals silicon and germanium, which are essential for the electronics industry, and red phosphorus, boron, selenium, and tellurium. In addition to these elemental semiconductors, the binary III-V semiconductors made of group 3 to 5 elements are of particular importance: nitrides, phosphides, and arsenides of gallium and indium, e. g., GaAs or InN, are required in large quantities for LED-based light fixtures and display panels.

In painting, on the other hand, we have to consider oxides, sulfides, and selenides of group 12, 14, and 15 metals, such as Pb, As, Zn, Cd, Hg, and Fe. These II-VI semiconductors have band gaps in the optical range so that they induce color. It was essential for the development of painting that many of these semiconductors occur naturally in the form of minerals: up to the nineteenth century, cinnabar (mercury sulfide) represented an essential pure red pigment, and lead oxides supplied crucial white and yellow ones (lead white, lead-tin yellow, lead yellow, Naples yellow).

Today, in addition to binary compounds, ternary compounds are also used, in which a third element replaces some atoms of the primary binary bodies. These ternary semiconductor alloys yield an extended color spectrum; examples are cadmium orange/cadmium red Cd(S, Se) and cadmium yellow lemon (Cd, Zn)Se.

A significant problem with traditional and modern semiconductor pigments is that they comprise toxic metals (lead, mercury, cadmium). Consequently, some prominent pigments (lead white, lead-tin yellow, Naples yellow) have already disappeared from the palette, or their use is declining (cadmium pigments). However, nontoxic new developments such as the oxynitrides (►p. 91) form a gapless series from $CaTaO_2N$ (yellow) to $LaTaO_2N$ (red).

Example: Titanium white, zinc white, cadmium yellow, vermilion

Titanium white, zinc white, cadmium yellow, and vermilion are examples of white, yellow, and red semiconductor-based colorants. The atomic orbital energies from ►Table 2.2 allow a first rough estimation of band gaps. However, atomic orbitals can seldom be compared directly with bands as bands are usually composed of several types of atomic orbitals. Therefore, we refer to the band structure diagrams published in the indicated sources to determine realistic band gaps for a more detailed analysis.

Table 2.2: Atomic orbital energies in eV for some elements from SC-based colorants for a rough estimation of band gaps [367].

	O	S	Se	Ti	Zn	Cd	Hg
6s							−8.93
5d							−17.68
5p							−96.27
5s						−7.66	−138.95
4d						−20.09	−402.61
4p			−10.93			−89.00	−710.87
4s			−23.48	−6.09	−8.13	−129.17	−834.43
3d			−71.30	−10.82	−20.98	−435.40	−2433.58
3p		−11.65	−187.26	−49.60	−107.70	−679.05	−3337.17
3s		−24.14	−250.95	−79.31	−157.83	−797.92	−3623.95
2p	−16.78	−182.87	−1516.00	−489.15	−1081.37	−3782.94	−14336.45
2s	−34.08	−246.57	−1697.17	−589.59	−1233.92	−4076.49	−14980.77
1s	−563.23	−2512.30	−12744.96	−5019.98	−9734.80	−26868.48	−83705.32

For example, consider titanium white TiO_2 with electron configurations $3d^2 4s^2$ for titanium and $2p^4$ for oxygen, or $3d^0 4s^0$ for $Ti^{4\oplus}$ and $2p^6$ for $O^{2\ominus}$. The valence band is formed by $2p$ electrons of the oxide anions, while the empty $3d$ orbitals of $Ti^{4\oplus}$ represent the conduction band, ►Figure 2.9(a) [636, 637]. The essential transition ν_1 in titanium white is thus O($2p$) → Ti($3d$). A more precise calculation shows that the densities of states for the O($2p$) band are also not zero in the Ti($3d$) band region and vice versa. To some extent, a mixing of d and p characters takes place. Indeed, the assignment of bands to atomic orbitals is generally only an approximation caused by the simple flat-band model used here. Furthermore, the bands show a LF splitting: the $3d$ band of the (approximately) octahedrally coordinated titanium consists (approximately) of the subbands $3d(t_{2g})$ and $3d(e_g)$.

In zinc white with configuration $3d^{10}4s^2$ for zinc or $3d^{10}4s^0$ for $Zn^{2\oplus}$, the valence band is formed by both the $2p$ electrons of the oxide anions and the $3d$ electrons of $Zn^{2\oplus}$, while the conduction band corresponds to the empty $4s$ and $4p$ orbitals of $Zn^{2\oplus}$, ►Figure 2.9(b) [324, 325, 639, 640]. The transition ν_2 is O(2p) → Zn(4s), Zn(4p). A rough calculation with ►Table 2.2 yields −8.13 − (−16.78) = 8.66 eV for the width of the band gap, corresponding to white.

Cadmium yellow possesses the configuration $4d^{10}5s^2$ for cadmium, $4d^{10}5s^0$ for $Cd^{2\oplus}$, and $3s^23p^6$ for the sulfide ions. The valence band includes the $3p$ electrons of the sulfide anions and the $4s$, $4p$, and $4d$ electrons of $Cd^{2\oplus}$, which have a similar spatial distribution. The empty $5s$ orbital of $Cd^{2\oplus}$ acts as a conduction band, ►Figure 2.9(c) [324, 325, 641]. A rough calculation with AO energies yields a smaller band gap of −7.66 − (−11.65) =3.99 eV, which is closer to yellow but still a very rough approximation.

Vermilion possesses the configuration $5d^{10}6s^2$ for mercury, $5d^{10}6s^0$ for $Hg^{2\oplus}$ and $3s^23p^6$ for the sulfide ions. The rough calculation with ►Table 2.2 yields an even smaller band gap of −8.93 − (−11.65) = 2.71 eV. The valence band comprises the $3p$ electrons of the sulfide anions and the $6s$ and $5d$ electrons of $Hg^{2\oplus}$. The empty $6s$ orbital of the mercury cation forms the conduction band, ►Figure 2.9(d) [643, 644].

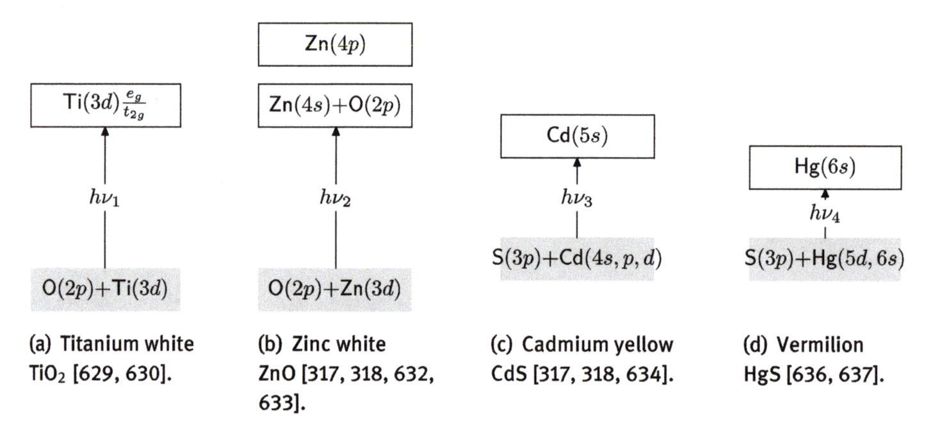

(a) Titanium white TiO_2 [629, 630].

(b) Zinc white ZnO [317, 318, 632, 633].

(c) Cadmium yellow CdS [317, 318, 634].

(d) Vermilion HgS [636, 637].

Figure 2.9: Flat-band model of classical SC-based pigments. The color changes due to the decrease of band gaps, caused by an increase in electronegativity from titanium to mercury and a decrease in electronegativity from oxygen to sulfur. It is visible that in actual colorants, each band is composed of several types of atomic orbitals (e. g., the O(2p) band also have Ti(3d) character).

The decrease of the band gap from zinc white to vermilion, perceived as a color sequence of white → yellow → red, is due to the increase in electronegativity from titanium and zinc to cadmium to mercury (values according to Pauling: 1.54, 1.65, 1.69,

1.9). Electrons are more tightly bound with increasing electronegativity, and the band's energy decreases. The same is true for the anions (electronegativity for oxygen, sulfur, and selenium: 3.44, 2.58, 2.55).

Oxynitride pigments

Aguiar et al. [445], Janson and Letschert [446], Günther et al. [447, 448] present an example of new inorganic pigments development. Here, oxynitrides with the formula $ATa(O,N)_3$ (perovskites with alkaline earth or rare earth metals as A) are examined, showing a wide variety of colors caused by solid solution formation and manipulation of the band gap, ►Table 2.3. Especially calcium-lanthanum-tantalum perovskites are very interesting as nontoxic and color-equivalent substitutes for cadmium sulfoselenide pigments such as cadmium yellow, cadmium orange, or cadmium red. Their hues vary from yellow → red → brown according to the fraction *x* of lanthanum and the oxygen/nitrogen ratio.

Table 2.3: Colors of oxynitrides of the form $ATa(O,N)_3$ and isoelectronic compounds [445–447].

Yellow	Orange	Red
$CaTaO_2N$ Calcium tantalum yellow	$SrTaO_2N$, $Ca_{1-x}La_xTaO_{2-x}N_{1+x}$	$BaTaO_2N$, $Ca_{0.25}La_{0.75}TaO_{1.25}N_{1.75}$ lanthan tantalum red
Brown	**Black**	**Green**
$SrNbO_2N$, $LaTaON_2$, $LaTiO_2N$	$BaNbO_2N$	$Yb_2Ta_2O_5N_2$

The color series shows different influences, ►Figures 2.10(a) and 2.10(b). Most metal oxides are colorless (white) because they have a large band gap. By lowering the electronegativity of the anion, its electrons become less firmly bound by the nucleus, the upper edge of the valence band increases, and thus the band gap decreases. We can change the EN by (partially) exchanging oxide anions for nitride anions in the perovskite and preserving charge neutrality by simultaneously exchanging calcium for lanthanum:

$$3\,O^{2\ominus} = 2\,N^{3\ominus}, \quad M^{n\oplus}O = M^{(n+1)\oplus}N$$

Günther et al. [448] discusses the nitride crystal chemistry and properties of oxynitrides in detail. Compared to pure oxides, the oxynitrides thus obtained are colored because the band gap is not between O(2*p*) and *d* bands but between N(2*p*) and *d* bands. Due to the lower EN of nitrogen than oxygen, its 2*p* band is higher, and thus reduces the band gap into the optical range.

Electronegativity also plays a role in the color differences between niobates and tantalates: $SrTaO_2N$ vs. $SrNbO_2N$ (orange vs. brown) and $BaTaO_2N$ versus $BaNbO_2N$ (red vs. black). Again, the higher electronegativity of niobium decreases the band

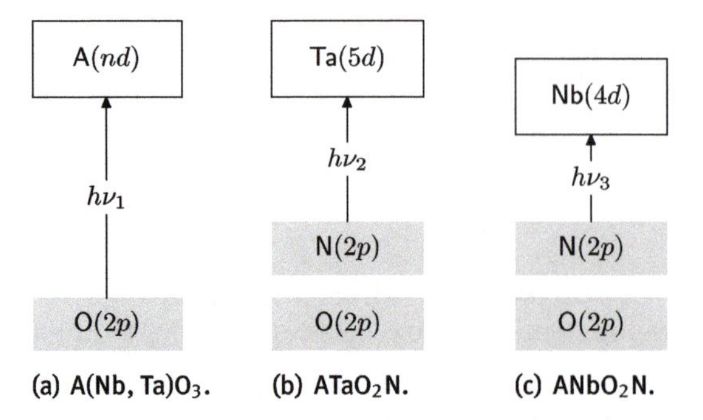

Figure 2.10: Flat-band model of oxynitrides [445]. (a): General band structure of oxynitrides. (b) and (c): Color change by forming a solid solution and adopting the band gap according to the composition. The lower electronegativity (EN) of nitrogen (3.04) increases the energy of the N(2*p*) band in comparison to oxygen (EN: 3.44), decreasing the band gap. Similarly, the lower EN of tantalum (1.5) increases the energy of the Ta(5*d*) band in comparison to the Nb(4*d*) band (EN of niobium: 1.6). All EN values according to Pauling.

gap because now the *d* bands are lower due to the tighter electron bonding, ►Figures 2.10(b) and 2.10(c).

Further differences in color result from different crystal lattices regarding dimensions and deformations and the degrees of purity.

2.2.4 Influence of lattice width and crystal structure, thermochromism

We have seen that the combination of many atoms into a solid-state body reveals properties caused by the formation of bands and the gaps between them. An important factor influencing the size of these band gaps is the atomic distance or the *lattice width* in the crystal. We can understand this influence by looking at the formation of the hydrogen molecule H_2: two spatially separated hydrogen atoms are brought together and begin to interact via their 1s electrons. As a result, a bonding and an antibonding molecular orbital form. Their energies depend on the spatial distance between atoms, and thus on the ability of atomic orbitals to overlap.

The same happens when atoms in a crystal lattice approach each other: if the lattice constant increases, the band's energies shift, and the fundamental band gap E_g generally decreases, ►Figure 2.11 [335, Chapter 6]. The change may reach the order of 4 eV/Å. The plot E_g versus a illustrates this trend using the example of (Zn, Cd)S and Cd(S, Se), ►Figure 2.12(a), data from ►Table 2.4.

The following factors affect the lattice width:

- Pressure. An increase in pressure generally decreases the lattice width.
- Temperature. A temperature rise generally increases the lattice width. For example, thermochromic zinc white reversibly turns yellow with a temperature rise.

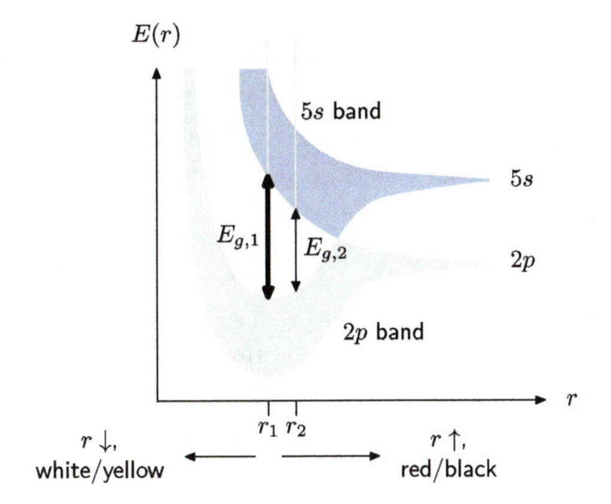

Figure 2.11: Dependence of the band gap on the lattice constant, respectively, the distance r between two lattice points ([335, Chapter 6], [210, p. 315], drawn after [321, Chapter 8.4.3, p. 300]). The diagram depicts a fictitious semiconductor's p and s bands whose atoms have the equilibrium distance r_1. The band gap 2p–5s for a lattice of width r_2 is smaller than that for the equilibrium width r_1, i. e., an increase of the lattice constant leads to a color shift to red, a decrease to a color shift to yellow.

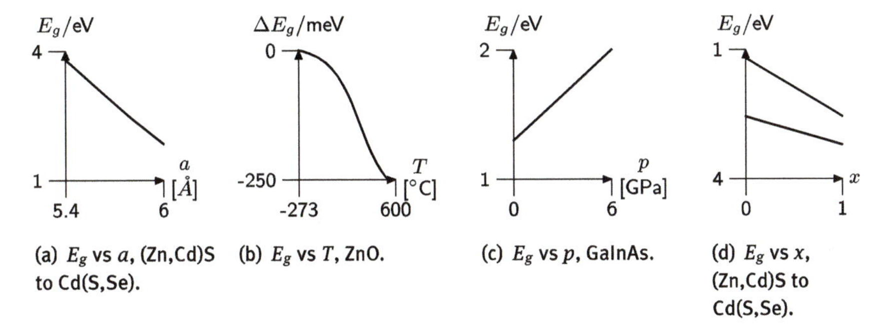

(a) E_g vs a, (Zn,Cd)S to Cd(S,Se). (b) E_g vs T, ZnO. (c) E_g vs p, GaInAs. (d) E_g vs x, (Zn,Cd)S to Cd(S,Se).

Figure 2.12: Schematic dependence of band gap E_g on the lattice constant a, temperature T, pressure p, and composition of an alloy x. Data from ►Table 2.4, [319, Chapter 6.7], [319, Chapter 6.10], ►Table 2.4.

- Crystal structure. The crystal structure determines the arrangement of atoms, which atoms are adjacent, and the distance between them. For example, lead(II)-oxide can crystallize in two crystal structures, which leads in the case of litharge to red color (tetragonal structure), in the case of massicot to yellow color (hexagonal structure).
- Composition (in the case of a semiconductor alloy). The lattice width of a ternary semiconductor $A_xB_{1-x}C$ changes continuously with x, exemplified by the cadmium pigments $Zn_xCd_{1-x}S$ and CdS_xSe_{1-x}. We will discuss the influence of composition on the band gap in ►Section 2.2.5 since profound changes also occur in the electronic band structure in addition to the change in lattice width.

We will see that the scientific application of these relationships led to a series of new pigments. The simplified band model used here helped us to understand these dependencies. Among others things, it assumes that bands can indeed be assigned to individual atomic orbitals. In reality, the composition of the bands and their energies depends in a complicated way on lattice constants and other factors.

2.2.4.1 Influence of pressure on lattice width

The described influence of the lattice width on the band gap E_g under pressure is shown for GaInAs in the plot E_g versus p, ►Figure 2.12(c). Greenaway and Harbeke [315] and Bassani and Brust [359] depict the shift of the lower conduction band edge Γ'_2 (s-like state) concerning the upper valence band edge Γ'_{25} (p-like state): at low pressure (when the lattice is wide), E_g is small, and at high pressure (when the lattice narrows), E_g increases. It is assumed that a core atom, including the inner electrons, mainly remains unchanged under pressure while the valence electrons move toward the core atom. Since s states have higher electron densities near the nuclear center, these states are more sensitive to pressure.

2.2.4.2 Influence of temperature on lattice width

The lattice width increases by raising the temperature due to the thermal movement of atoms. An example of such *thermochromism* is zinc white, being white at room temperature. However, at temperatures of a few hundred degrees Celsius, the pigment assumes a yellow color because the increased thermal motion of the atoms expands the crystal lattice, and the band gap E_g consequently decreases. The crystal lattice shrinks again on cooling, E_g increases, and the pigment becomes white again. Such thermochromic substances are used as robust passive temperature indicators, e. g., on pans or pots. The plot ΔE_g versus T depicts the described change for ZnO, ►Figure 2.12(b).

2.2.4.3 Influence of crystal structure on lattice width

The crystal structure of a semiconductor determines the positions of all of its atoms and their mutual distances. It thus has a fundamental influence on the chemical environment of each atom and the spatial distribution of the electric field, i. e., the semiconductor's electronic conditions and band structure.

Example: Massicot, lead yellow pigments, litharge

A review of crystallographic and electronic structures [650–652] shows that the color of red, tetragonal α-lead(II) oxide (litharge) originates in a band gap of 1.9 eV. Litharge crystallizes in the CsCl structure, where four oxygen atoms surround each lead atom at a distance of 0.232 nm. In contrast, yellow orthorhombic metastable β-lead oxide (massicot) has a band gap of 2.4–2.7 eV. Massicot consists of lead and oxygen layers with Pb–O chains and an average atomic spacing of 0.2358 nm.

In both modifications, Pb(6s) and Pb(6p) states form the valence band as well as the conduction band [653, 654]. In red lead oxide, O(2p) states participate at both band edges and decrease the band gap so that the red color results in contrast to the yellow color of massicot, ►Figures 2.13(a) and 2.13(b).

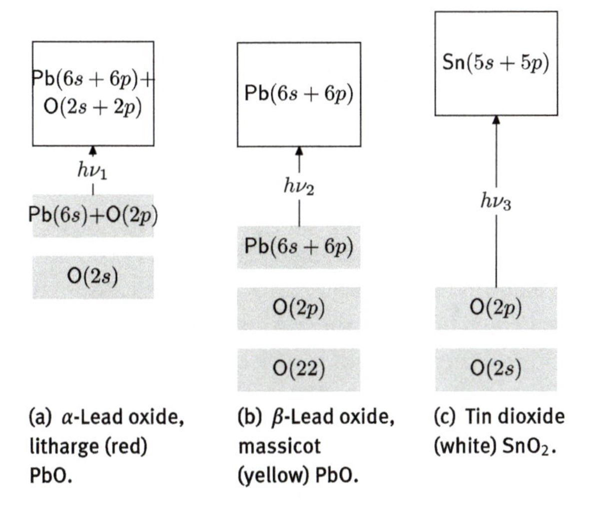

(a) α-Lead oxide, litharge (red) PbO.

(b) β-Lead oxide, massicot (yellow) PbO.

(c) Tin dioxide (white) SnO_2.

Figure 2.13: Flat-band model of classical, SC-based pigments. (a) and (b): Red tetragonal lead oxide (litharge) and yellow hexagonal lead oxide (massicot). The admixture of O(2p) states to the Pb(6s) and Pb(6p) bands decreases the band gap so that, in comparison to massicot, we perceive red color [650, 652–654]. (c): Tin dioxide occurs in lead-tin yellow but is itself white due to its large band gap between O(2p) states and Sn(5s + 5p) states [637, 655].

Naples yellow (lead antimonate) and lead-tin yellow (lead stannate) are other lead compounds classically used as pigments. They are closely related to yellow lead oxide and can be regarded as mixed oxides of lead oxide with antimony oxide or tin dioxide. As with the massicot itself, the yellow color originates in a transition from a Pb(6s + 6p) band to another Pb(6s + 6p) band. Tin dioxide itself exhibits a white color since transitions from its O(2p) band to its Sn(5s+5p) band require more energy than optical absorptions can deliver, ►Figure 2.13(c) [655]. The following section will discuss the exact processes within a solid solution.

2.2.5 Alloys, solid solutions, and color

The possibility of changing a semiconductor's band gap, and consequently the color, by applying high pressure or high temperature is intriguing in theory. Fortunately, for art museums, chemists have found more practical ways to cause such changes.

Several essential colorants consist of two different semiconductors, which appear in the pigment as *mixed crystals* or *alloys*. When the crystal structures of two semiconductors differ little or are compatible, we obtain solid solutions whose band gaps smoothly change between the pure substances' values, depending on the mixing ratio. We then observe a continuous color transition (E_g vs. x plot, ►Figure 2.12(d)). Examples are:

- Cadmium yellow, cadmium orange, and cadmium red. Cadmium yellow citron $Cd(S, SO_4)$ or $(Zn, Cd)S$ is a solid solution of cadmium sulfide and sulfate or zinc cadmium sulfide. Pure cadmium sulfide yields cadmium yellow medium, while solid solutions $Cd(S, Se)$ become redder with increasing selenium content (cadmium yellow dark, cadmium orange, cadmium red). Pure cadmium selenide is brownish-black but is not employed as a pigment.
- Cadmium cinnabar $(Cd, Hg)S$ is orange to red depending on the mercury content.
- Naples yellow $2\,PbO \cdot Sb_2O_5$ and lead-tin yellow $2\,PbO \cdot SnO_2$. Theoretically, they possess a fixed composition; practically, however, the mixing ratio differs due to manufacturing, especially in historical times.

Adachi [317], Grundmann [319], Sapoval and Hermann [320], Hunklinger [321] consider the properties of semiconductor alloys in detail and show that in semiconductor alloys, two influences superimpose and determine lattice width and band gap:

- the dependence of lattice width on the composition of the alloy;
- the reshaping of the electronic band structure during the transition from one pure alloy partner to the other.

2.2.5.1 Lattice width and composition

The change of lattice width in ternary semiconductor alloys $A_xB_{1-x}C$ with the mole fraction x can be modeled as incorporating larger or smaller atoms of an alloy partner B into the lattice of an initial binary compound AC [319, Chapter 3.7], causing the lattice to expand or collapse. The direct consequence is a displacement of atoms from their equilibrium positions, a change in bond angles and lengths, and correspondingly a change in the band structure and band gap, according to ►Section 2.2.4. A numerical example shall illustrate this: the band gap of quartz in feldspar changes with a ratio of 0.054 eV/degree (in case of changing the bond angle) or 35 eV/Å (in case of changing the bond length) [361].

Studies [317, 323], [326, Chapter 4.6], [319, Chapter 3.7] have shown that in ternary semiconductors $A_xB_{1-x}C$, all bond lengths, and thus the lattice width a change proportionally to the mole fraction x between the values of the pure substances AC and BC (law of Vegard):

$$a = a(x) = xa_{AC} + (1 - x)a_{BC} \tag{2.1}$$

This relation is idealized but comes close if ionic radii and electronegativity of the atoms involved differ little from each other, and the crystal lattices of the pure substances are largely compatible. If these conditions are not fulfilled, we observe a more complex progression of $a(x)$.

According to ►Section 2.2.4, the lattice width a and the band gap E_g ideally change proportionally with the mole fraction x [317]. In practice, deviations from linearity occur; we can express this by introducing a bowing parameter b:

$$E_g(x) = xE_{g,AC} + (1-x)E_{g,BC} - x(1-x)b \tag{2.2}$$

By forming a solid solution or alloying, it is possible to continuously change $E_g(x)$ of a material in an energy range limited by the band gaps of the pure materials. The change in electronic structure, i. e., the shift in bands and band edges can be seen in [365]. ►Figure 2.14 shows this schematically for the ternary semiconductors mentioned above, essential in painting.

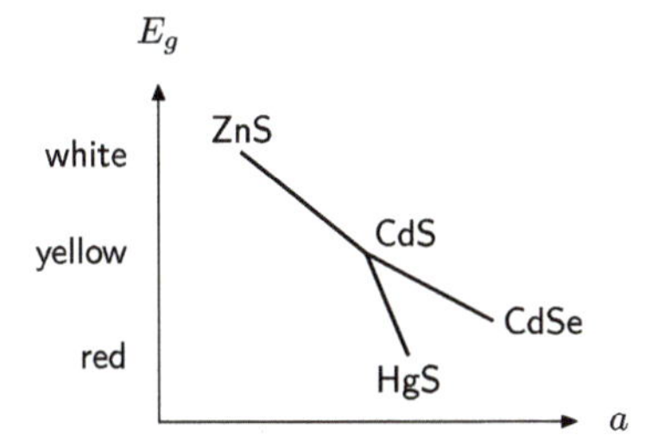

Figure 2.14: Schematic dependence of band gap E_g and color on the lattice constant a in ternary semiconductor alloys concerning painting (cadmium yellow, cadmium red, cadmium cinnabar).

2.2.5.2 Band structure and composition

The change of the band structure of AC is characterized by the appearance of bands of alloy partner B and by the transition to the band structure of BC [320, 357, 361–366]:

- In the pure semiconductor AC, its delocalized states concentrate up to energy E_V in its valence band and start from energy E_L in the conduction band.
- Single impurity atoms B form isolated new levels in the band gap, corresponding to localized states with a low density of states. However, this leads only to local deformation of the band structure. (Strictly speaking, we observe BC states, but since the counter ion C is the same, we can designate the new states to B for simplicity.)
- Increasing the concentration of B leads to increasing delocalization of B electrons throughout the whole crystal; the discrete B levels broaden and develop into *impurity bands*.

  Their interaction with the band structure of AC leads to a broadening and flattening of the AC bands and the development of *band tails*. These are extensions of the bands that reach from both sides into the band gap and form new band edges E'_V and E'_C, respectively, ►Figure 2.15. They decrease the band gap from $E_g = E_C - E_V$ to $E'_g = E'_C - E'_V$.

- AC- and B-bands merge. After [364], bands corresponding to the respective pure substances coexist, the contribution of AC bands decreasing and that of BC bands increasing.
- With an ever-decreasing contribution of AC, the band structure approaches that of BC with a final band gap $E''_g = E''_L - E''_V$. Now, A atoms represent an impurity and a perturbation to BC.

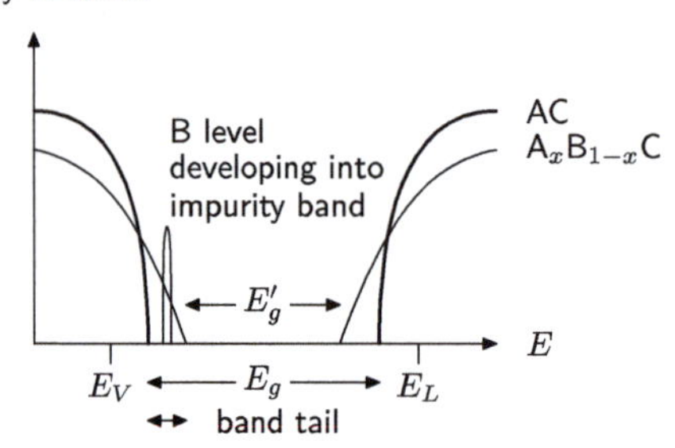

Figure 2.15: Dependence of an alloy's band gap (A, B)C on composition [326, 361]. At low concentrations of BC, B first forms isolated levels within the band gap, then *impurity bands*. Due to the interactions, the initial band structure of AC develops *band tails* and broadens; the band gap E_g decreases, and the color changes. The decrease is not symmetrical on both band gap edges, depending on the composition.

Naturally, the transformation of the band structures reflects not only the electronic but also all physical changes, like those of the lattice constant or of the crystal structure. As a consequence, also the optical absorption spectra change, ►Figure 2.16 [358, 360–363]:

- For AC, a steep absorption edge appears at the energy of the band gap $E_g = E_L - E_V$.
- Band tails develop and broaden the absorption edge, shifting it toward lower energies E'_g.
- The change in composition continuously shifts the absorption edge from E_g to E''_g.
- For BC, the absorption edge is again steep and located at the energy of the new band gap $E''_g = E''_L - E''_V$.

In the spectrum, we observe an overlay of all effects: the continuous transition from one band gap into the other and the complex modifications by band tails and impurity bands. Further explanation can be found in [320, 321, 323], [326, Chapter 5.8], [355, 356, 364–366].

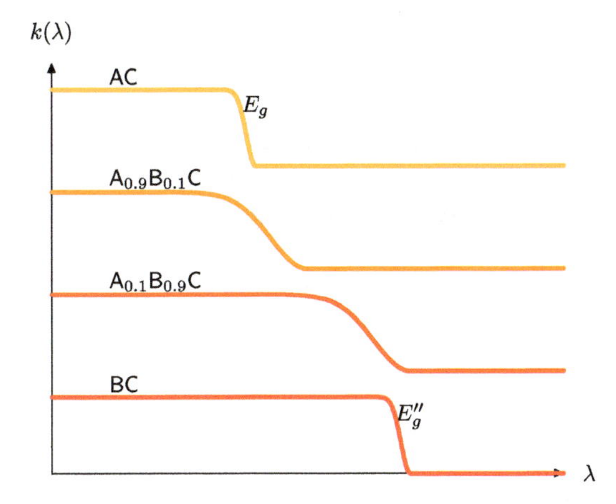

Figure 2.16: Schematic absorption spectra for an alloy $A_xB_{1-x}C$ [358, 360–363]. In the pure compounds, steep edges occur at the energies of the respective band gap E_g and E_g''. They are broadened by band tails in the alloy and develop into each other. Colors schematic for AC=CdS, BC=CdSe.

Example: Cadmium pigments

Pure yellow cadmium sulfide shows a color corresponding to the large gap between the S(3*p*) and Cd(5*s*) bands, ►Figure 2.17(b). Using a rough calculation and ►Table 2.2, we estimate the value to be −7.66 − (−11.65) = 3.99 eV. Pure brown selenide shows a smaller band gap because its more loosely bound 4*p* electrons form a Se(4*p*) valence band of higher energy, ►Figure 2.17(e). The estimate yields −7.66 − (−10.93) = 3.27 eV. The sulfide-selenide alloy shows a band gap between these values, corresponding to the mole fraction and colored dark yellow, orange, or red [323, 365].

The transition between the two compounds is smooth since cadmium sulfide and selenide have mutually compatible crystal lattices, which differ mainly in lattice width, ►Table 2.4. Therefore, larger selenide anions can replace sulfide anions to any extent. The continuous change of band gap from v_1 to v_{12a}/v_{12b} to v_2 is due to the change in lattice width and the band structure's transition. As a result, initially isolated selenium levels evolve into Se(4*s*) bands (►Figure 2.17(c)) and merge with S(3*s*) band into a band containing the combined electron density of sulfur and selenium, ►Figure 2.17(d). The visible result is the color change yellow → orange → red.

Since the crystal lattices of zinc sulfide and cadmium sulfide are also compatible (►Table 2.4), the preparation of a light shade of cadmium yellow is possible by admixture of the smaller zinc cations. The shade varies between a light yellow and yellow, ►Figure 2.17(a). Chen and Sher [323], Zhong et al. [645, 646], Zhong and Feng [647], Noor et al. [364], Weia et al. [366] present how the band gap increases with increasing zinc content and decreasing lattice constant a, resulting in the lighter pigment. The addition of colorless cadmium sulfate to cadmium sulfide also reduces a, consequently changing the color to lighter shades.

Table 2.4: Lattice constants *a* and band gaps for various binary semiconductors in cadmium pigments [323]. The semiconductors can crystallize in two modifications, the zincblende, and the wurtzite structures. Two lattice constants are therefore given.

Semiconductor	*a* (zincblende) [Å]	*a* (wurtzite) [Å]	Band gap [eV]
ZnS	5.406	3.811	3.8
(Zn, Cd)S	values in between		
CdS	5.835	4.137	2.45
Cd(S, Se)	values in between		
CdSe	6.05	4.30	1.85

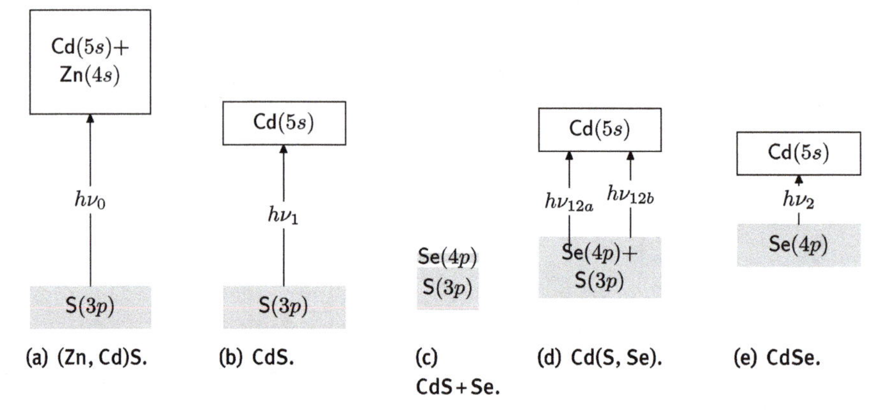

Figure 2.17: Flat-band model of cadmium yellow light, cadmium yellow medium, and cadmium red [324, 325, 364, 366, 641]. The color changes are caused by increasing the lattice width, going from (Zn, Cd)S to CdS, and by increasing the valence band's energy, going from S(3*p*) to Se(4*p*), reducing the band gap. In the alloys, the transition of the band structures changes the perceived color: isolated selenium levels broaden into bands and merge with the valence band, contributing more and more to the shade.

2.2.6 Manufacture of semiconductor alloys

►Section 2.2.4.3 introduced yellow lead oxide and its close relatives, lead-tin yellow and Naples yellow, which are alloys of lead yellow and tin(IV) oxide or antimony(V) oxide. Numerous recipes have survived for the production of Naples yellow, ►Section 3.4.1 at p. 234. It is interesting to note that the final hue depends on the reactants' proportions and the temperature; the same is valid for lead-tin yellow. Preparation at higher temperatures and with longer reaction times results in cool greenish or lemon yellow pigments, while lower temperatures yield reddish or orange-yellow pigments.

Two effects are responsible for this. On the one hand, different compounds form at different temperatures, which a phase diagram can confirm [321, Section 3.1.2]. Unfortunately, such is not known for lead yellow alloys, but [48, Volume 1] find experimentally that yellow lead oxides are formed instead of true Naples yellow at low temperatures. In the presence of halide anions, lead oxychlorides form.

On the other hand, manufacturing the alloy does not occur as a molten mass of the reactants but by a solid-state reaction, i. e., they are sintered below the melting point of any component. Apart from mechanical mixing while grinding the reactants, their bonding is achieved only by diffusion of the alloying partners in the solid, the rate of which is proportional to the temperature [321, Section 5.1]. Only by higher temperatures and sufficiently long reaction times reproducible, homogeneous products can be obtained. Otherwise, the material is far from equilibrium.

Specific examinations on this subject are not available. However, we can speculate that many structural defects occur at low temperatures, frozen so that the lead oxide crystal lattice expands slightly on average, ►Section 2.2.4.2. According to ►Figure 2.11, this leads to warmer redder colors.

At high temperatures and with a longer reaction time (a manufacturing process also followed in newer recipes of paint manufacturers), the defects equalize, and the crystal lattice approaches its equilibrium position. This lattice compaction is equivalent to a color shift to cooler, more whitish-yellow shades.

2.2.7 Doping and blue diamonds

Introducing traces of a second material into a solid, i. e., replacing only a few atoms of a semiconductor with other elements, is called doping and can have dramatic consequences for chromaticity. At the minute amounts used for doping, the physical crystal lattice and the electronic band structure are only locally modified, so there is no shift in hue perceptible. However, the dopant may create new energy levels within the band gap. Compared to the original band gap, these levels significantly decrease the energy of an electronic transition if they can accept or provide electrons. ►Figure 2.7 illustrates this: Rather than $h\nu_{\mathrm{min}}$, we only need to apply $h\nu_1$ or $h\nu_2$. Examples are doped rutile (DR) pigments.

In addition to producing artists' pigments, this fact explains why we sometimes find blue or yellow diamonds. A level of a naturally doped boron or nitrogen atom lying in the band gap lowers the initially large band gap to such an extent that visible light is absorbed. The absorption is sharp since the new level is strictly localized and not part of a band. Practically, however, it is broadened by the thermal excitation of electrons.

Example: Complex rutile pigments, DR pigments

Besides producing the well-known titanium white, rutile is also the starting material for producing DR or *doped rutile* pigments, ►Section 3.4.3.3. In these, metal cations are incorporated into a rutile matrix to impart colors to white (colorless) rutile. As coloring mechanisms, LF splitting of metal *d* orbitals and charge transfer are discussed [14, 648]. The metal cations introduce isolated states within the band gap between the O(2*p*) band and the Ti(3*d*) band. Depending on whether the metal is an electron acceptor A or an electron donor D, transitions from the valence band to the new states take place (O(2*p*) → A) or from the new states to the conduction band (D → Ti(3*d*)), ►Figure 2.18.

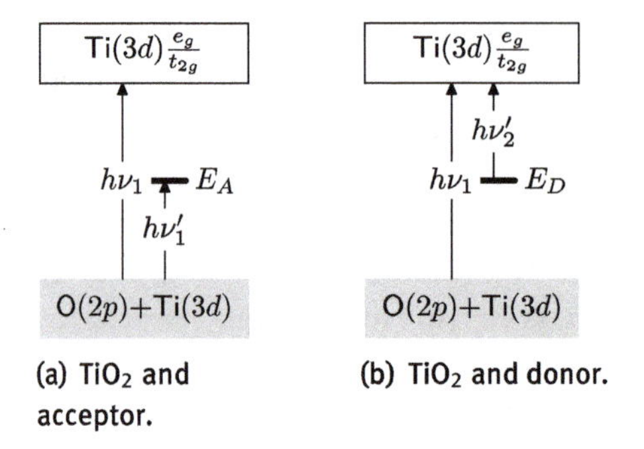

(a) TiO_2 and acceptor.

(b) TiO_2 and donor.

Figure 2.18: Flat-band model of doped rutile (titanium white) as present in the DR pigments. Metals used as dopants introduce isolated states in the band gap [14, 648]. Both CT transition energies $h\nu'_1$ and $h\nu'_2$ are smaller than the rutile band gap $h\nu_1$. Therefore, the doped rutile can be colored even if pure rutile is white. (a): Metals acting as electron acceptors enable charge transfer transitions from the valence band (O(2*p*) and Ti(3*d*) states) into the new state. (b): Metals acting as electron donors enables CT transitions from the new states into the Ti(3*d*) conduction band.

2.3 LF: splitting *d* orbitals in a ligand field

LF-based pigments consist of chromogenic metal ions as parts of a crystal lattice, as are their counteranions. For the subject of this text, we exclusively discuss *d* block metals as chromophores, more specific metal cations with a $3d^n$ electron configuration. In this setup, the metal is affected by the electric field of all crystal constituents, summarized as "crystal field." Due to the symmetry of the crystal, it is sufficient to describe the field and its influence on the metal only locally for representative positions, i. e., only the metal and its direct neighbors, the anions or "ligands." Additionally, the field's symmetry must be specified.

All *d* block elements have partially occupied *d* orbitals degenerated in the free atom or ion, i. e., equal in energy. Due to electron-electron interaction, intraatomic transitions of high energy are possible that are subject to atomic spectroscopy. Placing the metal atom or ion in an electric field lifts the degeneracy, making possible additional electronic transitions between *d* orbitals in the visible spectral range. The electric field can originate, e. g., in a crystal lattice, if the metal is part of a crystalline solid or in ligands, if the metal is part of a complex (which can be a constituent of a crystal). Consequently, the electric field is represented as a crystal field, or more generally, as a ligand field. The older *crystal field theory* describes the crystal lattice as a purely ionic crystal field. The succeeding *ligand field theory* (LFT) depicts it as a *ligand field* that includes covalent contributions to bonding.

Splitting *d* and analogously *f* orbitals by crystal or ligand fields is crucial for understanding coordination numbers and complex chemistry. We can therefore limit ourselves to an overview of the subject and refer to [224, 226–230, 234] and any other textbook of inorganic chemistry for detailed information. Applying LFT to minerals are [223] and [233]. A valuable introduction to the underlying group theory is available in [221].

Characteristics of colorants based on ligand field transitions are:

– The absorption is, in principle, narrow-banded, and the perceived colors pure. However, multiple transitions are possible depending on the chemistry and symmetry around the metal cation. They form a complex spectrum by superimposing, and the compound's color may get duller.
– The ligand field strength strongly influences the transition wavelength, which may span the entire color range. However, due to the limited number of actual ligands and their chemistry, not all colors are practically achievable. In natural pigments, ligands are preferably oxides, hydroxides, carbonates, sulfates, and halides. The strong influence of the ligand field also means that the color is sensitive to the purity of the colorant, i. e., colorants from natural sources are sensitive to admixtures of dirt and companion minerals.
– Transitions are often forbidden by spectroscopic selection rules and, therefore, not very intense. We see this by taking a mineral that is, at first glance, intensively colored. By dragging it across a rough surface like a piece of unglazed porcelain (a so-called "streak plate"), we frequently find that the streak is scarcely colored or even white and by no means shows the expected color intensity.
– Because of the low transition intensities, only the more abundant *d* block elements are considered chromophores. Therefore, according to the elemental abundance, the classical natural LF-based pigments contain Mn^{IV}, Fe^{III}, and Cu^{II} as chromophores. After the dawn of scientific chemistry, discoveries employed Co^{II} and Cr^{III} in artificial chromophores.
– The ligand field's origin can be an ordered crystal lattice (as in natural minerals or many synthetic pigments) or an amorphous substance (as in glasses). In the latter case, the ligand field is an average field. LF transitions colorize most stained glasses.
– Numerous anions can play a role as ligands. In practice, oxides, hydroxides, sulfates, and carbonates are frequent constituents in natural pigments. Oxides, in particular, provide pigments of high-temperature stability and are therefore used in ceramic paints and stained glasses. For synthetic pigments, anions such as acetate were used, in industrial times, even arsenite.

2.3.1 Crystal field theory and ligand field theory

Ligand field transitions are explained in terms of *crystal field theory (CF theory)* or *ligand field theory (LF theory)*. CF theory describes the effect of a static electric field, the crystal field (CF), on a metal atom or ion in a transition metal complex. The electric field origins in anionic point charges or a charge distribution surrounding the metal, i. e., from the ligands. Regarding pigments, the crystalline matrix comprises metal salts such as copper carbonate, and both metal and anions form the crystal lattice. The most notable effect of the CF is to lift the degeneracy of *d* orbitals of the metal depending on the CF's symmetry, i. e., geometric properties of the anion's arrangement.

Going from purely ionic complexes to such with covalent contributions to bonding, i. e., overlap or mixing of AOs of metal and ligands forming MOs, we arrive at LF theory. LF theory describes bonding in complexes by applying MO theory to transition metal complexes. It considers the interaction of the metal's nd, $(n+1)s$, and $(n+1)p$ AOs with ligand MOs, forming new MOs such as $d - p_\sigma$.

For the topic of this text, we do not strictly distinguish between CF theory and LF theory and speak of ligands and ligand fields.

2.3.2 Splitting of degenerated *d* orbitals

To understand why the set of five *d* orbitals, degenerated in a free atom, can support electronic transitions, we must take a closer look at the quantum mechanical description of the colorant. Gispert [229, Chapter 8], Marfunin [233, Chapter 1] give a more detailed derivation.

Hamiltonians, states and energy levels

The cause of chromogenic transitions is the presence of electronic states of different energies within the metal atom or ion. In this section, we discuss how energy differences arise in a $3d^k$ configuration, although in a free atom or ion, $3d$ AOs are fivefold degenerated and equal in energy. To do so, we need to describe electronic states uniquely.

States are sets of quantum numbers, assigning each electron values for properties such as shell, orbital angular momentum, and spin. Each state has a specific energy, and different states can have different energies, enabling transitions between them. The Schrodinger equation is solved to derive energies from states, yielding eigenvalues representing the states' energies.

The Schrodinger equation of a $3d$ metal in a crystal can be written using the following Hamiltonians:

$$\mathbf{H} = \mathbf{H}_0 + \mathbf{H}_{\mathrm{ee}} + \mathbf{H}_{\mathrm{LS}} + \mathbf{H}_{\mathrm{LF}} \tag{2.3}$$

Each term describes a specific interaction:
- $\mathbf{H}_0$ represents the fundamental Hamiltonian, describing the electron's kinetic energy and the electrostatic electron-nucleus interaction (Coulomb interaction), i. e.,

$$\mathbf{H}_0 = -\frac{\hbar^2}{2m}\sum \nabla^2 - Ze^2 \sum \frac{1}{r_i} \tag{2.4}$$

- $\mathbf{H}_{ee}$ is the electrostatic electron-electron interaction (Coulomb repulsion), i. e.,

$$\mathbf{H}_{ee} = e^2 \sum \frac{1}{r_{ij}} \tag{2.5}$$

- $\mathbf{H}_{LS}$ describes the interaction of orbital and spin angular momenta, $\mathbf{H}_{LF}$ the interaction of metal and ligand field.

Hamiltonian ordering

The contributions of individual Hamiltonians differ by magnitudes, depending on Z of the metal and the ligand field strength. For practical calculations, one solves the Schrodinger equation, employing the Hamiltonians of strongest influence, and treats weaker Hamiltonians as perturbations:
- $\mathbf{H}_0$ is always the most significant Hamiltonian, yielding "one-electron" energies, i. e., it treats the atom or ion as if it comprises Z independent electrons. Each electron is described by its quantum numbers n, l; the metal ion is described by an electronic configuration such as $3d^k$.
- $\mathbf{H}_{ee}$ is the next strong Hamiltonian, yielding "terms" such as ^3H, i. e., atomic states including electron-electron interaction, resulting in different energies. The existence of terms means that a configuration has several energy values, the lower being preferred to others. As we will see below, terms include microstates of equal energies.
- $\mathbf{H}_{LS}$ is, in general, the next strong Hamiltonian, yielding "multiplet levels" such as $^3H_6$, describing the splitting of terms into several levels in an external magnetic field.
- The strength of $\mathbf{H}_{LF}$ varies and depends on the composition of the crystal lattice. For rare earths and actinides ($4f$ and $5f$ ions), $\mathbf{H}_{LF}$ is weak, and $\mathbf{H}_{ee} > \mathbf{H}_{LS} > \mathbf{H}_{LF}$ applies. A strong spin-orbit coupling occurs, known as *jj* coupling because for each electron i, its orbital angular momentum l_i and its spin s_i couple to its total angular moment j_i. One would calculate AO occupation (e. g., $4f^2$), terms (e. g., ^3H) and multiplets (e. g., $^3H_6$) and split them according to the ligand field into irreducible representations (e. g., A_{1g}).
  For heavy transition metals ($4d$ and $5d$ metals), $\mathbf{H}_{LF}$ is strong, and $\mathbf{H}_{LF} \approx \mathbf{H}_{ee} > \mathbf{H}_{LS}$ applies. One calculate energies of AO configurations split first by LF, then by electron-electron interaction, ►Figure 2.20 from the right. Splitting by LF depends

on the symmetry of the field; thus LF states are denoted by group theory symbols such as t_{2g} or e_g, and then $^1T_{2g}$ or E_g, the so-called *irreducible representations* or *symmetry types*.

The last and for our topic, most important case is a medium LF, occurring for $3d$ metals. Here, $\mathbf{H}_{ee} > \mathbf{H}_{LF} > \mathbf{H}_{LS}$ applies. One calculates first terms for an AO configuration, then the irreducible representations due to CF splitting, ►Figure 2.20 from the left. In this case, all individual orbital angular momenta m_l and spins m_s couple to a total orbital angular momentum L and a total spin S (LS coupling). Using this knowledge, we will understand the coloration of chromium, copper, and cobalt pigments in more detail.

The correspondence of Hamiltonians, configurations, terms, and irreducible representations is as follows:

$$\begin{aligned} \mathbf{H} = \mathbf{H}_0 \quad & \text{AO configuration, e. g.} 3d^3 \\ + \mathbf{H}_{ee} \quad & \text{term, e. g.,} {}^4F, {}^4P, {}^2G \quad (\rightarrow \text{microstates}) \\ + \mathbf{H}_{LS} \quad & \text{multiplet level, e. g.,} {}^4F_7, {}^4F_6 \\ + \mathbf{H}_{LF} \quad & \text{irreducible representation, e. g.,} {}^4A_{2g}, {}^2E_g \end{aligned}$$

One-electron states of a free atom or ion

States of an isolated electron in a free atom or ion ($\mathbf{H} = \mathbf{H}_0$) can be described by four quantum numbers. The first two are:

– the *principal quantum number n*, $n = 1, 2, 3, \ldots$
– the *azimuthal or angular orbital momentum or orbital quantum number l*, $l = 0, 1, \ldots, n-1$, or, in spectroscopic designation, $l = s, p, d, f, \ldots$

Both quantum numbers describe the shape and size of AOs. The last two quantum numbers are:

– the *magnetic orbital quantum number* m_l, $m_l = l, \ldots, -l$, or, in symbols for $l = p$: $m_l = x, y, z$, and for $l = d$: $m_l = xy, xz, yz, x^2 - y^2, z^2$
– the *magnetic spin quantum number* m_s, $m_s = \pm\frac{1}{2}$

Both magnetic quantum numbers indicate the orientation of $\mathbf{l}$ (orbital angular momentum) and $\mathbf{s}$ (spin angular momentum) relative to a particular axis, e. g., the direction of an external magnetic field (denoted as z-axis).

In this description, e. g., a $3d$ electron with $m_l = -2$, $m_s = \frac{1}{2}$ has the quantum state $(3, 2, -2, \frac{1}{2})$. Regardless of its magnetic quantum numbers, the $3d$ electron has the same energy in this description; the d orbitals are degenerated.

Multielectron states of a free atom or ion

To describe the states of free atoms or ions, i. e., multielectron systems, electron-electron interaction (interelectronic repulsion) and electron-nucleus attraction must be taken into account ($\mathbf{H}_0 + \mathbf{H}_{ee}$). Two cases can be distinguished, depending on the Hamiltonian ordering:

- For $Z < 30$, the electron-electron interaction is described by the quantum numbers $L = \sum l_i$ and $S = \sum s_i$, describing orbital and spin angular momenta, and $J = L + S$, describing the total angular momenta. This summation of L and S is called LS- or Russel–Saunders coupling. The actual distribution of electrons to quantum numbers $L, S, \{m_l\}, \{m_s\}$ is called *microstate*. In contrast, a *term*, denoted by ${}^{2S+1}L$, describes only L and S of the atom. Each term summarizes all microstates of equal total energies.
- For heavier atoms, the resulting momenta $j_i = l_i + s_i$ are calculated first for each electron, then the total momentum $J = \sum j_i$ is calculated. This summation is called jj coupling.

One-electron and atomic quantum numbers are summarized in ►Table 2.5.

Table 2.5: Quantum numbers for descriptions of one-electron and atomic states [233, Chapter 1]. For simplicity, the vector character is omitted, so, e. g., $\mathbf{l}$ is written l.

Name	Electronic quantum numbers	Atomic quantum numbers
Principal	$n \in \{1, 2, 3, \dots\}$	–
(Total) azimuthal, angular orbital momentum, orbital	$l \in \{0, \dots, l-1\}$ or $\{s, p, d, f, \dots\}$	$L = \max M_L \in \{0, 1, 2, \dots\}$ or $\{S, P, D, \dots\}$
(Total) magnetic orbital	$m_l \in \{l, \dots, -l\}$	$M_L = \sum m_l \in \{L, \dots, -L\}$
(Total) spin angular	$s_i = \frac{1}{2}$	$S = \max M_S$ for a given L
(Total) magnetic spin	$m_s \in \{s, \dots, -s\}$	$M_S = \sum m_s \in \{S, \dots, -S\}$
Total angular momentum	–	$J \in \{L + S, \dots, L - S\}$
One-electron state	(n, l, m_l, m_s)	–
State/term, multiplet	–	${}^{2S+1}L$, ${}^{2S+1}L_J$
Microstate	–	$L, S, \{m_l\}, \{m_s\}$

LS or Russel–Saunders coupling, terms of a free atom or ion

Since we discuss LF-based pigments containing $3d$ cations, we follow the LS coupling scheme and include $\mathbf{H}_{LS}$ at this point. For the lighter elements in question, each electron distribution or *microstate* is characterized by their magnetic quantum numbers $\{m_l\}$ and $\{m_s\}$, and by the quantities $M_L = \sum m_l$ and $M_S = \sum m_s$, i. e., by coupling all orbital angular momenta and spins separately (LS coupling). In general, several microstates belong to the same values of L and S, whereby L and S are derived from the values M_L and M_S. These microstates are equal in energy; thus, they are characterized as *terms*, symbolized by a *term symbol*:

$$^{2S+1}L \quad \text{or} \quad ^{2S+1}L_J$$

L is the total orbital angular quantum number of the atom or ion and can assume integer values 0, 1, 2, ..., also denoted as letter S, P, D, F, ... The quantity S is the total spin quantum number of the atom or ion. The quantity $2S+1$ is called *spin multiplicity* and its values are given special names: 1 = singlet, 2 = duplet, 3 = triplet, etc. The latter notation includes the multiplet levels due to LS coupling, expressed as J. Here, J is the total angular quantum number of the atom or ion.

Ground state terms

The *ground state terms* can be derived from Hund's rules, aiming first at as many parallel electron spins as possible, and then at maximum L, ►Table 2.6. For complete subshells such as ns^2, np^6, nd^{10}, or nf^{14}, the ground state term is always $^1S_0$, because $M_L = L = S = 0 = M_S = J$. Thus, to determine atomic states, only incomplete subshells are considered and denoted the same way as excited states.

Table 2.6: d^n Configurations and ground state terms for atomic states, considering electron interaction and LS coupling [227, p. 101], [229, Chapter 8.3.1], [233, Chapter 1]. Also, term symbols for the excited states are given. Unfortunately, we cannot make valid statements about the general ordering of excited terms; they must be looked up in textbooks, calculated, or measured.

$3d^k$	Example	$M_L = \sum m_l =$ max $M_L = L$	$M_S = \sum m_s =$ max $M_S = S$	^{2S+1}L	Excited state terms
0, 10		0, S	0	^1S	
1, 9	$Cu^{2\oplus}$	2, D	$\frac{1}{2}$	^2D	
2, 8		3, F	1	^3F	$^3P, {}^1G, {}^1D, {}^1S$
3, 7	$Cr^{3\oplus}$, $Co^{2\oplus}$	3, F	$\frac{3}{2}$	^4F	$^4P, {}^2H, {}^2G, {}^2F, {}^2D, {}^2D, {}^2P$
4, 6		2, D	2	^5D	$^3H, {}^3G, {}^3F, {}^3F, {}^3D, {}^3P, {}^3P, {}^1I, {}^1G, {}^1G, {}^1F, {}^1D, {}^1D, {}^1S, {}^1S$
5	$Fe^{3\oplus}$	0, S	$\frac{5}{2}$	^6S	$^4G, {}^4F, {}^4D, {}^4P, {}^2I, {}^2H, {}^2G, {}^2G, {}^2F, {}^2F, {}^2D, {}^2D, {}^2D, {}^2P, {}^2S$

Excited state terms

To determine *excited state terms*, the total number of states belonging to a nl configuration must be taken into account, employing m_l and m_s of each electron. To each nl configuration belongs several states and, consequently, terms. For example, in a

$3d^1$ configuration ($n = 3, l = 2$), the electron can occur in 10 microstates of varying m_l ($m_l = 2, \dots, -2$) and m_s ($m_s = \frac{1}{2}, -\frac{1}{2}$). From $3d^1$ follows the maximum value $M_L = m_l = 2, M_S = m_s = \pm\frac{1}{2}$, and we immediately get $L = 2, S = \frac{1}{2}, J = \frac{5}{2}, \frac{3}{2}$, and thus the state or term ^2D and the multiplets $^2D_{5/2}$, $^2D_{3/2}$. The number of microstates in each multiplet is 6 for $^2D_{5/2}$ ($J = \frac{5}{2}, \frac{3}{2}, \frac{1}{2}, -\frac{1}{2}, -\frac{3}{2}, -\frac{5}{2}$), and 4 for $^2D_{3/2}$ ($J = \frac{3}{2}, \frac{1}{2}, -\frac{1}{2}, -\frac{3}{2}$). For $3d^1$, all 10 microstates belong to the same term since no electron-electron interaction occurs. However, they are denoted by two multiplet levels due to LS coupling.

For the following discussion, we will need the terms of some $3d^n$ configurations, ►Table 2.6. Unfortunately, we cannot make valid statements about the general ordering of excited terms. Consequently, we must look it up in textbooks, perform measurements, or calculate the details.

As an example, ►Table 2.7 depicts microstates that electrons can assume in typical cations of LF pigments. However, it contains only some of the many possible microstates. In general, for a $3d^k$ configuration, there are $\binom{10}{k} = \frac{10!}{k!(10-k)!}$ microstates. Gade et al. [230, p. 171] describes the procedure for determining all terms and their relation to microstates:

- Calculate M_L and M_S for all microstates and sort them into a table by M_L and M_S.
- Proceed as long as microstates remain in this table:
  - Find the maximum value $L = \max(M_L, m_l)$ and $S = \max(M_S, m_s)$. The configuration will then have a term denoted by $^{(2S+1)}L$.
  - Delete one microstate from each table cell within the rectangle $-L \dots L$, $-S \dots S$, which then belongs to this term.

Table 2.7: Examples of terms and microstates of typical elements contained in inorganic colorants. ↑ and ↓ symbolize electrons with spin quantum number $m_s = \pm\frac{1}{2}$ or $m_s =$↑↓.

Ion	m_l					$M_L = \sum m_l$	$M_S = \sum m_s$	Term
	2	1	0	−1	−2			$L = \max(M_L, m_l)$
	d_{xy}	d_{xz}	d_{yz}	$d_{x^2y^2}$	d_{z^2}			
$Cr^{3\oplus}$, $3d^3$	↑	↑	↑	–	–	3	$\frac{3}{2}$	^4F
	–	–	↑	↑	↑	−3	$\frac{3}{2}$	^4F
	–	↑	↑	↑	–	0	$\frac{3}{2}$	^4P
	↑	↑	–	–	↑	1	$\frac{3}{2}$	^4P
	↑↓	–	↑	–	–	4	$\frac{1}{2}$	^2G
$Cu_2^{2\oplus}$, $3d^9$	↑↓	↑↓	↑↓	↑↓	↑	−2	$\frac{1}{2}$	^2D
$Co_2^{2\oplus}$, $3d^7$	↑↓	↑↓	↑	↑	↑	−3	$\frac{3}{2}$	^4F

Since a term only specifies the total values of orbital angular momentum and spin, several microstates belong to it. Possible microstates belonging to, e. g., the ^4F term of $Cr^{3\oplus}$ ion are (↑, ↑, ↑, –, –) or (–, –, ↑, ↑, ↑), or written differently:

$$^4F: \quad (d_{xy})^1(d_{xz})^1(d_{yz})^1$$
$$^4F: \quad (d_{yz})^1(d_{x^2-y^2})^1(d_{z^2})^1$$
$$\dots$$

Significance of terms

Why do we need to know about terms? Terms explain atomic spectra, i. e., transitions between terms in free atoms or ions. Furthermore, they explain why several states differ in energy even if the underlying *d* orbitals are degenerated, i. e., have the same energy.

The configurations shown in ►Table 2.7 differ only in the distribution of electrons among individual *d* orbitals. Even if *d* orbitals are degenerated, the *total energy of a configuration* or the *energies of the related terms* differ due to electron-electron interaction, orbital geometries, and coupling of angular momenta. Thus, it is, e. g., more advantageous for two electrons to occupy different orbitals than to share an orbital (see the rule of Hund). Terms summarize all microstates of the same total energy, and electronic transition become possible between terms (►Figure 2.19(b)), not *d* orbitals (►Figure 2.19(a)).

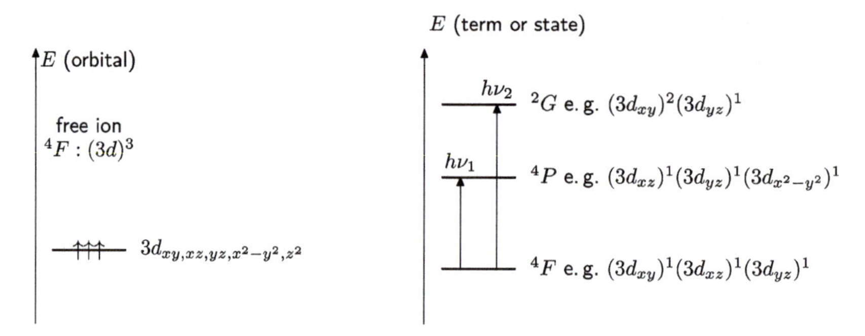

(a) Plot of the degenerated *d* orbitals vs. energy.

(b) Plot of the terms or states related to a configuration vs. energy, taking into account electron-electron interaction.

Figure 2.19: Origin of atomic spectra, exemplified by a $3d^3$ configuration in a free $Cr^{3\oplus}$ ion. (a): In the free ion, all *d* orbitals are degenerated. Consequently, no electronic transition is possible. (b): Due to the electron-electron interaction, each microstate (distribution of electron to specific *d* orbitals) possesses a specific energy out of a set of several fixed energies or *terms*. Electronic transitions between terms are possible. In inorganic colorants, they fall into the optical spectral range.

Ions in ligand field

If we now consider a free ion in a pigment crystal, the perturbation due to the ligand field $\mathbf{H}_{LF}$ finally enters the Schrodinger equation. Metal electrons "feel" the influence of all the ion's other electrons, the coupling of angular momenta, and the influence of electrons of all other atoms in the crystal. While distant atoms contribute to a static mean-field, the influence of the metal cation's nearest neighbors is significant. These neighbors are the metal's direct binding partners (ligands). They are arranged around the metal in a characteristic manner described by a *coordination number* and a *coordination geometry* typical for the metal. For example, the six hydroxyl ligands in the compound $Cr(OH)_6^{3\oplus}$ are arranged octahedrally around the chromium ion (coordination number: 6, coordination geometry: octahedrally).

The coordination situation is not immediately evident for compounds like $CuCO_3 \cdot Cu(OH)_2$: which ligands surround the copper ions, and in what arrangement? Crystal chemistry can answer this question in general, but the coordination geometry around cations essential for LF pigments is mostly octahedrally, distorted-octahedrally, or tetrahedrally. Primarily copper is known for strong distortions caused by the Jahn–Teller effect, leading to more complicated coordination geometries.

What consequences follow from coordination geometry? Metal electrons are influenced differently by the ligand field, depending on their quantum state. From a geometric point of view, the influence depends on an orbital's orientation to the ligand. Formally, the influence of ligands expresses itself as the splitting of terms depending on the ligands' geometric arrangement. Mathematically, electron-ligand interaction depends on coordination geometry and the geometry of all participating quantum states, expressed in their symmetry properties. Therefore, the splitting pattern of terms can be determined by group theory, as we will see in ►Section 2.3.4 for the significant case of octahedral ligand fields, frequent in inorganic colorants. However, we cannot make a general statement about energies and the sequence of the resulting levels.

2.3.3 Spectroscopic selection rules

Now that we know the origin and nature of LF transitions, we will briefly discuss why they are inherently weak. We will also need to clarify why some theoretical transitions do not qualify as LF transitions.

Electrons cannot be transferred arbitrarily from one quantum state to another. *Spectroscopic selection rules* govern electromagnetic transitions in molecules, atoms, or ions [229, Chapters 9.3, 9.4], [233, Chapter 6.2]. They constrain possible transitions and, therefore, influence features of the resulting electronic absorption spectra. Many LF transitions are forbidden or only weakly allowed, and their transition intensities are low compared to other transition types, e. g., charge transfer or interband transitions. Consequently, most LF transitions colorize minerals so weakly that they do not qualify as pigments. The bright colors of minerals, gemstones, and precious stones are quickly lost when ground into a fine powder, generally white or barely showing a hint of color. A few minerals have been used in painting only because no alternatives of comparable shade were available at that time: think of the weakly colored pigments azurite, cobalt violet, and yellow ocher (LF) in contrast to the intense Prussian blue (CT), quinacridone violet (MO), and cadmium yellow (SC). Of course, coloring power is only one criterion of many on an excellent pigment. Yellow ocher and cobalt violet are still enticing today due to their subtle colors. However, if alternatives existed earlier, especially in the blue/green range, copper minerals probably would not have been used as pigments.

Important electromagnetic selection rules are:

- Number of involved electrons rule. One-electron transitions are allowed.
- Preservation of spin multiplicity rule. Transitions changing the multiplicity (i. e., the number of unpaired electrons) are forbidden. Written as a mathematical condition: $\Delta S = 0, \Delta J = 0, \pm 1$.
  This rule applies less strictly when spin-orbit coupling occurs, particularly pronounced for *f* block elements.
- Laporte rule. Transitions maintaining parity are forbidden. Written as a mathematical condition: $\Delta L \neq 0$.
  The parity of an orbital describes its symmetry concerning inversion centers. *s* and *d* orbitals are symmetric concerning the nucleus and, therefore, have g-parity, while *p* and *f* orbitals are antisymmetric and have u-parity. (g- and u-parity means even- and odd-parity, respectively, for the german symmetry notation *gerade* and *ungerade*.)
  Application of the Laporte rule to a metal-ligand complex involves the following questions:
  - Is the metal cation an inversion center? A transition is allowed with high intensity if no (as in the tetrahedron). If yes (as in the octahedron):
  - Is the electron transferred to an orbital with different parity upon excitation? If yes ($g \rightarrow u$ or $u \rightarrow g$), the transition is allowed with high intensity. If no ($g \rightarrow g$ or $u \rightarrow u$), the transition is forbidden.

  The Laporte rule applies less strictly if the symmetry of the complex is broken by molecular vibrations or by mixing orbital types. Molecular vibrations must last long enough for an electron transfer to occur during this period. Due to the intermittent loss of symmetry, weak LF transitions become possible. Mixing occurs, e. g., when the bond covalency increases. Then, pure metal *d* AOs mix with ligand *p* AOs, forming a *dp* MO. g-Parity is lost, and *d-d* transitions become more intense. Mixing of states also occurs if a forbidden transition is close to an intense allowed transition, thus gaining "borrowed intensity."

These rules allow some general statements about LF colorants:

- Due to the preservation of spin multiplicity, $3d^5$ ions like $Fe^{3\oplus}$ are colorless because any excitation would decrease the total spin of $\frac{5}{2}$. However, LF transitions in iron oxides are more intense due to magnetic coupling, ►p. 118.
- The Laporte rule forbids transitions from *d* to *d* orbitals for octahedral complexes. However, octahedrally coordinated $Co^{2\oplus}$ is weakly colored because molecular vibrations break the symmetry. Similarly, most LF colorants with octahedral complexes are weakly colored.

- The Laporte rule allows transitions from *d* to *d* orbitals for tetrahedral complexes. Tetrahedrally coordinated cobalt possesses a widely known, deep blue color.
- Since only one-electron transitions are allowed, we are not observing multielectron transitions in LF colorants. An exception are electron-pair transitions (EPT) in iron oxides due to magnetic coupling, ►p. 118.

2.3.4 Ligand-field splitting in octahedrally coordinated complexes

So far, we have examined the influence of ligand fields only superficially. We now look into octahedrally coordinated 3*d* metal cations, the chromophores of many LF pigments. Adding the Hamiltonian $\mathbf{H}_{LF}$ to the Schrodinger equation considers the influence of ligand fields. In this setup, six anions or ligands located at the six vertices of the coordination polyhedron around the metal create the ligand field. Thus, the metal possesses coordination number 6, and the coordination polyhedron is an octahedron.

Actions of ligand fields on a 3*d* metal

As we have seen when discussing Hamiltonian ordering, depending on its strength, $\mathbf{H}_{LF}$ can be regarded as a perturbation to $\mathbf{H}_0 + \mathbf{H}_{ee}$, or $\mathbf{H}_{ee}$ can be regarded as a perturbation to $\mathbf{H}_0 + \mathbf{H}_{LS}$:

- For a small $\mathbf{H}_{LF}$, we first determine the *terms* under the influence of $\mathbf{H}_{ee}$ and then calculate an additional splitting due to $\mathbf{H}_{LF}$ [229, Chapter 8.3]. This procedure is the so-called *weak-field approximation*. It assumes that the principal force splitting orbital configurations into terms is the electron-electron interaction and that the ligand field's contribution is only small. We already considered the case of an infinitely weak ligand field ($\mathbf{H}_{LF} = 0$) in ►Section 2.3.2, deriving terms indicating the total energy of a particular distribution of electrons on degenerated *d* orbitals. Group theory tells us that an octahedral ligand field splits the terms into so-called *irreducible representations*, according to ►Table 2.8. For example, it splits an *F* term of a free ion into three irreducible representations, A_{2g}, T_{1g}, and T_{2g}. We cannot foresee the energetic ordering of these irreducible representations.
- Starting with atomic orbitals, we determine the level scheme under the influence of $\mathbf{H}_{LF}$ (►Table 2.9) and then calculate an additional splitting caused by $\mathbf{H}_{ee}$ [229, Chapter 8.4]. This procedure is called *strong-field approximation* since the ligand field is assumed to be stronger than the electron-electron interaction. Further analysis by group theory reveals that t_{2g} and e_g finally split into the same irreducible representations as to the weak-field approximation.

Ultimately, we get the same irreducible representations both ways. If we start with a term sequence of a free ion (no, i. e., infinitely weak ligand field) and increase the ligand field strength from small values (weak-field approximation) to maximum, we end

Table 2.8: Splitting of terms into irreducible representations in an octahedral (O_h symmetry) and tetrahedral (T_d symmetry) ligand field using the weak-field approximation [229, Chapter 8.3.3].

Term ($H_{LF} = 0, H_{ee} \neq 0$)	O_h ($H_{LF} \neq 0, H_{ee} \neq 0$)	T_d ($H_{LF} \neq 0, H_{ee} \neq 0$)
S	A_{1g}	A_1
P	T_{1g}	T_1
D	$T_{2g} + E_g$	$T_2 + E$
F	$A_{2g} + T_{1g} + T_{2g}$	$A_2 + T_1 + T_2$
G	$A_{1g} + E_g + T_{1g} + T_{2g}$	$A_2 + E + T_1 + T_2$
H	$E_g + 2\,T_{1g} + T_{2g}$	$E + T_1 + 2\,T_2$
I	$A_{1g} + A_{2g} + E_g + T_{1g} + 2\,T_{2g}$	$A_1 + A_2 + E + T_1 + 2\,T_2$

Table 2.9: Splitting of atomic orbital wave functions in octahedrally ligand fields (possessing O_h symmetry) and tetrahedrally ligand fields (possessing T_d symmetry) into orbital sets or irreducible representations (strong-field approximation, no electron-electron interaction) [229, Chapter 1.2].

Orbital	Splitting in O_h symmetry (irreducible representations)	Energy difference	Splitting in T_d symmetry (irreducible representations)	Energy difference
s	a_{1g}		a_1	
p	t_{1u}		t_2	
d	$t_{2g} + e_g$	$\Delta_o = E(e_g) - E(t_{2g})$	$t_2 + e$	$\Delta_t = \frac{4}{9}\Delta_o$
f	$a_{2u} + t_{1u} + t_{2u}$		$a_2 + t_1 + t_2$	

up with the strong-field approximation. The irreducible representations are the same in both approximations and only change their energies on the way from one to the other. This essential transition is recorded in *correlation diagrams*, e. g., ►Figure 2.20 for a $3d^3$ configuration in an octahedral ligand field. We can see that irreducible representations can occur several times, having different energies, as T_{1g} illustrate. We distinguish them by adding the original terms in parentheses: $T_{1g}(^4F)$, $T_{1g}(^4P)$.

While correlation diagrams schematically show relationships between terms and irreducible representations, *Tanabe–Sugano diagrams* depict precise energetic relationships. We find these essential tools in essays [340, p. 753] and tabular works [231].

Strong-field approximation acting on 3*d* metals

Many textbooks for coordination chemistry introduce the topic using the strong-field approximation. Although this is not the primary splitting model for 3*d* metals, we will outline the procedure since it provides insight into LF splitting: the principal action of a LF on the metal ion is to lift the *d* orbital's degeneracy, i. e., to split the five-fold degenerated *d* AOs into two or more sets. (It also acts on all other AOs with lower strength, but we are only interested in *d* orbital behavior.) These sets' exact number and charac-

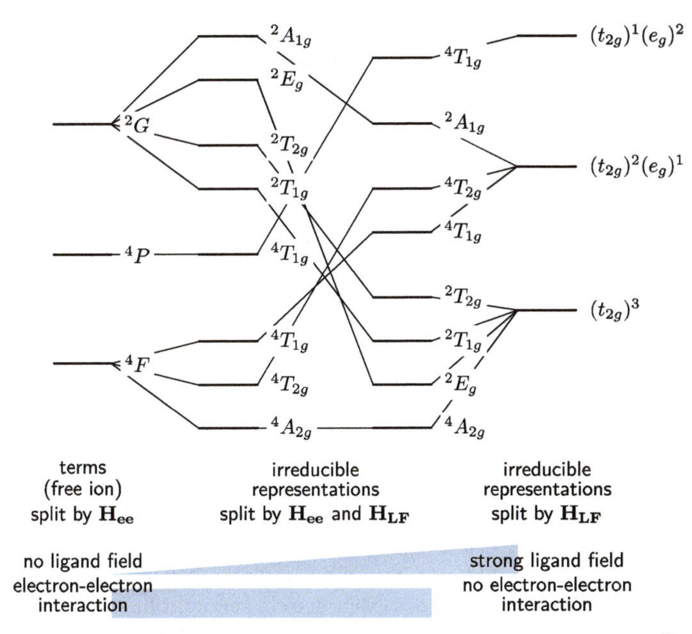

Figure 2.20: Correlation diagram for an octahedral ligand field and a $3d^3$ configuration [227]. Blue: strength of interaction. Left: electronic configurations split into terms through electron-electron interaction of the free ion in an infinitely weak ligand field. Right: electronic configurations of orbital sets (irreducible representations) resulting from splitting d orbitals in a strong ligand field without electron-electron interaction. Middle: splitting in both Hamiltonians into irreducible representations, their exact energies depending on ligand field strength.

ter depend on the spatial arrangement of ligands, expressed as *symmetry of the ligand field.*

In the strong-field approximation for octahedral ligand fields, $\mathbf{H} = \mathbf{H}_0 + \mathbf{H}_{LF}$ applies. In this setup, the five d AOs split into two sets of three and two orbitals, denoted by the symbols t_{2g}, and e_g, ►Figure 2.21 on the left. The sets are called *irreducible representations*, and the labels are group theory symbols describing particular symmetry properties.

The splitting pattern can be determined using group theory. When asked mathematicians, we might get statements like this: "d orbitals have the irreducible representations t_{2g} and e_g under an O_h symmetry." Chemically expressed, the degenerated d orbitals of a free ion in octahedral coordination O_h are split into two sets of orbitals, designated e_g and t_{2g}, ►Table 2.9 and ►Figure 2.21. Labels denote symmetry properties of d orbitals in each set, e. g., "e" stands for a pair of degenerated orbitals, "t" for a triplet. Dresselhaus et al. [221], Gispert [229, Chapter 1.2] explain details of the application of group theory to our question at hand. Such a splitting can be performed mathematically for each type of coordination polyhedron, ►Table 2.9 for O_h and T_d coordination.

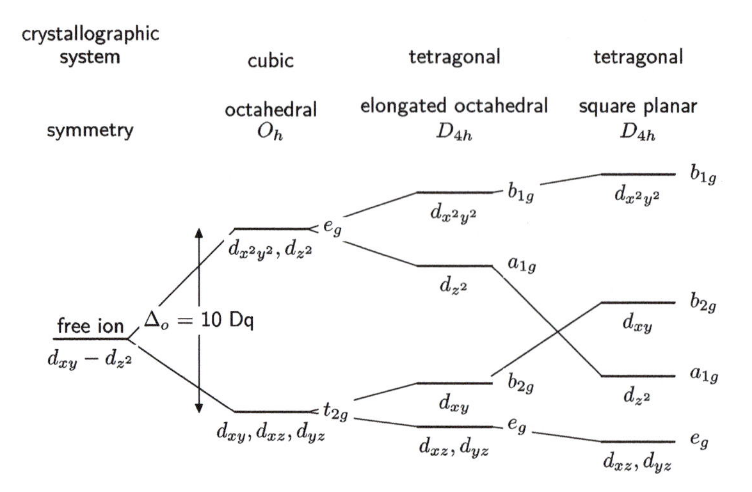

Figure 2.21: Level diagram of a cationic metal center in an octahedral ligand field without considering $\mathbf{H}_{ee}$ [226, Chapter 1.3, p. 26], [224, Figure 2.8, p. 35], [233, Figure 113a, p. 235]. A ligand field of O_h symmetry (a coordination octahedron) lifts the degeneracy of d orbitals and forms two sets of orbitals or *irreducible representations*, opening up the possibility of electronic transitions between them. Distortion of the coordination octahedron by stretching along its z-axis leads to elongated tetragonal 4+2 coordination (D_{4h} symmetry), then to square planar coordination (also D_{4h} symmetry). The diagram depicts orbital energies, not terms. In D_4 symmetry, the d_{z^2} level also can be even lower in energy than the d_{xz} and d_{yz} levels. To the right of each orbital, lowercase symbols such as e_g or a_{1g} denote the symmetry of the respective orbitals using group theory symbols.

The splitting can also be explained physically: the $d_{x^2-y^2}$ and d_{z^2} orbitals extend toward the ligands, and their electrons are subject to greater electrostatic repulsion than electrons from d orbitals extending between the axes. The e_g set is therefore higher in energy than the t_{2g} set. We provide a more detailed discussion of this orbital interaction using MO theory in ►Section 2.4.1 in the context of ligand-metal charge-transfer transitions. At this point, lifting the degeneracy of the d orbitals due to the ligand field opens the possibility of electronic transitions between the two orbital sets.

The energy difference between the sets e_g and t_{2g} is called Δ_o and is directly related to the ligands. In order to address the energy difference in general terms, a quantity "10 Dq" is used instead of actual energy units such as eV or J. In the case of octahedral splitting, set t_{2g} has an energy of –4 Dq relative to the free ion, set e_g of +6 Dq.

Δ_o and the visible light

After this theoretical digression, we return to the actual topic of color. By lifting the degeneracy of metal d orbitals in a crystal, i. e., the formation of distinct orbital sets, we can transfer electrons from one set to the other by supplying energy Δ_o. In other words, we transfer an initial configuration of lower energy to a final configuration of higher energy. If Δ_o corresponds to the energy of visible light, we can irradiate a metal complex or crystal with light, and the complex will appear colored due to this absorption. Especially for 3*d* metal complexes, Δ_o frequently matches the energy of visible light, so they are constituents of pigments and colored minerals.

Let us look at this process using a $3d^3$ configuration like in $Cr^{3\oplus}$ as an example, ►Figure 2.22. In the strong-field approximation, the *d* orbitals are split into two sets t_{2g} and e_g. Irradiation can transfer an electron from t_{2g} to e_g so that the ground state configuration $(t_{2g})^3$ changes to the excited-state configuration $(t_{2g})^2(e_g)^1$ (left). Without electronic interactions, the energy required would be $h\nu_1 = h\nu_2 = \Delta_o$.

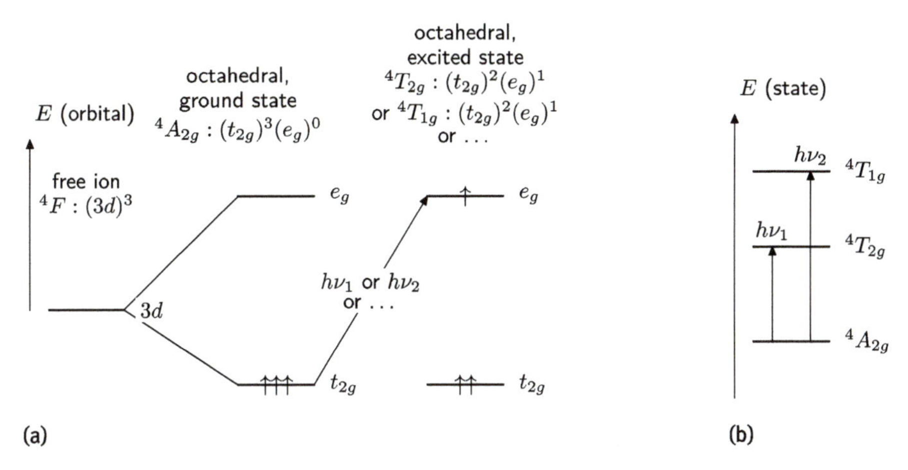

Figure 2.22: (a): The basic process of absorption by ligand field splitting, exemplified by a three-electron configuration $3d^3$ in $Cr^{3\oplus}$ using strong-field approximation [233, p. 210], [223, Chapter 3.2.1, p. 40], [224, Chapter 5.10.3, Figure 5.24]. *d* orbitals are split by the ligand field into two sets or *irreducible representations*. Lowercase symbols e_g and t_{2g} denote the symmetry of these sets. By irradiation with light of suitable energy, an electron can be transferred from t_{2g} to e_g. $h\nu_1 \neq h\nu_2 \neq \Delta_o$ due to electronic interactions (see states in (b)). (b): Plot of the terms (states or configurations) vs. energy, taking into account electron-electron interaction and coupling of angular momenta. Capitalized are the *term symbols* of states.

In reality, both electron-electron interaction and coupling of angular momenta occur, so we have to consider electronic transitions in terms of configurations, terms, and irreducible representations, ►Figure 2.20. The ground state (a $(t_{2g})^3$ configuration) yields a irreducible representation $^4A_{2g}$. The excited state (a $(t_{2g})^2(e_g)^1$ configuration) is represented by *two* irreducible representations, $^4T_{1g}$, and $^4T_{2g}$, with different energies. $^4T_{2g}$, of lower energy, represents the state wherein the former e_g electron occupies the d_{z^2} orbital, which interacts with the ligands along only one axis. $^4T_{1g}$, in contrast, represents the state wherein the former e_g-electron ends in the $e_{x^2-y^2}$-orbital, which interacts with the ligands along two axes and is, therefore, higher in energy. As a result, we observe two transitions with energies $h\nu_1$ and $h\nu_2$, respectively.

Example: Iron oxide pigments

As a first, though not simplest, example, let us examine the color of iron oxide pigments [65, Chapter 2, 6, 7], [223], [233, p. 217], [224, Chapters 3.7.2, 3.7.3, 10.8.1]. These oxides include α-Fe_2O_3 (red ocher, as mineral: hematite), α-FeOOH (yellow ocher, as mineral: goethite), γ-Fe_2O_3 (brown ocher, as mineral: maghemite), and γ-FeOOH (orange ocher, as mineral: lepidocrocite). ►Figure 2.23 schematically depicts reflectance spectra for these important earth colorants, ►Figures 3.12 and 3.14 show spectra of actual pigments.

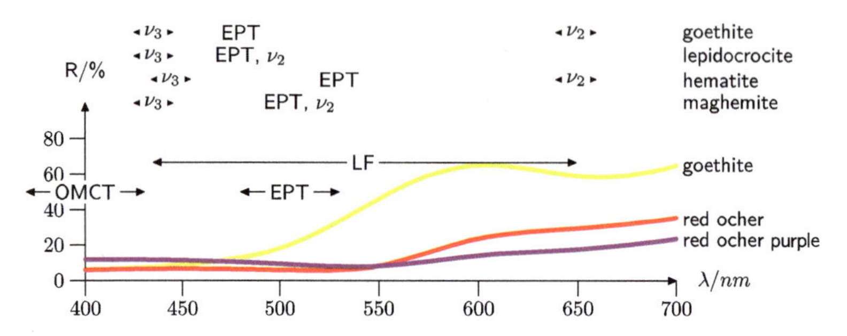

Figure 2.23: Schematic reflectance spectrum of yellow ocher (goethite), red ocher (hematite), and a purple-red ocher (spectra drawn after [48, Volume 4, p. 55], [65, Chapters 2, 6, 7]). Ligand field transitions ν_2 and ν_3 in iron are only faintly visible, e. g., the ${}^6A_1 \rightarrow {}^4T_2$ transition in goethite. Distinctly visible is the absorption in the blue range by the edge of a very intense OMCT transition around 250 nm in the UV, ►Section 2.4.1 at p. 136. Also clearly observable is the absorption in the blue and green spectral range due to an EPT (electron pair transition), supporting the distinct red color of hematite at 530 nm. Arrows labeled as "OMCT," "LF," and "EPT" indicate typical wavelength ranges of these features. ►Figure 3.12 depicts actual spectra.

Trivalent $Fe^{3\oplus}$, configuration $3d^5$, is incorporated into a high-spin complex and octahedrally surrounded by $O^{2\ominus}$ or $OH^{\ominus}$ ligands in all oxides. Building units are thus $Fe(O)_6$ or $Fe(O)_3(OH)_3$ octahedra, respectively. In the α phases, the ligands are hexagonal close-packed; in the γ phases, they are cubic closed-packed, ►Section 3.4.2.1.

The distribution of five d electrons to the two sets of d orbitals t_{2g} and e_g (►Figure 2.21) yields terms according to ►Figure 2.24. It allows multiple LF transitions, their energies ranging from near-infrared (NIR) to UV:

$$\begin{array}{lll}
{}^6A_1 \rightarrow {}^4T_1({}^4G) & (t_{2g})^3(e_g)^2 \rightarrow (t_{2g})^4(e_g)^1 & \nu_1 \text{ NIR} \\
{}^6A_1 \rightarrow {}^4T_2({}^4G) & (t_{2g})^3(e_g)^2 \rightarrow (t_{2g})^4(e_g)^1 & \nu_2 \text{ VIS} \\
{}^6A_1 \rightarrow {}^4E({}^4G), {}^4A_1({}^4G) & (t_{2g})^3(e_g)^2 \rightarrow (t_{2g})^3(e_g)^2 & \nu_3 \text{ VIS} \\
{}^6A_1 \rightarrow {}^4T_2({}^4D) & (t_{2g})^3(e_g)^2 \rightarrow (t_{2g})^3(e_g)^2 & \nu_4 \text{ UV} \\
{}^6A_1 \rightarrow {}^4E({}^4D) & (t_{2g})^3(e_g)^2 \rightarrow (t_{2g})^3(e_g)^2 & \nu_5 \text{ UV}
\end{array}$$

All LF transitions are spin forbidden as the spin multiplicity changes from 6 to 4 and are therefore inherently weak. However, due to the magnetic coupling of iron cations,

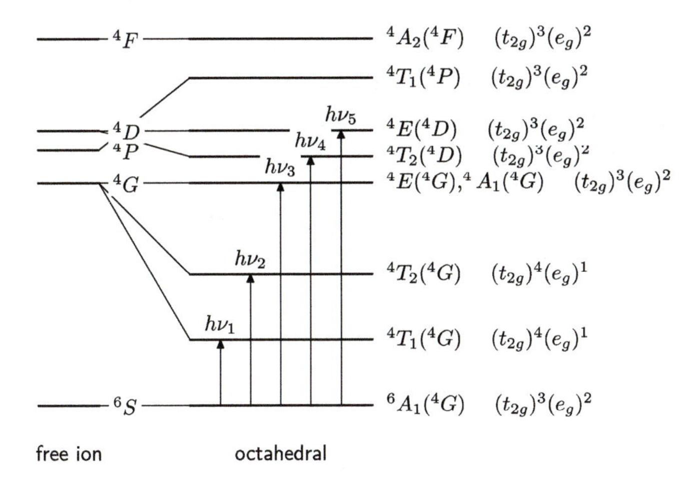

Figure 2.24: Term diagram of a $3d^5$ cation like $Fe^{3\oplus}$ in an octahedral ligand field [233, p. 219], [223, Chapter 3.2.1, p. 52], [224, Chapter 5.10.7, Figures 3.10, 3.16]. The electronic configuration (distribution of electrons to orbital sets) is indicated right to the term symbol in capitals.

LF transitions in iron oxides are stronger than expected. As ►Table 3.2 shows, the structural differences between α and γ phases and between oxides and hydroxides express themselves in ligand fields of varying strength, term splitting, and transition energies, ►Table 2.10.

Table 2.10: Transition wavelengths of color-inducing transitions in iron oxide pigments in nanometers [224, Chapters 3.7.2, 3.7.3, 10.8.1], [65, Table 7.5, p. 150], [631]. CT: charge transfer, LF: ligand field transition, EPT: electron pair transition.

Type	Transition	Hematite	Maghemite	Goethite	Lepidocrocite
CT	$O \rightarrow Fe^{III}$ $(2t_{1u} \rightarrow 2t_{2g}^*)$	270	250	250	240
CT	$O \rightarrow Fe^{III}$ $(1t_{2u} \rightarrow 2t_{2g}^*)$			250	210
LF	$^6A_1 \rightarrow {}^4T_1(^4G)$	885	935	917	960
LF	$^6A_1 \rightarrow {}^4T_2(^4G)$	649	510	649	485
LF	$^6A_1 \rightarrow {}^4E(^4G), {}^4A_1(^4G)$	444	435	435	435
LF	$^6A_1 \rightarrow {}^4T_2(^4D)$	405			
LF	$^6A_1 \rightarrow {}^4E(^4D)$	380	370	365	360
EPT	$^6A_1 + {}^6A_1 \rightarrow {}^4T_1 + {}^4T_1$	530	510	480	485

To make complicated things more difficult, distortions of the structure in actual iron-based pigments reduce the octahedral symmetry, lift any orbital degeneracies, and thus result in numerous LF transitions [630–633].

However, differences in LF transition energies between red to brown oxides (hematite, maghemite) and yellow to orange hydroxides (goethite, lepidocrocite) are not distinct enough, and intensities of these forbidden transitions are usually too low to explain the pronounced differences in the perceived colors. A closer analysis reveals that two more processes cause the distinct coloring of iron oxide pigments:

- A strong OMCT transition O(2*p*) → $Fe^{III}(3d)$ between oxide anions and iron is centered in the near-UV range around 210–270 nm, ►Section 2.4.1 at p. 136. It is so intense that even at iron contents as low as 0.5 %, an edge of it extends into the visible range, absorbing the blue parts of the spectrum. Many iron minerals, therefore, exhibit reddish-brown to dark colors.
  This absorption is even more intense in pure compounds such as iron oxides. It absorbs the blue range of the visible light entirely, resulting in an intensive yellow color, only slightly modified by LF transitions or exact composition.
- The $Fe^{III}O_6$ or $Fe^{III}O_3(OH)_3$ units occur in different arrangements (face-, edge-, or corner-sharing), changing Fe^{III}-Fe^{III} and Fe^{III}-O distances and lifting the magnetic equity of the iron cations. $Fe^{3\oplus}$ can then widely couple magnetically via electrons of the oxide and hydroxide bridges. Details can be found in [630, 631], [65, Chapters 6, 7].

While the first process explains the yellow colors, the second has two effects, one yielding intensive red colors:

- Coupling relaxes the spectroscopic selection rules that apply to isolated ions rather than pairs of ions. This relaxation can go as far as allowing LF transitions, otherwise forbidden in isolated iron ions. Therefore, they gain considerable intensity and become observable.
- In hematite and maghemite, Fe^{III} ions are close neighbors due to a face-sharing arrangement of FeO_6 octahedra (about 0.29 nm). This spatial proximity facilitates magnetic coupling across the electron-rich oxide bridge (Fe–O–Fe). It is so intense that a new absorption called *electron pair transition* (EPT) appears. The EPT is located at 530 nm in hematite and yields the red color of the red ocher by absorbing the spectrum's green parts (the blue parts already extinguished by the flank of the OMCT transition in UV).
  Magnetic coupling is weaker in goethite and lepidocrocite due to the different geometry of their crystal lattices (Fe^{III} distances about 0.3–0.35 nm in edge- or corner-sharing arrangements). Therefore, the EPT appears at shorter wavelengths at about 470 nm in goethite, only slightly absorbing green and yellow components. Together with the absorption of blue components by the OMCT transition in UV, these compounds appear yellow.

The magnetic coupling can be described mathematically by a Hamiltonian that combines the electron spins of the coupling partners Fe_a and Fe_b, J being the exchange integral [631]:

$$\mathbf{H} = J\mathbf{S}_a \cdot \mathbf{S}_a \tag{2.6}$$

Performing a perturbation calculation with this Hamiltonian as a perturbation to the ligand field, we obtain the energies of the coupled system for

$$E = \frac{J}{2}(S(S+1) - S_a(S_a+1) - S_b(S_b+1)) \tag{2.7}$$

The total spin S of the coupled system can take on the values $|S_a + S_b|$, $|S_a + S_b - 1|$, ...$|S_a - S_b|$. For both cations in the ground state ${}^6A_1 + {}^6A_1$, $S_a = S_b = \frac{5}{2}$ and $S_{GS} = (0, 1, 2, 3, 4, 5)$ applies.

For a *single* LF transition of energy $h\nu_1$ or $h\nu_2$ to excited states such as ${}^6A_1 + {}^4T_1$ or ${}^6A_1 + {}^4T_2$, $S_a = \frac{5}{2}, S_b = \frac{3}{2}$, and $S_{ES} = (1, 2, 3, 4)$ applies. We see that by coupling two iron cations, four transitions from the ground state to an excited state become possible, for which the selection rule $\Delta S = 0$ applies. Transitions such as ${}^6A_1 + {}^6A_1 \rightarrow {}^6A_1 + {}^4T_1$ are now spin-allowed in the coupled system and correspondingly become more intense.

Similarly, in hematite or maghemite, having short Fe^{III}-Fe^{III} distances due to the face-sharing FeO_6 octahedra, *two* LF transitions of energy $h\nu_1$ can strongly couple. This *electron pair transition* or EPT is described by ${}^6A_1 + {}^6A_1 \rightarrow {}^4T_1 + {}^4T_1$, of about twice the energy of a single transition. It, therefore, moves from the near-infrared to the visible range. For the EPT, $S_a = S_b = \frac{3}{2}$ and $S_{ES} = (0, 1, 2, 3)$ applies. Again, this time due to the EPT, there are four spin-allowed transitions from the ground state to the excited state with $\Delta S = 0$. EPTs are, therefore, so intense that they absorb significant fractions of the green and yellow spectral range, leading to a distinct red color of hematite or red ocher.

Coupling in iron oxide hydroxides is less efficient due to different crystal lattices and, therefore, iron-oxygen-iron angles and longer Fe^{III}-Fe^{III} distances of the edge- and corner-sharing $FeO_3(OH)_3$ octahedra. Since the EPT transition is located at higher energies in these iron hydroxides, the green and yellow spectral range of goethite and lepidocrocite is less affected, so they retain their yellow color.

Magnetic coupling is sensitive to a change in particle size. A reduction in particle size reduces coupling, leaving a higher proportion of yellow light in the reflectance spectrum and causing the particle to show a warmer or more orange hue. Crystallinity also plays a substantial role in the intensity of the coupling, and thus affects the color. In addition, the development of surface plasmons plays a significant role in the dependence of color on particle size. We discussed plasmons in more detail in ►Section 1.6.4.

2.3.5 Influence of ligand field strength

Chromium oxide is an example of the relationships just described. Its green color is determined by ligand field splitting in the Cr^{III} ion, as is the violet color of chromium alum and the red of ruby [616]. How can this be?

Example: Chromium oxide green

$O^{2\ominus}$ ligands octahedrally surround the ion $Cr^{3\oplus}$ ($3d^3$) in chromium oxide green. Its *d* orbitals split into two orbital sets, t_{2g} and e_g (►Figure 2.21), and the states are ordered according to ►Figure 2.25 [223, 224]. In the ground state, the electron configuration $(t_{2g})^3(e_g)^0$ is present, from which the following spin-allowed transitions are possible:

$$^4A_{2g} \rightarrow {}^4T_{2g}(^4F) \quad (t_{2g})^3(e_g)^0 \rightarrow (t_{2g})^2(e_g)^1 \quad \nu_1 \text{ VIS}$$
$$^4A_{2g} \rightarrow {}^4T_{1g}(^4F) \quad (t_{2g})^3(e_g)^0 \rightarrow (t_{2g})^2(e_g)^1 \quad \nu_2 \text{ VIS}$$

We also observe a transition to the higher ^4P state. This transition is weaker because it is a two-electron transition with lower probability:

$$^4A_{2g} \rightarrow {}^4T_{1g}(^4P) \quad (t_{2g})^3(e_g)^0 \rightarrow (t_{2g})^1(e_g)^2 \quad \nu_3 \text{ UV}$$

Transitions like $^4A_{2g} \rightarrow {}^2T_{2g}$ are spin-forbidden (the spin changes from $\frac{3}{2}$ to $\frac{1}{2}$) and therefore (theoretically) do not occur. Since the $Cr^{3\oplus}$ cation is octahedrally coordinated, all LF transitions are forbidden by the Laporte rule (g-g transitions from *d* to *d* orbitals) and are, therefore, correspondingly weak in intensity.

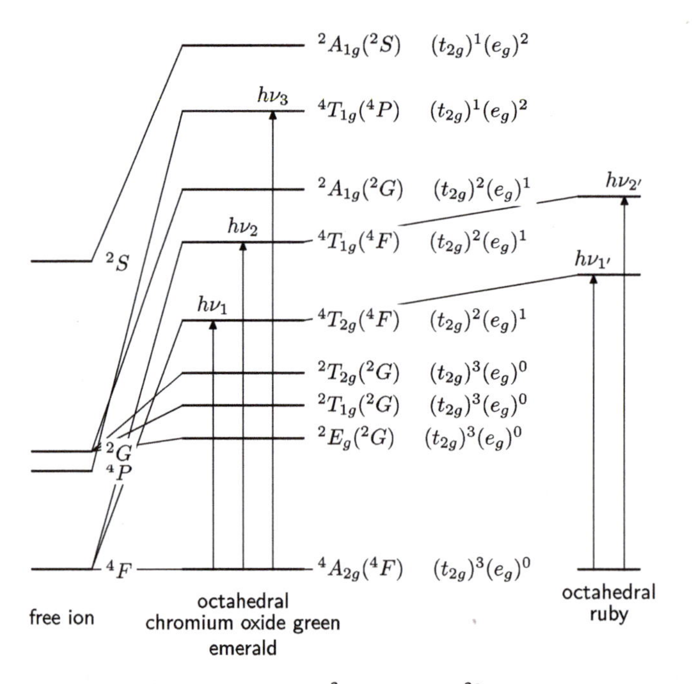

Figure 2.25: Term diagram of a $3d^3$ cation like $Cr^{3\oplus}$ in an octahedral field [233, p. 210], [223, Chapter 3.2.1, p. 40], [224, Chapter 5.10.3, Figure 5.24]. The electronic configuration (distribution of electrons to orbital sets) is indicated right to the term symbol in capitals.

In many chromium compounds, ν_1, as well as ν_2, correspond to energies of visible light so that the compounds are colored. This constellation provides a rare chance to isolate the green spectral range by absorbing the red and blue parts, like chromium oxide green or emerald. In these, the absorption around ν_1 removes the red spectral part, that around ν_2 the blue one, leaving only the green one, ►Figure 3.28. In the case of ruby, however, the states corresponding to the irreducible representations $^4T_{2g}$ and $^4T_{1g}$ has higher energy, hence $\nu_{1'} > \nu_1, \nu_{2'} > \nu_2$. Consequently, both absorptions occur at lower wavelengths in the yellow and blue-violet spectral range. A significant amount of red light is transmitted in addition to a blue one, yielding the bluish-red color of ruby, ►Figure 2.26.

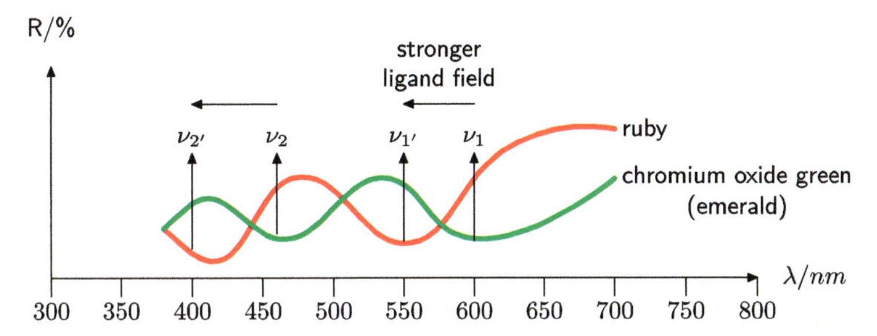

Figure 2.26: Reflectivity of two minerals with a composition of chromium oxide (ruby and chromium oxide green or emerald). In chromium oxide green and emerald, absorptions at ν_1 and ν_2 remove red and blue spectral parts, letting green light pass. In ruby, the ligand field is more potent, so the energy of the states shifts to shorter wavelengths ν_1' and ν_2', absorbing yellow and blue-violet spectral regions. As a result, ruby exhibits red as well as blue-green light and, therefore, displays cool red color [223] (spectra schematically drawn after [223]).

Ligand field strength

Why do the terms behave in the manner described? The energy difference Δ_o between orbital sets t_{2g} and e_g is directly related to the chemical environment of the metal cation, i. e., to its ligands. Systematic studies helped establish the *spectrochemical series*, which arranges ligands according to their "ligand strength," i. e., the magnitude of the ligand field splitting they cause, ►Figure 2.27:

$$(\Delta \text{ small})\ I^{\ominus} < Br^{\ominus} < S^{2\ominus} < Cl^{\ominus} < NO_3^{\ominus} < F^{\ominus} < OH^{\ominus} < O^{2\ominus} < H_2O < NH_3 < NO_2^{\ominus} < CN^{\ominus}\ (\Delta \text{ large}).$$

The influence of ligands on Δ can be studied by successively exchanging one or more ligands in a complex under otherwise identical conditions. If the new ligands are of higher ligand strength, we observe an increase in Δ in the electronic spectra. In pigments, oxide and sulfide ligands often produce a strong field, so working in the strong-field approximation is justified in these cases.

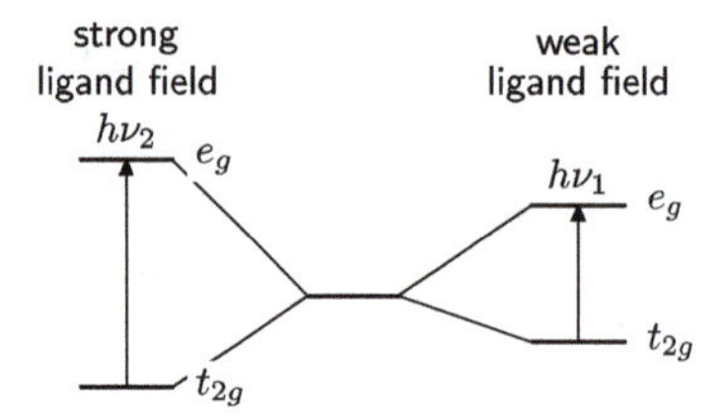

Figure 2.27: Influence of ligand field strength on the splitting of d orbital sets, shown for a $3d^1$ configuration in an octahedral ligand field [224, Figure 3.11, p. 60]. A strong ligand field causes a large splitting, increases the transition energy between orbital sets (corresponding to light of shorter wavelengths), and shifts the color impression to red. In contrast, a weak ligand field decreases the transition energy (corresponding to light of longer wavelengths) and shifts the color impression to blue.

Furthermore, there is a proportionality between Δ_o and the distance a between the metal cation and ligand [227, Chapter 2]:

$$\Delta_o \propto a^{-5} \tag{2.8}$$

Depending on the underlying computational model (ligands as point charges or as dipoles), the exponent varies somewhat; a dependence of a^{-6} applies for the dipole approximation. Apart from these details, we recognize that the size of the crystal's elementary cell (the lattice constant) determines a, and thus strongly influences d orbital splitting. Schmitz-DuMont et al. [628] and Reinen and Schmitz-DuMont [629] illustrate how an expansion of various crystal lattices of oxides shifts the color of cobalt and chromium dopants to blue because an increase in lattice constant decreases the ligand field and the orbital splitting energy and increases the wavelength of the absorption linked to this electronic transition. Longer wavelength absorptions signify color impressions toward the blue end of the spectrum.

To get an impression, we calculate this dependence approximately, using an example from the world of gemstones: both corundum Al_2O_3 (colorless) and eskolaite Cr_2O_3 (green) crystallize in the corundum lattice and form a gapless series of solid solutions with increasing Cr^{III}/Al^{III} ratio when mixed. The minerals' colors change with chromium concentration as follows: colorless (corundum, no chromium), red (ruby, some chromium), green (more chromium), green (eskolaite, chromium completely replaced aluminum). We see how the color of the same crystal lattice and an identical chromophore change with chromophore concentration. More obvious for painters is a similar series of corundum (colorless), aluminum chromium pink (red, but sphene lattice instead of corundum), chromium oxide green (green), ►p. 263.

The ligand-metal distance O–Cr is about 191 pm in ruby with an approximate octahedral splitting of $\Delta = 535$ nm. With these values, we can calculate the proportionality constant and estimate $\Delta \approx 668$ nm for an atomic distance of 199 pm in eskolaite. ►Figure 2.26 shows that this rough estimation is not so far off from the actual value. At least we discern how a slight change in ligand spacing results in a significant color shift.

To get back to LF-based artists' paint, it becomes clear that large amounts of crystal defects yield a wide variety of atomic distances, resulting in broad absorptions and dull, muddy shades. Severe and frequent crystal defects occur, e. g., during the hydrothermal formation of minerals or their weathering.

Again, we can explain thermochromism with this strong distance dependence: increasing temperatures expand the crystal lattice, a increases, and thus Δ becomes smaller. With the fifth power, the change in a is amplified.

2.3.6 Distortion of the octahedral field, Jahn–Teller effect

►Figure 2.21 depicts the splitting of d orbitals in an octahedral ligand field and in ligand fields created by distorting the octahedron: if we pull two ligands outwards along the z-axis, we end up with an elongated octahedron and tetragonal symmetry. If we continue to pull the two ligands outwards infinitely (actually removing them), we yield quadratic-planar coordination. This distortion results in altered coordination geometry and lifts nearly all orbital degeneracy. The more the symmetry decrease from O_h to D_{4h}, the more degeneracy is lifted, in terms of group theory: "The correlation diagram for O_h shows that lowering the symmetry from O_h to D_{4h} causes a splitting of t_{2g} into b_{2g}/e_g and of e_g into a_{1g}/b_{1g}." As a result, the two orbital sets (irreducible representations) in an octahedral field split further into four sets e_g, a_{1g}, b_{2g}, and b_{1g}.

Since this splitting may reduce the total energy if d orbitals are partially occupied, we often observe the distortion of crystal lattices. We can notice this interaction of metal with crystal lattice in copper pigments. On the one hand, the crystal lattice (which in turn depends on the chemical composition) determines the coordination polyhedron (the spatial position of the ligands); on the other hand, the metal cation interacts with the ligands, favoring a certain degree of distortion of the crystal lattice to minimize the total energy.

This so-called *Jahn–Teller distortion* occurs whenever degenerate orbitals of high energies are partially filled, as are orbitals from the e_g set in $Cu^{2\oplus}$, having the electronic configuration $3d^9$. e_g is degenerated twice and must accommodate three electrons. In ►Figure 2.21, we can see what happens: a slight distortion, i. e., elongation of the octahedron in the z-direction, lowers its symmetry, and lifts the degeneracy of the e_g set. Due to the decreased interaction of electrons in d_{z^2} with the ligand along the z-axis, d_{z^2} decreases in energy. It is now energetically favored to accommodate *two* electrons in d_{z^2} (lowered in energy), while only *one* electron occupies $d_{x^2y^2}$ (raised in energy). This net gain in energy yields a distortion of the crystal lattice, which is counteracted by repulsive forces from other lattice atoms until a stable equilibrium is reached.

Example: Copper pigments (malachite, azurite, Egyptian blue)

Ligand field splitting of $Cu^{2\oplus}$ is the coloring principle of all copper-based pigments. The free ion has the electron configuration $3d^9$ (state ^2D). It is octahedrally coordinated in the colored compounds relevant to us and has the ground state $^2E_g(^2D)$ and the excited state $^2T_{2g}(^2D)$, ►Figure 2.28 [223, 224]. Therefore, we would expect only the one transition:

$$^2E_g(^2D) \rightarrow {}^2T_{2g}(^2D) \quad (t_{2g})^6(e_g)^3 \rightarrow (t_{2g})^5(e_g)^4$$

However, the Jahn–Teller effect allows an energy gain by reducing the coordination symmetry and elongating the octahedron. The broken symmetry lifts the degeneracy of t_{2g} and e_g orbital sets, and the following three transitions become possible:

$$\begin{array}{lll} ^2B_{1g}(^2D) \rightarrow {}^2A_{1g}(^2D) & (e_g)^4(b_{2g})^2(a_{1g})^2(b_{1g})^1 \rightarrow (e_g)^4(b_{2g})^2(a_{1g})^1(b_{1g})^2 & \nu_1 \text{ IR} \\ ^2B_{1g}(^2D) \rightarrow {}^2B_{2g}(^2D) & (e_g)^4(b_{2g})^2(a_{1g})^2(b_{1g})^1 \rightarrow (e_g)^4(b_{2g})^1(a_{1g})^2(b_{1g})^2 & \nu_2 \\ ^2B_{1g}(^2D) \rightarrow {}^2E_g(^2D) & (e_g)^4(b_{2g})^2(a_{1g})^2(b_{1g})^1 \rightarrow (e_g)^3(b_{2g})^2(a_{1g})^2(b_{1g})^2 & \nu_3 \text{ VIS} \end{array}$$

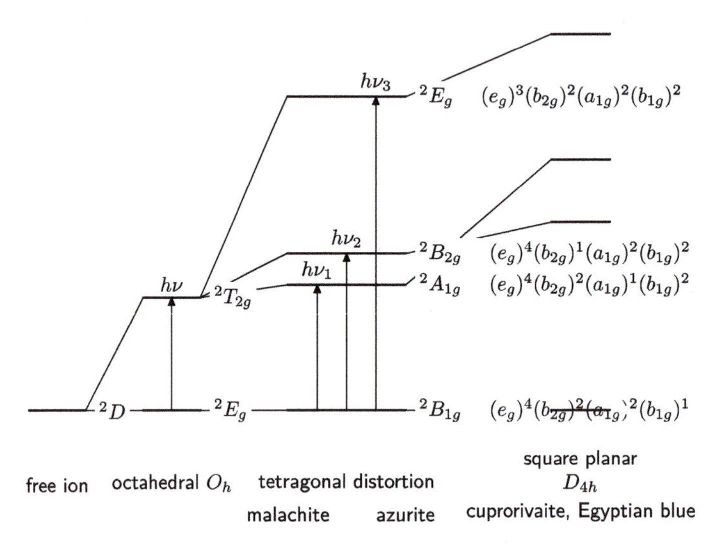

Figure 2.28: Term diagram of a $3d^9$ cation like $Cu^{2\oplus}$ in the octahedron field [233, p. 235], [223, Chapter 3.2.1, p. 37], [224, Chapter 5.10.12, Figure 3.5]. From left to right, the tetragonal distortion of the coordination octahedron increases up to the point of square-planar coordination. The distortion corresponds to the minerals malachite, azurite, and cuprorivaite (Egyptian blue). The electronic configuration (distribution of electrons to orbital sets) is indicated right to the term symbol in capitals.

The distortion of the octahedron increases from malachite (six ligands, strongly distorted octahedron) to azurite (four ligands square planar, two bipyramidal) to cuprorivaite (only four ligands, square-planar). This distortion increases the energy of the final ^2E_g state and, therefore, the transition energy $h\nu_3$ of the optical transition. Malachite induces a green color impression ($h\nu_3$ small), azurite a blue-green one ($h\nu_3$ higher), and cuprorivaite and Egyptian blue a blue one ($h\nu_3$ even higher, absorption of yellow spectral parts). ►Figure 2.29 illustrates how the absorption shifts from ν_3 to higher energies, ending at $h\nu_3''$. The interplay of two phenomena causes the overall color impression by creating a distinct reflection maximum in the green to the blue spectral range:

- the LF transition at $\nu_3/\nu_3'/\nu_3''$, absorbing red to yellow spectral parts
- a strong OMCT (ligand-copper) transition at high energies in the UV, absorbing purple spectral parts

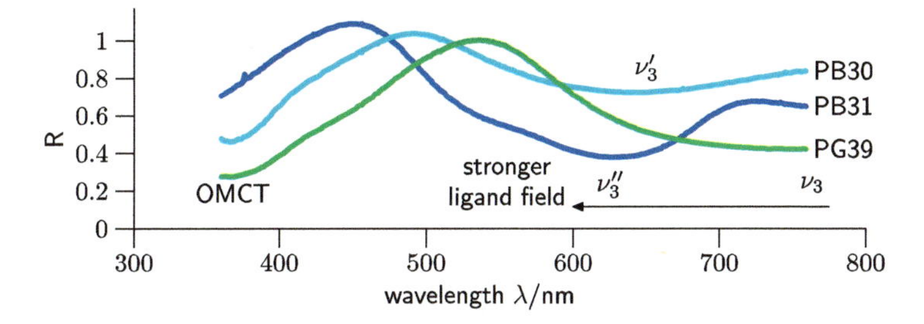

Figure 2.29: Influence of Jahn–Teller effect on the color of copper pigments. Reflectance spectra of various copper minerals and pigments normalized to an arbitrary unit: PB30 (azurite, Kremer no. 102078, $Cu(OH)_2 \cdot 2\,CuCO_3$), PB31 (Egyptian blue, Kremer no. 10060 $CaCu[Si_4O_{10}]$), and PG39 (malachite, Kremer no. 103458, $Cu(OH)_2 \cdot CuCO_3$). The spectra show a shift of the optical transition ν_3 in copper from lower to higher energies. It is caused by an increasing distortion of coordination octahedra: malachite (weak ligand field, distorted octahedron, ν_3), azurite (stronger ligand field, square-planar and bipyramidal coordinated copper, ν_3'), and cuprorivaite (strongest ligand field, square planar coordinated copper, ν_3'') [223]. In all cases, ν_1 is an IR transition. In the near-UV, strong OMCT transitions shape a pronounced reflection maximum located in the green spectral range in the case of malachite and in the blue-green to blue spectral range in the case of azurite and cuprorivaite.

As we have seen, OMCT transitions strongly influence a copper pigment's color. The composition of copper compounds influences the energy of their OMCT transition, and it varies, especially in the case of natural copper pigments from ores. Their formation in the weathering zone of ore deposits or hydrothermal precipitation results in various mixtures of carbonates, hydroxides, and other anions, even at small deposit scales. A change of OMCT energy toward the blue spectral range strongly influences the color of malachite and other green copper carbonate colorants, determining the perceived color's bluish tint. The broad reflection peak in copper carbonate colorants is observed between 505 and 525 nm [48, Volume 2].

2.3.7 Tetrahedral coordination

After the discussion of color and octahedral ligand fields, we look at tetrahedral ligand fields, essential for cobalt pigments. ►Figure 2.30 depicts the typical splitting of the

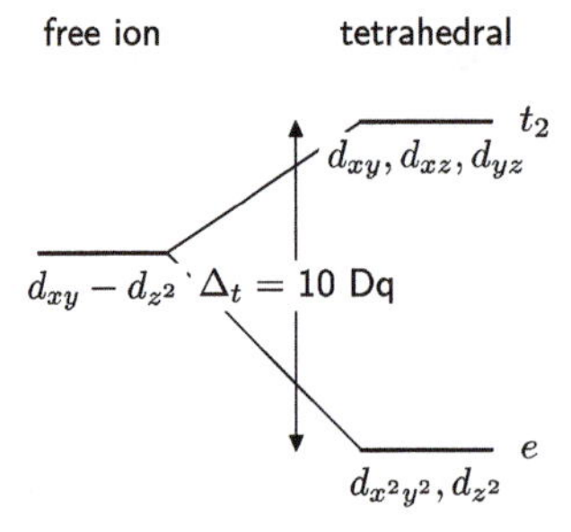

Figure 2.30: Splitting of *d* orbitals in two sets (or irreducible representations) of different symmetry in a tetrahedral ligand field (without considering terms or LS coupling) [224, Figure 2.7, p. 21]. The symmetry is inverted compared to octahedral ligand field.

d orbitals into two sets, t_2 and e. The energy difference Δ_t between the sets is smaller compared to Δ_o, and the symmetry of the sets is interchanged compared to an octahedral field.

Different coordination polyhedra result in significant differences in the electronic structure, exemplified by cobalt-containing colorants. In smalt, a deep blue cobalt glass, Co^{II} is tetrahedrally coordinated, Δ_t is small, and absorption occurs in the red spectral range, inducing a blue color. In the red-violet cobalt violet, cobalt is square planar coordinated, Δ_q is large, the absorption occurs at higher energies in the green spectral range, and induces a reddish violet color. Finally, in aureolin, cobalt is octahedrally coordinated by six cyano ligands, which cause a ligand field so intense that absorption occurs in the blue spectral range, resulting in yellow color. Colored glasses also provide examples: Adding cobalt salts to ordinary silicate glass turns it into splendid blue cobalt glass due to tetrahedrally coordinated Co^{II} ions. In contrast, phosphate glass is pink due to octahedrally coordinated cobalt.

The color diversity of cobalt pigments originates in the ability of cobalt to occur in different valences and coordination geometries. $Co^{2\oplus}$ ($3d^7$) forms high-spin complexes that are either tetrahedrally coordinated with configuration $(e)^4(t_2)^3$ and predominantly blue, or octahedrally coordinated with configuration $(t_2)^5(e)^2$ and primarily red.

Example: Cobalt blue

Cobalt blue is an excellent example of blue tetrahedrally coordinated cobalt. It is a normal spinel $A^{II}B_2^{III}O_4$ in which divalent cations A occupy tetrahedral gaps, and trivalent cations B occupy octahedral gaps. The ground state is $^4A_2$. We observe the following transitions, ►Figure 2.31 [223, 224]:

$$
\begin{array}{lll}
{}^4A_2({}^4F) \rightarrow {}^4T_2({}^4F) & (e)^4(t_2)^3 \rightarrow (e)^3(t_2)^4 & \nu_1 \text{ IR} \\
{}^4A_2({}^4F) \rightarrow {}^4T_1({}^4F) & (e)^4(t_2)^3 \rightarrow (e)^3(t_2)^4 & \nu_2 \text{ NIR} \\
{}^4A_2({}^4F) \rightarrow {}^4T_1({}^4P) & (e)^4(t_2)^3 \rightarrow (e)^2(t_2)^5 & \nu_3 \text{ VIS}
\end{array}
$$

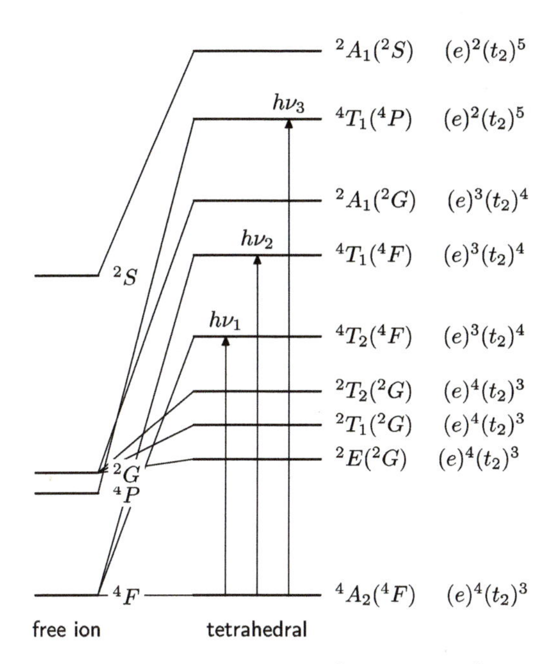

Figure 2.31: Term diagram of a $3d^7$ cation like $Co^{2\oplus}$ in a tetrahedral ligand field [233, p. 229], [223, chapter 3.2.1, p. 40], [224, Chapter 5.10.10, Figure 3.12]. The electronic configuration (distribution of electrons to orbital sets) is indicated right to the term symbol in capitals.

Due to the tetrahedral symmetry, each transition is allowed by the Laporte rule and, therefore, an intense one. In cobalt blue, the intense optical transition ν_3 occurs in the green to the red spectral range, causing the pigment to appear deep blue, ►Figures 3.22 and 3.25.

Compared to octahedral chromium, ν_3 in the chromium ion occurs in the UV, while ν_1 and ν_2 are in the optical spectral range. This shift to higher transition energies illustrates that the splitting energy Δ_o in an octahedral field is larger relative to Δ_t in a tetrahedral field.

Example: Aureolin

In aureolin, trivalent cobalt $Co^{3\oplus}$ ($3d^6$) is present, which almost always forms octahedral low-spin complexes $(t_{2g})^6(e_g)^0$ with a ^1I ground state term [224]. Ligand field splitting results in the following transitions, ►Figure 2.32:

$^1A_{1g}(^1I) \rightarrow {}^3T_{1g}(^3H)$	$(t_{2g})^6(e_g)^0 \rightarrow (t_{2g})^5(e_g)^1$	ν_1 IR (spin-forbidden)
$^1A_{1g}(^1I) \rightarrow {}^3T_{2g}(^3H)$	$(t_{2g})^6(e_g)^0 \rightarrow (t_{2g})^5(e_g)^1$	ν_2 IR (spin-forbidden)
$^1A_{1g}(^1I) \rightarrow {}^1T_{1g}(^1I)$	$(t_{2g})^6(e_g)^0 \rightarrow (t_{2g})^5(e_g)^1$	ν_3 VIS (spin-allowed)
$^1A_{1g}(^1I) \rightarrow {}^1T_{2g}(^1I)$	$(t_{2g})^6(e_g)^0 \rightarrow (t_{2g})^5(e_g)^1$	ν_4 VIS (spin-allowed)

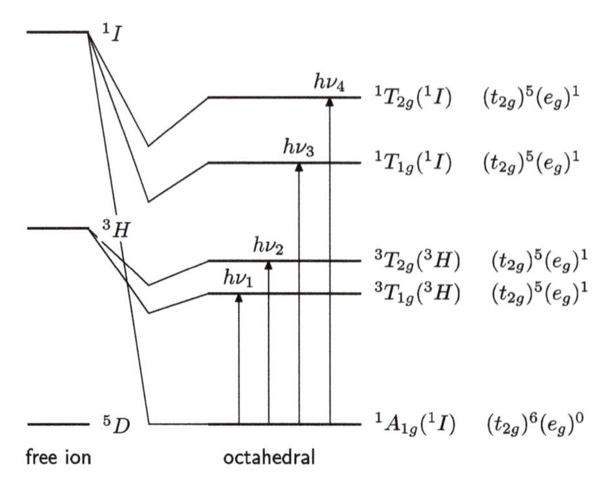

Figure 2.32: Term diagram of a $3d^6$ cation like $Co^{3\oplus}$ in an octahedral field (low-spin complex) [224, Chapter 5.10.9, Figure 5.26]. The kink symbolizes the change from energy levels of high-spin complexes (left from kink) to those of low-spin complexes (right from kink). For our subject, we are interested in low-spin complexes. The electronic configuration (distribution of electrons to orbital sets) is indicated right to the term symbol in capitals.

The optical ν_3 and ν_4 transitions are located in the blue and red spectral ranges to reflect the yellow and yellow-green spectral ranges, giving aureolin its pure yellow color.

2.3.8 LF-based chromophores

Pigments based on ligand field transitions contain a complex of a *d* block metal as the chromophore. Therefore, the ligands must build a ligand field sufficiently strong for the intended transition energy, determining absorption wavelength and color perception. Due to the high natural abundance of iron, manganese, and copper, we find them among the classic metals included in LF pigments. With the development of inorganic chemistry in modern times, Cr, Co, and V also became standard metal components.

Crystal lattices and periodic ligand fields

One of the crystals' characteristics is a high degree of near and far order, i. e., the ligand field structure is the same in all directions for all metal cations. As a result, we observe one or more distinct absorption bands, resulting in a pure color perception when located in the optical spectral range.

The metal must be the main constituent of a crystal lattice to obtain a high color intensity. It is often contained only as an impurity in natural minerals, so many are only weakly colored. Precious examples are gemstones such as ruby or emerald, composed of colorless aluminum oxide or silicate. Oxide ions octahedrally coordinate the aluminum atoms, and a color impression is induced by partly replacing aluminum atoms with chromium atoms. Since chromium is present only in small amounts, the mineral powder is nearly colorless and appears white due to scattering phenomena.

Frequent ligands in minerals of plutonic origin (i. e., formed underground under heat and pressure) are oxide and sulfide anions. They induce a middle to strong ligand field and form chemically stable compounds used as pigments. Nowadays, natural crystal lattices act as role models for the formation of new mineral phases in which cations are entirely or partially replaced by another metal, thereby changing their color or inducing color in the first place. Frequent host lattices are corundum Al_2O_3 and rutile TiO_2. Since these mixed oxide pigments represent a high proportion of color-stable and refractory pigments today, we will deal with them in more detail in ►Section 3.4.3.

Common ligands in minerals hydrothermally deposited or formed by weathering are hydroxide ions, carbonate ions, and water. These minerals are formed by spontaneous precipitation from aqueous solutions or piecemeal weathering and often have significantly disturbed crystal lattices and inhomogenous ligand fields that reduce color quality. We have already addressed this problem in ►Section 1.7.2.

Amorphous materials and statistical ligand fields

No crystal lattices exist in amorphous materials such as glasses and glazes, but they nevertheless show a specific order on average. Each metal cation is surrounded by its unique chemical environment. However, statistically, most of these environments are similar, and in a first approximation, can be combined into a mean lattice (near order). A predefined far order beyond this does not exist, but the contribution of remote regions to the local mean-field is, on average, the same everywhere for homogeneous composition. Combined, both effects create an average local ligand field, which varies to some degree, so that we can observe broadened absorption peaks in the absorption spectrum. Glasses can thus also be relatively pure in color.

Since the composition of amorphous substances is not governed by stoichiometric rules but only by electrical charge neutrality and the solubility of one mineral phase in another, the ligand fields in glasses can vary considerably with composition. The colorations of glasses are accordingly very sensitive to a change in composition and not easily grasped by general rules.

2.4 CT: Charge transfer transitions

We saw how electronic transitions between *d* orbitals induce the color in the previous section. A ligand field lifts the degeneracy of *d* orbitals so that an electron transition between them becomes possible when the *d* orbitals are only partially occupied. Surprisingly, compounds in which the metal has a d^0 or d^{10} configuration can also be intensively colored, although no *d* electron transition can occur here; examples include HgI_2 (red) or $MnO_4^{\ominus}$ (purple).

Instructive examples also exist in the world of gemstones: Corundum Al_2O_3 is colorless in its pure state (white sapphire). Doped with less than 1 % Ti^{IV}, it is still colorless, while 1 % Fe^{III} induces a faintly yellow color, and small fractions of Cr^{III} produce red ruby. These colors can be traced back to LF transitions of iron and chromium. Accordingly, in 1902 the French chemist, Verneuil, successfully created artificial rubies with relative ease by adding small amounts of chromium salts to pure corundum. However, a subsequent attempt to produce similar sapphires failed: neither induced cobalt salts the sought-after sapphire-blue color nor did any tested metal salts achieve the desired result. To his surprise, Verneuil was successful only when he added *two impurities simultaneously*: the combined application of a few parts per thousand of Ti^{IV} and Fe^{II} produced a magnificent sapphire-blue color.

Charge transfer

In HgI_2 and $MnO_4^{\ominus}$, we observe the (partial) transfer of charge from a ligand to the metal. In the molecular orbital model, electrons move from a MO with ligand character (e. g., $I^{\ominus}$ or $O^{2\ominus}$) to a MO with metal character (e. g., $Hg^{2\oplus}$ or $Mn^{7\oplus}$). This mechanism is generally called *charge transfer*, and the transfer can be initiated by irradiation with light. Formally, the transfer lowers the metal's oxidation number, i. e., the metal is (partially) reduced. Therefore, we can regard CT transitions as photoinduced redox reactions, like that of silver iodide in photopaper. Here, a ligand ($I^{\ominus}$) also reduces a metal cation ($Ag^{\oplus}$) in a light-induced electron transfer. In photography, the reduction continues to black metallic silver. In general, CT processes transfer charge from a donor to an acceptor, so we can address it as an internal redox reaction [343], [229, Chapter 13], [226].

The following section will deal with electronic transitions between MOs in general. What characterizes CT is that we can assign both the initial and the final MO to *distinct parts* of the molecule. Both examples illustrate that the initial MO is a ligand MO, and the final MO is a metal MO. Consequently, the CT is accurately called *ligand-metal charge transfer* (LMCT). In contrast, MOs in MO-MO transitions are generally *delocalized over large molecule regions*. The assignment of MOs to distinct molecule parts in a CT also distinguishes CT transitions from semiconductor transitions. In SC, both the initial and final bands are *collective states (bands)*. Finally, the distinction between CT and LF transitions is that LF transitions involve electrons changing orbitals *within the metal.*

The example of sapphire illustrates another type of CT, the *intervalence charge transfer* (IVCT) or *metal-metal charge transfer* (MMCT). $Ti^{4\oplus}$ and $Fe^{2\oplus}$ ions replace some $Al^{3\oplus}$ ions in the corundum lattice in sapphire. By irradiation with light, an electron changes from $Fe^{2\oplus}$ to $Ti^{4\oplus}$. Again, unlike LF transitions, this transition involves orbitals of two reactants.

The process is not restricted to d^0 or d^{10} configurations but occurs rather more generally, as [343] states:

- in metal-ligand complexes, e. g., of *d* and *f* block metals and inorganic or organic ligands
- in compounds containing different metal cations
- in compounds containing a multivalent metal in several oxidation states

The gemstone examples show that a bond need not directly connect the metals involved; a spatial arrangement in a crystal lattice, allowing an electron transfer, is sufficient. Usually, the observed absorptions are located in the UV spectral range and therefore induce no color. However, suppose one of the partners has a high oxidation potential (high electron affinity) and the other is a reducing agent. In that case, the transition energy may be lowered so strong that the absorption takes place in the blue spectral range (yielding a yellow color) or even in the red one (yielding a green to blue color).

As we already noted, the chemical nature of the redox partners determines the type of CT transition:

- Ligand metal CT (LMCT): electrons flow from ligand to metal orbitals. Often the oxide anion acts as the ligand, this particular significant case is called *oxygen metal CT* (OMCT). Examples include chromate anions (yellow), permanganate anions (purple), and iron(III) oxide (hematite, red).
- Intervalence CT homonuclear (IVCT): electrons are transferred from a metal of a lower oxidation state to the same metal in a higher oxidation state. An example is a Prussian blue or green earth with a $Fe^{II} \rightarrow Fe^{III}$ transition.
- Intervalence CT heteronuclear or metal-metal CT (MMCT): electrons are transferred from one metal of a low oxidation state to another metal of a high oxidation state. An example is a blue sapphire with a $Fe^{II} \rightarrow Ti^{IV}$ transition.
- Metal-ligand CT (MLCT): electrons flow from the metal to the ligand. This type does not occur in colorants but, e. g., in nickel carbonyl.
- Donor–acceptor CT intramolecular and intermolecular: Electrons move from an (often organic) donor to an (often organic) acceptor. This type is also not relevant for colorants; an example is a quinhydrone with an electron transfer from hydroquinone to a quinone.

The characteristics of the charge transfer mechanism are:

- CT transitions are allowed by selection rules and are very intense ($\epsilon > 10^3\,mol^{-1}\,cm^{-1}$).
- As a consequence of the high intensity, the absorption band is very broad.
- The transition could, in principle, induce any color but is practically limited by the finite number of suitable redox partners.

- Often LMCT transitions are located in the blue or near-UV spectral range. The edge of the intense peak absorbs a significant part of the neighboring wavelengths, especially violet and blue, or even green, ►Figure 2.2. This wide-range absorption is typical for OMCT transitions near-UV, imparting a yellow to red color to a long series of oxide minerals or chromates. Examples are chrome yellow (lead chromate) and yellow and red ochers (iron oxide).
  IVCT transitions, on the other hand, are often located in the VIS or near IR spectral range. Hence we observe a blue color (Prussian blue, blue, and green minerals containing mixed-valent iron).
- The color induced by a CT transition depends on pressure and temperature. An increase in pressure shortens all bonds and increases the overlap of the participating orbitals, facilitating the transfer of electrons. The result is a more intensive color. In contrast, an increase in temperature increases the bond lengths, and thus decreases the overlap integral and the color intensity.
- CT transitions often overlay the much weaker LF absorptions.

In some compounds, CT transitions are not limited to two partners. Due to the crystal structure, electrons can change from one partner to the next in a chain reaction. Consequently, the absorption bands are intensified, broadened, and extended over the entire VIS range. This unspecific absorption imparts a brown to black color to the compounds, an example being black magnetite.

Sulfides are not similar to oxides in this respect, as sulfur tends to a covalent rather than ionic bonding, and thus electron transfer is reduced. In addition, the electrical and optical properties of solid sulfides indicate electrons delocalized throughout the crystal. This delocalization results in a molecular orbital or band structure, and thus electrons alternate between MOs and bands rather than between metal or ligand orbitals. Sulfides, unlike oxides, therefore show resemblance to metals, and they often possess high luster and metallic color (examples are FeS_2 pyrite or PbS galenite). We have already examined sulfide colorants while discussing the SC mechanism, ►Section 2.2.

2.4.1 Ligand-to-metal transition and oxygen-to-metal transition

We focus first on ligand-metal charge transfer, or LMCT, crucial to colorants. Charge flows from ligands to a central metal atom. Examples of LMCT are the intensively colored anions permanganate $MnO_4^{\ominus}$ (purple) and chromate $CrO_4^{2\ominus}$ (yellow). Due to the $3d^0$ configuration of the central metal cations, no *d-d* LF transition is possible. However, fractions of the charge of the oxo ligands can be transferred to the highly charged metal cation when irradiated with light, ►Figure 2.33(b). Since oxo ligands frequently occur in complexes, this particular case of LMCT is called OMCT (oxygen metal charge

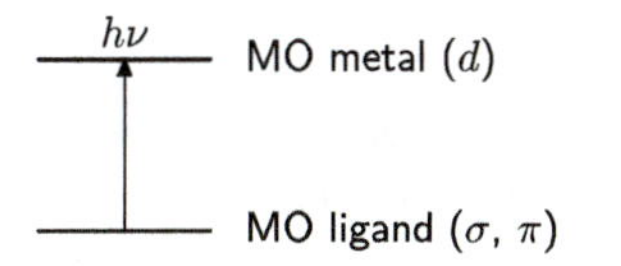

(a) The general transition from a MO located at the ligand to a MO located at the metal [226].

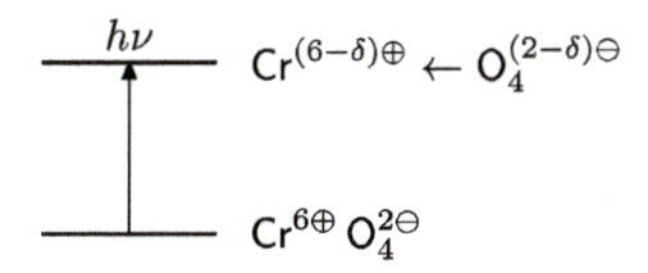

(b) OMCT transition in the chromate ion from the oxide ligand to the chromium (metal). There may be s, p, or d orbitals involved and mixtures such as d-π.

Figure 2.33: Schematic illustration of an LMCT transition.

transfer). OMCT can also occur for metal cations of smaller oxidation numbers as Fe^{III}, Mn^{IV}, or Cu^{II}.

CT can well be described by molecular orbital theory, ►Figure 2.33(a) [229, Chapter 9.5.2, 13], [221, Chapter 6], [232], [226]. As we will see below, a d block metal in a complex anion shares its $3d$, $4s$, and $4p$ orbitals. These orbitals form the symmetry sets t_{2g}/e_g, a_{1g}, and t_{1u} (in octahedral complexes), or e/t_2 and a_1 (in tetrahedral complexes). Ligands participate in the complex, e. g., with $2s$ and $2p$ orbitals of varying symmetry.

Occurrence of mechanism, LMCT colorants

CT can occur for transition metals with considerable ionization energies or in a high oxidation state, i. e., oxidizing agents having unoccupied orbitals of low energy. As a ligand, we can use nonmetals with low electron affinity (potent reducing agents) having occupied orbitals of high energy, which can easily be oxidized. These conditions fulfill complex anions with metal cations in a high oxidation state (V^V, Cr^{VI}, Mo^{VI}, Mn^{VII}, Ti^{IV}, Fe^{III}, Cu^{II}) and oxide ligands. They frequently lead to an absorption located in the UV or blue spectral range, yielding yellow color impression:

- copper compounds: blue to green (due to additional absorptions in the red and yellow spectral range)
- iron(III) oxide hydrate FeOOH (goethite), yellow ocher: yellow
- iron(III) oxide Fe_2O_3 (hematite), red ocher: red (additional absorptions in the yellow spectral range)
- lead chromate $PbCrO_4$ (crocoite), chrome yellow: yellow
- lead molybdate $PbMoO_4$ (wulfenite): yellow; molybdate red $Pb(Cr, S, Mo)O_4$: red
- bismuth vanadate $BiVO_4$ and bismuth vanadate molybdate $4\,BiVO_4 \cdot 3\,Bi_2MoO_6$: yellow
- potassium permanganate $KMnO_4$: purple

Except for potassium permanganate, all the compounds mentioned were or are valued pigments because of their intensive color. Some occur in nature as minerals; others are produced synthetically only in modern times. The heavy halides such as HgI_2, BiI_3, or

PbI_2 can also act as ligands in compounds, but only HgI_2 was shortly employed as a pigment (iodine scarlet in pictures by J. W. M. Turner). In addition, OMCT transitions impart dark red to brown colors to common iron magnesium silicates such as biotites and hornblende [224, p. 132].

Fe^{III} in oxidic environments, such as in hematite and goethite, is also subject to OMCT transitions and is of utmost importance for the coloration of our entire environment. The absorption corresponding to the intense transition of $O^{2\ominus} \rightarrow Fe^{3\oplus}$ occurs in the near-UV spectral range, and its edge absorbs up to the blue spectral range, inducing the observed yellow and red colors. Since iron oxides are present in almost all minerals and rocks as impurities, these are often tinted yellow, red, or reddish-brown. For example, the northern half of Africa, the Sahara, is colored yellow thanks to OMCT transitions in Fe^{III} compounds; the other half, Namib, Kalahari, and all laterite soils, is colored red due to additional EPT transitions in iron oxide. This variability makes iron a valued element for producing pigments, ►Section 3.4.2. We have already discussed LF and EPT transitions in iron in detail in ►Section 2.3.4 at p. 118.

Example: Octahedral complex ML_6, iron oxide pigments

In iron oxide pigments, the coloring cation of iron is octahedrally coordinated. The iron complex can be expressed as either FeO_6 (hematite) or as $FeO_3(OH)_3$ (goethite) [65, Chapters 2, 6, 7], and it possesses a high-spin configuration. Electron counting for $FeO_6^{9\ominus}$ yields $6 \times 8 = 48$ electrons from six oxide ligands and 5 electrons from the iron cation.

►Figure 2.34 [233, p. 116] shows the theoretical MO scheme; a more precise derivation can be found, e. g., in [210, Chapter 15.2], [229, Chapter 1.3.1]. In reality, conditions for iron oxides are much more complicated due to distortions of the crystal lattices [630–633]; in general, the crystal symmetry is lowered in many iron minerals. Consequently, this largely lifts the degeneracy of MOs, and many closely spaced transitions become possible, leading to complicated fine structures in the spectra.

σ System

In addition to the *s* orbital, each of the six oxide ligands contributes one orbital to the σ system, which, e. g., is formed by a linear combination of *p* orbitals (symmetry-adopted linear combination, SALC, [210, Chapter 14.1]). The six *s* and six p_σ orbitals exhibit symmetries a_{1g}, e_g, and t_{1u} in the octahedral point group. Thus, the σ system consists of fully occupied σ and empty σ^* orbitals. σ MOs exhibit a strong ligand character, while σ^* MOs exhibit a strong metal character.

π System

If the ligand can form π bonds to the central metal, e. g., through the two remaining *p* orbitals, these ligand orbitals have symmetry types t_{1g}, t_{1u}, t_{2g}, and t_{2u} in an octahedral field. t_{1u} and t_{2g} orbitals interact with metal *d* and *p* orbitals, and t_{1g} and t_{2u} orbitals

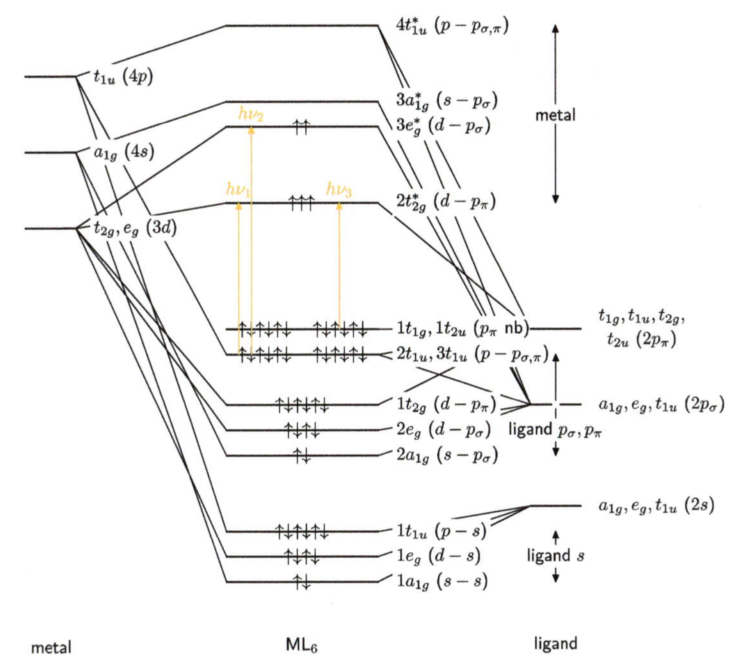

Figure 2.34: Schematic MO diagram of an ML_6-complex [233, p. 116], [234, p. 248], [210, Chapters 14.1, 15.2], [229, Chapter 1.3.1]. Marked in yellow are electronic transitions, qualifying as LMCT transitions. They frequently are of high energy, located in the blue spectral range, and induce a yellow color impression. The diagram shows d^2sp^3 hybrid orbitals typical of *inner orbital complexes* of *d* block metals and describes approximately d^5 high-spin configurations such as Fe^{III} (hematite, iron oxide pigments). The MOs of highest energy possess strong metal character, the ones of lower energy predominantly ligand character.

remain nonbonding. If the ligand is incapable of forming π bonds, the t_{2g} orbitals also remain nonbonding.

Transitions

For ligands like $O^{2\ominus}$, ligand electrons occupy the MOs up to and including the nonbonding MOs. The metal *d* electrons occupy the t_{2g}^{*} MO (possibly also e_g^{*}) and can be transferred to e_g^{*} by incident light, corresponding to a normal ligand-field transition. However, the actual ligand-metal transition occurs with electrons localized in MOs of strong ligand character:

$$
\begin{aligned}
2t_{1u} &\rightarrow 2t_{2g}^{*} && \nu_1 \text{ VIS} \\
2t_{1u} &\rightarrow 3e_{g}^{*} && \nu_2 \text{ VIS} \\
1t_{2u} &\rightarrow 2t_{2g}^{*} && \nu_3\ n \rightarrow \pi^{*} \text{ CT, VIS} \ldots
\end{aligned}
$$

In actual iron oxides, crystal lattices are distorted and level splitting, electronic configuration, and transitions are far more complicated, as shown in detail in [630–633].

Example: Tetrahedral complex ML_4, chrome yellow, chromate, permanganate

The discussion is similar for tetrahedral complexes occurring, e. g., in chromate or permanganate anions For example, after correlating the orbitals of the σ and π systems, we obtain the MO diagram shown in ►Figure 2.35 [233, p. 116]; a theoretical derivation is contained [229, Chapter 1.3.1]. The following transitions are crucial for an intensive color:

$$\begin{array}{ll} 1t_1 \rightarrow 2e^* & \nu_1\ n \rightarrow d\pi^* \text{ CT, VIS} \\ 1t_1 \rightarrow 4t_2^* & \nu_2\ n \rightarrow d\pi^* \text{ CT, VIS} \\ 3t_2 \rightarrow 2e^* & \nu_3\ d\pi \rightarrow d\pi^* \text{ CT, VIS} \\ 3t_2 \rightarrow 4t_2^* & \nu_4\ d\pi \rightarrow d\pi^* \text{ CT, VIS} \end{array}$$

Ballhausen [617], Hillier and Saunders [618], Johnson and Smith [619], Johansen [620], Mortola et al. [621], Hsu et al. [622], Connor et al. [623], Johansen and Rettrup [624], Jitsuhiro et al. [625], Wolfsberg and Helmholz [626], Millier et al. [627] describe actual level splittings and electronic configurations in the chromate or permanganate anion, essentially corresponding to the scheme. Due to the slight energy difference between the HOMO group ($1t_1$, $2t_2$, $3t_2$, $2a_1$) and the LUMO group ($2e^*$, $4t_2^*$, $3a_1^*$), almost all other HOMO-LUMO transitions can occur in addition to the four transitions mentioned above.

2.4.2 Metal-to-metal transition (MMCT), intervalence transition (IVCT)

Metal-to-metal transitions (MMCT) are electron transitions taking place between two *different metal* cations. In the case of intervalence transition (IVCT), the same metal is present in *various oxidation states*, and electrons can move from the lower to the higher oxidized metal cation [229, Chapter 13].

Heteronuclear intervalence transition or metal-to-metal transition

The charge transfer occurs between two different metal cations. The classic example is sapphire, in which charge flows from Fe^{II} to Ti^{IV}:

$$Fe^{2\oplus} + Ti^{4\oplus} \xrightarrow{2.11\,\text{eV}} Fe^{3\oplus} + Ti^{3\oplus}$$

In sapphire, $Fe^{2\oplus}$ and $Ti^{4\oplus}$ replace some $Al^{3\oplus}$ cations in the crystal lattice of corundum. They can occupy adjacent positions in the octahedral coordination sphere around alumina (which become somewhat distorted as a result) and approach up to a distance of about 0.265 nm below which the d_z orbitals of both cations overlap. This overlap facilitates an electron transfer when the energy of 2.11 eV is applied, e. g., by the yellow light of 588 nm (photochemical oxidation). The perceived color is a deep sapphire blue.

The pair Fe^{II}/Ti^{IV} also gives color to other blue, brown, or black minerals, such as extraterrestrial pyroxenes $(Mg, Fe^{II}, Mn, Ti^{IV})(Mg, Mn, Fe^{II})(Si, Al, Fe^{III})_2O_6$ or kyan-

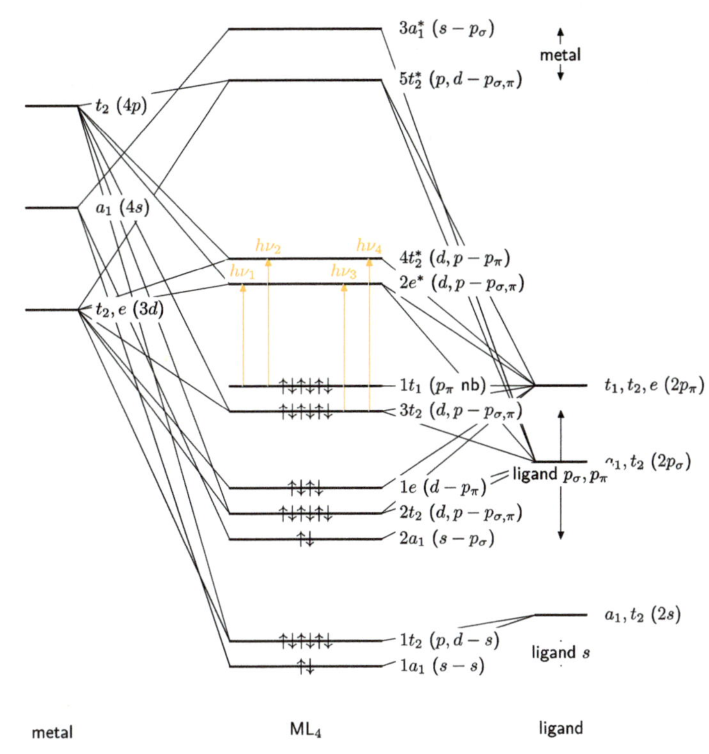

Figure 2.35: Schematic MO diagram of an ML_4 complex [233, p. 116], [234, p. 248], [229, Chapter 1.3.2]. Marked in yellow are electronic transitions, qualifying as LMCT transitions. They frequently are of high energy and located in the blue spectral range, inducing a yellow color impression. The diagram is applicable for d^0 configurations such as Cr^{VI} (chromate) and Mn^{VII} (permanganate).

ite Al_2SiO_5 [224, p. 115], [223, p. 62]. $Fe^{2\oplus}$ and $Ti^{4\oplus}$ are present as admixtures in these minerals. Since CT transitions are intense, even small quantities of iron and titanium produce a distinct color. For example, compared to LF transitions, sapphires require only a per mil admixture to exhibit an intensive blue color. At the same time, chromium in the percentage range is necessary to color a ruby via LF transitions intensively. Nevertheless, the color of the ground stones is too weak to have made an appearance as colorants.

Homonuclear IVCT transition

In homonuclear CT, the charge is exchanged between two cations of the same element but in different oxidation states. A classic example is the pair $Fe^{2\oplus}/Fe^{3\oplus}$ in Prussian blue and magnetite:

$$Fe_A^{2\oplus} + Fe_B^{3\oplus} \longrightarrow Fe_A^{3\oplus} + Fe_B^{2\oplus}$$

Different chemical environments A and B must surround the cations to induce chromaticity. Otherwise, initial and final states are indistinguishable, and they would be equal in energy; hence no transition would occur. There are numerous examples of compounds in which mixed-valent iron Fe^{II}/Fe^{III} is responsible for intense colors [224, p. 115], [223, p. 58]:
- Prussian blue (intensive blue)
- brown bottle glass (well-known brown)
- allochromatic (colorless) minerals colored by iron

The latter are of interest since some colorants are among them:
- Blue alkali amphiboles such as glaucophane or riebeckite of general composition $Na_2(Mg,Fe^{II})_3(Al,Fe^{III})_2[(OH)_2|Si_8O_{22}]$ served as blue pigments in early times, ►Section 1.2.1 and below.
- Green earth, a mixture of celadonite and glauconite of varying composition $(K,Na)(Al,Fe^{III},Mg)_2(Si,Al)_4O_{10}(OH)_2$, has been a green pigment since early times, ►Section 1.2.1, ►Section 3.4.2.3 at p. 248 and below. The green color origins by mixing a blue color (caused by IVCT) with a yellow one (caused by LF transitions in $Fe^{3\oplus}$).
- Vivianite $Fe_3^{II}(P_1O_4)_2$, lazulite, and chlorite $(Fe^{II},Fe^{III},Mg,Al)_6(Si,Al)_4O_{10}(OH)_8$ were sporadically used as a blue pigment in ancient times.
- Aquamarine $Be_3Al_2Si_6O_{18}$, blue and green tourmalines (blue).

Cation pairs such as Fe^{II}/Fe^{III} are constituents of the mineral, impurities, or minor components. In the latter case, the achieved color intensity is usually low. In the former case, various polyvalent metals are responsible for the intensive color of some idiochromatic oxidic minerals:
- magnetite (iron(II,III) oxide) Fe_3O_4 or $Fe^{II}Fe_2^{III}O_4$ (black)
- red lead (lead(II,III) oxide) Pb_3O_4 or $Pb_2^{III}Pb^{II}O_4$ (orange-red)
- manganese(II,III) oxide Mn_3O_4 or $Mn^{II}Mn_2^{III}O_4$ (black)

Although not used as colorants, other examples are molybdenum blue and tungsten blue. They are colloidal, hydrated, partially reduced oxides of molybdenum and tungsten, in which the various oxidation states of the metals cause an IVCT transition according to

$$Mo_A^{V} + Mo_B^{VI} \longrightarrow Mo_A^{VI} + Mo_B^{V}$$

A and B are crystallographic different chemical environments for the metal cations. The transition occurs in the near-infrared; therefore, intensive blue color results. The arrangement of metal oxide polyhedra ensures the necessary spatial proximity of the metal atoms.

The enumeration shows that the IVCT mechanism provides natural colorants with blue and green shades that have been in use since ancient times, along with copper pigments. Prussian blue was among the first and one of the most important synthetic blue pigments of modern times.

An exciting and colorful example is the mineral vivianite, which corresponds to a colorless iron(II) phosphate when freshly precipitated. In the air, Fe^{II} is partially oxidized gradually to Fe^{III}. Since this process takes some time, the conditions for IVCT are fulfilled as long as both oxidation states coexist, and a color change from green to blue to black-blue and back takes place until the final color of Fe^{III} salts is perceived:

$$\underset{\text{fresh colorless}}{Fe_3^{II}(PO_4)_2 \cdot 8\,H_2O} \xrightarrow{O_2} \text{green}\rightarrow\text{blue}\rightarrow\text{blackblue}\rightarrow\text{blue}\rightarrow\text{green} \xrightarrow{O_2} \underset{\text{aged yellow}}{Fe_3^{III}(PO_4)_2(OH)_3}$$

Since hydroxide anions are available in the oxidized form, both oxidation states of iron are present in a different chemical environment. Thus, the diversity condition also is fulfilled. A similar process occurs during manganese(II) sulfide oxidation. Freshly precipitated, it shows a pink color, then evolves to brown.

Another noteworthy example is magnetite, which absorbs light so intensely that it appears black and opaque. Magnetite consists of endless chains of edge-linked iron oxide octahedra. Electrons in these chains can not only switch between two particular iron ions but also move to the neighboring ion in the chain at a time. This movement establishes an electron system delocalized across the chain, having a variety of excited states. Thus, magnetite absorbs light of any wavelength, rendering it black. Similarly, many minerals with chain- or band-linked iron oxide octahedra are brown to black.

Example: Prussian blue

From 1700 onward, a cheap blue pigment of intensive color expanded the artist's palette radically, colored by IVCT. This blue pigment was Prussian blue or Berlin blue, being very popular at that time and named after the location of its discovery, ►Section 3.7.

$$Fe_4^{III}[Fe^{II}(CN)_6]_3 \cdot 16H_2O$$

has an intensive blue color because it contains divalent and trivalent iron in the same molecule. In the crystal lattice, the chromogen $Fe^{II}{-}C{-}N{-}Fe^{III}$ is present. Both $Fe^{2\oplus}$ and $Fe^{3\oplus}$ cations are octahedrally coordinated so that a ligand field splitting of iron *d* orbitals occurs [423], [385, p. 298]. While $Fe^{2\oplus}$ is surrounded by six cyanide carbon atoms and forms a low-spin complex (environment A, ►Figure 2.36 on the left), $Fe^{3\oplus}$ is surrounded by cyanide nitrogen and crystal water oxygen (environment B, ►Figure 2.36 on the right). B creates a weaker ligand field and smaller ligand field splitting than A, forming high-spin complexes.

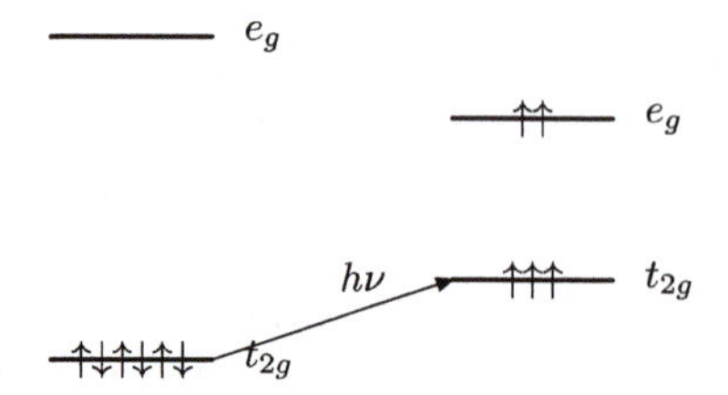

Figure 2.36: Level diagram of iron *d* orbitals in Prussian blue [423], [385, p. 298]. Left: Low-spin complex formed by Fe^{II} and cyanide ligands bonded via carbon. Right: High-spin complex formed by Fe^{III} and cyanide ligands bonded via nitrogen.

In the arrangement described above, both kinds of iron cations are placed in different chemical environments A and B. However, they are spatially close enough to interact: irradiation with low-energy light, e. g., red light initiates the transfer of an electron from the low-spin t_{2g} state to the high-spin t_{2g} state:

$$(t_{2g})^6_A \cdot (t_{2g})^3_B (e_g)^2 \rightarrow (t_{2g})^5_A \cdot (t_{2g})^4_B (e_g)^2$$

This intense transition imparts a distinct blue color to the compound.

Example: Green earth

Like Prussian blue, green earth (►p. 248) appears green due to an IVCT transition from the t_{2g} state of Fe^{II} to the t_{2g} state of Fe^{III}, which is of higher energy:

$$t_{2g}(\mathrm{Fe^{II}}) \rightarrow t_{2g}(\mathrm{Fe^{III}})$$

Although this CT transition also induces a blue color, the overall color impression of green earth is green since the blue color mixes with a yellow color caused by LF transitions within the Fe^{III} cation. Depending on the ratios of Fe^{II} and Fe^{III}, the color tends to be more blueish or yellowish-green. Hence, frequently the plural “green earths” is used since there is no particular green earth or composition as reference.

Example: Blue amphiboles

In early times, especially in the eastern Mediterranean, the Aegean, Minoans, and Mycenaeans used blue to blue-gray amphiboles as blue fresco pigments [482–486]. Alkali amphiboles have the general composition $\mathrm{Na_2(Mg, Fe^{II})_3(Al, Fe^{III})_2[(OH)_2|Si_8O_{22}]}$, and the family includes some blueish minerals:

Ferro-glaucophane	$Na_2Fe^{II}_3Al_2[(OH)_2|Si_8O_{22}]$
Glaucophane	$Na_2Mg_3Al_2[(OH)_2|Si_8O_{22}]$
Riebeckite	$Na_2Fe^{II}_3Fe^{III}_2[(OH)_2|Si_8O_{22}]$
Magnesio-riebeckite	$Na_2Mg_3Fe^{III}_2[(OH)_2|Si_8O_{22}]$

SiO_4 octahedra are edge-linked to double rows of infinite length in the amphibole structure. Mg and Al, respectively, are located in typical octahedrally coordinated environments, crystallographically called M1, M2, and M3. In particular, we can immediately see that the presence of mixed-valent iron Fe^{II} and Fe^{III} (which are regular constituents of riebeckite) and its distribution to M1, M2, and M3 causes coloration by IVCT transitions, inducing a blue color impression:

$$Fe^{II}_{M1} + Fe^{III}_{M2} \xrightarrow{\beta} Fe^{III}_{M1} + Fe^{II}_{M2} \quad \text{and} \quad Fe^{II}_{M3} + Fe^{III}_{M2} \xrightarrow{\beta} Fe^{III}_{M3} + Fe^{II}_{M2}$$

$$Fe^{II}_{M1} + Fe^{III}_{M2} \xrightarrow{\gamma} Fe^{III}_{M1} + Fe^{II}_{M2}$$

In fact, along with Egyptian blue, riebeckite was part of the basic inventory of Minoan painters. The other members of the amphibole family mentioned above never occur naturally in their ideal chemical composition but always contain Fe^{II} or Fe^{III} as impurities or minor constituents. Thus, glaucophane and magnesio-riebeckite also may exhibit blue colors caused by IVCT. The older literature identified glaucophane as another blue pigment besides Egyptian blue, [486] argues it is perhaps rather magnesio-riebeckite than glaucophane.

The IVCT transitions responsible for the blue color mentioned above occur at about 540 nm (β) and 620 nm (γ) in the yellow and red spectral range, respectively [223], [224, p. 124], [345]. The intensity of the color is proportional to the concentrations of Fe^{II} and Fe^{III}, respectively, so that the color intensity depends, as expected, on the degree of iron admixture.

Interestingly, the mentioned minerals show pleochroism, i. e., their color depends on the polarization of light. The explanation is that the crystal structure spatially orients each transition, and absorption depends directly on the direction of polarization of the incident light. The transitions labeled β orient along the b-axis (mineral appears purple), and the γ transitions along the c-axis (mineral appears blue). Since no transition can be excited along the a-axis, the mineral appears colorless.

2.5 MO: molecular orbital transitions

The mechanisms of chromaticity discussed so far have been characterized by the fact that they occurred primarily in inorganic materials. They were essentially confined to one or a few atoms or—in the case of CT—parts of an organic molecule, making it relatively easy to identify and describe the chromophore. In contrast, we can frequently describe organic colorants and some inorganic ones better by including many, if not all, atoms in the molecule. Color then results from the transition between *molecular orbitals*. Unfortunately, we cannot provide a precise mathematical description in closed form, and numerical approximations require a vast amount of computational power. Therefore, we frequently work with models, of which there is quite a number, and we can provide only a glimpse on this subject. In particular to [3–5, 203, 204, 206, 207, 209, 211, 368] provide further details.

Colors caused by MO transitions have the following characteristics:

- The (theoretically) narrow-band absorptions lead to pure hues.
- The absorption bands' position is freely selectable by suitable chemistry, i. e., all colors besides white are possible. However, *two* absorption bands are required to obtain middle green colorants.
- The absorption can be very intense and induce intensive colors.

MO transitions occur in the most organic and some inorganic colorants, e. g., in the sulfide anions $S^{3\ominus}$ or $S^{4\ominus}$ (ultramarines), but there is no sharp distinction. Other transition types such as CT can occur in organic compounds besides the MO ones. In metal complexes, MOs appear next to inner-metal LF transitions. The complete molecule or only parts of it can participate in MOs, while other parts of the electron system can be more confined, localized, or isolated and subject to CT. From the MO point of view, some MOs are confined to parts of a molecule (or localized) while others are not, or they primarily have σ-character and others mainly *d* orbital character, or some MOs are predominantly ligand MOs or metal MOs.

Molecular orbitals

The concept of *molecular orbitals* is an approach in theoretical chemistry that extends the notion of "orbital" from a single atomic orbital (AO) to an orbital in which large parts or the whole molecule participate. This approach was already developed in the first half of the last century, starting with the LCAO one. Since then, theoretical and numerical methods have evolved considerably [5].

To see why the formation of molecular orbitals (MO) can cause chromaticity, consider a highly simplified picture in which two atoms with *s* and *p* AOs form a molecule. The electrons can move in regions that include *both* atoms, ►Figure 2.37.

Due to a sufficient overlap of the s orbitals in the molecular axis direction, they form a binding σ_s MO and an antibinding σ_s^* MO with higher energy, the energy difference being $h\nu_{s1}$. Also, the one *p* orbital propagating along the molecular axis can form a σ bond, from which another pair of binding σ_p and antibinding σ_p^* MOs follows, this time with an energy difference of $h\nu_{s2}$. Since the overlap integral (the common region

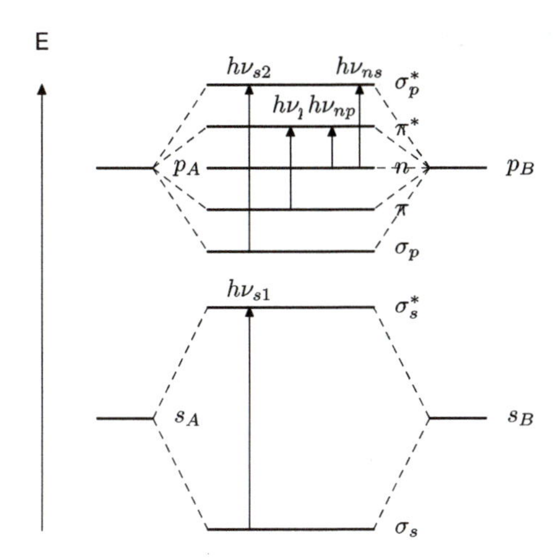

Figure 2.37: Formation of MOs from s and p orbitals of two atoms A and B, and the possible radiation-induced transitions between the σ and π MOs. In this example, A and B form a diatomic molecule with a double bond between them, consisting of σ_p and π, e. g., from p_z and p_x. However, the remaining p orbital is not involved in bonding and remains a nonbonding MO.

between the atoms) is smaller for p orbitals, the splitting energy $h\nu_{s2}$ between σ_p MOs is smaller than $h\nu_{s1}$, the energy difference between σ_s MOs.

The remaining p AOs can also overlap, but not along the molecular axis, but within a plane containing this axis. If a double bond links the atoms, only one of the remaining p AOs forms a π bond, and we obtain one binding π and one antibinding π^* MO. The overlap integral for π bonds is much smaller than for σ bonds, so the energy difference $h\nu_p$ between the π MOs is smaller than for the σ_p MOs.

In the example of a double bond, the remaining p orbital of each atom does not participate in bonding at all, and they form a nonbonding MO n. Often, these are free electron pairs of oxygen, sulfur, or nitrogen. Therefore, nonbonding orbitals are not lowered in energy.

The situation described in ►Figure 2.37 provides several possibilities for light-induced electronic transitions. Depending on the electronic occupation of the MOs, we can transfer electrons from the binding to the antibinding MOs ($\sigma \rightarrow \sigma^*$ transitions with energies $h\nu_{s1}$ and $h\nu_{s2}$ as well as $\pi \rightarrow \pi^*$ transition with energy $h\nu_p$). Nonbonding electrons can also be transferred to σ^* or π^* orbitals ($n \rightarrow \sigma^*$ or $n \rightarrow \pi^*$ transition with energies $h\nu_{ns}$ and $h\nu_{np}$). Since n orbitals occur between binding and anti-binding MOs, the necessary energies for transitions to σ^* or π^* orbitals are smaller than the other energies. ►Figure 2.38 depicts the typical locations of all these electronic transitions in a spectrum. The energy difference is lowest between HOMO (highest occupied MO) and LUMO (lowest unoccupied MO), so HOMO-LUMO transitions are natural candidates for optical transitions.

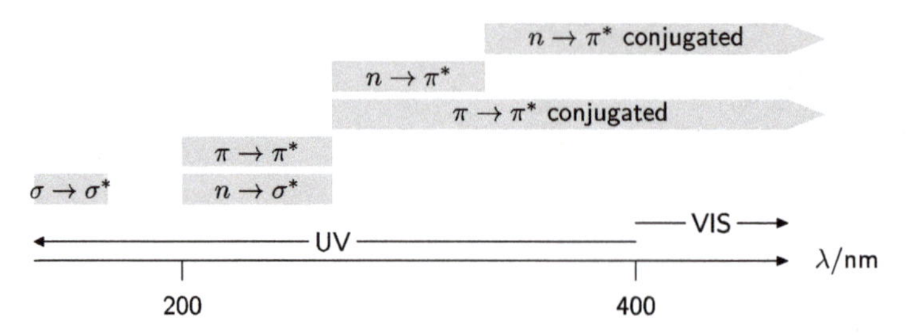

Figure 2.38: Schematic location of main types of electronic transitions between molecular orbitals (drawn after [211]). Most of these transitions are too high in energy to cause absorptions in the optical spectral range. The transitions can be shifted to lower energies (higher wavelengths) by conjugation, the so-called bathochromic shift. One goal for color chemists is locating intense transitions in the visible spectral range to induce intensive colors.

$\sigma \to \sigma^*$ transition

The energy difference between the σ MOs is so large that the $\sigma \to \sigma^*$ absorption band occurs in the UV, and we do not perceive any color. For this reason, most organic compounds are colorless. However, the absorption causes them to be sensitive to UV light, which destroys them by cleavage of σ bonds.

A $\sigma \to \sigma^*$ absorption band located in the UV can be so intense, i. e., its bandwidth can be so large, that the absorption extends into the visible spectral range, reducing violet and blue spectral parts. Thus, many organic compounds are perceivably yellowish tinted even if they do not contain chromophores.

$n \to \pi^*$ transition

Due to the small energy difference between n and π^* MOs, the $n \to \pi^*$ transition absorbs in the optical or near-UV spectral range. Known $n \to \pi^*$ chromophores are:

- Carbonyl groups $R_2C{=}O$, the nonbonding electron pairs of oxygen reducing the energy gap between π and π^* MOs.
- Azo groups NR=NR, the nitrogen provides nonbonding electron pairs, as does the doubly bonded nitrogen of the imino group =C=N–.

We encounter these chromophores in numerous organic compounds, some of which exhibit beautiful colors. However, their extinction coefficients are low due to the low transition probability of the $n \to \pi^*$ transition. The carbonyl chromophore unintentionally develops in varnish aging, painting materials, and paper, imparting a yellow tint to the material, ►Sections 6.7.10 and 7.4.8.3.

The carbonyl and the azo group are essential components of colorants, ►p. 156 and 161. However, in these cases, they do not impart color through the $n \to \pi^*$ transition but via their ability to act as electron acceptors in $\pi \to \pi^*$ transitions.

$\pi \rightarrow \pi^*$ transition

Due to the relatively small splitting of the π MOs, a $\pi \rightarrow \pi^*$ absorption band appears in the near-UV spectral range. Molecular structures with such an absorption only nearly miss being colored. Of particular importance is the double bond $R_2C{=}CR_2$, acting via a $\pi \rightarrow \pi^*$ transition. Either it is the principal chromophore in a colorant in its own right, or it shifts the absorption bands of another chromophore to longer wavelengths by creating or extending their conjugated electronic systems. To entirely shift the absorption of the $\pi \rightarrow \pi^*$ chromophore into the visible range, we have two options: adding auxochromes or enlarging the conjugated electronic system, gaining enough resonance energy and decreasing the energy difference between π and π^* MOs.

Auxochrome substituents

Auxochromes change the electronic structure of a molecule and the transition energies between some MOs. *Auxochromes* are electron donors whose free electron pair interacts with the HOMO of a chromophore and raises its energy therefore decreasing the transition energy between HOMO and LUMO from $h\nu_1$ to $h\nu_2$, ►Figure 2.39.

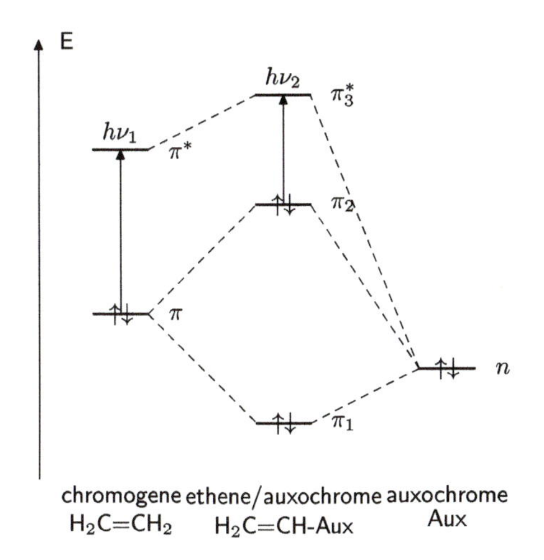

Figure 2.39: Bathochromic shift of $\pi \rightarrow \pi^*$ electronic transitions by auxochromes –OR, –SR, $-NR_2$, –Hal (drawn after [211]). The auxochromes contribute a nonbonding MO n, which interacts with a π MO from the chromophore, yielding two new MOs, π_1 and π_2. π_2 has higher energy than the original π MO, and therefore the transition energy $h\nu_2$ is smaller than $h\nu_1$.

Conjugation

When enlarging a π system with more and more conjugated π bonds, we increase the energy of its HOMO, decrease the energy of its LUMO, and thus also decrease the energy of the $\pi \rightarrow \pi^*$ transition to $h\nu_4$, ►Figure 2.40. Designers of chromophores frequently use this possibility (see, e. g., polyenes and polymethines) ►Sections 2.5.4

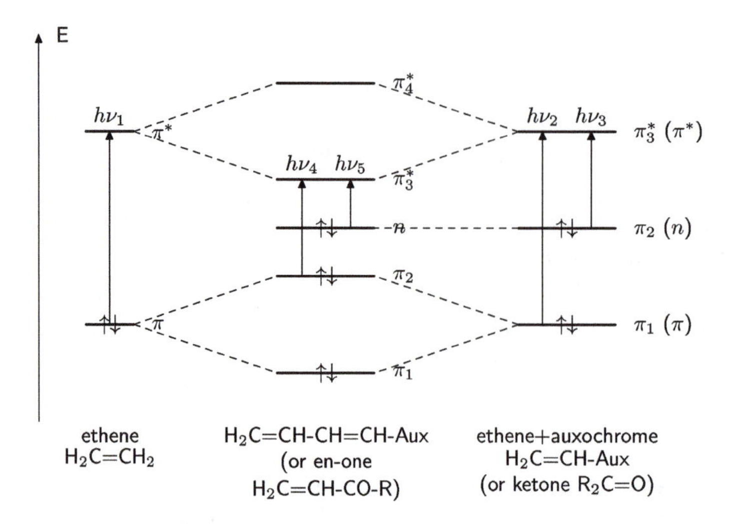

Figure 2.40: The additional bathochromic shift of $\pi \rightarrow \pi^*$ transitions (ν_1, ν_2) and $n \rightarrow \pi^*$ transitions (ν_3) by conjugated double bonds (drawn after [211]). Each additional ethene unit increases the HOMO (from π to π_2) and decreases the LUMO (from π^* to π_3^*). ν_4 and ν_5 are smaller than the original frequencies.

and 2.5.5. Thus, large conjugated π systems are common structural elements in organic colorants.

2.5.1 VB and MO model, resonance structures

We have so far considered MO diagrams of colorants. The valence bond (VB) theory discusses chromophores in terms of resonance structures. VB theory is, besides MO theory, the other theory of explaining chemical bonding in quantum mechanical terms. It focuses on AOs, combining and forming chemical bonds. According to VB theory, covalent bonds form by overlapping two half-filled AOs, each contributing one electron to the bond. The bonds, along with lone electron pairs, are depicted as lines in the classical Lewis structures such as H–$\underline{\text{O}}$–H. Each line depicts a shared pair of electrons or a lone electron pair.

VB theory describes compounds that a single Lewis structure cannot express by employing *resonance* or *mesomerism*. Resonance must be applied when electrons cannot be located between particular atoms, such as the double bond in the carbonate anion $CO_3^{2\ominus}$. VB theory uses two or more *contributing*, *canonical*, or *resonance structures*, forming a *resonance hybrid*. Their contribution to the hybrid can vary, depending on stabilizing and destabilizing factors in each resonance structure.

Ultimately, both Mo and VB theory describe different aspects of the same problem, ►Figure 2.41. Colorants, having resonance structures of similar energy, are colored because the resonance structures can be transformed into each other by irradiation

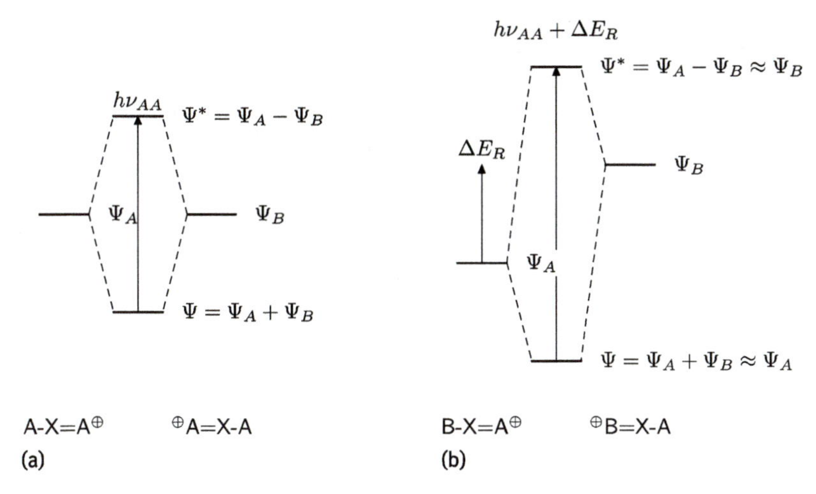

Figure 2.41: Correlation between resonance structures (VB theory) and MOs (drawn after [4, Chapter 4.2]). (a): Both resonance structures are equal in energy. In MO theory, two new MOs, Ψ and Ψ^*, evolve from the interaction of the MOs Ψ_A and Ψ_B, describing individual resonance structures A and B. The energies of the new MOs differ by an amount of $h\nu_{AA}$, allowing an electronic transition. (b): The resonance structure with cationic A is preferred, i. e., more stable, and the energy of its MO Ψ_A is reduced by an amount ΔE_R. The energy difference between the new MOs increases by that amount.

with light. The incident light provides the energy, and the absorption induces color. Depicted in VB theory, the process looks like this:

$$\underset{\text{resonance structure 1}}{\text{A—X=A}^{\oplus}} \quad \xleftrightarrow{h\nu_{AA}} \quad \underset{\text{resonance structure 2}}{\text{A}^{\oplus}\text{=X—A}}$$

►Figure 2.41(a) shows the MO representation. The compound is not represented by either A–X=A$^{\oplus}$ or A$^{\oplus}$=X–A but by a superposition of both structures, i. e., the formation of new MOs, Ψ and Ψ^*. An electronic transition of energy $h\nu_{AA}$ becomes possible between them. The more resonance structures a compound possesses, the smaller the energy differences between Ψ (the HOMO) and Ψ^* (the LUMO) becomes, i. e., the transition is shifted toward the optical spectral range; the compound gains color.

In the example above, both resonance structures were equal in energy since A–X=A$^{\oplus}$ is identical to A$^{\oplus}$=X–A. In reality, however, one of them is almost always preferred by an energy ΔE_R, assuming B–X=A$^{\oplus}$ compared to B$^{\oplus}$=X–A. In this case, ΔE_R must be applied additionally to $h\nu_{AA}$. Therefore, the absorption is shifted toward the UV range and appears at higher energies (shorter wavelengths). ►Figure 2.41(b) depicts the corresponding MO representation.

2.5.2 Chromophore enlargement, bathochromic shifts

As we have seen, the presence of a single $\pi \rightarrow \pi^*$ chromophore is usually not sufficient to yield colored compounds. In most cases, a bathochromic shift of the UV absorption band into the optical spectral range is necessary. We can combine auxochromes with conjugation, achieving bathochromic shifts large enough to place the resulting absorption in the optical spectral range. In this way, we obtain typical structural units that frequently occur as chromophores in MO-based colorants,►Figure 2.42:

- Donor–acceptor systems comprising electron donors, acceptors, and conjugated bridges are fundamental structural units of many natural and synthetic colorants. The chromogenic effect origin in the transition between donor and acceptor, the bridges contribute to a bathochromic shift.
  We can distinguish simple, localized acceptors from complex acceptors. For example, the isolated carbonyl group is a simple acceptor in many colorants, e. g., indigoid colorants. Complex acceptors include donor-substituted quinones and diazo compounds. In the case of quinones, the carbonyl group is part of a more extensive conjugated system and, therefore, not classified as "simple."
- Compounds with an odd number of atoms and MOs in a conjugated π system possess transitions between a nonbonding MO(NBMO) and the LUMO. This structural unit includes the essential natural and synthetic colorants based on oxonol and cyanine systems.
- Compounds with an even number of atoms and MOs in a π system belong to the polyenes, in which the HOMO-LUMO transition induces color. In addition to linear polyenes, this group includes cyclic annulenes and polycyclic hydrocarbons.

In all of these classes, $n \rightarrow \pi^*$ transitions in carbonyl groups do not contribute to the coloration.

2.5.3 Donor–acceptor chromophores

The most common chromophores in colorants include the types:

$$\text{D–B–A (simple)}, \quad \text{D–A (complex)},$$

which comprise an electron donor D, an electron acceptor A and a bridge B of multiple conjugated bonds [4, Chapter 6, 7]. The excitation energy of a donor electron determines the color of this chromophore:

$$\underset{\text{simple acceptor A}}{\text{D–B–A}} \xrightleftharpoons{h\nu} \text{D}^{\oplus}\text{=B=A}^{\ominus} \qquad \underset{\text{complex acceptor A}}{\text{D–B–A}} \xrightleftharpoons{h\nu} \text{D}^{\oplus}\text{=B}^{\delta-}\text{=A}^{\delta-}$$

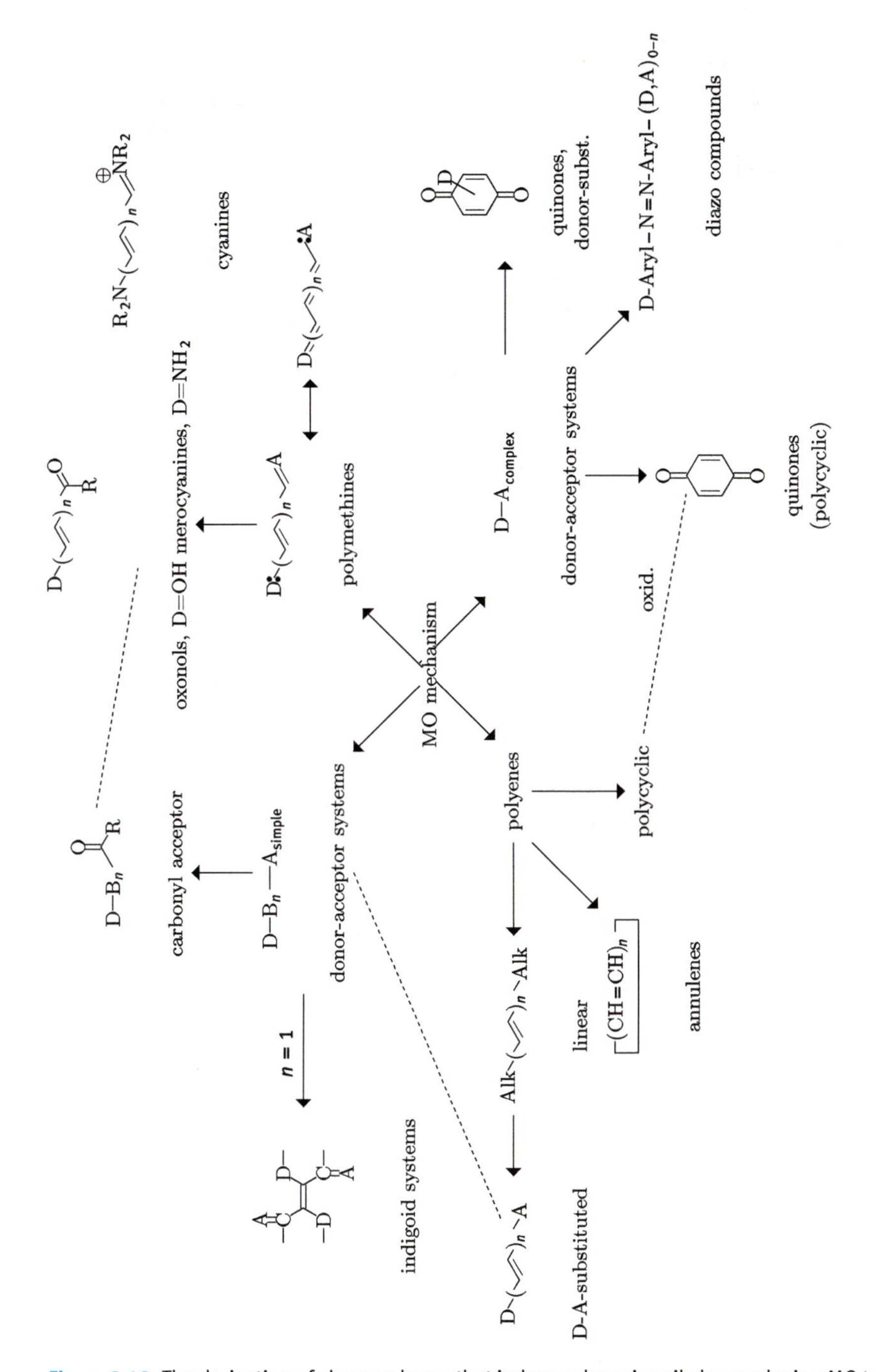

Figure 2.42: The derivation of chromophores that induce color primarily by employing MO transitions. The prominent representatives are π systems, systems with azo groups, and carbonyl groups with π systems. The dashed lines mark equivalent chromophores. B = bridge, B_n = bridge, which is equivalent to n $CH_2{=}CH_2$ units.

If A is a simple acceptor, we can designate distinct parts of the molecule as being this acceptor, absorbing the charge in the excited state. However, suppose this designation is not possible. In that case, we are dealing with a complex acceptor, distributing the charge over extended parts of the molecule (possibly including the bridge) in such a complicated way that we cannot depict it as a simple negative charge.

The excitation frequency of the chromophore D–A usually appears in the UV; therefore, the conjugated system of the bridge and the complex acceptor is essential. Both contribute to a bathochromic shift, moving the absorption bands of the chromophore into the optical spectral range, as we can see from this series of compounds:

H_2N-NO_2 nitroamine colorless

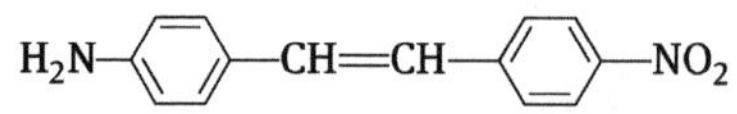

H_2N–C₆H₄–CH=CH–C₆H₄–NO_2 1-p-aminophenyl-2-p-nitrophenyl-ethene, simple acceptor

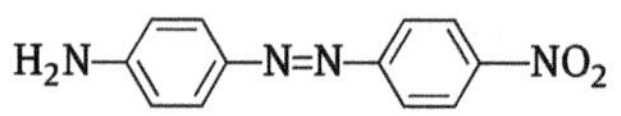

H_2N–C₆H₄–N=N–C₆H₄–NO_2 p-amino-p'-nitro-azobenzene complex acceptor

Although nitroamine possesses a strong donor (the amino group) and a strong acceptor (the nitro group), the absorption does not appear in the visible spectral range. Incorporating a polyene bridge comprising two phenyl rings and a vinyl unit shifts the absorption into the optical spectral range. In both cases, the nitro group is a simple electron acceptor. In the case of the azobenzene, we encounter a complex acceptor since both nitrogen atoms of the azo bridge and the nitro group participate in the charge distribution in the excited state.

Donors are atoms or atom groups with a lonely electron pair in a nonbonding orbital of high energy. This orbital can overlap with the π system and is therefore involved in the formation of MOs. (In contrast, the lonely electron pair of an isolated nonbonding AO would only give rise to the localized $n \rightarrow \pi^*$ transition discussed above). Frequently oxygen, sulfur, or nitrogen atoms deliver the lonely electron pair. Important donors, ordered by strength, are [4, Table 6.1]:

$$-O^{\ominus}, -NR^{\ominus}, -S^{\ominus} > -N(CH_3)_2 > -NHCH_3 > -SCH_3 > -NH_2 > -SH > -OCH_3$$
$$> -NHCOCH_3 > -OH > -OCOCH_3$$

A system of multiple conjugated bonds arranged in chains or rings plays the role of the *bridge*. The wider this system is, the larger the achieved bathochromic shift. The absolute magnitude depends on the location of donors and acceptors and on the arrangement of the bridge itself. For example, the longitudinal extension of the bridge is more effective than the lateral arrangement:

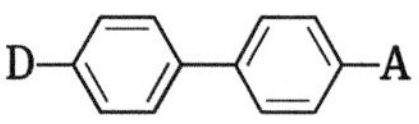

D–C₆H₄–C₆H₄–A more effective than D–C₁₀H₆–A (naphthalene)

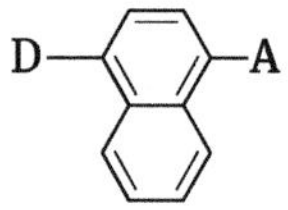

When isolated, the parts of the bridge that are not participating in aromatic systems show distinct differences between single and double bonds. However, any differences in bond lengths will be equalized as soon as the bridge is part of alternating resonance structures. In VB theory, each bond would be a single or double bond, yielding hybrid bonds with aligned lengths. The polyenes discussed below illustrate this.

The *acceptor* is composed of two atoms that are connected by multiple bonds and of which the terminal atom has a higher electronegativity than carbon. Important acceptors in falling strength are [4, Table 6.2]:

$$-SO_2CF_3 > -NO_2 > -SO_2CH_3 > -CN > -COCH_3 > -COOH > -SOCH_3 > -COOCH_3$$
$$> -CONH_2 > -CHO > -NO > -COO^{\ominus}$$

In the colorants we consider in this text, simple donor–acceptor systems have either carbonyl groups as acceptors or are indigoid chromophores. Donor-substituted quinones and donor-substituted diazo compounds are, in contrast, complex acceptors. Both types of acceptors are of great importance for natural and synthetic colorants.

2.5.3.1 Simple D-A chromophore with carbonyl acceptor

The carbonyl group is a common acceptor in donor–acceptor systems, frequently combined with a polyene as the conjugated bridge [4, Chapter 6.2]:

R, D, O — $h\nu$ ⇌ — R, $\oplus$ D, $O^{\ominus}$

I
D=OH: oxonol,
D=NH_2: merocyanine

II

In this system, an asymmetric charge distribution exists. Already in the ground state, structure II with a concentration of negative charge at the carbonyl oxygen is present to a certain amount. In the excited state, the share of structure II distinctly increases. We still observe an alternation between single and double bonds in the bridge with increasing bridge length, but the bond orders show much greater uniformity in the excited state. We also recognize a convergence of the bathochromic shift instead of a steady rise, even when we employ strong donors. However, depending on the bridge length and the donor, this system can induce an absorption over the whole optical spectral range, i. e., yield any color from yellow to blue-green.

We obtain the maximum bathochromic shift of the HOMO-LUMO transition with symmetrical electron distribution in the ground state, i. e., a certain amount of the neutral structure I and the ionic structure II. We do not achieve maximum shift by employing the most potent donor and carbonyl acceptor since this would only establish structure II as a strongly ionic ground state.

Instead, the strength of the donor and carbonyl acceptor must match so that we achieve a high degree of electronic symmetry. The maximum bathochromic shift we can reach is about 80 nm per vinyl unit in the bridge. (The highest symmetry, for the shown example, is obtained with an oxonol, i. e., when the donor atom matches the carbonyl acceptor. However, then donor and acceptor are no longer distinguishable, and we are discussing a polymethine. Its electronic structure differs strongly from donor–acceptor systems, e. g., it comprises a nonbonding MO and maintains equivalence of single and double bond lengths, ►Section 2.5.5.)

2.5.3.2 Indigoid chromophores

Indigo is a classic colorant, having a frequently discussed chromophore. The starting point was the observation that the small indigo molecule shows a remarkable massive bathochromic shift up to the blue spectral range, which is otherwise only achievable with considerably more extended molecules or strong donor groups.

The history of this debate is fascinating and spans many decades of chemical research, during which numerous theories tried to explain the cause of the intensive color, ►Section 4.7.4 [450–452]. Chemists reduced the chromophore more and more until they identified structure I as chromophore, the so-called *H-chromophore* [4, Chapter 7.9]:

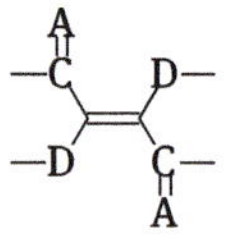

I, H-chromophore 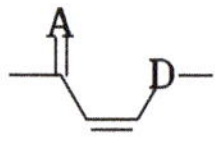   II, pentacyanine 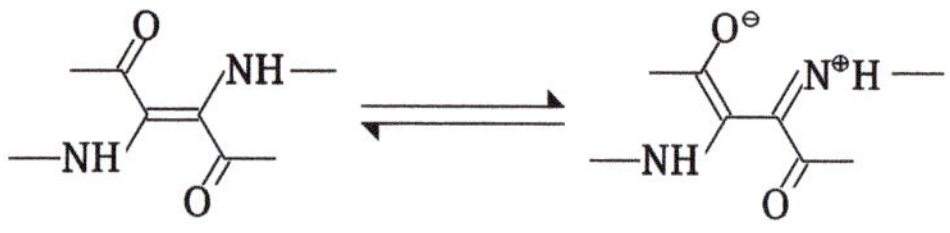   III   IV

D represents electron donors (–O–, –S–, –NH–, –Se–, –Te–), A represents acceptors (=O, =NH, =NR).

Formula I can be viewed as a simple donor–acceptor system with two donor–acceptor pairs connected by a single vinyl group as a bridge. Each pair consists of the structural element –CO–CH=CH–NH–. The secondary amino group is the electron donor, while the carbonyl group is the acceptor in each pair. However, I can also be considered a double merocyanine (pentacyanine, II) with one double bond, the shortest possible carbon chain. VB theory depicts the chromophore as resonant structures such as III and IV. MO theory, however, describes the actual electronic situation better.

However, even merocyanines substituted with donor and acceptor groups, such as in II, do not yet cause any color impression, having such a short bridge. The strong bathochromism results from substituting II with a second donor, X, and another acceptor, Y, ►Figure 2.43 [5]. The substitution with a donor happens at the center carbon

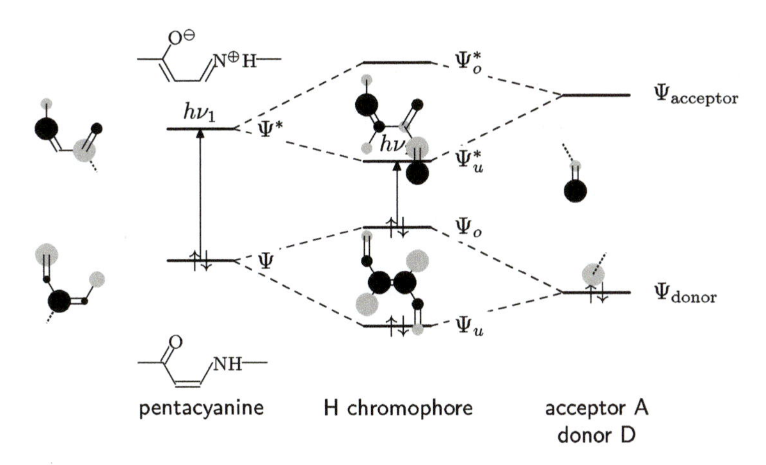

Figure 2.43: MO scheme of the pentacyanine structure underlying indigoid colorants (drawn after [5, Chapter X.1]). Left: MOs of pentacyanine. The spheres indicate the value of orbital coefficients of the MO's wave function at each atom; thus, they also indicate the share of each AO on the MO. Right: MOs of the additional donor and acceptor groups (amino, resp., carbonyl groups). Middle: In the pentacyanine substituted with donor and acceptor, HOMO and LUMO approach, the absorption energy of a transition decreases from $h\nu_1$ to $h\nu_2$.

atom, which has a large orbital coefficient in the HOMO. Donor and structure II, therefore, couple strongly; Ψ and Ψ_{donor} form the MO pair Ψ_u and Ψ_o, and the energy of the new HOMO Ψ_o raises. The LUMO Ψ^* does not couple with the donor since it has a vanishing orbital coefficient at the center atom.

The acceptor, on the other hand, substitutes an atom with a large orbital coefficient in the LUMO so that Ψ^* and $\Psi_{acceptor}$ couple and form the MO pair Ψ^*_u and Ψ^*_o. The energy of the new LUMO Ψ^*_u decreases. Due to the combined effects, the HOMO-LUMO gap of structure I considerably decreases compared to that of II, and thus the transition energy $h\nu_2$ is also reduced.

We can derive this result by applying the Dewar rules to structure II, interpreting it as a heterosubstituted pentacyanine with nitrogen as D and oxygen as A, ►Section 2.5.5 at p. 186. The donor (another nitrogen) then substitutes a starred position of the polymethine, the acceptor an unstarred position. Both changes yield a bathochromic shift of the transition.

The H chromophore causes a large bathochromic shift, but the actual color of the indigo is not blue but red, so the actual shift is lower than expected [453]. The well-known blue color is an artifact of indigo being solid, and it is not observable in the vapor phase. Regarding the solid state, further interactions between carbonyl groups and hydrogen atoms of the amino group are discussed, enlarging the conjugated system considerably (theoretically to the complete solid). As we have seen (and will continue to see), the larger the conjugated system is, the smaller the HOMO-LUMO gap and the lower the transition or absorption energy.

2.5.3.3 Donor-substituted quinones

Essential natural colorants, classical red pigments such as madder lake (►Section 4.1), as well as synthetic anthraquinone colorants (►Section 4.2), are based on donor-substituted quinone chromophores. The base structures are benzoquinone, naphthoquinone, and especially anthraquinone. The $n \rightarrow \pi^*$ transition (localized within the carbonyl group) may induce color, but its intensity is not sufficient for commercial use. Therefore, color-intensive quinones are substituted with electron donors and, consequently, exhibit an intense $\pi \rightarrow \pi^*$ transition based on the transfer of electrons from the donor to the carbonyl oxygen. Adding electron-rich substituents, mono- and disubstituted anthraquinones produce any hue from yellow to blue.

In combination with the π system of the aromatic ring structure, the carbonyl oxygen represents a powerful, complex electron acceptor. We can recognize this by employing a schematic depiction of charge distribution after excitation, using VB theory [3] [4, Chapter 7.2]:

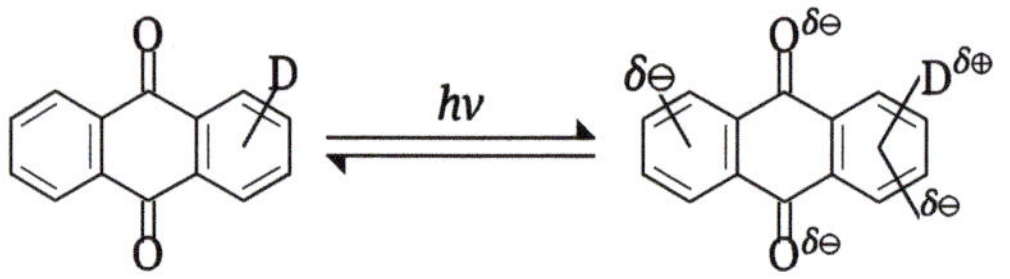

The formula implies that charge is removed from the donor and applied primarily to the ring system's carbonyl oxygen and other positions. According to VB theory (see below), this charge redistribution yields intensive absorption bands and more or less distinct ionic states. Donors in decreasing strength are [4, Table 6.1]:

$$-O^{\ominus}, -NR^{\ominus}, -S^{\ominus} > -N(CH_3)_2 > -NHCH_3 > -SCH_3 > -NH_2 > -SH > -OCH_3 \\ > -NHCOCH_3 > -OH > -OCOCH_3$$

We can observe several bands in the absorption spectrum. These can be assigned to structural elements if we first identify the different decomposition possibilities of the colored molecule [5]. For example, we can decompose quinones into two carbonyl groups, each having a conjugated π system. We can regard *p*-quinone as a "double en-one" and anthraquinone as a "double acetophenone." Aromatic hydrocarbons are also possible constituents. Example decompositions of p-quinone and naphthoquinone are

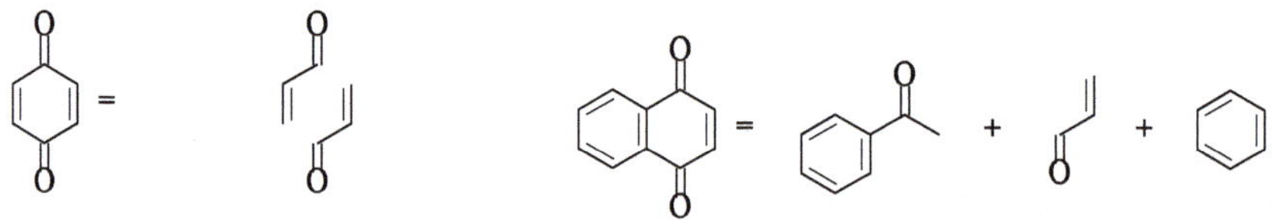

en-one (polyene) fragment

acetophenone en-one benzene

Each of these decompositions explains certain aspects of the spectrum:
- Bands in the UV belong to $\pi \rightarrow \pi^*$ transitions within benzene rings.
- Intense bands in the optical spectral range belong to $\pi \rightarrow \pi^*$ transitions of the quinoid system.
- Weak absorptions at long wavelengths belong to $n \rightarrow \pi^*$ transitions of the conjugated carbonyl group.

Etaiw et al. [683] and Peters and Sumner [684] illustrate this for anthraquinones.

The absorption bands of $\pi \rightarrow \pi^*$ transitions are shifted to higher wavelengths (lower energies) by fusing quinones with benzene rings, enlarging the conjugated electronic system. However, anthraquinone hardly differs from naphthoquinone. The spectroscopic properties are already determined in naphthoquinone, possessing a single benzene ring only. The second benzene ring in anthraquinone does not shift the absorption to higher wavelengths but doubles its intensity. In this respect, we can divide anthraquinone into two overlapping naphthoquinone structures. As a rule of thumb for polycyclic aromatic hydrocarbons, the more extensive ring system determines the bathochromic shift; all other rings increase the absorption's intensity.

VB theory, donor-substituted quinones

VB theory describes $\pi \rightarrow \pi^*$ transitions of quinones by the equilibrium between a neutral and several ionic resonance structures, ►Figure 2.44 [3]. The stability of ionic structures depends on the quinone's substitution pattern with electron donors so that we obtain the following series of increasing bathochromism:

1-substitution < 1,5-disubstitution < 1,8-disubs. < 1,2-disubs. < 1,4-disubs.
< 1,4,5,8-tetrasubs.

►Table 2.11 gives examples of resulting colors.

Unsubstituted anthraquinone

In the transition from the ground state of the unsubstituted anthraquinone I to its excited state (described by resonance formulas like II and III), we lose the benzene ring's resonance stabilization energy (RSE). Therefore, the transition energy is high, and the compound is only faintly yellowish, ►Figure 2.44(a). As a result of the charge redistribution, a carbocation forms.

1-Donor substitution

Introducing an electron-donating substituent at position 1 stabilizes the carbocation III by forming an ionic structure V. In V, the oxidized donor possesses greater stability

Table 2.11: Colors of variously substituted anthraquinones [3, Chapter 4.6] (the color column shows the theoretical color of the compound as a complementary color to the indicated wavelength of the absorption band).

Substitution	Color	$\lambda(\pi \rightarrow \pi^*)$
2-Hydroxy-anthraquinone	Yellow-green	368 nm
1-Hydroxy-anthraquinone	Yellow	402 nm
1,5-Dihydroxy-anthraquinone	Yellow	425 nm
1,2-Dihydroxy-anthraquinone	Yellow	430 nm
1,8-Dihydroxy-anthraquinone	Yellow	430 nm
2-Amino-anthraquinone	Yellow	440 nm
1,4-Dihydroxy-anthraquinone	Orange	470 nm
2-Dimethylamino-anthraquinone	Orange	472 nm
1-Amino-anthraquinone	Orange	475 nm
1,2-Diamino-anthraquinone	Orange	480 nm
1,5-Diamino-anthraquinone	Red	487 nm
1-Dimethylamino-anthraquinone	Red	504 nm
1,8-Diamino-anthraquinone	Red	507 nm
1,4-Diamino-anthraquinone	Blue	550 nm, 590 nm
1,4,5,8-Tetra-amino-anthraquinone	Blue	610 nm

than the carbocation. Due to this, the transition energy decreases, ►Figure 2.44(b). The magnitude of this bathochromic shift is directly related to the donor strength of D:

$$NHCH_3 > NH_2 > NHCOCH_3 > OCH_3 > OH$$

Due to the symmetry of anthraquinone, the same situation exists for 1,5-disubstituted anthraquinones. The spatial proximity of the donor cation to the oxide anion at C^9 in the excited state gains electrostatic energy. Donors linked to hydrogen can additionally form hydrogen bonds to the carbonyl oxygen and move electron density to the carbonyl oxygen:

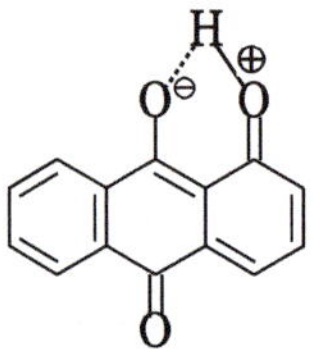

2-Donor substitution

A similar situation exists for a donor substitution at position 2, but in structure VII, the electrostatic stabilization between donor and O^9 is missing. Therefore, the transition energy ν_3 is higher than ν_2, ►Figure 2.44(c).

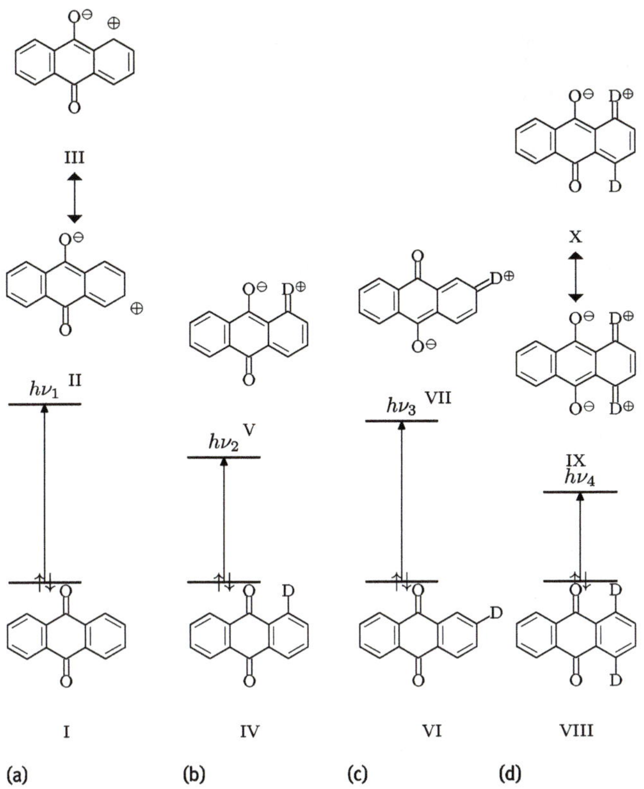

Figure 2.44: Influence of location and number of electron-donating substituents D on transitions in anthraquinones in VB representation [3, Chapter 4.6]. The ground state is depicted at the bottom, and the excited state and its resonance structures are at the top. The charge of the electron donor can be taken over and stabilized by carbonyl groups, acting as electron acceptors. As a result, a carbocation or positively charged donor and a hydroxylate anion form. From (a) to (d), the increasing resonance stabilization yields a decrease in transition energy from ν_1 to ν_2, ν_3, or ν_4.

1,4-Donor substitution

We observe an exceptionally high stabilization by substituting *two* electron donors at positions 1 and 4, as in structure VIII: resonance structures like IX gain the RSE of a naphthene system, ►Figure 2.44(d).

MO representation, donor-substituted quinones

In the MO representation, the change in electron density during the transition from the ground state to the excited state is more complex [3, 4]. We also clearly observe that the donor is the principal supplier of electron density (density decreases about −0.38 and −0.43 units, see figure below), which built up at the carbonyl oxygen (density increases about +0.19, +0.11, +0.08, and +0.13 units) and at the carbonyl carbon (density increases about +0.13 and +0.26 units):

+0.19 O −0.38 N +0.26 + + −0.37 +0.13 O +0.11

+0.08 −0.43 O N + + A B − + + + O +0.11 +0.13

This charge distribution imparts a charge-transfer character to the transition, in which charge moves from donor to oxygen (D → O).

It is interesting to note that, in contrast to the simple VB representation, MO theory predicts an increase in electron density at the respective other carbonyl oxygen and various carbon atoms. Substituting these positions with electron acceptors can cause a further bathochromic shift. The evaluation of individual changes in electron density is complicated and outlined in [5] in more detail. Using ►Figure 2.45, we can understand the substituent's influence in the context of MO theory. It shows the magnitude of orbital coefficients for the highest occupied MOs, including HOMO and LUMO of some basic quinone structures. At the positions already recognized as donor positions by VB theory, orbital coefficients of occupied MOs are significant, and the influence of a donor is destabilizing, i. e., increasing the MO energy and decreasing the energy difference of the MO-LUMO transition (bathochromic shift). Affected are positions 5 and 8 of naphthoquinones and positions 1, 4, 5, and 8 of anthraquinone. We will learn more about this in the context of Dewar rules, ►p. 186.

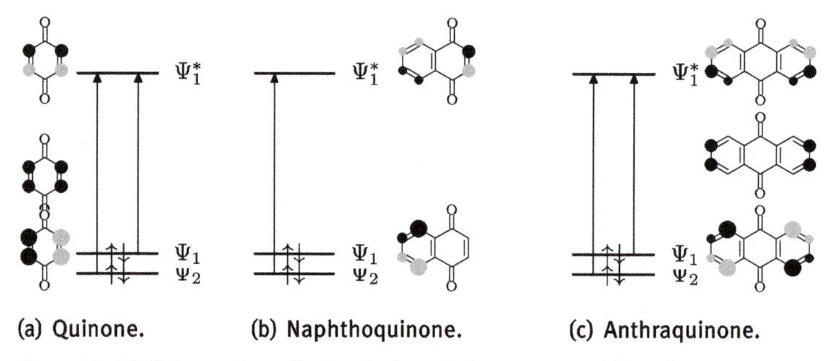

(a) Quinone. (b) Naphthoquinone. (c) Anthraquinone.

Figure 2.45: Schematic radiation-induced electron transitions between MOs of quinone, naphthoquinone, and anthraquinone (drawn after [5]). Only allowed transitions are shown. Orbitals labeled Ψ_1 and Ψ_2 are the two highest occupied MOs, including HOMO Ψ_1, whereas Ψ_1^* denotes the LUMO. The disk size symbolizes the value of orbital coefficients. For the HOMOs, it also indicates the influence of donor substituents on a bathochromic shift.

Using MO calculations, we can more precisely determine the character of transitions. ►Figure 2.46 depicts the four highest occupied MOs Ψ_1 to Ψ_4 and the LUMO of anthraquinone and their orbital coefficients. We see that, during the transition, the orbital coefficients at donor positions decrease for MOs Ψ_1 and Ψ_2, while they increase

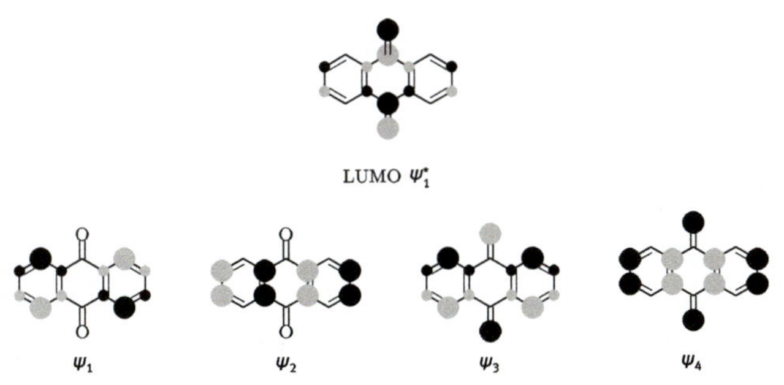

Figure 2.46: Orbital coefficients of the four highest occupied MOs Ψ_1–Ψ_4 and the LUMO Ψ_1^* of anthraquinone, calculated with ArgusLab [334] (default settings). The exact energetic position or ordering of Ψ_1 to Ψ_4 is not depicted. During transitions $\Psi_1 \rightarrow \Psi_1^*$ and $\Psi_2 \rightarrow \Psi_1^*$, electron density at donor positions decreases and increases at carbonyl positions, imparting a donor–acceptor character to these transitions. In contrast, transitions $\Psi_3 \rightarrow \Psi_1^*$ and $\Psi_4 \rightarrow \Psi_1^*$ are MO-MO transitions without the participation of the carbonyl acceptor.

at the carbonyl oxygen atoms in the LUMO. Therefore, $\Psi_1 \rightarrow \Psi_1^*$ and $\Psi_2 \rightarrow \Psi_1^*$ possess the quality of donor–acceptor transitions. In contrast, $\Psi_1 \rightarrow \Psi_1^*$ and $\Psi_2 \rightarrow \Psi_1^*$ are MO-MO transitions in which oxygen atoms are not significantly involved. This type of transition becomes dominant in polycyclic quinones.

2.5.3.4 Diazo chromophore

The azo group –N=N– constitutes an $n \rightarrow \pi^*$ chromophore, employing its free electron pair of nitrogen, and as such, is colored (yellow to orange). The transition, however, is not suitable for colorants due to its low intensity. However, by adding electron donors and a conjugated system, the azo group becomes a vital $\pi \rightarrow \pi^*$ chromophore [4, Chapter 7.3, 7.4] with the basic structures

D–C₆H₄–N=N–C₆H₅ D–C₆H₄–N=N–C₆H₄–D

In a first approximation, the excitation can be described by charge transfer from donor D to acceptor A (structure II):

$$\text{D–C}_6\text{H}_4\text{–N}_\alpha\text{=N}_\beta\text{–C}_6\text{H}_4\text{–A}_{0,1} \quad \overset{h\nu}{\rightleftharpoons}$$

$$\text{D}^\oplus\text{=C}_6\text{H}_4\text{=N–}\overset{\ominus}{\text{N}}\text{–C}_6\text{H}_4\text{–A}_{0,1} \longleftrightarrow \text{D}^\oplus\text{=C}_6\text{H}_4\text{=N–N=C}_6\text{H}_4\text{=A}^\ominus$$

I II

Diazo chromophores possess a complex acceptor because the acceptor can take over charge, as depicted in structure II, but also nitrogen atoms of the azo bridge (structure I) and large parts of the molecule even if no acceptor A is present at all. VB theory only predicts charge transfer to the β-nitrogen, but we also observe significant charge built up on the α-nitrogen, which MO theory describes well.

Classical donors and acceptors used in earlier times could shift the absorption band of a diazo chromophore from UV to the blue spectral range, imparting azo colorants the well-known yellow, orange, and red hues. Blue and especially green azo colorants were rare and required modern substituents to induce large bathochromic shifts. We will look into this in more detail when discussing azo colorants, ►Section 4.11.5. ►Table 2.12 lists magnitudes of bathochromic shifts achievable by simple, classical azo colorants.

Table 2.12: Bathochromic shifts of the $\pi \rightarrow \pi^*$ transition in simple, classical azo colorants [5].

Compound	Color	$\lambda_{\pi\rightarrow\pi^*}$ [nm]
$C_6H_5{-}N{=}N{-}C_6H_5$		316
$HO{-}C_6H_4{-}N{=}N{-}C_6H_5$		339
$H_2N{-}C_6H_4{-}N{=}N{-}C_6H_5$ (SY1, CI 11000, Sudan Yellow R)	Yellow-green	364
1-Naphthyl-N=N-1-naphthyl	Yellow-green	371
$HO{-}C_6H_4{-}N{=}N{-}C_6H_4{-}NH_2$	Yellow	399
$(H_3C)_2N{-}C_6H_4{-}N{=}N{-}C_6H_5$ (SY2, CI 11020, butter yellow)	Yellow	400
$(H_3C)_2N{-}C_6H_4{-}N{=}N{-}C_6H_4{-}N(CH_3)_2$	Orange	460

VB model

►Figure 2.47 depicts the influence of substitution using VB theory [3, Chapter 3.5]. ►Figure 2.47(a) shows azobenzene I (yellow), the fundamental structure of an azo colorant. After irradiation with light, its excited state is approximately represented by the ionic structure II. Due to the loss of aromaticity in one benzene ring, this transition is not favored and requires high excitation energy $h\nu_1$ (located in the UV). Substitution of azobenzene with an electron acceptor A, e. g., a nitro group, does not significantly improve the situation since the azo group is already a suitable acceptor (structure II). In contrast, even more aromaticity is lost in the excited state III.

Electron donors in o- or p-position to the azo bridge stabilize all resonance structures with a carbocation, as structure V, ►Figure 2.47(b). Due to the amino group's ability to deliver electrons while forming an ammonium ion, the energy of the excited state V decreases, and $h\nu_2$ is correspondingly lower, i. e., the absorption band is distinctly shifted into the optical spectral range.

►Figure 2.47(c) depicts the combined effect of simultaneous o- or p-substitution with a donor and an acceptor. The energy difference $h\nu_3$ between ground state VI and excited state VII is even smaller because also the acceptor takes over the negative charge from the azo group.

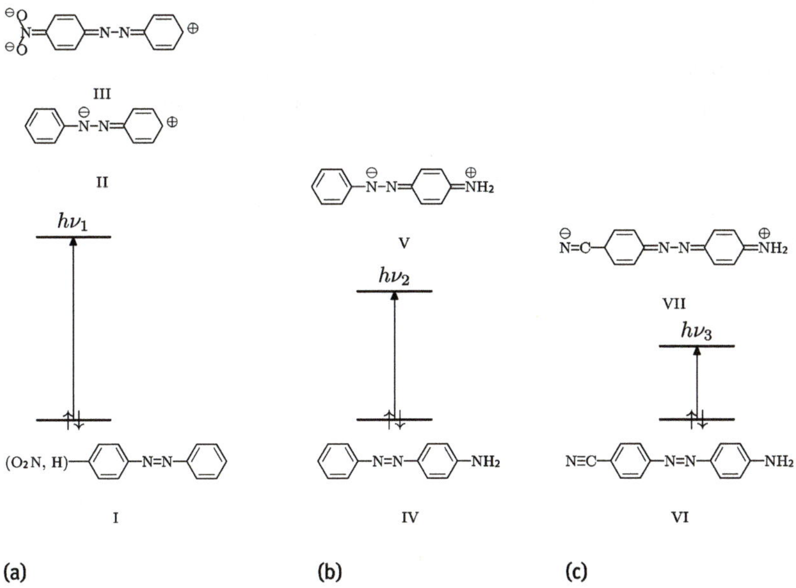

Figure 2.47: Effects of substitution on transition energies of a diazo compound using VB theory [3]. (a): Transition from unsubstituted azobenzene I to the excited state II with transition energy ν_1. (b): An electron donor in structure V can compensate for the charge at the carbocation, accompanied by a decrease in transition energy from ν_1 to ν_2. (c): An electron acceptor can take over the negative charge off the azo bridge, accompanied by a further decrease in transition energy to ν_3.

MO model

MO calculations for p-amino-azobenzene show that during the transition from the ground state to the excited state, electron density moves from the amine nitrogen to the nitrogen atoms of the azo group, as expected [3, Chapter 3.5], [5, Chapter VII]:

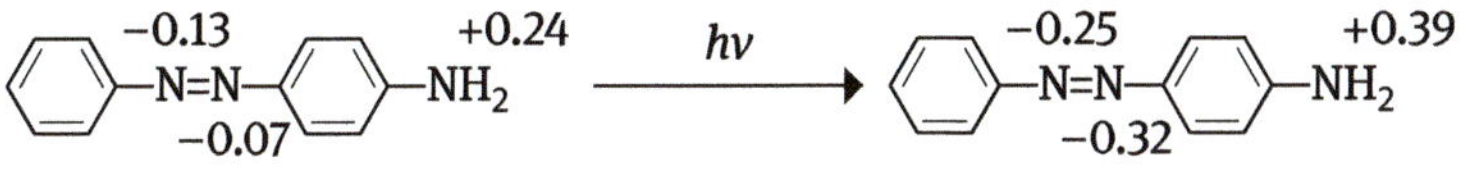

The increase in electron density at the β-nitrogen roughly corresponds to the azo anion, as shown in the introduction above (structure I). However, in contrast to VB theory, we observe a significant increase in electron density at *both* nitrogen atoms. This density built-up is due to an interaction of nonbonding AOs of both nitrogen atoms: the free electron pairs with sp^2-character, located in the σ-bonding plane, combine to form a binding and an antibonding MO n_+ and n_-. Therefore, the charge can no longer be assigned to a single nitrogen atom. The change in electron density at the various carbon atoms is complicated, [5] discusses details and presents other possible MO-MO transitions.

Metallization

With metals, the diazo chromophore can form complexes. One uses metallization to increase the lightfastness of colorants and to change their hue, as metallization yields a bathochromic shift of $\pi \rightarrow \pi^*$ transitions [4, Chapter 7.8]. Stable complexes form especially with o,o′-substituents next to the azo group, ►Section 5.5. The bathochromic shift is caused by the exact mechanism as the color deepening when forming color lakes, ►Section 2.6.3.

2.5.4 Polyene chromophore

Polyene chromophores [4, Chapter 8] do not possess recognizable donor or acceptor groups that could support a charge transfer upon excitation. Instead, they consist of n conjugated double or multiple bonds, forming a system of $2n$ π electrons in various arrangements. Depending on the structure of the double bond skeleton, we distinguish three types that frequently impart color to colorants:

– *Linear polyenes* represent the simplest form of organic colorants: a chain with n conjugated double bonds or vinyl units hosting $2n$ π electrons. They carry aliphatic or alicyclic groups at both ends of the conjugated system, and these groups do neither participate in the conjugated system nor contribute to color.

alkyl $-(\text{CH=CH})_n-$ alkyl

One or several methine groups of the conjugated system may be replaced by heteroatoms or part of a carbo- or heterocyclic ring system.

– *Annulenes* are cyclic polyenes with n vinyl units, forming a single ring with $2n$ π electrons.

$[-(\text{CH=CH})_n-]$ (cyclic)

Again, individual methine groups may be part of a ring system, frequently an aromatic ring.

– In *polycyclic hydrocarbons*, the n vinyl units form polycyclic ring systems of complex and varying geometry. They are constructed from benzene rings (comprising three vinyl units) and fused along one or more vertices in a linear or angular way. Polycyclic hydrocarbons possess aromatic or olefinic character.

In contrast to σ systems, isolated double bonds show relatively small energy differences between HOMO and LUMO in the near-UV spectral range. The more extended the conjugated π-electron system is, the smaller this energy difference becomes. In the case of polyene chromophores, it falls into the optical spectral range.

Linear polyenes, annulenes, and many (but not all) polycyclic hydrocarbons belong to the so-called "even alternant hydrocarbons," or alternating hydrocarbons. They possess an even number of atoms in the π system; the name indicates a specific geometric property: We can mark their carbon atoms alternately, one with a star and the other without any mark. In the end, each atom with a star is only linked to atoms without a star and vice versa (alternation principle):

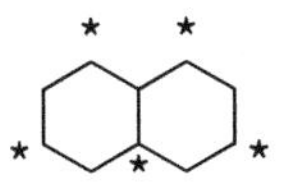

even alternant hydrocarbon

odd alternant hydrocarbon

nonalternant hydrocarbon

Alternating hydrocarbons have a fundamentally different electronic structure than non-alternating compounds [208, Chapter 5.3], [207, Chapter 3.2]:

- The MOs always occur in pairs, i. e., for every Ψ with energy $E = \alpha + i\beta$, there is a Ψ^* with energy $E = \alpha - i\beta$. The MO scheme is constructed symmetrically to an energy α. We can see this in ►Figure 2.48(a), the MOs Ψ_{n-1} and Ψ_n, as well as Ψ_{n-2} and Ψ_{n+1}, are such pairs.
- For each of the paired MOs, the orbital coefficients have equal magnitudes.
- The electron density q_r for the atom r across all MOs j adds up to unity, $q_r = 2\sum_j c_{rj}^2 = 1$.

For discussions of the optical spectra, only the transitions with the lowest energy (longest wavelength) are relevant; thus, we limit ourselves to the four MOs nearest to the HOMO-LUMO boundary, the so-called *four orbitals*. Between them, the HOMO-LUMO transition $\Psi_{n-1} \rightarrow \Psi_n$ (energy difference $h\nu_1$) and the transition $\Psi_{n-2} \rightarrow \Psi_{n+1}$ (energy difference $h\nu_3$) are possible. As a consequence of the MO pairing, the energy difference of the two remaining transitions $\Psi_{n-1} \rightarrow \Psi_{n+1}$ and $\Psi_{n-2} \rightarrow \Psi_n$ have the same value, $h\nu_2$. Thus, in a one-electron approximation, the transitions between the *four orbitals* are those with the lowest energies, namely ν_1 to ν_3. Up to now, we would expect three absorption bands in a spectrum, ►Figure 2.48(b) (left).

The so-called configuration interaction lifts the two-fold degeneracy, giving rise to four distinct absorption bands at long wavelengths, called p, α, and β, and β' (notation following Clar [373]), ►Figure 2.48(b) (right) [208, Chapter 11.2]. p corresponds to the HOMO-LUMO transition, α and β to the transitions from previous degenerated states, and β' corresponds to the transition between the outer MOs $\Psi_{n-2} \rightarrow \Psi_{n+1}$ (HOMO-1 and LUMO+1). The exact ordering of p, α, β, and β' depends on the splitting of the former degenerate states, which are denoted according to Platt as ${}^1L_a$ or ${}^1B_b$ [371, 372]. We expect the sequence α, p, β, β' for large splitting and p, α, β, β' for small splitting.

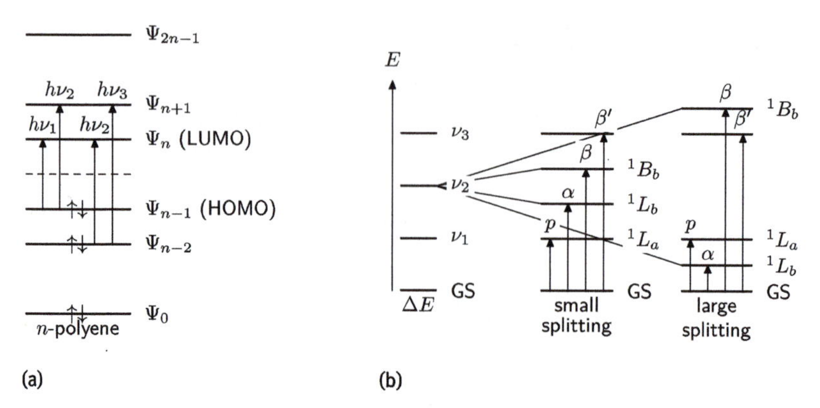

(a) (b)

Figure 2.48: Schematic electronic transitions between molecular orbitals of an alternating hydrocarbon with an even number of MOs (even-alternating hydrocarbon) induced by radiation. GS = ground state; p, α, β, β' are the names of the transitions after Clar [373], ^1L_a, ^1L_n, and ^1B_b are the names of the states, according to Platt [371, 372], ν_1–ν_3 are the transition energies. (a): The MO scheme includes the four orbitals at the HOMO-LUMO gap, the imaginary centerline located around $E = \alpha$ (dashed). Also included are the four possible transitions between the four orbitals with the smallest possible energies. (b) left: Energetic location of these transitions with the degenerate transition ν_2. (b) right: Energetic location of the states after lifting the degeneracy for small (middle) and large splittings (right) [208, Chapter 11.2].

2.5.4.1 Linear polyenes

The simplest polyene colorants are *linear polyenes* consisting of n conjugated double bonds [4, Chapter 8.2]. This chromophore is at both ends linked to groups R, which do not affect the electronic excitation of the molecule, such as alkyl groups or alicyclic residues:

alkyl $(\;)_n$ alkyl

Linear polyenes have $2n$ π electrons and $2n$ MOs in the chromogenic chain. For example, butadiene ($n = 2$) possesses a π system with four electrons; the MOs are denoted by Ψ_0–Ψ_3, ►Figure 2.49(a). Here, Ψ_1 is the HOMO, and Ψ_2 is the LUMO, so these four MOs also represent the *four orbitals*. The transition with the smallest transition energy is the HOMO-LUMO transition $\Psi_1 \to \Psi_2$, followed by the $\Psi_1 \to \Psi_3$ transition, which is equal in energy to the $\Psi_0 \to \Psi_2$ transition into the LUMO. The transition $\Psi_0 \to \Psi_3$ possesses the highest energy (the shortest wavelength for the absorption band):

$$(\Psi_0)^2(\Psi_1)^2 \to (\Psi_0)^2(\Psi_1)^1(\Psi_2)^1 \quad \nu_1$$
$$(\Psi_0)^2(\Psi_1)^2 \to (\Psi_0)^2(\Psi_1)^1(\Psi_3)^1 \quad \nu_2$$
$$(\Psi_0)^2(\Psi_1)^2 \to (\Psi_0)^1(\Psi_1)^2(\Psi_2)^1 \quad \nu_2$$
$$(\Psi_0)^2(\Psi_1)^2 \to (\Psi_0)^1(\Psi_1)^2(\Psi_3)^1 \quad \nu_3$$

As described, the degeneracy of ν_2 is lifted by configuration interaction, resulting in two transitions so that we can expect the transitions p (ν_1), α, β, and β' (ν_3), sorted by increasing transition energy.

By elongation of polyenes, we push $h\nu_1$ toward optical energies. ►Figure 2.49(b) schematically shows that with increasing n, e. g., for hexatriene or octatetraene, the absorption moves toward longer wavelengths since the HOMO energy increases and the LUMO energy decreases. Consequently, the transition frequencies ν_3 and ν_4 decrease and approach energies in the visible spectral range at about $n = 7$. Illustrative natural examples are carotenes, ►Section 4.3. To be colored, i. e., to possess an absorption band in the optical spectral range from about 400 nm onward, 14 methine groups are necessary, which are equivalent to $n = 7$ double bonds. The shift of the transition toward longer wavelengths becomes evident in the series bixin (yellow $n = 7$), α-carotene (orange, $n = 10$), and lycopene (red, $n = 11$).

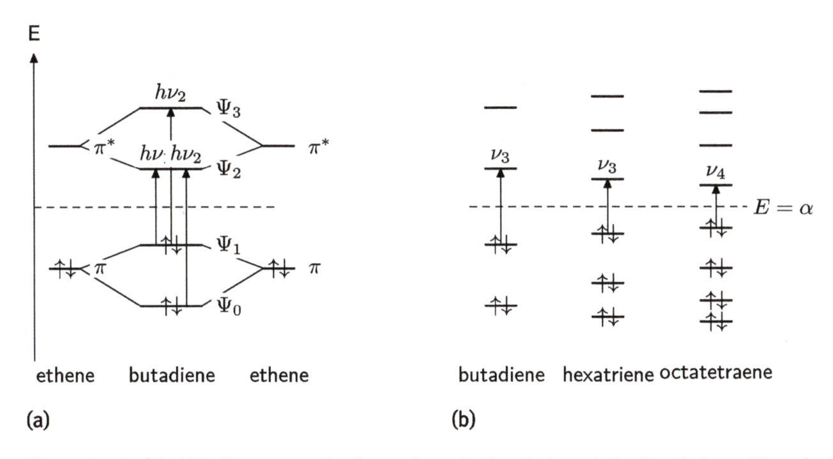

Figure 2.49: (a): MO diagram and schematic radiation-induced electronic transitions between MOs of a four-atom molecule with two conjugated double bonds, e. g., butadiene [208, Chapter 5.1], [206, 207]. The MOs Ψ_0 to Ψ_3 originate in π and π^* MOs of the ethene units. (b): MO level diagrams for the homologous hexatriene and octatetraene, their HOMO and LUMO approaching $E = \alpha$, their HOMO-LUMO gap decreasing.

The described behavior originates in theoretical models. For polyenes with $2n$ carbon atoms, [4, Chapter 8.2], [206, 207] describe the calculation of MO energies for their Ψ_i and their transition frequency, using the Hückel approximation:

$$h\nu = \Psi_n - \Psi_{n-1} = -4\beta \sin \frac{\pi}{2n+2} \tag{2.9}$$

More intuitive than the mathematical derivation of the orbital energies as eigenvalues is a graphical method known as a *Frost's Circle* or *Musulin-Frost diagram*: we draw a semicircle along the energy axis and divide it into $n + 1$ equal sectors. The projections of the vertices of the sectors onto the energy axis indicate the MO energies. (Of course, this is a severe simplification; nevertheless, it illustrates some fundamental aspects of the electronic structure of polyenes.)

In reality, we observe a convergence of transition frequencies to a specific value with increasing n, about 600 nm [4, Chapter 8.2]. Therefore, the polyene chromophore cannot produce colors beyond red. This behavior testifies that some preconditions of the Hückel theory are not strictly obeyed, e. g., the equivalence of bonds: Polyenes show distinct differences between single and double bonds [4, Chapter 8.2], [207, Chapter 3.9]. However, the simple theory correctly states that substituting chain atoms with heteroatoms does not influence the position of absorption bands.

2.5.4.2 Polyenes substituted with donors and acceptors

If we terminate a polyene not with alkyl groups, which are inert regarding π electrons, but with groups acting as electron donors and acceptors instead, we obtain a chromophore whose transition frequency decreases with increasing chain length n. Both donor and acceptor support the delocalization of π electrons throughout the polyene chain, and thus better satisfy the conditions of the Hückel model. VB theory illustrates how donors and acceptors in II support electron delivery into the π system, compared to the carbocation of an unsubstituted polyene I:

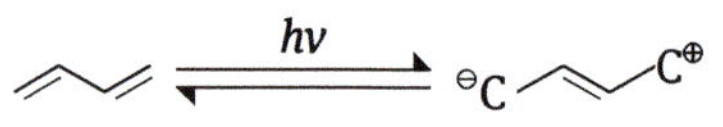

$h\nu$ ⊖C C⊕

I

A D A⊖ D⊕

II

O NH_2 ⊖O $NH_2^{\oplus}$

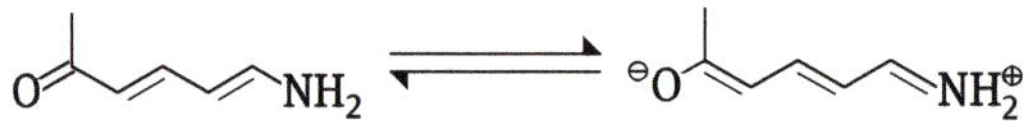

III

MO theory states that the terminal atoms of a polyene chain possess significant orbital coefficients in both HOMO and LUMO, ►Figure 2.50 (left). So, donor and acceptor substituents at these locations strongly couple with the frontier orbitals. According to the Dewar rules, ►p. 186, substitution with a donor increases the HOMO energy, while substitution with an acceptor decreases the LUMO energy, ►Figure 2.50 middle. Both trends reduce the energy gap and the transition energy, shifting the absorption to higher wavelengths up to the optical spectral range. The resonance structures are classically interpreted as ground state and excited state.

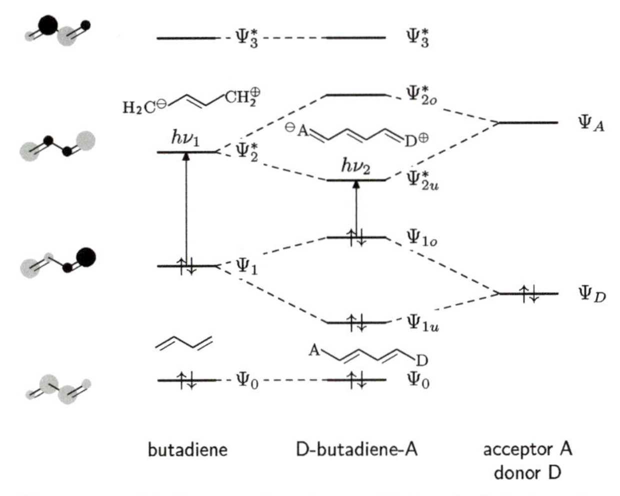

Figure 2.50: MO diagram of a polyene with terminal electron donor and acceptor groups [5, Chapter VI.1]. The absorption energy decreases from $h\nu_1$ for the pure polyene to $h\nu_2$ for the substituted polyene. The magnitude of orbital coefficients at the atoms in the respective MO are indicated on the left by disks of the corresponding size.

We may prefer to consider the substituted π system as a polyene, substituted with a donor and an acceptor, or instead as a donor–acceptor chromophore with the polyene as the bridge, ►Section 2.5.3. All groups presented there can be employed as donors or acceptors.

Some acceptors, such as the carbonyl group, extend the carbon chain of the polyene by one carbon atom. Including the two heteroatoms of the donor and acceptor groups, we formally obtain a structure known as *polymethine*, comprising an odd number of chain carbon atoms. They are classified depending on which donor and acceptor groups terminate the chain, yielding oxonols, streptocyanins, or other types:

$^{\oplus}H_2N$ … NH_2 ⟵ polyene ⟶ O … OH

streptocyanin, $n = 3$ polyene, $n = 2$ oxonol, $n = 3$

Polymethines will be the subject of ►Section 2.5.5.

2.5.4.3 Polycyclic polyenes

n vinyl units or double bonds can merge to form mostly six-membered (aromatic) rings, and these can then be joined and fused in many ways [305–307]. They are collectively called polycyclic aromatic hydrocarbons (PAH). Possible arrangements are linear series of rings, angular series of fused rings with one or more bends, and planar systems of rings annulated in many ways. They can be entirely aromatic throughout the molecule, i. e., π electrons are delocalized over the complete molecule. They can also contain smaller aromatic units, e. g., benzene units (aromatic electron sextets), naphthalene units, and sections of olefinic character (isolated double bonds).

Unfortunately, it is impossible to make general statements about the MO diagram of PAH due to their structural diversity. It is equally impossible to derive generalized formulas for the HOMO-LUMO-difference as it is possible for monocyclic polyenes. ►Figure 2.51 depicts examples of MO diagrams of some higher annulated PAH, which occur in colorants (MO energies were obtained by simple Hückel-MO calculation). We see from this that the patterns known from polyenes and polymethines are no longer applicable to these polycyclic compounds.

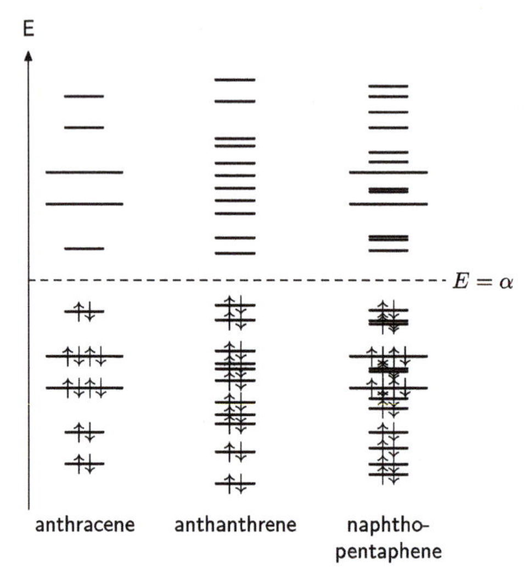

Figure 2.51: MO level diagrams of some polycyclic aromatic hydrocarbons occurring as parent compounds of colorants. Level energies are derived from simple Hückel calculations according to [208, Chapter 4]. The familiar, symmetric, or regular patterns of MOs from polyenes and polymethines are no longer applicable.

As a group, polycyclic hydrocarbons are only weakly colored; smaller systems with four rings such as pyrene or chrysene are colorless. However, as with all polyenes, the HOMO-LUMO gap and, therefore, the transition energy decreases until it enters the visible spectral range for sufficiently large n. Higher representatives show intensive colors in all hues, e. g., perylene (yellow), diindenoperylene (red). The highest-condensed PAH are black, such as graphite and its smaller fragments, ►Section 3.1.

Because of their diversity, general statements about the color of PAH are only possible for individual structure classes. In [307], this has been done for numerous classes. It is sufficient to know that a bathochromic shift of similar magnitude occurs for each ring fused to the system for a given structural class.

The most uncomplicated cases are aromatic, linearly fused benzene rings or *acenes*. The first and smallest representatives, benzene, naphthalene, and anthracene, are colorless. Color occurs from the fourth ring onwards: tetracene is orange (λ_{max} 470 nm), pentacene purple (λ_{max} 580 nm), and hexacene green (λ_{max} 690 nm). However, as the length of the acenes increases, the aromatic character is lost in favor of a polyolefinic character (see below), the higher representatives becoming reactive and unstable compounds not suitable as colorant bases.

As described initially, there are four long-wavelength transitions, two of which are degenerated originally, ►Figure 2.48. Configuration interaction lifts the degeneracy, and we observe four absorption bands p, α, β, and β'. In the series of acenes, the wavelength of the p transition (HOMO-LUMO) increases strongly with n, which manifests itself in the appearance of color from tetracene onwards. For the first members of the acene series, transitions occur in the order β, p, α, and they are located in the UV to the visual spectral range, while for higher acenes from tetracene onwards, the order is β, α, p. Due to their intensity, the p and the β transitions dominate the induced color.

In the series of *phenes*, all rings are linearly fused with one exception so that an angular arrangement results. In this series, the wavelengths of β, α, and p only increase weakly; intensive colors appear only in more extensive systems than acenes. In general, angularly fused compounds undergo a hypsochromic shift, as the following examples illustrate [4, p. 219]:

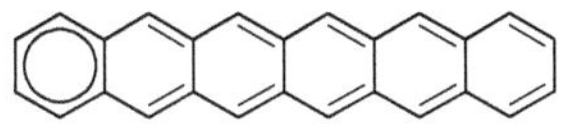

I, hexacene, green (λ_{max} 690 nm)

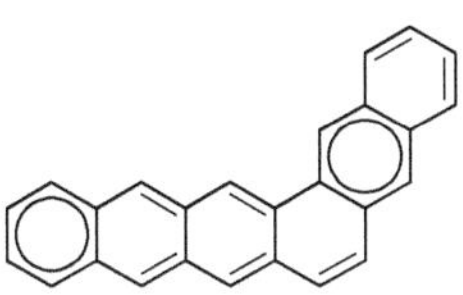

II, purple (λ_{max} 550 nm)

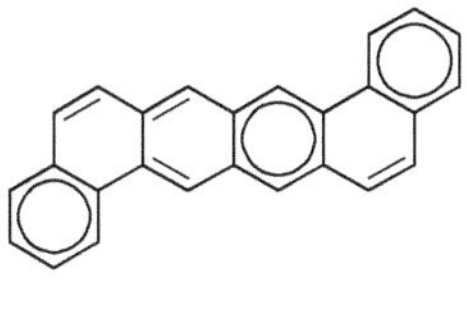

III, yellow (λ_{max} 438 nm)

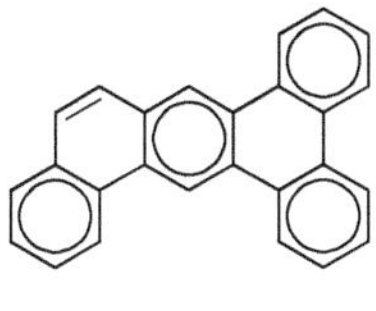

IV, colorless(λ_{max} 388 nm)

We can interpret this color series if we draw polycycles after Clar in such a way that complete π electron sextets are denoted with a circle, the so-called *Robinson symbol* [306, 307]. The Robinson symbol was introduced by Robinson in 1925 and symbolized the property "6 π electrons," i. e., the presence of an aromatic electron sextet with benzene-like properties and relative stability [369]. The rules of Clar and Zander for drawing the structural formula from 1958 are [306, 370]:

- Draw as many Robinson circles as possible.
- Never draw Robinson circles in adjacent six rings.
- Draw a Kekule structure with double bonds for the molecule residue after removing the six-membered rings marked with Robinson circles.

If we compare isomeric benzenoid hydrocarbons, we find that the stability of polycycles increases with the number of Robinson circles. The absorption wavelength decreases to the same extent: In hexacene I, there is one Robinson ring and ten double bonds, formula II comprises two rings and seven double bonds, structure III has three rings, and the colorless compound IV has a total of 4 aromatic, closed rings. With an increasing number of Robinson ring symbols, the resonance stabilization energy gained by an aromatic configuration increases. To a growing extent, this uncouples aromatic subunits from the overall π system. The disintegration attenuates the large-scale electronic delocalization. Consequently, we observe an increase in the HOMO-LUMO energy difference and transition energy. ►Figure 2.52 depicts these trends for typical hydrocarbon structures. Linear annulation generally leads to a redshift of the absorption band due to the enlarged π system. In contrast, angular annulation yields a blueshift due to the formation of new isolated electron sextets and disruption of the overall conjugated systems.

Many of the larger colored systems are strongly olefinic. The Clar representation of the formulae depicts this clearly; see, e. g., hexacene. Double bonds, which do not belong to a benzoic subunit marked by a Robinson circle symbol, remain as olefinic double bonds. Structures with many reactive olefinic bonds are hardly suitable as colorants. However, some polycycles, like quinones, possess so many aromatic substructures that they are stable and serve as indispensable vat dyes, as we will see in the next section.

2.5.4.4 Polycyclic quinones

We derive quinones from polycyclic compounds by replacing two methine groups with carbonyl groups [4, Chapter 8.5]. The carbonyl groups can be located in the same ring or separated from each other by several fused rings, as supported by the phenylogy principle. Due to this dicarbonyl structure, polycyclic quinones are both vat dyes (►Section 4.6.1) and pigments (►Section 4.6.4.3), e. g., indanthrene blue.

The quinoid structure is crucial for dissolving polycycles into a vat. It also increases the stability of polycycles, as we can see by hexacene:

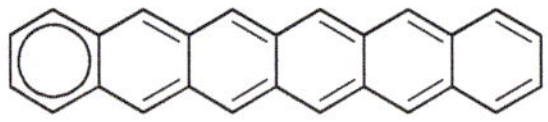

hexacene

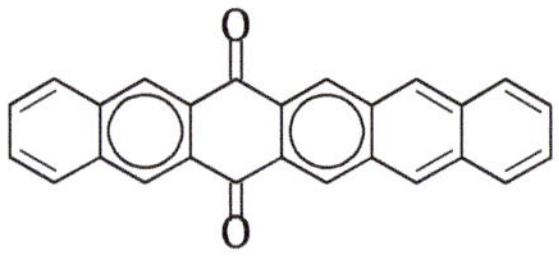

hexacene-dione

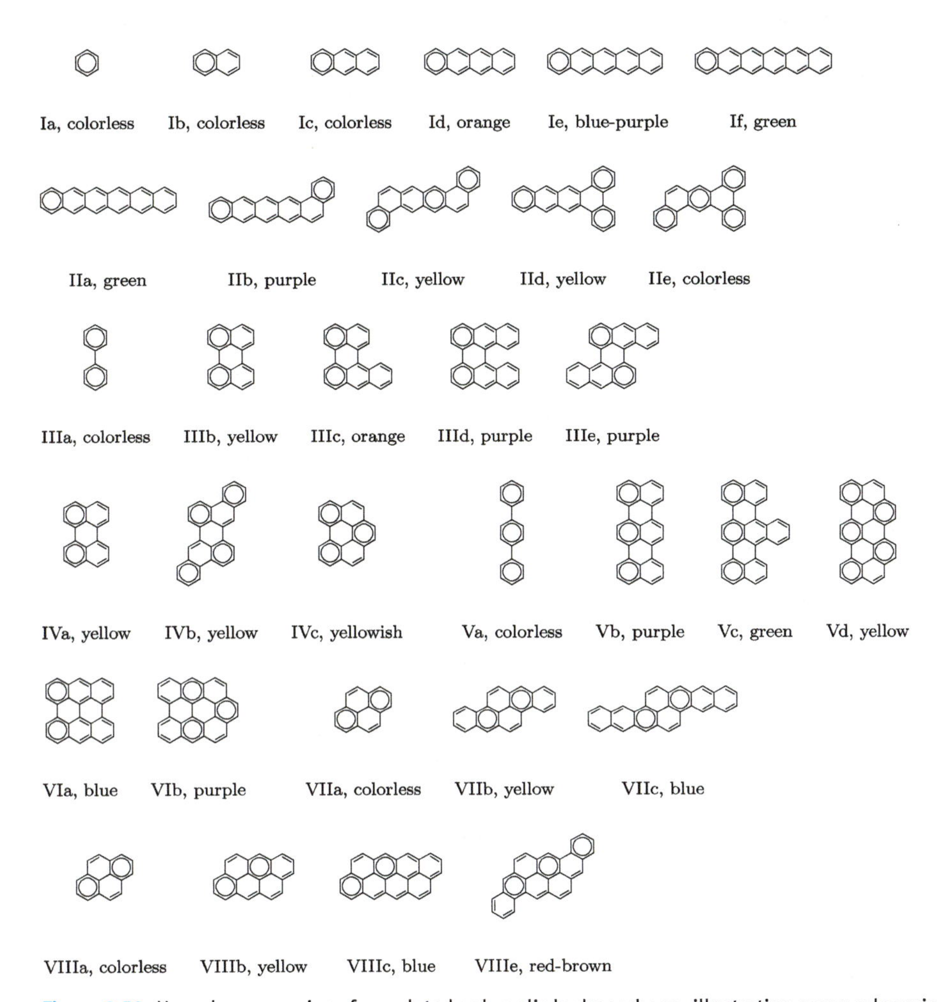

Figure 2.52: Homologous series of annulated polycyclic hydrocarbons, illustrating some color principles [307]. Linear annulation leads, in general, to a bathochromic shift. Angular annulation, in contrast, yields a hypsochromic shift due to the formation of isolated aromatic electron sextets, disintegrating the entire π system.

The quinone structure significantly reduces the contiguous π system, thus shifting the absorption toward the short-wavelength or UV spectral range. Simultaneously, the number of reactive olefinic double bonds decreases in favor of a new aromatic sextet (marked by a Robinson ring). In commercial polycyclic quinones, the number of six-membered rings increases, or electron donors are incorporated to compensate for the loss of conjugation.

Polycyclic quinoid hydrocarbons such as anthanthrene (►Section 4.6.4.3) without electron donors (►Section 2.5.3.3) must provide extended π systems to be colored. Therefore, we can imagine that the extended aromatic system acts as a donor. In the case of anthanthrene, this would be the central naphthalene fragment [5]. The carbonyl oxygen is then part of a complex acceptor, and the π system stabilizes the positive charge by resonance:

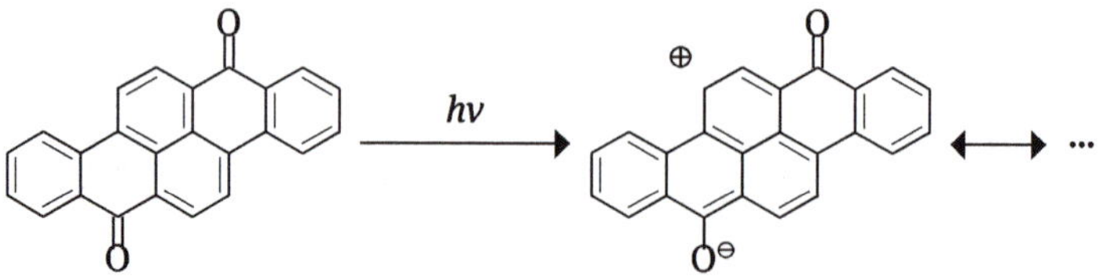

$h\nu$

However, studies of the spectra show that higher polycyclic quinones increasingly resemble pure hydrocarbons and that the carbonyl groups are only small perturbations of an aromatic system, ►Figure 2.53. The top row illustrates that large amounts of charge move from the outer rings to the central ring and the carbonyl oxygen atoms (transition with a donor–acceptor character) during the HOMO-LUMO transition in anthraquinone. In the middle row, we see that in anthanthrone, the HOMO-LUMO transition also involves an accumulation of charge in the central region between the carbonyl oxygen atoms but no longer focuses on the oxygen atom itself. In the bottom row, a distinct charge shift is not visible for anthanthrene, the pure hydrocarbon. The donor–acceptor character of the transition has given way to a MO-MO character. The coloration in this and similar cases is due to $\pi \to \pi^*$ transitions between MOs of the entire hydrocarbons.

This discussion uses a highly simplified model. We have not considered if transitions between MOs are allowed by selection rules and symmetry. Furthermore, we must note that in addition to the HOMO, lower MOs (HOMO-1, HOMO-2, ...) can also be involved in transitions, whose donor–acceptor character varies considerably. Thus, transitions in anthraquinone occur between deeper MOs and the LUMO without the participation of carbonyl oxygen atoms, i. e., transitions of pure MO-MO character.

2.5.4.5 Annulenes

If we link both ends of a polyene chain with n conjugated double bonds together, forming a ring, we obtain a cyclic system with $2n$ π electrons, a so-called *[2n]annulene*:

$$\ulcorner(\mathrm{CH{=}CH})_n\urcorner$$

The most prominent representative of this type is [3] annulene or benzene. Annulenes possessing $2n = 4m+2$ electrons obey the Hückel rule and are supposed to be aromatic according to Hückel theory. This manifests itself, among other things, in the equivalence of single and double bonds [4, Chapter 8.2, 8.6], [207, Chapter 3.11]. Regardless of that, they, like polyenes, induce color [4, Chapter 8.6] when $n > 6$ is sufficiently

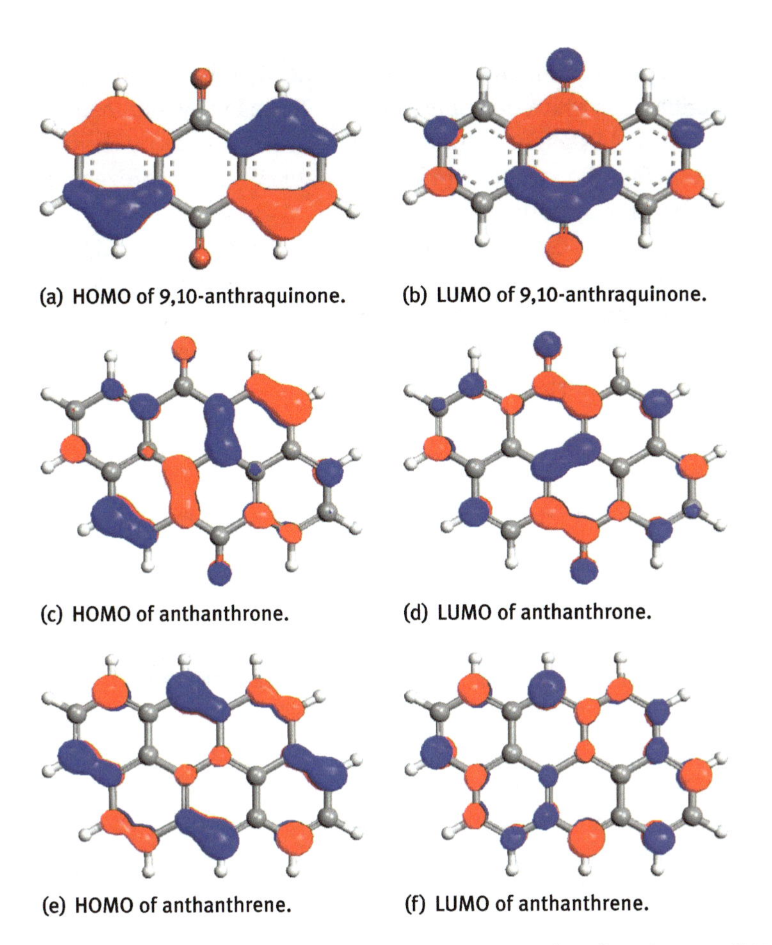

(a) HOMO of 9,10-anthraquinone. (b) LUMO of 9,10-anthraquinone.

(c) HOMO of anthanthrone. (d) LUMO of anthanthrone.

(e) HOMO of anthanthrene. (f) LUMO of anthanthrene.

Figure 2.53: Schematic changes in electron density in polycyclic quinones. All MOs were calculated with ArgusLab [334] (default settings) and are shown without taking into account whether transitions between them are allowed by selection rules or not. For quinones, charge accumulates in the central region around the acceptor axis, but higher quinones like anthanthrone (middle) increasingly resemble pure hydrocarbons such as anthanthrene (bottom).

large. As described on ►p. 165, the degeneracy of transitions between the *four orbitals* is lifted by configuration interaction so that between HOMO-1 and LUMO+1, four transitions p, α, β, and β' occur. Since annulenes belong to the same series as benzene, their spectra resemble benzene; they all possess a long-wavelength α transition, followed by the p transition and the short-wavelength β band.

2.5.4.6 Porphines and phthalocyanines

In nature, [16] annulenes occur as fourfold nitrogen-substituted tetraaza-annulenes, better known as tetrapyrroles. The simplest one is porphine I, the core element in crucial natural dyes such as hemin. The pure linear polyene structure is extended by two-membered bridges, yielding four five-membered rings. (Following biosynthesis, one could alternatively describe them as five-membered rings interlinked by four methine bridges yielding a tetrapyrrole.) Different natural hydrogenation patterns on the five-membered rings classify them further. For example, one double bond is hydrogenated in chlorins (represented by chlorophyll).

Regarding colorants, we focus on the fully unsaturated porphines (porphyrins) and their derivatives, all having 16 π electrons. Further substituting the *meso*-carbon atoms (the bridge atoms) with nitrogen yields tetraazaporphines (tetraazaporphyrins) (TAP) II. Finally, annulation of benzene rings leads to phthalocyanines III (Pc), the parent of vital synthetic colorants, ►Section 4.10:

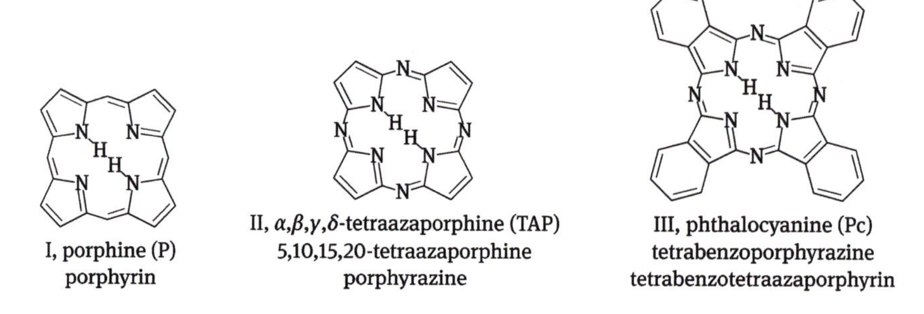

I, porphine (P)
porphyrin

II, $\alpha,\beta,\gamma,\delta$-tetraazaporphine (TAP)
5,10,15,20-tetraazaporphine
porphyrazine

III, phthalocyanine (Pc)
tetrabenzoporphyrazine
tetrabenzotetraazaporphyrin

Porphines are characterized by two regions of intense absorption located near the UV and the IR spectral region. The strongest band lies near the UV and is called the B or Soret band; the next most intensive and color-determining band, called the Q band, is located near the IR spectral region and frequently split into two bands, depending on molecule symmetry.

Observation data

Depending on which of the types I–III is present, the position of the B and Q bands, and thus the color changes in a characteristic way, ►Table 2.13 [376–380]. The intense B band is located in the near-UV/VIS spectral range in the parent body I. If the B band is located below 400 nm and a weak Q band around 550 nm, red color results (hemin: B < 400 nm, Q = 550 nm = red). The hue becomes yellowish if the B band is shifted to 400 nm. In asymmetric porphines, the Q bands can extend beyond 600 nm, and the overall color impression is green as in chlorophyll (B = 400 nm = yellow, Q = 600 nm = blue), ►Figures 2.54(a) and 2.54(b).

Replacement of the four methine groups of the bridges with nitrogen in tetraazaporphine II decisively changes the spectrum: the B band retains its intensity and moves to the UV spectral range. The Q bands gain considerably in intensity but maintain their position. The color impression is pure blue.

Table 2.13: Location of the main absorption bands in some porphines and phthalocyanines [4, Chapter 8.6f], [5, Chapter XIII], [376]. Listed are the band position (schematic) and the absorption coefficient log ϵ.

Class	B	Qy	Qx
[16] Annulene dianion	412/5.2	560/4.1	600/4.2
Porphine (general)	around 400	4x between 500–700 nm	
Me_4Et_4-porphine	398/5.1	529/3.8	619/3.6
Copper-porphine	387/10	531/1	566/2
Tetraazaporphine	340/4.9	556/4.7	624/4.9
	(Q more intense, B UV-shifted, constant intensity)		
Copper-tetraazaporphine	334/5.5	578/6	
Phthalocyanine	330	556/3.8 576/4.0 604/4.5 636/4.7 668/5.2 703/5.2	
Copper-phthalocyanine	325	620/5	657/10

The annulation of four benzene rings yields phthalocyanines III. It does not significantly alter the location or intensity of the B band. However, the Q band now extends into the IR spectral region, ►Figure 2.54(c), blue phthalocyanine. Substitutions in the benzene ring or replacement, e. g., with naphthalene, do not lead to distinct changes.

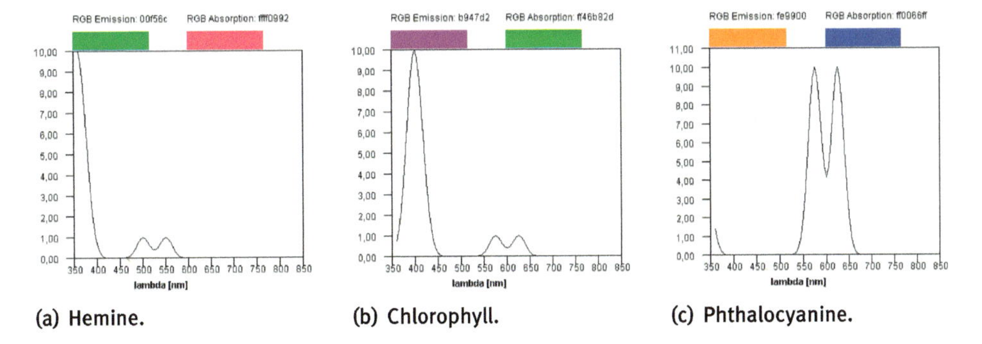

(a) Hemine.

(b) Chlorophyll.

(c) Phthalocyanine.

Figure 2.54: Schematic influence of B and Q band location on the color of porphines and derivatives. The absorption bands are shown, and only the main bands B and Q are included in the calculation. (a): Hemine (a porphine, red absorption of blue-green and yellow spectral range by B and Q bands). (b): Chlorophyll (a porphine, green absorption of blue, yellow, and red spectral range by B and Q bands, Q bands shifted toward longer wavelengths). (c): Phthalocyanine (turquoise, absorption of yellow and red spectral region by Q bands, B band located in UV spectral range).

Furthermore, we observe the splitting of the bands depending on molecule symmetry: Metal complexes of all three types, as well as their dianions, belong to the highly symmetric point group D_{4h}, in which the absorption bands B and Q are degenerated. In the neutral dihydrogen compounds, the presence of two amino and two imino groups reduces symmetry to D_{2h}. This reduction lifts electronic degeneracy, and we observe splitting the absorption bands into an x and a y component. The same happens when the central metal cation is so large that it does not fit into the ring plane, lowering the symmetry to C_{4v}.

Metallization of the neutral dihydrogen compounds leads to a slight blue shift of the B band. The blue shift increases with the electronegativity of the metal. The Q band is shifted toward blue by dimerization or aggregation. This dependence is critical in the production of blue pigments for four-color printing since, in this way, crystal structure influences the hue of the final colorant.

Electronic structure

Studies have shown that the chromogenic electronic structure of porphine is the inner 16-membered ring, corresponding to a [16] annulene but comprising 18 π electrons [4, Chapter 8.7], [5, Chapter XIII], [376, 377, 380]. The same situation is present in phthalocyanines, being aza-substituted porphine derivatives. The number 18 results from the following calculation: each methine group (12 or 8), each imino group =N– (2), and each aza group =N– (0 or 4) provide one π electron. The tertiary nitrogen atoms –NH– provide two π electrons, employing their lone electron pair (4). The imino or aza nitrogens lone pair cannot participate in such a way since it extends into the ring plane, not perpendicular. We are thus dealing with a structure isoelectronic to the dianion of [16] annulene.

There has been no lack of attempts to interpret the spectra with band location, splitting, intensity, and shift, among other things, by derivation from annulenes and by HMO calculations. Gouterman [376] and Stillman and Nyokong [377] provide an overview of this. A series of articles published by Gouterman discussing the so-called *four orbital model* for the description of porphines, tetraazaporphines, and phthalocyanines, explain the basic features nicely [381–383].

The simple annulene model by Simpson (summarized in [382]) considers porphines as [4n]annulenes and [4n + 2]annulenes, their degenerated HOMOs $\Psi_{\pm 4}$ being fully occupied. ►Figure 2.55(a) depicts the MO scheme of the [16] annulene dianion, $n = 4$ applies. Due to the degeneracy of HOMO and LUMO, theoretically, only a single but four-fold degenerated transition $\Psi_{\pm 4} \rightarrow \Psi_{\pm 5}$ into a four-fold degenerated excited state is possible.

A magnetic quantum number ν distinguishes the MOs, for the HOMOs apply $\nu =$ ± 4, and for the LUMOs apply $\nu = \pm 5$. We expect strong configuration interaction due to the severe degeneracy in this simple MO approach. The configuration interaction split the single excited state into two, ${}^1E_{7u} < {}^1E_{1u}$, each two-fold degenerated and enabling two transitions:

Ground state	$\rightarrow {}^1E_{7u}$	Q	forbidden, weak, VIS, $\Delta\nu = \pm 9$
Ground state	$\rightarrow {}^1E_{1u}$	B	allowed, intense, UV/VIS, $\Delta\nu = \pm 1$

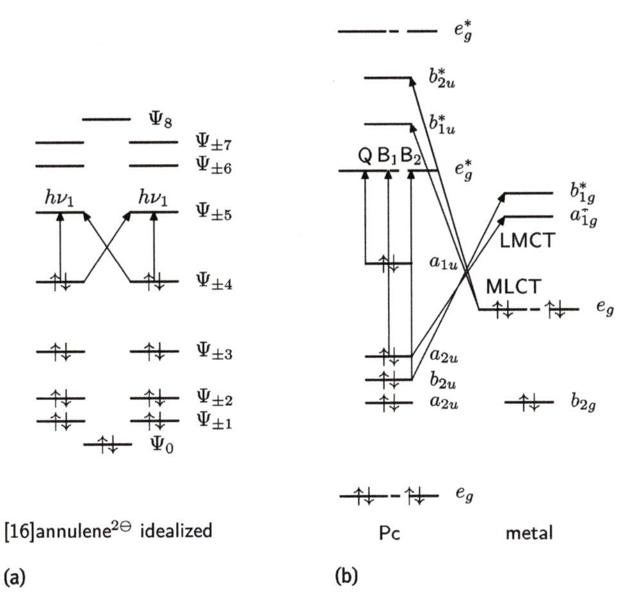

Figure 2.55: Origin of B and Q bands in porphines and phthalocyanines. (a): Schematic MO diagram of the π system of [16] annulene's dianion, isoelectronic to porphine with 18 π electrons. It shows the two-fold degenerated HOMO, LUMO, and the four-fold degenerated HOMO-LUMO transition [208, Chapter 5.2], [377, p. 166], [382]. (b) left: Actual MO diagram of a phthalocyanine, based on calculations, depicting the four central MOs of the *four orbital* models and the origin of B and Q bands (drawn after [377, 380]). (b) right: Typical location of metal orbitals in phthalocyanine-metal complexes with MLCT and LMCT transitions in addition to Q and B bands [380].

The transition GS $\rightarrow$ $^1E_{7u}$, $\Delta\nu = \pm 9$ causes a significant change in the magnetic quantum number and is symmetry-forbidden and weak. Therefore, literature associates it with the Q band. In contrast, the transition GS $\rightarrow$ $^1E_{1u}$, $\Delta\nu = \pm 1$, is symmetry-allowed and correspondingly intense. Accordingly, it belongs to the B band. Thus, the simple model reproduces the existence and intensity of two absorption bands approximately correctly for porphines and phthalocyanines of D_{4h} symmetry.

The model can also explain splitting the Q bands into two peaks, assuming that, in the free porphine base H_2P_1, the imino hydrogen atoms inhibit electron flow through the [16] annulene. Instead, electrons flow through the outer bonds of the two pyrrole rings, effectively yielding an [18] annulene. This [18] annulene belongs to the type $4n + 2$, lifting the degeneracy of one excited state so that the Q band split into two:

Ground state $\rightarrow$ $^1B_{2u}$	Q	forbidden, weak, VIS, $\Delta\nu = \pm 9$
Ground state $\rightarrow$ $^1B_{1u}$	Q	forbidden, weak, VIS, $\Delta\nu = \pm 9$
Ground state $\rightarrow$ $^1E_{1u}$	B	allowed, intense, UV/VIS, $\Delta\nu = \pm 1$

The transitions GS $\rightarrow$ $^1B_{2u}$ and GS $\rightarrow$ $^1B_{1u}$, with $\Delta\nu = \pm 9$, again cause a significant change in magnetic quantum number; they are symmetry-forbidden and weak

(Q band split into two peaks). The transition GS → $^1E_{1u}$ ($\Delta\nu = \pm1$) is symmetry-allowed, intense, and belongs to the (single) B band.

Gouterman and his group summarized the merits of the polyene and all the following models (electrons on wire, HMO calculations): essentially, the two HOMOs and two LUMOs mentioned at the beginning, the *four orbitals*, determine the spectra of porphine derivatives. This realization created a sound long-term basis for further investigations.

In ►Figure 2.55(b), we see a part of an actual calculated MO diagram depicting significant differences from that of annulene, ►Figure 2.55(a). The HOMOs of the *four orbitals* are of the symmetry types a_{1u} and a_{2u}. The LUMOs are two-fold degenerated and of symmetry type e_g^*. ►Figure 2.56 compares the *four orbitals* of porphine structures I–III, thus explaining the different visual behavior of these similar porphines derivatives.

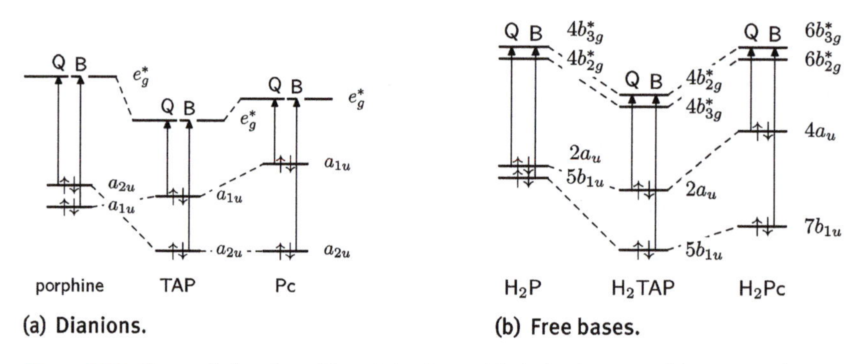

Figure 2.56: Change in band position of the *four orbitals* for three porphine types, the associated bathochromic shift of the Q band, and the hypsochromic shift of the B band according to MO calculations. (a): MO diagram and transitions for porphine, tetraaza-porphine (TAP), and phthalocyanine (drawn after [383]). (b): The same for the free bases (drawn after [375]).

Porphines

In the case of porphines, the HOMOs a_{1u} and a_{2u} are closely spaced and quasi-degenerated, leading to strong configuration interaction. Instead of simple transitions like $a_{1u} \rightarrow e_g^*$, the excited states contain mixtures of different transitions with various intensities. Actual calculations for the free porphine bases are available in [374, 375]. Accordingly, the free porphine base allows the following transitions:

$$\begin{aligned}
&Q_x\!: && k_1(2a_u \rightarrow 4b_{3g}) + k_2(5b_{1u} \rightarrow 4b_{2g})\\
&Q_y\!: && k_3(2a_u \rightarrow 4b_{2g}) + k_4(5b_{1u} \rightarrow 4b_{3g})\\
&B_x\!: && k_5(2a_u \rightarrow 4b_{3g}) + k_6(5b_{1u} \rightarrow 4b_{2g}) + k_7(4b_{1u} \rightarrow 4b_{2g})\\
&B_y\!: && k_8(2a_u \rightarrow 4b_{2g}) + k_9(5b_{1u} \rightarrow 4b_{3g})
\end{aligned}$$

Due to similar energies of all HOMOs, the contributions k_1–k_4 of the Q transitions are of a corresponding magnitude, and the transition dipole moments add up to approximately zero. Thus, the Q transitions are weak, as we can observe. The contributions of the B bands are such that a large net transition moment and intense B bands results.

Tetraazaporphines (TAP)

In the transition from porphine to tetraazaporphine (TAP), one must consider the influence of the substitution of bridge carbon by nitrogen (*meso*-substitution). Nitrogen possesses higher electronegativity than carbon and, therefore, stabilizes MOs with large orbital coefficients at the bridge position, i. e., the MO energy decreases. This stabilization is pronounced for the second-highest MO, HOMO-1, and somewhat less so for the LUMOs, ►Figure 2.57. The HOMO has nodes at the bridge positions and hardly changes. ►Figure 2.56 show the resulting band energies for TAP and H_2TAP. As we can see, the B band moves toward the blue spectral range, while the Q band shifts toward the red spectral range.

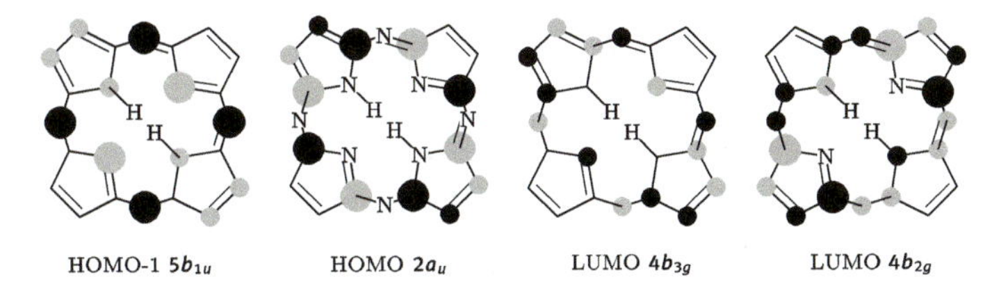

Figure 2.57: Depiction of HOMO, LUMO, and the two adjacent MOs of H_2TAP and magnitude of orbital coefficients (symbolized by the size of the disks, drawn after [374]).

Lifting the degeneracy of HOMOs reduces configuration interaction. Still, there are no simple transitions but mixed ones. However, the contributions of individual transitions are now such that the transition dipole moments no longer compensate for the Q bands. Therefore, they gain about the same intensity as the B bands.

Phthalocyanines (PC)

The annulation of four benzene rings to TAP, yielding phthalocyanines, changes the energy of the *four orbitals*, as depicted in ►Figure 2.56 for Pc and H_2Pc. We can model this step as the interaction of TAP with four butadiene entities, ►Figure 2.58. HOMO $2a_u$ interacts with a $2a_u$ MO of butadiene entities from which $4a_u$ emerges as a new HOMO that is destabilized and raised in energy. TAP's second-highest HOMO $5b_{1u}$ does not change because the counterpart in butadiene is too low; instead, the lower $4b_{1u}$ MO interacts with butadiene. The counterparts of TAP's LUMOs in butadiene ($3b^*_{3g}$, $3b^*_{2g}$) are at such high energies that the benzo annulation has little effect on the LUMO.

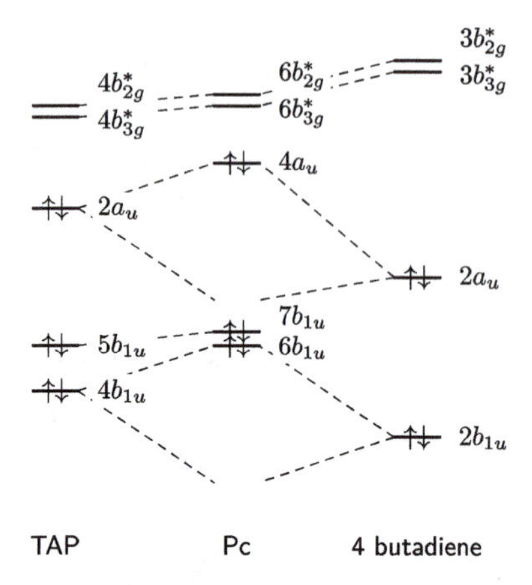

Figure 2.58: Transition of the free bases of tetraazaporphines H_2TAP to phthalocyanines H_2Pc, shown as the interaction of H_2TAP with four butadiene units (drawn after [374, 375]). The position of the LUMO remains unchanged because there is no matching counterpart in butadiene. The interaction of HOMO $2a_u$ with butadiene destabilizes the new HOMO $4a_u$ and lowers the energy of the Q band. HOMO-1 remains unchanged, and so does the position of the B band.

The result of the interaction is a further bathochromic shift of the Q band, which is moved so far to the IR spectral range that a clear green tint adds to the color induced by the phthalocyanine. Due to the even larger imbalance of HOMO-LUMO energies, the transition dipole moments of the Q band add up even more than in the case of TAP, thus increasing the intensity of the Q band. The B band remains essentially unchanged.

Other components of the spectra

In reality, the situation is complicated because MOs below the HOMO also contribute to transitions in the near-UV below the B band. Lone electron pairs of nitrogen atoms can cause $n \rightarrow \pi^*$ transitions. Finally, in some metallo-phthalocyanines, metal d orbitals are located near π and π^* MOs and can act as either donors or acceptors in LMCT and MLCT transitions, ►Figure 2.55(b). These CT bands generally lie between the B and Q bands, broadening them or appearing as isolated peaks. Toyota et al. [374] provides information about the MO diagrams of some metallo-phthalocyanines. Toyota et al. [375] partially reinterprets the bands based on recent SAC-CI calculations. An example may show the complex configurational mixtures discussed in recent literature: $0.57(4a_{2u} \rightarrow 7e_g) + 0.31(2a_{1u} \rightarrow 7e_g) - 0.55(3b_{2u} \rightarrow 7e_g) - 0.2(3a_{2u} \rightarrow 7e_g) + \cdots$ [377].

Dimerization and aggregation

The planar molecular shape of phthalocyanines allows dimerization and dense stacking (aggregation). Dimerization leads to characteristic changes in the spectra, particularly a broadening and shifting of the Q bands.

Dimerization can be interpreted as an interaction of the e_g MOs of two phthalocyanine molecules that form a new set of excited states (*exciton splitting*), ►Figure 2.59 [377, 380]. A suitable spatial arrangement (degree of aggregation, spacing, angle) is required.

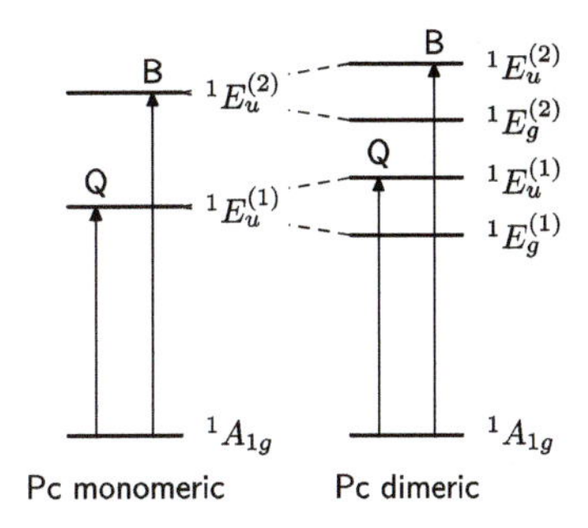

Figure 2.59: Origin of the hypsochromic shift of the B and Q bands in phthalocyanine dimers (drawn after [377, 380]). Depending on the mutual arrangement, transitions into the new E_g states (bathochromism) or as shown into E_u (hypsochromism) become possible.

In the dimer, the excited ^1E_u states each form a $^1E_g^{(1)}$ or $^1E_g^{(2)}$ state (lower in energy) and a $^1E_u^{(1)}$ or $^1E_u^{(2)}$ state (higher in energy). Depending on the mutual orientation of the dimers, transitions to these new states are allowed or forbidden. Transitions into the E_g states that are lower in energy correspond to a redshift of the absorption band, yielding greenish-cool blue hues. Transitions into the E_u states higher in energy correspond to a blue shift, leading to reddish-warm blue hues. In [384], the phenomenon of exciton splitting in detail is discussed.

Ring substitution

B and Q band transitions cause an electron rearrangement from the molecule's center toward its perimeter [4, Chapter 8.6f]. Therefore, electron-withdrawing substituents at the ring periphery stabilize the excited state and lead to a bathochromic shift. In agreement with theory, halogenated phthalocyanines such as PG7 show a distinctly green tint. In the same way, an electropositive metal in the center destabilizes and causes a hypsochromic shift (tetraphenylporphyrin: B = 440 nm, Q = 635 nm; copper-tetraphenylporphyrin: B = 420 nm, Q = 580 nm). However, metallization is not exploited technically; metal-free phthalocyanines or copper-phthalocyanines are almost exclusively used.

2.5.5 Polymethine chromophores

In the previous section, we considered polyenes, i. e., systems comprising n vinyl units or $2n$ π electrons. Suppose we add one methine unit CH_2 with an sp^2-hybridized carbon to such a system. In that case, its p orbital overlaps with the π electrons of the polyene, and the electronic structure of the resulting compound changes considerably. The polyene, a hydrocarbon of the "even alternant" type consisting of localized single and double bonds, develops into a polymethine, an "odd alternant" type. In the polymethine, electrons are widely delocalized along the chain, and a new, central nonbonding MO (NBMO) appears [4, Chapter 9]. The resulting compounds of type I

$$H_2\dot{C}\text{–}(\text{CH=CH})_n\text{–} \underset{}{\overset{h\nu}{\rightleftharpoons}} \text{–}(\text{CH=CH})_n\text{–}\dot{C}H_2$$

I II

$$^{\ominus}H_2C\text{–}(\text{CH=CH})_n\text{–} \overset{h\nu}{\rightleftharpoons} \text{–}(\text{CH=CH})_n\text{–}CH_2^{\ominus}$$

III

$$^{\oplus}H_2C\text{–}(\text{CH=CH})_n\text{–} \overset{h\nu}{\rightleftharpoons} \text{–}(\text{CH=CH})_n\text{–}CH_2^{\oplus}$$

IV

exhibit π systems with an odd number $2n + 1$ of atoms and MOs, called *polymethines*. The neutral hydrocarbon I has an odd number of π electrons and is a radical if the methine group $-CH_2\cdot$ donates a single p electron. There are also anionic polymethines III comprising $2n + 2$ π electrons, to which the methine fragment $-CH_2$**:** donates two p electrons. Polymethine cations IV, having $2n$ π electrons, result if the methine fragment $-CH_2$ only donates sp^2 electrons.

Let us move from a given polyene to the next higher polymethine, one carbon atom longer. The new MO appears as nonbonding MO (NBMO) between the HOMO and LUMO of the polyene, ►Figure 2.60 [208, Chapter 5.1].

The NBMO considerably reduces the energy difference between HOMO and LUMO. Polymethine chromophores with as few as five to seven methine groups ($n = 2$ or 3) possess an intense absorption in the visible spectral range if dimethylamino groups are used as donors and acceptors. Comparable polyenes appear colorful only from 14 methine groups ($n = 7$) onward.

Compared to polyenes, the simple Hückel rule can describe polymethines even at large n because the electrons are delocalized evenly over the entire chain. As an expression for the transition energy, we obtain after Hückel [206–208]:

$$h\nu = -2\beta \sin \frac{\pi}{n+1} \tag{2.10}$$

Replacement of terminal methine groups with heteroatoms leads to compounds that can be formally regarded as donor–acceptor-substituted polymethine and also

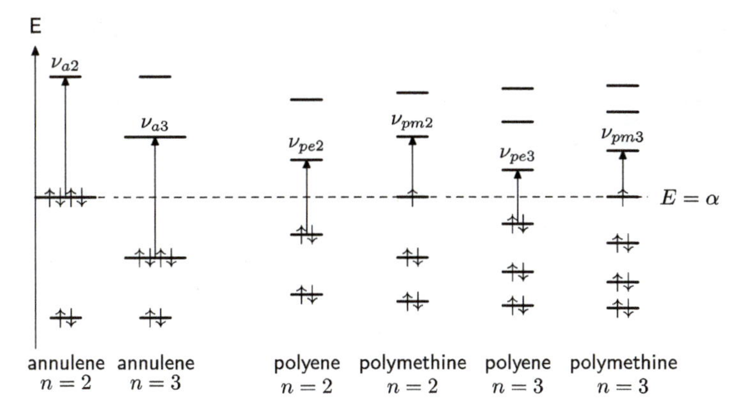

Figure 2.60: Comparison of the MO structure of conjugated π electron systems. All MOs are located symmetrically to an energy α if using the simple Hückel model [208, Chapter 5]. In all cases, the HOMO-LUMO energy difference decreases with n. Left: Annulenes [208, Chapter 5.2]. In theory, their HOMO is two-fold degenerated. Configuration interaction lifts the degeneracy, and multiple transitions become possible. Right: Polyenes and polymethines [208, Chapter 5.1]. Due to the odd number of MOs in polymethines, a nonbonding MO is located between their bonding and antibonding MOs, considerably decreasing the transition energy ν_{pm2}, compared to the corresponding polyene ν_{pe2}.

possess at least two identical or similar resonant structures:

$$\text{D:}\!\diagup\!\!\diagdown\!\!\diagup\!\!\diagdown\text{A} \overset{h\nu}{\rightleftharpoons} \text{D}\!=\!\cdots\text{:A}$$

$$^{\ominus}\text{O} \cdots \text{O} \overset{h\nu}{\rightleftharpoons} \text{O} \cdots \text{O}^{\ominus}$$

$$\text{H}_2\text{N} \cdots \overset{\oplus}{\text{N}}\text{H}_2 \overset{h\nu}{\rightleftharpoons} \text{H}_2\overset{\oplus}{\text{N}} \cdots \text{NH}_2$$

As heteroatoms, oxygen or nitrogen play the most crucial role. Based on the nature of donor and acceptor, we can classify polymethines. ►Table 2.14 shows types based on oxygen and nitrogen, which are very important for natural and synthetic colorants, ►Chapter 4.

It is interesting to see how the properties of polymethines change as a result of this substitution. If heteroatoms or substituents are of the same type, electronic symmetry and properties of the carbon chain, in particular, are preserved. In contrast, different substitution leads to asymmetry and, ultimately, loss of polymethine properties. The systems become donor–acceptor systems with a carbonyl acceptor, which we have already discussed in ►Section 2.5.3.1.

We take merocyanines as an example, exhibiting an electronic asymmetry due to substitution with oxygen and nitrogen. The equilibrium in the ground state then lies

Table 2.14: Important types of polymethines. For amino donors and acceptors, membership in a ring or the chain is essential [205, Chapter 34].

Acceptor	Donor	Class
=O	–OH	Oxonol 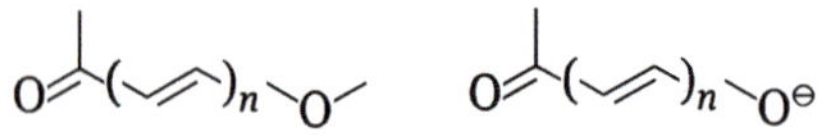
$=O^{\oplus-}$	–OH	
=X, =O	–Y:, $-NH_2$	Mero-, neutrocyanine 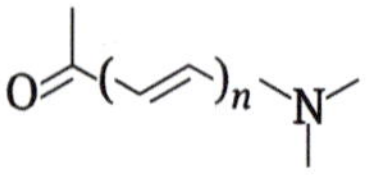
$=NH^{\oplus-}$	$-NH_2$	Streptocyanine 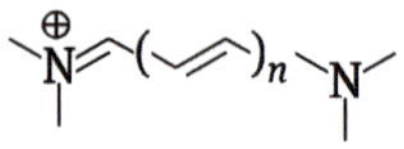
$=NH^{\oplus-}$	$-NH_2$	Hemicyanine 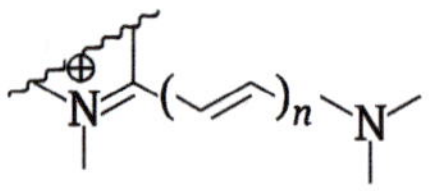
$=NH^{\oplus-}$	$-NH_2$	Cyanine 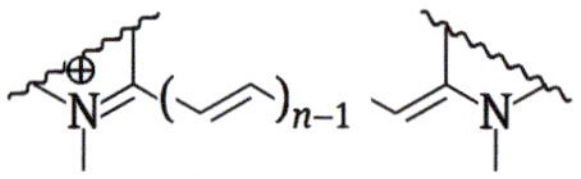

firmly on the side of the neutral form I:

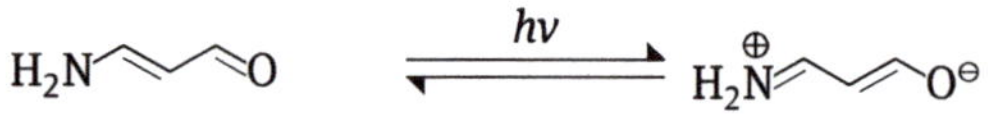

$h\nu$

I
donor-bridge B_n-acceptor

II

The compounds show alternating bond lengths and convergence of the absorption at long wavelengths with increasing chain length and significant differences between the ground and excited state. In the case of merocyanines, differences between nitrogen and oxygen are not yet so distinctive that we can no longer speak of polymethines. However, we can see that the essential properties of the polymethines attenuate.

Dewar rules

We can modify polymethines in various ways without changing their fundamental structure and thus optical properties. These modifications include replacing carbon atoms or methine groups of the chain with heteroatoms and substituting hydrogen atoms of the chain with side groups. The effects of such modifications on absorption bands can be attributed to altered electronegativities (EN) of heteroatoms and the ability of the substituents to supply or accept electrons. Dewar has described these effects in a set of rules for polymethines based on perturbation calculations [207, Chapters 6.9–6.11], [4, Chapter 4.4], [5, Chapter XII], [338, p. 26ff].

Understanding the rules summarized in ►Table 2.16 helps to know what effect occurs when replacing an atom of a π system with a heteroatom or a substituent. The effects only occur when the affected hetero- or substituted atom has large orbital coefficients in the concerned MO, in particular in the HOMO or LUMO, i. e., those MOs involved in the optical transition.

In general, replacing a carbon atom in the chain with a heteroatom of lower EN weakens the π electrons bond to the atom. The energies of all MOs rise. Similarly, substituting the chain atom with an electron donor has the same effect. In ►Figure 2.61 on the left, this is shown graphically.

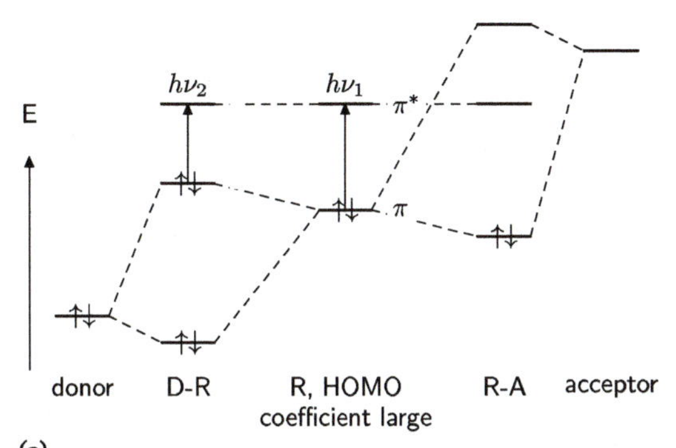

(a)
Donor substitution of a position with a large HOMO coefficient increases HOMO (bathochromism, $\nu 2 < \nu 1$). In contrast, acceptor substitution of such a position lowers the HOMO (hypsochromic shift).

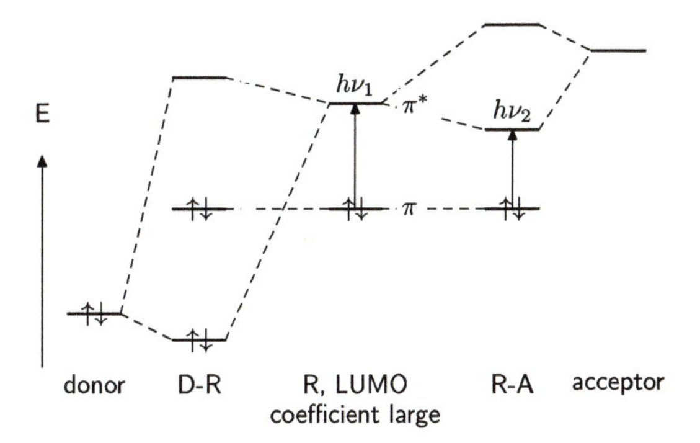

(b)
Acceptor substitution of a position with a large LUMO coefficient lowers the LUMO (bathochromic shift, $\nu 2 < \nu 1$). The donor substitution of such a position, on the other hand, increases the LUMO (hypsochromic shift).

Figure 2.61: Influence of donor and acceptor substitution on the position of HOMO and LUMO of a chromophore R (drawn after [338, p. 26ff], [4, Chapter 4.4], [5, Chapter XII], [207, Chapters 6.9–6.11]).

In contrast, replacing a chain atom with a heteroatom of higher EN tightens the bond of π electrons to the atom and the energies of all MOs decrease. The substitution of the carbon atom with an electron acceptor acts similarly, shown in ►Figure 2.61 on the right.

The effects are pronounced for those MOs to which the heteroatom or the substituted atom strongly contributes, i. e., where the heteroatom or the substituted atom possesses a large orbital coefficient, ►Table 2.15. If the orbital coefficient is large, the observable effect is also significant. If the orbital coefficient is small, the impact is also low and disappears entirely when the orbital coefficient becomes zero, which happens at the nodes.

Table 2.15: Effect of donor and acceptor substitution or change of EN on the location of HOMO and LUMO in polymethines.

Change	Orbital coefficient large at	Effect
Donor substitution, heteroatom with small EN	HOMO	HOMO ↑
Donor substitution, heteroatom with small EN	LUMO	LUMO ↑
Acceptor substitution, heteroatom with large EN	HOMO	HOMO ↓
Acceptor substitution, heteroatom with large EN	LUMO	LUMO ↓

The question now is at which points of the polymethine chain orbital coefficients are significant. To solve this, we need to know that polymethines belong to the so-called "odd alternant" compounds with an odd number of π centers. In the odd-alternant compounds, it is possible to mark the carbon atoms of the π system alternately with stars so that atoms with a star ("starred position") are only connected to atoms without a star and vice versa. We infer from MO calculations that starred positions have a sizeable orbital coefficient in the ground state (HOMO). In contrast, unmarked positions are nodes, i. e., have an orbital coefficient of zero. In the LUMO, all positions have a nonzero orbital coefficient, smaller than the HOMO coefficients since the electron density spreads across more atoms. The size of the disk indicates the magnitude of the orbital coefficient:

D–C–C*–C–A
C*–C–C*–C–C*
polymethine marker
* = starred position

polymethine ground state
(HOMO)

$h\nu$

excited state
(LUMO)

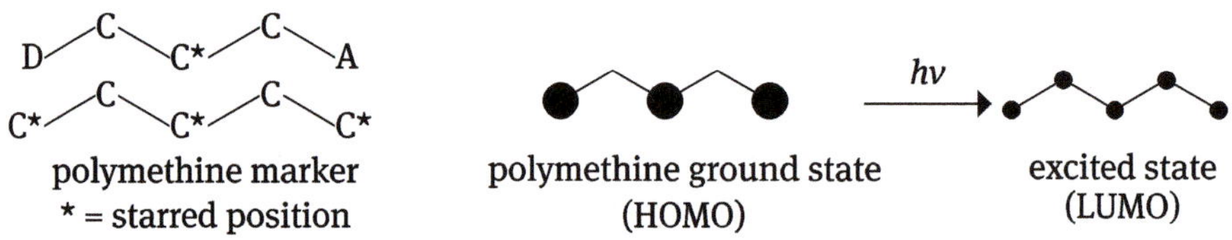

The outer positions are starred positions for a pure carbon chain. If these positions represent a donor D or an acceptor A, they are not subject to Dewar rules. Since the orbital

Table 2.16: The rules of Dewar for estimating substitutional influences on optical properties of polymethines (odd alternant hydrocarbons, the HOMO here corresponds to the NBMO) [207, Chapter 6.10]. "*" indicates a starred position, " " an unmarked position. ↑ means an increase in MO energy, ↑↑ a significant increase. The same is true for ↓. = symbolizes a position where the MO energy remains unchanged.

Substitution with	Pos.	Shift	HOMO position	LUMO position
Donor or heteroatom of lower EN	*	Bathochromic	HOMO ↑↑	LUMO ↑
Acceptor or heteroatom of higher EN	*	Hypsochromic	HOMO ↓↓	LUMO ↓
Donor or heteroatom of lower EN		Hypsochromic	HOMO =	LUMO ↑
Acceptor or heteroatom of higher EN		Bathochromic	HOMO =	LUMO ↓

coefficients represent the strength of the effect (how much a MO energy is increased or decreased), we can depict the following result for a polymethine chain using Dewar's rules, ►Table 2.16:

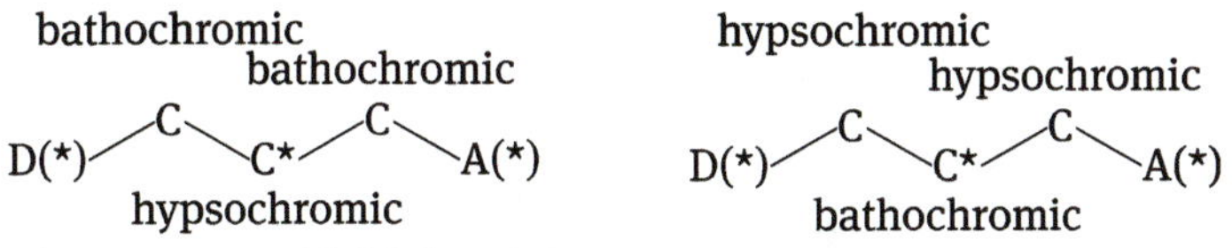

bathochromic
bathochromic
D(*)–C–C*–C–A(*)
hypsochromic
heteroatom with higher EN
or acceptor substitution

hypsochromic
hypsochromic
D(*)–C–C*–C–A(*)
bathochromic
heteroatom with lower EN
or donor substitution

Since polymethines are essential dyes, these rules are often applied, e. g., in colorants for color photography, printing, and particular subject domains. We have already encountered substitutional changes in the case of indigo ►Figure 2.43; we have formally described it either as polyene with a carbonyl acceptor or as polymethine.

(The MOs of polyenes react similar to substitutional changes, but as even-alternant hydrocarbons, polyenes are subject to the pairing theorem, ►Section 2.5.4. The orbital coefficients in the HOMO correspond precisely to those in LUMO, except for the sign. Subsequently, all shifts in the energy occur concordant for HOMO and LUMO. Therefore, neither bathochromic nor hypsochromic shifts result.)

2.5.6 Other chromophores: Sulfide radical ions

The radical trisulfide anion is an important chromophore, especially noticeable in bright blue Lapis Lazuli and ultramarine. We can understand trisulfide and similar polysulfide anions based on MO considerations.

The electronic structure of the disulfide radical anion $S_2^{\ominus}$, having a yellow color and influencing the color in ultramarine and other sulfur-bearing minerals, is shown in ►Figure 2.62 [460]. The allowed electron transitions are essentially the following:

$$^2\Pi_g \rightarrow {}^2\Pi_u \quad (1\pi_u)^4(1\pi_g)^3 \rightarrow (1\pi_u)^3(1\pi_g)^4 \quad \nu_1 \quad \text{VIS, 400 nm}$$
$$^2\Pi_g \rightarrow {}^2\Sigma_u \quad (1\pi_u)^4(1\pi_g)^3 \rightarrow (1\pi_u)^4(1\pi_g)^2(2\sigma_u)^1 \quad \nu_2 \quad \text{UV, 340 nm}$$

►Figure 2.62 also depicts the structure of the otherwise largely unexplored trisulfide radical anion $S_3^{\ominus}$ [460, 461]. This anion is the primary cause of the deep blue color of ultramarine and many other sulfur-containing minerals. The hue can be shifted toward green in the simultaneous presence of the yellow disulfide anion. The trisulfide anion, an angular molecule belonging to point group C_{2v}, enables the following essential electronic transitions:

$$^2B_2 \rightarrow {}^2A_2 \quad (4a_1)^2(1a_2)^2(3b_1)^2(2b_2)^1 \rightarrow (4a_1)^2(1a_2)^1(3b_1)^2(2b_2)^2 \quad \nu_1 \quad \text{VIS, 600 nm}$$
$$^2B_2 \rightarrow {}^2A_1 \quad (4a_1)^2(1a_2)^2(3b_1)^2(2b_2)^1 \rightarrow (4a_1)^1(1a_2)^2(3b_1)^2(2b_2)^2 \quad \nu_2 \quad \text{IR, 870 nm}$$
$$^2B_2 \rightarrow {}^2A_1 \quad (4a_1)^2(1a_2)^2(3b_1)^2(2b_2)^1 \rightarrow (4a_1)^2(1a_2)^2(3b_1)^2(2b_2)^0(5a_1)^1 \quad \nu_3 \quad \text{UV–VIS, 400 nm}$$

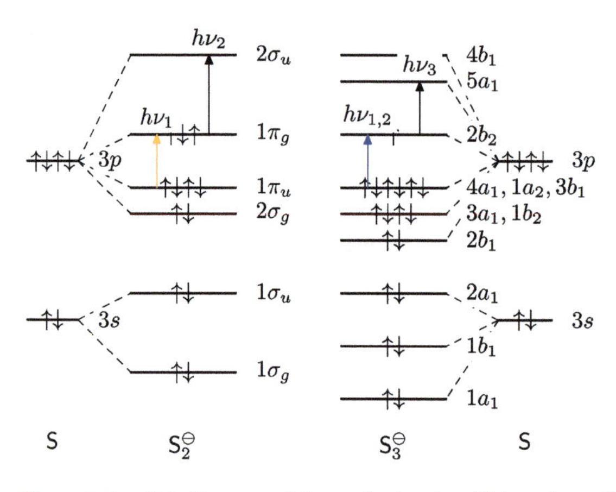

Figure 2.62: MO diagram of the radical polysulfide anions $S_2^{\ominus}$ (yellow) and $S_3^{\ominus}$ (blue), as they occur in ultramarines (drawn after [460]).

Compounds analogs to ultramarine containing selenium and tellurium instead of sulfur also exist. Instead of polysulfide centers, they possess red or blue-green diselenide and ditelluride color centers [462–465]. We return to this in ►Section 3.3.

2.6 Laking and colored lakes

Many colored compounds from either natural plant or animal sources or synthetic compounds are actually soluble dyes. To apply them as paint, we must decrease their solubility in binder and solvent and impart body to achieve a bodily texture. By producing *color lakes*, that are dyes deposited or adsorbed on insoluble inorganic material, we can accomplish both tasks. While the dye component induces the color of lake colorants, the carriers lend them body and texture, paintability, and opacity. Color lakes belong to the mordant or metal complex dyes, ►Section 5.5.

►Table 2.17 lists some color lakes well known to the painter. In [48, Volume 3], [67, 70, 394, 715] details of their spectra and production are given, Arendt [94] is devoted in detail to the in-house production of plant lakes. Bartl et al. [108] contains a large number of annotated medieval recipes. Kirby et al. [392] deals with the nature of precipitate and substrate of color lakes.

Table 2.17: Known color lakes made from the named dyes by laking with aluminum (alum), iron, and tin salts and historically used in panel painting [67, 70, 394, 490, 715]. Lakes used in book illumination are listed in ►Table 8.5 in ►Section 8.3.1.

Plant (animal)	Underlying dye	Proper name of the paint
Yellow		
Dyer's mulberry	Morin (flavonoid)	
Dyer's broom, reseda	Luteolin (flavonoid)	
Dyer's broom	Luteolin (flavonoid), genistein (isoflavonoid)	
Dyer's broom	Quercetin (flavonoid)	
Cross berries (unripe)	Rhamnetin (flavonoid)	Stil de Grain
Red		
Safflower	Carthamine (chalcone)	
Madder root	Alizarin (anthraquinone)	Madder lake, rose madder, turquoise red
Kermes lice	Kermic acid (anthraquinone)	Kermes lake, carmine
Cochineal lice	Carminic acid (anthraquinone)	Carmine red
kerria lacca lice	Laccaic acid A, B (anthraquinone)	Lac dye
Redwood, brazilwood	Brazilin (neoflavone)	
Green		
Cross berries (ripe)	Rhamnetin (flavonoid), anthocyanins	Sap green

Color lakes may cover the whole spectral range, but they possess, like the corresponding dyes, varying and mostly minor light fastnesses. Therefore, in panel painting, only yellow and red lakes of minimum light fastness were used; see the table. However, for book illumination, blue, green, and purple lakes of lower lightfastness could be used since those lakes are protected by closing the books, ►Table 8.5 in ►Section 8.3.1. ►Section 7.4.11.7 discusses damage patterns, attributed to fading lake pigments. In textile dyeing, multicolored lakes have been made for a very long time with the help of mordants [705].

Production of color lakes is a two-step process. In the first step, complexes of the dyes with polyvalent metal cations form. The complexes are precipitated or adsorbed on an insoluble carrier or substrate in the second step.

By complexation, the dyes are stabilized against light since metal-oxygen bonds are very strong and more robust against high-energy UV radiation than carbon bonds. As a result, the absorbed radiation energy spreads over a wider part of the molecule, and fewer bonds break. Most importantly, however, complexation reduces the dye solubility because the molecular weight increases significantly. Typically, two or three dye molecules are part of one complex molecule. Furthermore, the number of functional groups contributing to solubilities, such as hydroxyl, carbonyl, or carboxyl groups, is reduced since they participate in bonding to the metal cation [699, p. 727]. The molecular weight becomes exceptionally high when one or more metal cations bridge several dye molecules in a polynuclear complex, such as Turkish red. Dyes with several suitably arranged coordination sites can bond multiple times to the metal cation and form multidentate complexes of high stability (chelates).

Complexing metals

Metals capable of complexation are iron, aluminum, tin, lead, copper, and chromium. Historically significant is aluminum, as it was widely available in antiquity due to rich deposits of alum. Alum (potassium aluminum sulfate) is the most crucial aluminum salt $KAl(SO_4)_2 \cdot 12H_2O$ and produced from alunite $KAl_3[(OH)_6|(SO_4)_2]$. Known deposits of alunite, formed from potassium-rich feldspars, were found at Phocoea in the eastern Mediterranean and, from 1460 onwards, at Tolfa, Italy. (Its value for the medieval Christian dying industry was so high that quarrels and battels were fought about the deposits and their exploitation. Only after discovering deposits in Italy did the Turks' domination decline over delivery routes from the East.) Due to the crystallization required in alum production, the product was already relatively pure in the Middle Ages. Iron was also readily available, and around 1600 the tin lake was discovered.

Carrier materials

The carrier or *substrate* of a color lake was, from the twelfth century to the eighteenth century, usually produced from an aluminum salt and alkali in a precipitation reaction, roughly corresponding to the following equation:

$$\underset{\text{alum}}{KAl(SO_4)_2} \xrightarrow{HO^{\ominus}} \underset{\text{'Alumina'}}{\text{'}Al(OH)_3\text{'}\downarrow}$$

The alkali used was usually lye from wood ash (predominantly K_2CO_3) or plant ash (predominantly Na_2CO_3). Rotten urine (ammonium hydroxide NH_4OH), calcined tartar (potassium sodium tartrate, yielding K_2CO_3) or milk of lime from slaked lime ($Ca(OH)_2$) could also be used.

The resulting voluminous, gelatinous precipitate traps or coprecipitates the dye, possible accompanying substances, and ions from impurities. The precipitate is amorphous, variable in composition, and difficult to characterize. It does not correspond to any of the crystalline alumina hydrates γ-$Al(OH)_3{=}Al_2O_3 \cdot 3\,H_2O$ (gibbsite) or γ-$AlOOH{=}Al_2O_3 \cdot H_2O$ (boehmite). However, we can characterize the precipitates according to their elemental composition:

- Precipitates containing Al and O are best described as amorphous hydrated alumina $Al(OH)_3$.
- Precipitates containing Al, O, and much S as sulfate are known, especially from the nineteenth century onward.

In recipes for yellow lakes and pink or red brazilwood lakes, lime or other types of calcium carbonate are added, such as marble or eggshells.

The exact nature of precipitates depends on reaction conditions, i. e., pH value, temperature, the concentration of reactants, rate of addition, and crystallization speed. The decisive factor is the sequence of essential steps carried out. Until the eighteenth century, the usual sequence was:

- first, extracting the colorant from the raw material with alkali
- then adding alum for precipitation

In this way, a voluminous precipitate of the first type of hydrated alumina forms, which can coprecipitate any anions present, such as carbonate. The sequence mentioned in recipes of the nineteenth century was:

- first, extracting the colorant from the raw material with alum solution
- then adding alkali for precipitation

This sequence of steps yields precipitates of the second type, incorporating sulfate anions into the substrate. The procedure was used only rarely in the past and, if at all, for brazilwood lakes.

In addition to the alumina mentioned above, gypsum, chalk (calcium carbonate), and lead white (lead carbonate) were also used as neutral carriers and added as such. Soluble metal salts such as alum, stannous chloride, or lead acetate could produce insoluble carriers such as alumina, basic tin oxide, or lead sulfate:

$$2\,KAl(SO_4)_2 + 3\,K_2CO_3 \longrightarrow 2\,Al(OH)_3\downarrow + 3\,CO_2\uparrow + 4\,K_2SO_4$$

$$SnCl_2 + K_2CO_3 \longrightarrow SnO{\cdot}H_2O\downarrow + CO_2\uparrow + 2\,KCl$$

$$2\,KAl(SO_4)_2 + 3\,CaCO_3 \longrightarrow 2\,Al(OH)_3\downarrow + 3\,CO_2\uparrow + 3\,CaSO_4\downarrow + K_2SO_4$$

$$2\,KAl(SO_4)_2 + 3\,PbCO_3 \longrightarrow 2\,Al(OH)_3\downarrow + 3\,CO_2\uparrow + 3\,PbSO_4\downarrow + K_2SO_4$$

One could quickly obtain all the above ingredients in ancient times. The added chalk reacts in the manner shown as a precipitating reagent for an aluminum substrate. When added in large amounts, it serves as a substrate in its own right, being colored by the dye.

Example: Madder lake

In order to take a closer look at the process of laking, we will examine madder lake, a complex of alizarin with polyvalent metal cations. In the naturally occurring madder root, anthraquinones like alizarin act as chromophores. Depending on the predominant anthraquinone and the metal, colors ranging from light pink to dark brown or purple can be observed: Sn^{IV} orange, Al^{III} purple, Al^{III}/Ca^{II} purple, Ca^{II} blue-violet, Fe^{III} blue-black [711, Chapter 2.8.3, Appendix 1.4]. We encounter two questions:

- What is the structure of the resulting lake?
- Why do we perceive a deepening of the color after the metal complex formation?

2.6.1 Structure of the color lakes

The base of the color lakes is an aromatic quinoid structure generally comprising one, two, or three fused rings, carrying hydroxyl and carbonyl groups. They thus offer ideal conditions for assembling metal complexes and act as ligands to the metal cation.

Several factors determine the product of laking. On the one hand, the coordination number of the central metal cation determines the basic complex geometry and number of ligands. On the other hand, the substitution pattern of ligands determines in which orientation and position they bond to the metal center. Furthermore, the nature and substitution of ligands also influence how many ligands can be arranged around the metal simultaneously, respecting their size and geometry. Secondary ligands (solvent or anions of the original metal salt) join to achieve the maximum possible coordination number:

2 alizarin $\xrightarrow[-H_2O]{Al(OH)_3}$ I, alizarin-Al lake proposal 1963

alizarin

I, alizarin-Al lake
proposal 1963

Structure I for the aluminum lake was proposed in 1963 [707], [699, p. 727ff] and is only one milestone on a long (misleading) path. The problem is that aromatic ligands are frequently ambident, i. e., they possess several positions capable of chelating. For example, in the case of alizarin, aluminum can coordinate via oxygen atoms at positions 1,2 or 1,9. Baumann and Hensel [699] already established the 1,9-coordination, but it was confirmed only by modern solid-state spectroscopy [712]. Spectroscopy also showed that in contrast to I, the sixth coordination position is not occupied by the solvent (water) but the hydroxy group of a second aluminum complex, resulting in an oxygen-bridged, polynuclear complex II:

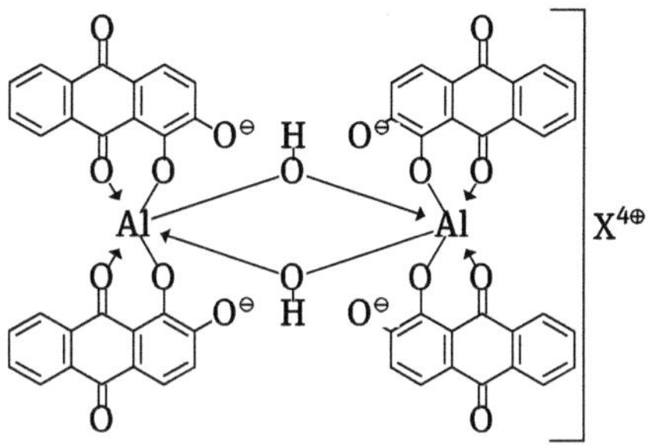

II, Al lake
proposal 1994–1996

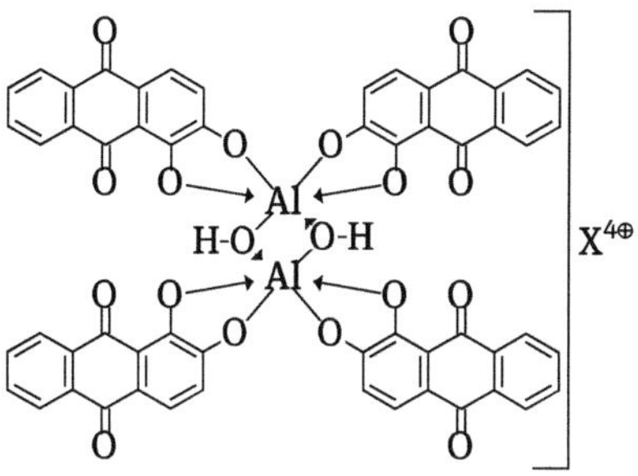

III, Al lake
proposal by Wunderlich, 1993

Wunderlich [711, Chapter 3.2.2.2], however, points out that the negative charge in alizarinate dianions preferentially concentrates at O^1 and O^2 so that for aluminum alizarinate, expecting structure III similar to that of Turkish red IV.

It becomes interesting when, as in the case of Turkish red, several metals are added. In this case, chelates form using positions 1,2 and 1,9. Their size and Lewis acid strength largely determine the distribution of metals among these positions; see below.

Historically, in the course of the structural elucidation of aluminum lakes, a structure IV was proposed for the calcium aluminum lake [706], [699, p. 727ff]. It has also been attributed to pigment PR83:1 (alizarin carmine), ►Section 4.6.4.1 [12]. According to [708], the formation of the calcium-aluminum lake starts by forming an acidic calcium salt at the 2-hydroxyl group since the 1-hydroxyl group forms a relatively strong hydrogen bond with carbonyl oxygen and is not very reactive. The acid salt can either react further to a pure calcium lake or, in the presence of a trivalent cation such as aluminum, form the dimeric structure IV, including the 1-hydroxyl group, ►Figure 2.63.

However, according to [710, 711], the calcium-aluminum lake is a polynuclear, aluminum-bridged complex V analogous to II. In V, two aluminum and two calcium cations form a tetrameric metal complex. Each metal cation coordinates two alizarin molecules by oxygen bridges, aluminum via the 1,2-positions (yielding a five-

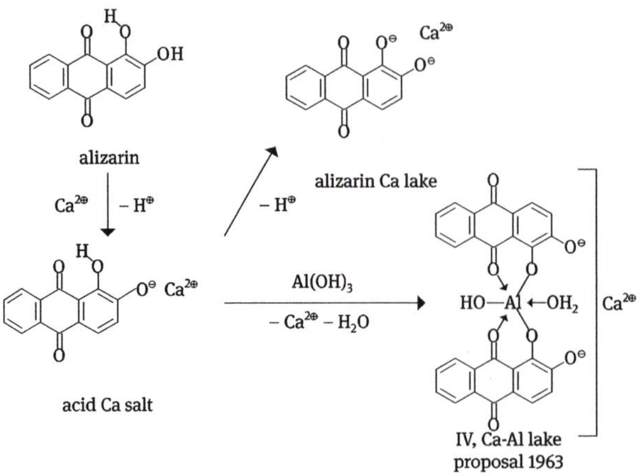

Figure 2.63: Formation of calcium and calcium-aluminum lakes of alizarin [708].

membered chelate ring), and calcium via the 1,9-positions (yielding a six-membered chelate ring). Altogether, this creates a tetramer of four alizarin molecules arranged around a metal-oxygen core:

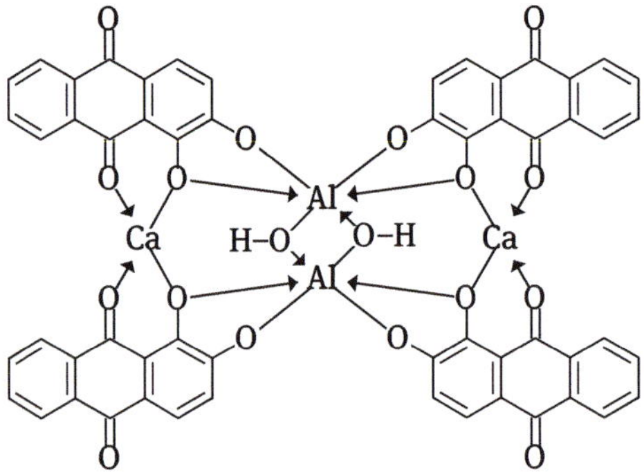

V, Ca-Al lake, Turkish red
1994

We can explain the preference for certain coordination positions on the ambident alizarin ligand by the Lewis acidity of the involved metal cations [711, Chapter 3.2.1.1]. In the alizarinate dianion, the negative charge concentrates on oxygen at positions 1 and 2. Therefore, we expect the cation with the highest Lewis acidity (the highest ability to accept electrons) to attach to these positions when different metal cations are present. In the case of the calcium-aluminum lake, aluminum has higher Lewis acidity and coordinates with alizarin at 1,2-positions, leaving 1,9-positions for calcium.

A series of similar lake structures (e. g., Ba-Al lake) is discussed in [712]. Relevant for the painter is that bridging several alizarin units yields an oligomer. The increase in molecular weight of the oligomer decreases its solubility significantly. This decrease explains the famous washing fastness of Turkish red and the desirable pigment properties of the calcium aluminum lake and lake pigments in general.

2.6.2 Practical procedure

Practically, laking can be achieved in various ways [48, Volume 3], [70, 108, 603, 604]. Natural raw materials such as alum and soda (potassium aluminum sulfate and sodium carbonate) are of paramount importance, as some historical recipes may show [48, Volume 3, p. 122]:

- *Madder lake* was formed by leaching madder roots with cold water, washing out accompanying soluble substances, while remaining alizarin and purpurin as insoluble pigments in the root. After squeezing to remove the cold water, the root is soaked for a while in a hot alum solution, allowing the complex to form. From the complex, the lake evolves. After filtration of the hot solution, the addition of sodium carbonate (soda), potassium arsenate, or borax (Az = alizarin) precipitates the lake:

$$4\,Az(OH)_2 + 3\,KAl(SO_4)_2 + 13\,NaOH \longrightarrow$$
$$[(AzO_2)_4Al_2(OH)_2]^{4\ominus} + Al(OH)_3 + 8\,H_2O + 3\,K^{\oplus} + 13\,Na^{\oplus} + 6\,SO_4^{2\ominus}$$

madder lake

In this and the following procedures, choosing the correct solution temperatures is crucial to accelerate the complex formation and keep alizarin in solution without producing brown decomposition products.

- *Rose madder* was produced by leaching madder root with dilute sulfuric acid. On heating, pseudopurpurine Psp was deposited, dissolved in alum solution, and precipitated with warm soda solution. Precipitation on chalk yields a lighter shade.

$$4\,Psp(OH)_2 + 3\,KAl(SO_4)_2 + 3\,NaOH + 2\,H_2O \longrightarrow$$
$$[(PspO_2)_4Al_2(OH)_2]^{4\ominus} + Al(OH)_3 + 3\,K^{\oplus} + 3\,Na^{\oplus} + 10\,H^{\oplus} + 6\,SO_4^{2\ominus}$$

rose madder

The precipitated flakes could also be converted into a concentrated lake by dissolving in sodium hydroxide solution and precipitating with aluminum sulfate.

$$4\,Psp(OH)_2 + Al_2(SO_4)_3 + 8\,NaOH \longrightarrow$$

$$[(PspO_2)_4Al_2(OH)_2]^{4\ominus} + 8\,Na^{\oplus} + 2\,H^{\oplus} + 3\,SO_4^{2\ominus} + 6\,H_2O$$

rose madder

- Pure aluminum alizarinate was formed by cold leaching of madder roots, extracting the dye with hot alum solution, and precipitating the sulfate ions by adding lead acetate. After filtration, the concentrated aluminum lake deposits.

$$4\,Az(OH)_2 + 2\,KAl(SO_4)_2 + 2\,H_2O + 4\,PbAc_2 \longrightarrow$$

$$[(AzO_2)_4Al_2(OH)_2]^{4\ominus} + 4\,PbSO_4 + 2\,K^{\oplus} + 2\,H^{\oplus} + 8\,HAc$$

alizarin carmine

- For manufacturing the brilliant Turkish red (calcium aluminum lake of alizarin), the substrate (aluminum hydroxide) was prepared first by mixing a hot solution of natural alum and sodium carbonate. After filtering, washing, and adding calcium chloride and madder, the calcium-aluminum lake of alizarin formed and settled on cooling.

$$4\,Az(OH)_2 + 2\,CaCl_2 + 2\,KAl(SO_4)_2 + 6\,NaOH \longrightarrow$$

$$[(AzO_2)_4Al_2(OH)_2]^{4\ominus} + 4\,Cl^{\ominus} + 2\,K^{\oplus} + 6\,Na^{\oplus} + 4\,H^{\oplus} + 4\,SO_4^{2\ominus} + 4\,H_2O$$

Turkish red

In modern times, pure substances such as aluminum sulfate and alizarin have been used instead of natural substances, increasing the purity of the product. Turkish red oil was added to improve the wetting of alizarin. Turkish red oil is a surfactant derived from castor oil treated with sulfuric acid and then neutralized with sodium hydroxide solution. It is a mixture of castor oil (ricinoleic acid glyceride), the sodium salts of ricinoleic acid sulfated at the hydroxyl group, polyricinoleic acids, and their anhydrides and lactones.

OH

COOH

ricinoleic acid

From the Roman epoch, we know about the use of color lakes as artists' materials only on a case-by-case basis. Color lakes are more often described in the Middle Ages, and we even dispose of recipes. Nevertheless, the principle of laking was already used in the second millennium before Christ for textile dyeing, ►Section 5.5. The textiles were impregnated with a solution of the complexing metal salt, and the metal hydroxide was precipitated on the fiber by adding alkalis. Subsequent soaking of the stained fabric in a dye solution led to laking directly on the yarn, fixing the dye on it.

2.6.3 Hue shift

The formation of the complex between a metal and an organic ligand gives rise to two types of color phenomena:
- The starting compounds consist of a *d* block metal salt and an organic ligand. They are colorless, the ligand absorbing in the UV, and only complexation induces color caused by ligand field absorption.
- The organic ligand is already colored, or it absorbs in the near-UV. By complexation, a bathochromic shift into the visible spectral range takes place. The metals involved do not have to be *d* block metals; e. g., also aluminum or tin will do. The optical transition possesses $\pi \rightarrow \pi^*$ or MO-MO character.

d Orbital bathochromism: ligand field and LMCT transitions

In the first case, metallization with a *d* block metal yields a metal cation as a chromophore: the organic molecule acts as a ligand and builds up a ligand field surrounding the metal. This lifts the degeneracy of the metal's *d* orbitals, making *d*-*d* transitions possible in the visible spectral range, ►Section 2.3.

Furthermore, in complexes with highly charged cations such as $Fe^{3\oplus}$ or $Ti^{4\oplus}$, LMCT transitions $\pi \rightarrow t_{2g}(t_2)/e_g(e)$ can occur. The metal accepts electrons from electron-rich ligands and their π systems, giving rise to color, ►Section 2.4.1.

$\pi \rightarrow \pi^*$ Bathochromism

In the second case, anthraquinones illustrate the bathochromic shift of $\pi \rightarrow \pi^*$ transitions from UV into the visible spectral range: anthraquinones are pale-red; in contrast, anthraquinone-metal complexes such as madder lake are bright red. Since the extent of bathochromic shifts depends on the metal, we can obtain several shades from one dye through laking with different metals. Besides the carmine madder lake, there is a scarlet and a violet-red lake.

The bathochromic shift can go so far as metal complex dyes are blue or even green, ►Section 5.5. Since these shades are difficult to achieve otherwise, metal complex dyes are often involved in dark colorations [700]. In addition to the bathochromic shift, a broadening of absorption bands occurs. We visually perceive it as dullness: Metal complexes show less pure colors but are muted, dull, brownish, or greyish. It thus opens up the possibility of producing brown and black dyes.

What is the cause of these phenomena? One factor is that the formation of a chelate ring favors planar molecular shapes, supporting extended conjugation of π orbitals. Essential to the bathochromic shift, e. g., in compounds of the types I to III, however, are two factors that are generally exemplified in [4, 710] and detailed in [703, 704]:

- Metallization increases the electron donor strength of groups such as the hydroxyl group. The free electron pair of the donor group is more easily released into the π system.
- The free electron pair of acceptor groups like the azo or the carbonyl group is shared with the metal, thus increasing the electronegativity of the acceptor.

Both factors enhance the substituent effects of donor–acceptor chromophores.

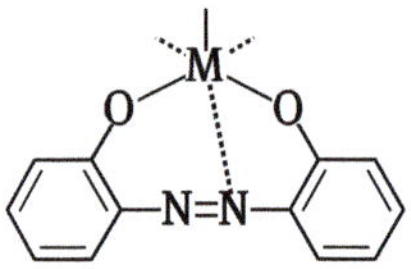

I, azo metalized 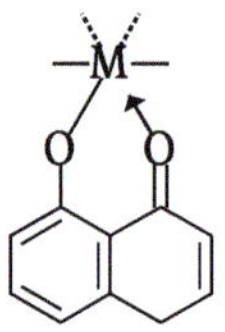   II, quinone metalized 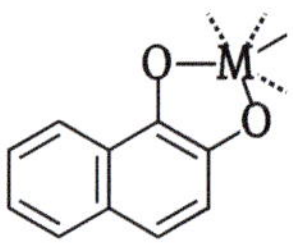   III, dihydroxylated aromatic compound metalized

Increase donor strength

The Lewis structure notation describes metal-ligand σ as covalent but, in fact, they have ionic components. The more electropositive the metal is concerning the ligand, the more pronounced the polarity is. Since the ligand is often oxygen (E_n 3.44) or nitrogen (E_n 3.04), which are comparatively highly electronegative elements, the negative charge concentrates at the ligand, increasing its donor capabilities:

$$\underset{\text{covalent bond}}{\underset{E_n(M)\text{ high}}{M{-}L}} \longleftrightarrow \underset{\text{partial ionic}}{M^{\delta\oplus}{-}L^{\delta\ominus}} \longleftrightarrow \underset{\text{ionic bond}}{\underset{E_n(M)\text{ small}}{M^{\oplus}\quad L^{\ominus}}}$$

Using a hydroxylated anthraquinone as an example, we can think of the metal's involvement in the excitation process as follows [711, Chapter 3.3.2]:

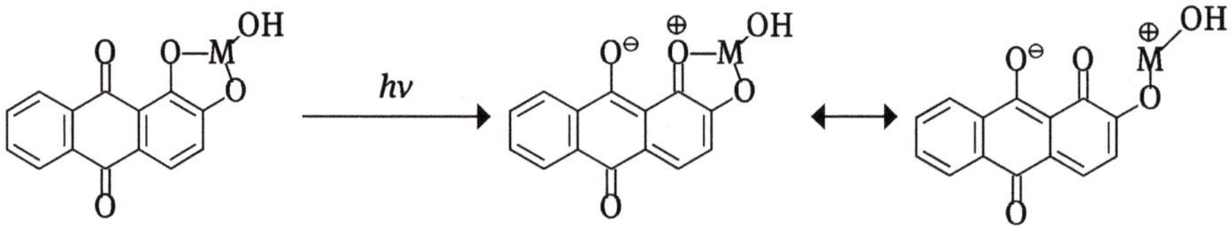

$h\nu$

For electropositive metals, the equilibrium is on the ionic side. In the case of alkali metals, the hydroxyl group –OH even becomes the powerful donor $-O^{\ominus}$:

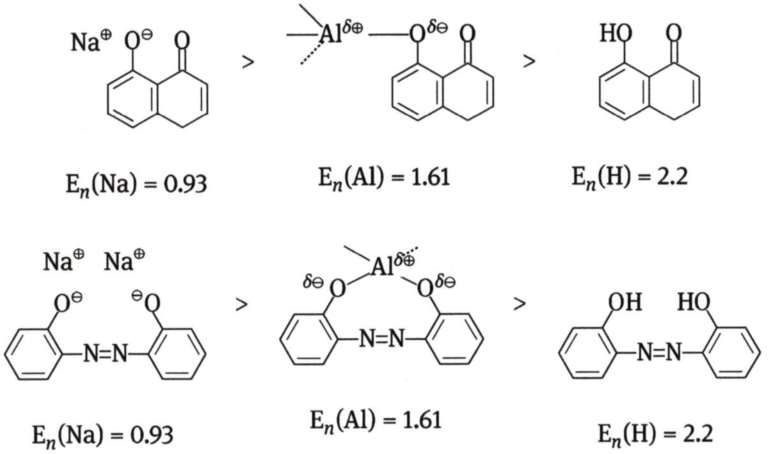

$Na^{\oplus}$ $O^{\ominus}$ O > $Al^{\delta\oplus}$ — $O^{\delta\ominus}$ O > HO O

$E_n(Na) = 0.93$ $E_n(Al) = 1.61$ $E_n(H) = 2.2$

$Na^{\oplus}$ $Na^{\oplus}$ $O^{\ominus}$ $^{\ominus}O$ –N=N– > $^{\delta\ominus}O$ $Al^{\delta\oplus}$ $O^{\delta\ominus}$ –N=N– > OH HO –N=N–

$E_n(Na) = 0.93$ $E_n(Al) = 1.61$ $E_n(H) = 2.2$

In the case of strong donors, the absorption of low-energy red light can be sufficient to initiate the electron transfer from the donor to the acceptor. The perceived color is then a shade of blue, while the weaker, nonmetalized donors induce colors in the yellow to red range. ►Figure 2.64 illustrates trends for anthraquinone lakes: electropositive metals such as calcium or strontium (E_n 1.0, resp., 0.95) lead to absorption above 550 nm, while moderately electropositive metals such as tin and aluminum (E_n 1.96 and 1.61, resp.) lead to absorption around 500 nm. The even more electronegative hydrogen (E_n 2.2) leads to absorption at 450 nm.

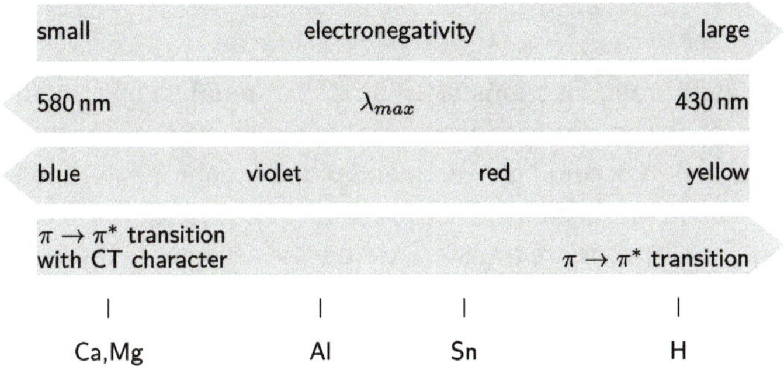

small electronegativity large

580 nm λ_{max} 430 nm

blue violet red yellow

$\pi \rightarrow \pi^*$ transition with CT character $\pi \rightarrow \pi^*$ transition

Ca,Mg Al Sn H

Figure 2.64: Influence of electronegativity (EN) of the metal on the color of lake colorants (anthraquinones) [710], [711, Chapter 3.3]. From left to right, the EN of the metal increases, the wavelength of the absorption band decreases, the induced color shifts from blue to yellow or colorless, and the transition loses any CT character. Ca, Mg, Al, Sn, and H are marked as examples.

Another consequence of a distinct polar metal-donor bond is that the $\pi \rightarrow \pi^*$ transition increasingly assumes a charge-transfer character: the high charge of the donor system can partially flow into an empty MO of strong metal character. This transition occurs with the CT-typical high intensity.

Increase in acceptor strength

The second effect of metallization does not increase the strength of donor groups but that of acceptor groups. The complexation of oxygen- or nitrogen-based acceptors partially transfers the free electron pair of carbonyl oxygen or azo nitrogen to the metal. As a consequence, the electron density at the acceptor decreases. The electron transfer from a donor through the π system to such an acceptor, which has effectively become more electronegative, can already occur by absorbing low-energy light. From another point of view, we can conclude that the metal or a metal chelate act as a suitable acceptor, increasing the conjugated π system. It thus represents an auxochrome, yielding a bathochromic shift. We illustrate this using Lewis structures:

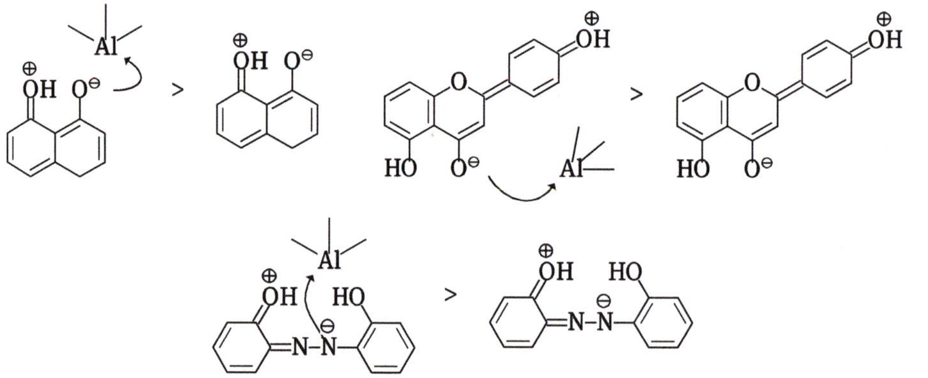

In p block metals such as aluminum, electrons of an acceptor are taken over by the metal's p orbitals and, therefore, contribute to acquiring the noble gas configuration. For d block metals, an interaction occurs between metal d and acceptor p_π orbitals, which we already have described in ►Section 2.4.1 and ►Figure 2.34, yielding d-p MOs of strong metal character. The electrons are transferred from the acceptor to these d-p MOs, and thus to the metal.

Aluminum alizarinate (madder lake) illustrates the ability of a complexed metal to act as an acceptor [711, Chapter 3.2.2.2]. MO calculations show that for a HOMO-LUMO transition, charge flows into a $3p$ orbital of aluminum, i. e., a $\pi \rightarrow 3p$ transition of distinct CT character takes place. In free alizarin, the charge is predominantly taken over by the carbonyl system in the quinone ring:

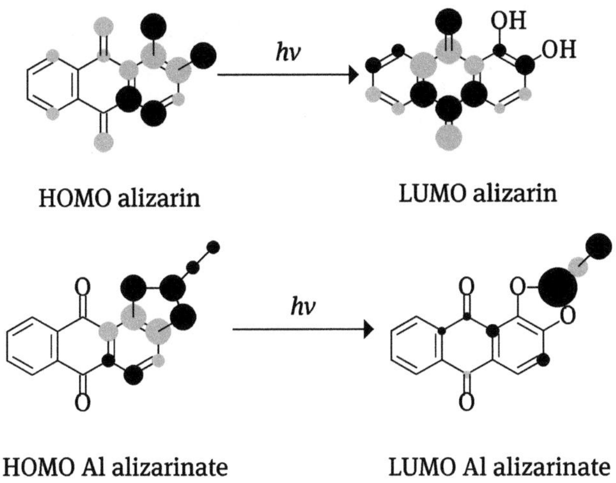
hν
OH
OH
HOMO alizarin
LUMO alizarin
O
O
hν
O
O
O
O
HOMO Al alizarinate
LUMO Al alizarinate

This intense LMCT transition dominates the induced color for some metals compared to the weaker $\pi \rightarrow \pi^*$ transitions. This dominance explains the frequent appearance of intensive colors in complexation reactions with, e. g., Fe^{III} cations, such as the classical iron-phenol reaction, ►Section 8.4.4.

3 Inorganic pigments

The history of painting, at least the part whose artworks have survived to this day, begins with inorganic colorants. They surround humankind from the beginning and build the world at large in the form of earths, minerals, and rocks. They are often present in large quantities and readily available. Their occurrence is not limited to the habitats of certain animals and plants, and the most important of them are available in most cultures.

The advantage of inorganic colorants is their high to very high resistance to a wide range of environmental conditions (light, air, temperature). Especially in the early days of painting, when chemistry did not even exist as alchemy, they offered a more or less complete color palette in a wide range of applications.

Humans painted caves with them over twenty thousand years ago; later, they created religious and decorative mural and panel paintings and decorated ceramic objects with the hot-painting technique. The main disadvantage is the limited color palette that can be composed of commonly found materials. In addition to various blacks and whites, primarily yellow, red, and brown earth colors (ochers and earths) are available, which show beautiful, but also dull colors. The reason for the dullness is that, chemically, the bulk elements of the geology (e. g., Si, O, alkaline earth elements such as Ca, Mg) are either colorless, or their color results from LF transitions (Fe, Mn). Since spectroscopic selection rules forbid many LF transitions, transitions have low intensity and produce dull colors. However, regarding their chemistry, the natural oxides of these elements are exceptionally stable.

Many elements that significantly enlarge the color space only occur as rare minerals. They extend the palette into the green, blue, and purple range (Cu, Co, Cr) or the yellow, red, and white one (Pb, As, Sn, Sb, Hg). They introduce new mechanisms of color formation (SC, CT), enabling intensive and pure colors, especially in the white, yellow, and red range. Admittedly, these minerals are not only rare, but their color-bearing elements were primarily toxic heavy metals (Pb, As, Sb, Hg) or mobile transition metals (Cu, Co, Cr^{VI}), which are also unhealthy or even toxic. In addition, natural hydroxides, carbonates, and sulfides often possess low stability against chemical influences. Light and temperature induce decomposition of the pigment or chemical conversion into differently colored or colorless compounds, i. e. blackening, browning, or bleaching are the consequences.

Humans wished to expand the color spectrum beyond the naturally available compounds; therefore, early in history, they carried out chemical syntheses of colored substances that did not exist in nature. The ancient Egyptians made the first known inorganic synthetic achievements and produced Egyptian blue and green as early as 5000 years ago. 2000 years later, white, yellow, and red derived from lead, and verdigris and vermilion followed. The Renaissance produced lead-tin yellow; with the advent of early alchemy and mineralogy in the seventeenth and eighteenth centuries, a whole series of pigments (lead-tin yellow, Naples yellow, Prussian blue) was

https://doi.org/10.1515/9783110777116-003

developed. From the nineteenth century onward, scientific and industrial chemistry provided numerous new inorganic pigments, which replaced unsatisfactory precursors and even possessed new hues, e. g., orange and purple. Most pigments of that time are no longer in use today for toxicological reasons, as they base on harmful heavy metals such as Pb, As, Sb, Cd, Co, Se, and Cr^{VI}. Examples of green arsenic compounds are emerald green and Scheele's green; examples of lead-containing pigments are chrome yellow and Naples yellow.

From twentieth century onward, the industry developed new chemical classes for pigment chemistry, e. g., mixed metal oxides, incorporating color-bearing elements into a mineral host lattice. Hence, they are no longer mobile, i. e., bioavailable. Consequently, it has become possible to use chromogenic elements such as Cd, Se, Co, Cr, or Mn without risk. However, the disadvantage is often a weaker coloring power and purity. Modern inorganic pigment chemistry aims to develop non-toxic yet economical pigments without the weaknesses mentioned above. Legislation is increasingly concerned with threats to people's health and the environment, placing more and more significant restrictions on the chemistry of new pigments.

As a consequence, progress tends to come on a small scale. For example, in the case of mixed metal oxides, new metal combinations provide yellow and orange-yellow pigments that are increasingly equivalent to the critically viewed brilliant chrome yellow in terms of color power and purity. Bismuth vanadate and cerium sulfides are representatives of this new chemistry. In addition, inorganic colorants have been developed in recent years based on a better understanding of solid-state structures and the consequences of manufacturing conditions and post-refinement. Therefore, the artists' palette slowly but steadily expands with high-quality nontoxic pigments.

The restrictions on pigment chemistry force manufacturers to leave familiar paths and enter new ones. Nowadays, chemists can delight in colored compounds such as $LaTaON_2$ or $YInO_3 : Mn^{3\oplus}$.

Is it worth the effort to further develop inorganic pigment chemistry? The answer is yes because inorganic pigments offer advantages:
- high refractive index, and thus a high opacity
- high light, temperature, and weather resistance
- high color fastness
- high resistance to solvents and bleeding
- economic production

What disadvantage remains after eliminating brilliant but dangerous pigments? After all, many have already vanished or will do so in the future. The disadvantage is frequently lower brilliance compared to organic colorants, and in some cases the use of toxic starting materials or intermediates.

Today, inorganic pigments are produced on an industrial scale but now under strictly controlled manufacturing conditions. By these means alone, today's pigments are more brilliantly colored, purer in color, and more reproducible than their predecessors. In addition, their optical, chemical, and application properties and their stability can be optimized by post-treatment even better to fulfill the intended use. These steps forward frequently also decrease toxicity by effectively removing hazardous intermediates or blocking conversion reactions into toxic products. Careful control of manufacturing also prevents hazards during production and application.

We will take a closer look at the following classes of preindustrial pigments:
- carbon (as soot, char, and coke), ►Section 3.1
- copper pigments such as carbonates and hydroxides, mainly from natural sources, ►Section 3.2
- natural ultramarine, ►Section 3.3
- oxides and sulfides of As, Sb, Pb, and Hg, mostly of natural origin, in some cases also products of early synthesis, ►Section 3.4.1
- natural iron oxides, ►Section 3.4.2

The chemical industry produces other classes by building on chemical research:
- oxides/hydroxides of iron and several other metals (Cr, Mn, Ni, Co, Cu, Pr, V) often present as mixed oxides and synthetically produced, ►Section 3.4.3
- cerium pigments, synthetic, ►Section 3.4.4
- chromium oxide, chromate, and molybdate pigments, some of natural origin, many synthetic, ►Sections 3.4.5 and 3.6
- titanium, zinc (►Section 3.4.6), cadmium (►Section 3.4.7), and bismuth pigments, synthetic, ►Section 3.5
- iron blue pigments, synthetic, ►Section 3.7
- other inorganic pigments, mainly synthetic, ►Section 3.8

►Figure 3.1 depicts schematically how these colorants cover the entire color spectrum. In addition, [9, Chapter 7, app. D] contains reflectance spectra for some of these pigments.

3.1 Carbon pigments

The Colour Index lists black pigments under the denominations PBk6 to PBk10, based on elemental carbon of varying purity [48, Volume 4], [49], [201, p. 358ff], [398, 399], [149, p. 101ff]. All but PBk10 (graphite) form by combustion or charring of carbonaceous organic material. The carbonization products can be classified as soot, chars, cokes, and coals:
- Soot or flame carbon forms in the gas phase of flames in an atmosphere that is poor in oxygen, and has a typical morphology of small spherical particles. The gas phase is created by burning fuels consisting of complex natural materials like oils, fats, and resins or pure materials like natural gas or petrochemical oils.

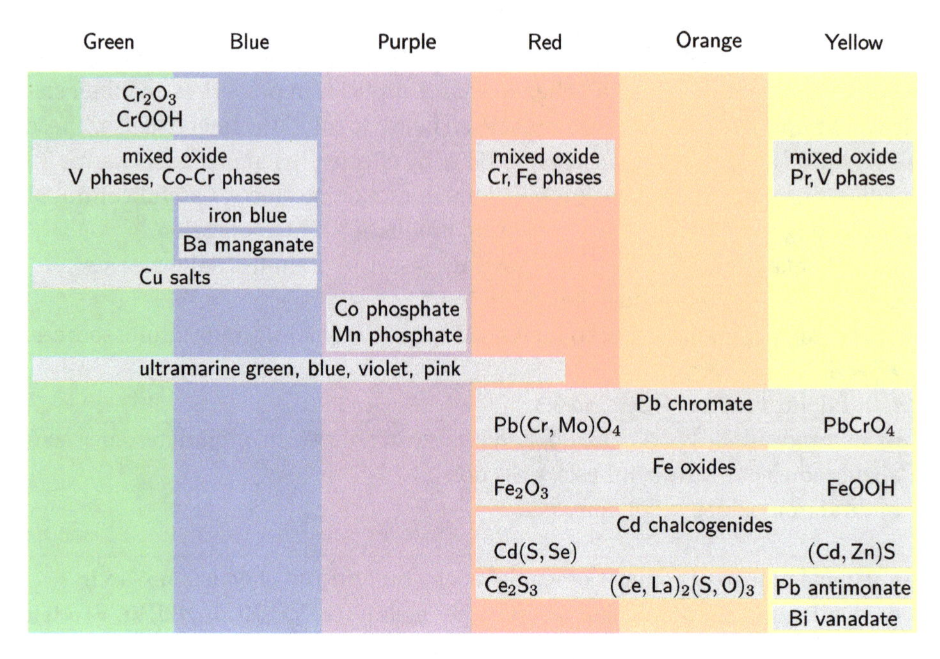

Figure 3.1: Coverage of the color spectrum by inorganic colorants.

- Chars form from solid precursors that remain solid during carbonization. Precursors are wood (►charcoal), cork, fruit stones, twigs, vines, and other plant materials. Chars retain the morphology of their precursor and have much porosity. Due to their immobility during carbonization, the fuel's carbon incompletely rearranges into a graphite structure or does not rearrange at all. Thus, chars are classified as "hard carbon."
- Cokes form by carbonization of liquid or plastic precursors. Precursors can be sugars, collagen, gelatine, bitumen, and other nonvolatile liquids. Since bones contain collagen (in conjunction with apatite), they are also precursors to coke formation. Due to the precursor's flexibility, a molecular rearrangement occurs during carbonization, reducing the overall energy. The precursors are convertible into graphite at a lower temperature than chars. Thus cokes are called "soft carbon." Furthermore, cokes do not retain their precursor's morphology and typically form lumps of high porosity.
- Coals are organic rocks formed by natural, slow carbonization of plant remains in the Earth's crust during some period. Coals are rarely used as pigments.

Since early times, naturally occurring products such as graphite or charcoal have been used as black pigments. Many were produced artificially early on (charcoal, charring products, and soot from fires and lamps). Control combustion processes increased purity in the nineteenth century.

Carbon pigments are characterized by high opacity and usually strong coloring power; moreover, they are resistant to light and other environmental conditions. Exceptions are products such as bister, which are only partially carbonized. Here, chemically unstable brownish tarry components determine the intended delicate hue.

Carbon occurs in several modifications, such as diamond, graphite, fullerenes, carbon-tubes, and possibly, a high-pressure modification [195, Chapter XV.1.1.2]. For the topic of this text, only ►graphite is of interest since carbonaceous black materials consist of "amorphous," i. e., microcrystalline graphite in various arrangements. Therefore, graphite represents the actual chromophore.

Graphite (PBk10, CI 77265)

Graphite is crystallized carbon, organized in infinite layers or sheets of annulated 6-membered rings called graphene [48, Volume 4], [49, 195, 197, 398, 399]. The C-C bonds possess σ character and are 0.142 nm long, close to the length in aromatic compounds, and equally strong. Within each carbon atom, 2*s* and two 2*p* AOs combine into sp^2 hybrid orbitals, and three of four valence electrons of carbon participate in three σ bonds to the next-neighboring carbon atoms, thus creating the hexagonal layer structure. The remaining $2p_z$ AO combine into a π valence band, occupied by the fourth valence electrons of each carbon, and an empty π^* conduction band, ►Figure 3.2.

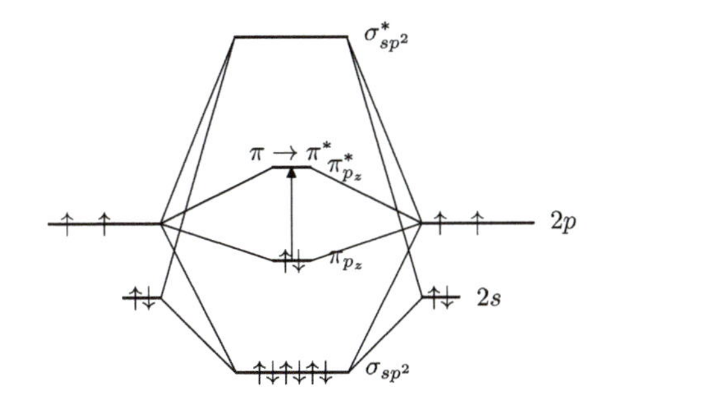

(a) Schematic development of bands in graphite or graphene

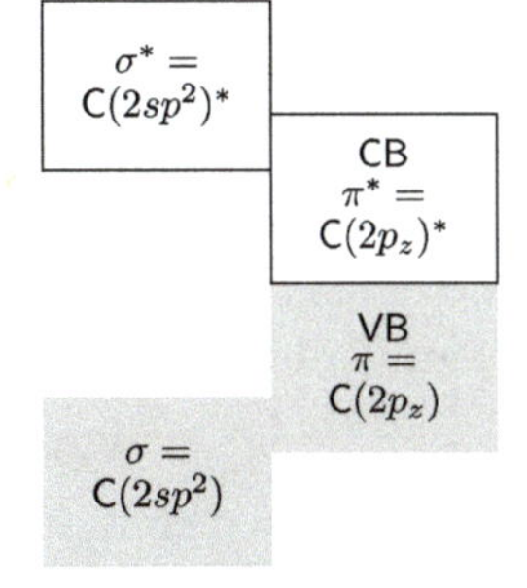

(b) Flat-band model of graphite or graphene [197], [339]

Figure 3.2: Origin of high absorption, luster, and grayish to black color of graphite or graphene [197, Chapter 7, p. 240], [346]. Hybridization of 2*s* and two 2*p* AOs of carbon form sp^2 hybrid orbitals, developing into a σ and σ^* band and causing covalent linkage of carbon in a flat hexagonal structure. The remaining $2p_z$ AOs combine into a filled π valence band (VB) and an empty π^* conduction band (CB). Due to a slight overlap of π and π^* bands, the transition $\pi \to \pi^*$ can be excited by light of every energy, responsible for electrical conductivity, high absorption and, therefore, luster and dark color.

Individual graphene layers are separated by a distance of 0.335 nm. In stable, hexagonal or α-graphite, layers are displaced so that each third layer is located precisely over its predecessor (ABAB sequence). In rhomboedric or β-graphite, each fourth layer possesses an equivalent position (ABCABC sequence). The layers are only weakly held together by van der Waals forces and weak π interactions.

Graphite shows a silvery gray color in large pieces, whereas it is gray as a powder. Papoular and Papoular [348] depicts graphite's absorption in the visible spectral range and beyond. Artists rarely use graphite as a pigment but often as a drawing material. While the coherence within covalently bonded graphene is high, the interaction between neighboring graphene layers is not, and graphite can easily be cleaved along with these layers. This fissility makes graphite suitable for drawing by facilitating the abrasion of layers on paper or other drawing surfaces. Decorative drawings that Greek artists applied to ceramic articles [48, Volume 4] around 500 BC are early witnesses of this use; discovery of pure graphite in Borrowdale parish in England in the sixteenth century started the pencil industry. Nowadays, graphite in the form of wood-cased pencils with which we are familiar has been available since the seventeenth century after manufacturing powdered graphite.

Cause of color

The optical properties of graphite are based on the electronic structure of the graphene layer. The individual graphene layers represent an infinitely extended aromatic system. $2p_z$ electrons are delocalized within the π band, imparting metallic properties to graphene such as electrical conductivity within graphene layers and high light absorption, associated with high luster, ►Section 1.6.7. As ►Figure 3.2(b) depicts, absorbing visible light of any energy can initiate $\pi \rightarrow \pi^*$ transitions since π, and π^* bands merge into one another. As a result, bigger lumps of graphite are highly absorbing, highly reflective, and shimmering gray. Powdered graphite is dark gray and matte, missing higher degrees of order required for geometric reflection.

Lamp black, Chinese ink, soot (PBk6, CI 77266)

The classic black carbon pigment is *lamp black* [48, Volume 4], [16, 49, 398, 399]. It is already mentioned in the fifth century as a component of Chinese inks, but was already used for black inks on Egyptian papyri 5000 years ago [771].

Lamp black consists of the soot of flames from burning oils, fats, waxes, resins, or resinous woods. The fuel burns at a wick (e. g., in a lamp) or in a shallow vessel. The soot forms out of the gas phase under restricted air supply and deposits on various surfaces. The surfaces are cooled plates, walls, rough tissues such as sheepskins, or other collection constructions. This way, the soot can be collected by either scraping or shaking. ►Section 8.1.3 outlines the formation of soot by fuel burning.

Strictly speaking, the soot constituting the finest grades of Chinese ink originated in burning pine wood and was therefore a carbon black [157, p. 236]. It was collected after drifting longways through tunnels and chamber systems made of bamboo or pots to reduce the tarry components (a defining part of ►bister). The tunnel and pot system also separated the soot into different grades, ►Chapter 8 and ►Section 8.1. Only from the Song dynasty (960 to 1279) on lamp black made from animal, vegetable, or mineral oils, often substituted pine soot. The recipe and procedure remained otherwise unchanged.

A specific particle morphology characterizes carbon particles for this pigment; i. e., small spherical particles form chains or clumps. The structure is described for ►carbon black. In addition to graphite-like carbon, polycyclic aromatic hydrocarbons, especially pyrene, perylene, fluoranthene, and acenaphthene, are present in significant quantities:

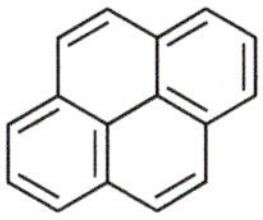

pyrene 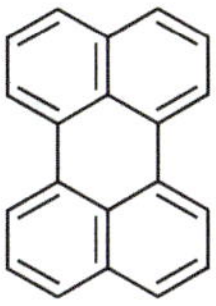 perylene 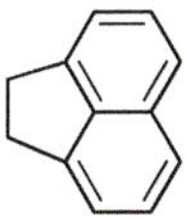 acenaphthene 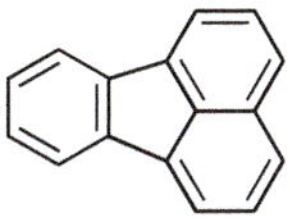 fluoranthene

Since these lower polycyclic aromatic hydrocarbons are colorless or yellow, they do not contribute to the dark gray to black color impression. The color originates in the minute, disordered ►graphite-like carbon crystallites and numerous tiny fissures, irregularities, and cavities that reflect the incident light multiple times within the structure, each time absorbing fractions of it, ►carbon black [238, p. 23].

Lamp black poses a problem as the liquid fuel must be completely evaporated and burnt with the precise amount of oxygen. Since earlier production methods could control this only incompletely, early lamp black contains tar-like organic products. These result from partial combustion, which considerably aggravates dispersing of the pigment in water.

Carbon black, channel black, gas black, acetylene black (PBk7, CI 77266)

The term carbon black describes different groups of black pigments:
- Any pigment based on carbon.
- Natural soots from burning oils, resins, and other materials. These are denoted more specifically as lamp blacks.
- Specific pigments from industrially burning or pyrolyzing natural gas, oil, or other fuels yielding soot. The fuel type or manufacturing conditions designate the exact soot type, e. g., channel black (soot deposited on iron channels) or gas black (soot from burning gas). Also, artificial lamp black is included here.

Today one generally assumes that carbon black denotes the latter group of industrially manufactured black pigments, and so do we in this text.

Carbon black was developed around 1864 and constituted the modern black pigment [48, Volume 4], [16, 49, 398, 399]. Its purity and homogeneous composition, achieved by a precisely controlled production, distinguish it from other carbon black types and make it crucial for the industry. Since it lends particular mechanical properties to rubber, about 90 % of carbon black is used in the rubber industry. Furthermore, carbon black is one of today's essential black pigments; it provides practically all the black needed by the printing and ink industry. The quantity needed for artists' paints is insignificant.

Composition

Carbon black consists of aggregates of tiny, mostly spherical particles [16, Chapter 5.1], [195, Chapter XV.1.1.2]. Primary manufacturing products are extremely small, spherical particles with diameters in the range 5–500 nm, comprising amorphous carbon and small graphite-like crystallites (size about 1.5–2.0 nm in length, and 1.2–1.5 nm in height, containing 4–5 graphene layers). At the high temperatures of combustion, polycyclic aromatic hydrocarbons (PAH) form from the fuel and attach to condensation nuclei. There the PAH condensate irregularly, yielding graphene fragments and graphite-like quasi-crystallites in spherical layers. The crystallites are loosely tied together by van der Waals forces, forming aggregates such as chains or clusters and impressing a particular structure to carbon black. These aggregates also tend to agglomerate, forming larger structures, and significantly influencing application properties. ►Section 8.1.3 outlines the formation of soot by fuel burning.

Byproducts of soot formation are lower PAHs, often with five-membered rings, such as pyrene, fluoranthene, benzo[j]fluoranthene, benzo[k]fluoranthene, perylene, coronene, or benzpyrene.

Color formation

Compared to grayish graphite, carbon black is dark black due to its extended system of minute fissures, irregularities, voids, and cavities. Incident light is reflected and scattered multiple time at these inhomogenities, and each time, a fraction of it is absorbed, until finally up to 99.5 % of incident light are absorbed [238, p. 23]. In case of carbon black with its loose structure, reflection and absorption processes happen many times, in contrast to crystalline, middle-gray, and shimmering graphite or middle-gray and sometimes shiny charcoal, all possessing a certain degree of ordering or larger structures.

Graphite oxides, hydrophilic carbon black

Carbon atoms at the borders of graphene layers or carbon atoms at the surface of amorphous parts of carbon black are highly disturbed, having not enough neighbors to form covalent bonds. They exhibit unpaired valence electrons and are subject to frequent attacks of hetero atoms, e. g., oxygen, forming a series of oxygen-containing functional groups [197, Chapter 7.3.5]. The oxidation can be controlled by applying oxidation agents, e. g., nitric acid or O_2. Depending on the reaction conditions, the surface consists of more or less pure carbon with functional groups such as alcohols, carboxylic acids, esters, or ketones resulting from surface oxidation reactions:

O_2, HNO_3

native carbon black
hydrophobic

graphite oxide
hydrophilic

Suppose we want to use the pigment to manufacture ink, often working with polar binders. In that case, these groups are desirable and allow us to disperse the carbon in aqueous solvents, which would be impossible without further treatment. Often, the oxygen amount is considerably increased by a controlled post-oxidation process [200, keyword "Pigmente, anorganische: Schwarzpigmente"], [16]. ►Section 8.5.4 explores possibilities of functionalization for dispersing in more detail.

Production

►Table 3.1 enlists the manufacturing processes used since industrial production started [16, Chapter 5.2], [201]. For all processes, the basic reaction is

$$C_nH_x \xrightarrow{\Delta} n\,C + \tfrac{x}{2}\,H_2$$

The pure thermal decomposition processes require an external energy supply to achieve and maintain the high temperatures needed for pyrolysis. In contrast, oxidative decomposition processes also use air for the combustion of parts of the raw material, providing pyrolysis energy.

A wide variety of fuels can be used. The selection influences process details and properties of the end product; therefore, raw materials, fuels, and process parameters must be balanced according to the intended application properties.

– The *furnace black process* is used in manufacturing all types of carbon black for plastics, rubbers, paints, and printing inks. It was developed around 1920 and is one of carbon black's most significant production processes. The raw material is injected as a spray into the high-temperature zone of a furnace. Burning fuels such as natural gas or oil maintain the temperature. Enough oxygen is provided to support the fuel's burning, but not enough for complete combustion of the raw materials so that they pyrolyze at 1200–1900 °C to carbon black. The product is

Table 3.1: Processes used for industrially manufacturing carbon black pigments [16, Chapter 5.2].

Type of process	Manufacturing process	Main raw material
Thermal oxidative decomposition, partial combustion of raw material	Furnace black process	Aromatic oils from coal tar or mineral oil, natural gas
	Gas black process, channel black process	Coal tar distillates
	Lamp black process	Aromatic oils from coal tar or mineral oil
Thermal decomposition, no combustion of raw material	Thermal black process	Natural gas, mineral oils
	Acetylence black process	Acetylene

filtered from the tail gas. Additives such as KOH or KCl control the carbon black's structure.

- The *channel black process* is the oldest manufacturing process, achieving yields of about 3–6 %. It was used in the late nineteenth century and is now abandoned. The raw material is natural gas. The *gas black process*, invented around 1930, uses coal tar oils as raw material and achieves higher yields up to 60 %. It is still in use today.
  The raw materials are vaporized, and a combustible carrier gas such as H_2 or CH_4 transports the vapor to the reaction vessel. The gases are mixed with air and burned, forming carbon black.
- The *lamp black process* is the oldest commercial process, ►lamp black. However, the modern process is not comparable to the classic one. The liquid raw materials are deposited on an iron pan, surrounded by a fireproof hood, and burned. The air is supplied from the sides of the pan through the slit between the pan and hood. The heated hood helps vaporize the liquid raw materials and directs the combustion gases and the soot formed into a cooled pipe to a collection system. In contrast to other processes, lamp black particles possess a broad size distribution of 60–200 nm.
- The *thermal black process* is nonoxidative, using primarily natural gas as raw material. The reactor is heated, and the starting material is injected, undergoing thermal decomposition without air, forming carbon black. The product comprises coarse particles up to a diameter of 500 nm, and its properties differ significantly from those obtained by oxidative processes.
- The *acetylene black process* was developed in the early twentieth century. Acetylene is fed into a preheated reactor, and its pyrolysis starts after ignition. It continues due to the decomposition heat released from the exothermic reaction. The process yields about 95 % carbon black, and the particle shape differs from those of other processes. Acetylene black is therefore limited to particular applications.

Ivory black, bone black (PBk9, CI 77267)

Bone black or ivory black, which has only been used for artists' paints since about 1500, is produced by charring bone-like substances [48, Volume 4], [49, 399]. Therefore, it consists of coke from burnt collagen, e. g., from ivory, bones, or horn, and of calcium phosphate (apatite), summarized in the composition $C{\cdot}Ca_3(PO_4)_2$. The vast black surface of the coke in the ramifications of fine holes in the apatite structure causes the deep black appearance of carbon black.

Charcoal black, vine black, vegetable black, plant black (PBk8, CI 77268)

These artists' pigments also belong to the classics [48, Volume 4], [49, 398, 399]. They are the carbonization products of various materials in closed vessels under careful temperature and oxygen or air control. Precursors may be wood, wine lees, cork, vine sprigs, twigs, almond shells, and pits of peaches or other fruit. The purpose of temperature control is to prevent the formation of tarry substances when the temperature is too low and the burn-up of the precursor when the temperature is too high. ►Section 8.1.3 outlines the formation of charcoal by pyrolysis of wood.

Charcoal black materializes not in powder or flake form but relates structurally to the original product. Charcoal provides a fine example: thin wood chips (grapevines, willow sticks, peach pits, almond shells, or pieces of oak) are heated in closed containers until charred but still retain their (stick-like) shape. Already 14000 BC, humans used charcoal pieces for drawings and charcoal powder for black surfaces while painting stone-age caves.

Charcoal black consists mainly of carbon. In addition, we find polycyclic aromatic hydrocarbons as byproducts, often with five-membered rings, as well as a large number of other hydrocarbons:

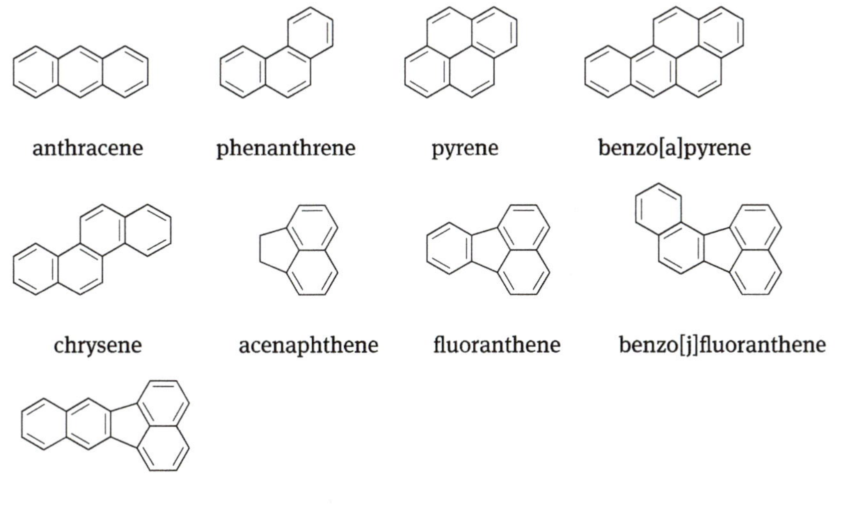

Color formation

Since these lower polycyclic aromatic hydrocarbons are colorless or yellow, they do not contribute to the dark gray color impression. The color originates in the small, disordered ►graphite-like carbon crystallites and numerous fissures and cavities that reflect the incident light a few times within the structure, absorbing fractions of it, ►carbon black. The starting material's structure governs the carbonization process; thus, charcoal possesses a higher degree of order than carbon black and is therefore less intensely black [238, p. 23].

False blue

For painting, the artist must grind the charcoal pieces, but this does not achieve the fineness of the particles of carbon black. We can find a fascinating application of this black pigment in medieval English wall paintings: a blue-gray hue sometimes occurs in them, sometimes called "false blue" and created by the painter mixing charcoal black with white or applying it to a white background. The charcoal particles are ground so that instead of the expected gray tone, a distinct blue color impression emerges by Rayleigh scattering, ►Section 1.6.6 at p. 55 [54]. Rubens, Van Dyck, and others knew how to apply this, demonstrated by the blue and purple colorations created with charcoal black in the paintings "Samson and Dalilah" and "Woman with child," [555, 558]. The effect can be enhanced by contrasting this false blue with warm tones.

Bister

Bister is a glazing paint used mainly in the eighteenth century in watercolors and for pen and wash [48, Volume 4], [49, 398, 399]. It shows a warm reddish-brown to brownish-black and is made from wood soot, the best varieties from beechwood soot. The soot originates, e. g., from fireplaces and stoves heated with a specific sort of wood. After burning, the deposited fat mass is scraped off the fireplace. Soluble components are dissolved in hot water, decanted, and the remaining sediment is dried.

Bister consists of products of more or less complete combustion of the wood: Flame soot (as in lamp black), charcoal (as in charcoal black), coke, tarry components, but predominantly, polycyclic aromatic hydrocarbons, e. g.,

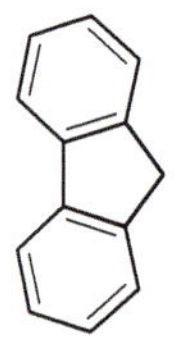
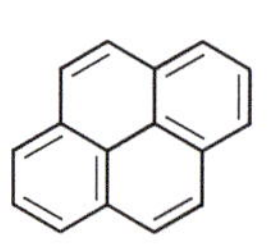
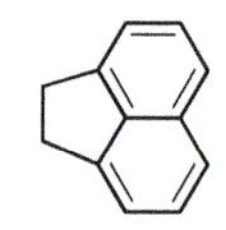
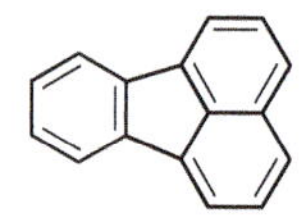

fluorene perylene pyrene acenaphthene fluoranthene

In addition, bister contains oxidation and decomposition products of these compounds.

While soot and charcoal have a dark gray or "black" color, many lower polycyclic aromatic hydrocarbons are colorless or yellow. The tarry component is responsible for the brown color, which has a yellow, red, or brown tinge depending on the type of wood used in the process. Since the tarry components agglomerate primarily at the point of combustion, soot for bister is scraped off near this point. In contrast, soot for Chinese inks is collected as far away as possible from the combustion site to reduce the tarry components (►lamp black).

3.2 Copper pigments

Copper was one of the essential elements for painting since it displays blue and green colors through the ligand field transitions of its *d* electrons. All durable green colorants and most blue ones from antiquity until the nineteenth century rely on these transitions. Therefore, it was crucial for paint development that copper, with its electron configuration d^9, is susceptible to Jahn–Teller distortion. More specifically, a LF band absorbs the yellow to the red spectral range, while a strong OMCT band $O^{2\ominus} \rightarrow Cu^{2\oplus}$ absorbs in the near-UV to the blue spectral range. The reflectance spectrum is thus limited to blue and green light, ►Section 2.3.6 and ►Figure 3.3.

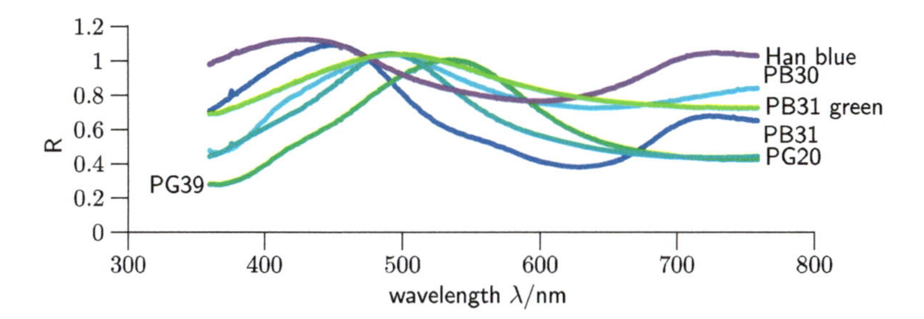

Figure 3.3: Reflectance spectrum of copper-based watercolors normalized to an arbitrary unit: PB30 (azurite, Kremer no. 102078, $Cu(OH)_2 \cdot 2\,CuCO_3$), PB31 (Egyptian blue, Kremer no. 10060, $CaCu[Si_4O_{10}]$), PB31 (Egyptian green, Kremer No. 10064, green copper silicate glass), PG20 (verdigris, own preparation, about neutral copper acetate $CuAc_2 \cdot H_2O$), PG39 (malachite, Kremer No. 103458, $Cu(OH)_2 \cdot CuCO_3$), and Han blue (Kremer No. 10072, $BaCu[Si_4O_{10}]$). They all have an absorption in the yellow to red spectral range in common, resulting in blue to green colors. The influence of the Jahn–Teller effect on the exact hue is described in ►Section 2.3.6 at p. 125.

The variations of color stem from different ligand field strengths of chloride, hydroxide, carbonate, and other ligands present in compounds with different crystal geometry, and thus different coordination polyhedrons affecting both types of absorption bands. Since the transitions due to the selection rules for octahedral and square complexes are forbidden, copper-based colorants are not distinct in color.

A favorable factor for copper as a colorant is that green and blue copper minerals occur in large quantities in the vicinity of copper ore deposits as superficial weathering products or products of hydrothermal precipitation. They could be collected, purified, and ground. Minerals of antiquity, such as atacamite (►Section 1.2.1), no longer played a role in European panel painting, but two significant pigments, malachite and azurite, survived into the second millennium. An early outstanding achievement in synthesis is Egyptian blue and green. Verdigris and copper resinates are further examples of early synthetic pigments. An overview, including modern recipes for synthesis, can be found in [416]. Scott [109] summarizes the state of knowledge on copper pigments. Ellwanger-Eckel [417] is devoted to less common pigments established on various basic copper chlorides, sulfates, and carbonates, such as e. g.,

$$Cu_4SO_4(OH)_6 \cdot H_2O \text{ (posnjakit)}, \; Cu_4SO_4(OH)_6 \cdot 2\,H_2O \text{ (langit)}$$
$$Cu_7Cl_4(OH)_{10} \cdot H_2O$$
$$Cu_7SO_4(OH)_6, Cu_7SO_4(OH)_6 \cdot H_2O, Cu_7SO_4 \cdot Cu(OH)_2$$
$$Na_2Cu(CO_3)_2 \cdot 3\,H_2O$$
$$Cu_2(OH)_3NO_3, Cu_5(PO_4)_2(OH)_4$$

The fact that copper, similar to lead, is a bivalent metal cation that catalyzes the drying of so-called “drying oils” is helpful to the oil painter. Book illuminators or graphic artists somewhat suffered from copper pigments. Some early copper pigments contributed considerably to the decay of the carrier paper used for graphic artworks, ►Section 6.7.8. In panel painting, their application is less problematic since a linseed oil film surrounds and isolates the pigments so that they no longer directly contact the carrier material (which is also more stable).

Antique and various green pigments (atacamite and others)

In antiquity, many more or less identifiable copper compounds served as green and blue pigments [49, 109, 414]. Besides the essential minerals malachite and azurite, painters employed basic chlorides of copper, copper hydroxychloride Cu(OH)Cl, and the two modifications atacamite (green) and paratacamite (pale green) $Cu_2(OH)_3Cl$. Atacamite and paratacamite were also used from the eighth to the nineteenth century. An antique green to turquoise pigment is chrysocolla $(Cu, Al)_2H_2Si_2O_5(OH)_4 \cdot n\,H_2O$, which was also used from the tenth to the sixteenth century; as watercolor predominantly in the sixteenth century.

Unfortunately, it is not always clear whether green copper pigments are conversion products of other copper pigments or intentionally applied.

Egyptian blue (PB31, CI 77437)

Already around 3000 BC, we find a bright blue in Egyptian wall paintings (►Figure 3.3), which consists of calcium-copper tetrasilicate $CaCu[Si_4O_{10}]$ and is considered to be one of the first synthetic pigments artificially produced [48, Volume 3], [49, 109, 429, 771]. In this compound, which occurs naturally as mineral cuprorivaite, copper is square-planar coordinated (for color formation, ►Section 2.3.6).

Depending on grain size, exact composition, and method of production, the blue hue obtained varies from whitish-blue to dark blue; small pigment grains result in a lighter color, as we have already seen in ►Section 1.7.1.

In addition to a beautiful pure blue tone, the pigment exhibits exceptional lightfastness and is stable in all standard painting techniques. Therefore, it is not surprising that since its discovery, it remained the most important ancient blue pigment until the ninth century, according to [109] until the fourth century. Its disappearance after the collapse of the Roman Empire forced European panel painters to rely on azurite and, in exceptional cases, ultramarine. Nevertheless, already in 1815, the chemist Davy successfully recovered the secret of producing Egyptian blue.

The raw materials lime, sand, and copper minerals were readily available in early Egypt [429]. After the reaction, excess copper oxide, silicon dioxide, sodium oxide, or calcium silicate remain. Since the composition of historical Egyptian blue has been relatively constant over four millennia, we may assume that the early craftspeople were aware of the need for defined proportions of ingredients. They were furthermore able to maintain temperatures of 800–900 °C over long periods (up to a 100 hours) without any real possibility of control.

Egyptian green, green frit

Besides Egyptian blue, a green pigment was in use in Egypt between 2300 BC and about 600 BC (►Figure 3.3), which was melted similar to Egyptian blue from quartz, lime, and copper minerals [49, 109]. Since none of the numerous contemporary or later recipe books mention this pigment, its production and composition were the subjects of numerous speculations and theories. Options ranged from a cuprorivaite rich or poor in calcium to copper-containing wollastonite $(Cu, Ca)SiO_3$ or a green-colored copper glass. Another opinion claimed that Egyptian green represented not a pigment but a failed Egyptian blue.

Only recently, several detailed investigations have been carried out on this pigment. According to [63], Egyptian green is silicate glass colored green by copper ions. The wollastonite embedded in the glass is colorless, contains practically no copper, and does not contribute to the color. According to [109], Egyptian green is either a glass phase with copper wollastonite as admixture or glass added to wollastonite. Similarly, [480, 481] also see admixtures of wollastonite to glass or glass to wollastonite, respectively. However, they assume that one or the other type was not purposefully produced but that the variations occurred unintentionally. In all cases, LF transitions in the copper ion induce the green color.

The color of the glass depends on the cooling rate of the product and ranges from green (rapid cooling) to turquoise to blue (slow cooling). According to [63], the measurable difference is only a higher calcium content of the green glass. Responsible for the green color are LF transitions within a copper complex formed by oxide anions of the glass, acting as ligands and surrounding the central copper ion. It is assumed that calcium ions change the Cu-O bond length, and thus the ligand field splitting of the complex so that blue and blue-green instead of green colors result. According to [109], the color depends on the calcium and copper content and manufacturing conditions.

Han blue, Chinese blue; Han purple, Chinese purple

During the Han Dynasty (200 BC to 220), the Chinese used a blue chemically similar to Egyptian blue. It consists of barium copper tetrasilicate $BaCu[Si_4O_{10}]$ and was used, e. g., to paint the Terracotta Army, ►Figure 3.3 [49, 109, 429]. The quartz-poor Han purple, a barium copper disilicate $BaCuSi_2O_6$, is similar to this blue and the only nonorganic purple pigment. Like Egyptian blue, LF transitions in the square-planar coordinated $Cu^{2\oplus}$ ion cause the Han color, ►Section 2.3.6.

Antique preparation

Remarkably, the change from calcium to barium leads to considerably more problematic chemistry [429]. The $BaO/CuO/SiO_2$ system exhibits many separate phases. For the intentional preparation of the desired Han blue phase, strict stoichiometry control, and thus control of the amounts of material used is necessary. High purity of starting products and adequate experience are prerequisites to achieving the desired color.

As a source of barium oxide, barium carbonate is especially suitable, naturally occurring as witherite $BaCO_3$. Unfortunately, witherite is rare; thus, ancient craftspeople resorted to barite $BaSO_4$, not providing satisfactory results. Admirably, they found a way to dissolve barite via lead salts (lead carbonate or oxide): from a lead salt and barite, lead sulfate and the desired barium oxide form in a circular process. Then the lead sulfate is converted back into lead oxide at high temperatures while sulfur is released as oxide:

$$PbO + BaSO_4 \longrightarrow PbSO_4 + BaO$$

$$PbSO_4 \longrightarrow PbO + SO_3 \text{ or } SO_2 + 1/2\,O_2$$

To produce Han blue, the artisans had to keep the high temperature of 1000 °C as stable as possible for a very long time. This stability is necessary because, initially, Han purple is produced as a kinetically controlled product and only slowly transforms into Han blue.

Because of the complications mentioned above, Berke [429] suspects that the Chinese were first familiar with Egyptian blue, which at the time was a standard trade product throughout the Mediterranean region, and that the subtleties of Han blue chemistry were only gradually discovered later.

Modern production

Production nowadays employs a solid reaction between barium carbonate, copper oxide, and silicon dioxide following the presumed ancient synthesis route. The mixture is sintered at about 1000 °C for one week [658]. At the beginning of the reaction, approximately equal amounts of Han blue $BaCu[Si_4O_{10}]$ and Han purple $BaCuSi_2O_6$ emerge. The equilibrium shifts to pure Han purple through successive grinding and sintering steps.

Excess copper is used for preparing Han blue and later removed with hot hydrochloric acid. The acid destroys the more sensitive Han purple simultaneously, leaving pure Han blue.

The bonding situation of copper explains the lability of Han purple [429]. Han purple contains more copper than Han blue, and in the purple's structure, two copper atoms are such close neighbors that we can speak of a Cu–Cu bond, which is unstable against acids.

Azurite, mountain blue, blue verditer, Cyprian blue, blue bice (PB30, CI 77420)

The beautiful blue azurite [48, Volume 2], [47, 49, 109], like malachite, is a basic copper carbonate $Cu(OH)_2 \cdot 2\,CuCO_3$, and occurs mainly in the upper oxidic areas of copper veins and only needs to be ground, washed, and sieved for use. The more thorough this sorting is, the more homogenous particle sizes become, and the purer the color obtained. Azurite becomes lighter with decreasing particle size and loses much of its color intensity. In contrast to ultramarine, azurite shows a green undertone. This undertone is a characteristic of all copper compounds, as these absorb in the red region of the spectrum, ►Section 2.3.6.

Azurite occurs in Neolithic decorations around 6000 BC in Catal Hüyük. Egyptian antiquity appreciated and used azurite from 2500 BC [23]; however, the ancient advanced civilizations already disposed of Egyptian blue as a good blue pigment for painting. Rich deposits in silver and copper mines of the German-speaking world and the lack of other blue pigments after the disappearance of Egyptian blue made azurite the most important blue pigment for European painters in the Middle Ages, especially from the fourteenth to the seventeenth century. Artisans often applied an azurite underpainting before glazing with ultramarine to save expensive natural ultramarines.

In Europe, azurite is displaced only from the eighteenth century onward due to the discovery and manufacture of Prussian blue and synthetic ultramarine. Attempts to produce azurite artificially from copper salts never achieved the color beauty of the natural mineral. The reaction is

$$CuSO_4 + Na_2CO_3 \longrightarrow CuCO_3 \downarrow + Na_2SO_4$$

Azurite was also produced by the reaction of copper salts with lime:

$$CuSO_4 + CaCO_3 \longrightarrow CuCO_3 \cdot Cu(OH)_2 \downarrow + CaSO_4$$

Artificial malachite and azurite were used excessively in the Middle Ages after [109], Azurite until the seventeenth century.

Azurite is stable to light and air as carbonate but decomposes with dilute acids. The inclusion in an oil film increases its resistance significantly, and azurite becomes even more durable in egg tempera since the tempering effect mentioned below occurs.

In azurite, the ligands surround copper in a strongly elongated octahedral shape, so a 4 + 2 square-planar bipyramidal coordination occurs. With this, azurite occupies a middle position in coordination and color. The former is between distorted octahedral coordination (malachite) and 4 + 2 square planar coordination (Egyptian blue), ►Section 2.3.6, the latter between green (malachite) and pure blue (Egyptian blue), ►Figure 3.3.

In linseed oil, the green tint can intensify to the point of greening. This may be related to the slow conversion of azurite into green organocopper compounds such as copper resinate or copper oleate. In paintings of the Van Eycks, unchanged azurite occurs. A fine protein film coats the azurite particles in these paintings. *Tempering*, i. e., the saturation of crystal defects with proteins, counteracts the process of greening [61]. Based on the available information, it is impossible to decide whether the protein film prevents the formation of organocopper compounds or whether the stabilization of the crystal lattice prevents the transition of the planar coordination around copper into an octahedral one. A marked color change to green (malachite) accompanies this transition, ►Section 2.3.6. Azurite can also turn green in mural paintings, which is explained by the transformation into green atacamite caused by salts and moisture in the masonry.

Malachite, mountain green, green verditer, mineral green, copper green, green bice (PG39, CI 77420)

Malachite [48, Volume 2], [47, 49, 109, 407], a basic copper carbonate $Cu(OH)_2 \cdot CuCO_3$, was an important blue to green copper pigment. To what extent all green pigments addressed as malachite are really copper carbonate is discussed in [408].

Malachite occurs in large quantities together with azurite in the upper oxidic parts of copper veins; the pigment is produced by grinding, washing, and sieving the mineral. Since malachite becomes lighter and weaker in color with decreasing grain size and copper pigments have a low coloring power, malachite is not finely ground; we find grain sizes up to 100 µm. Malachite was more often used in underpaintings or under glazings of verdigris or copper resinate. It does not possess high opacity in oil or watercolor; however, it proves itself in tempera. It was employed in Egyptian tomb paintings around 2500 BC [771], then frequently in medieval illuminations, less so in panel painting. Malachite gradually disappeared from the palette after more intensive application around the fifteenth and sixteenth centuries.

We can artificially produce malachite by precipitating copper nitrate or sulfate onto calcium carbonate or by precipitating soluble copper salts with alkali. One yields various basic but weakly colored copper carbonates. We find a current prescription in [416]. Artificial malachite and azurite were used excessively in the Middle Ages [109]; specifically, malachite also in the fifteenth century.

Malachite as carbonate is stable to light and air but unstable to hydrogen sulfide (browning with sulfide formation) and diluted acids. However, embedding the pigment in oil provides some protection against these influences.

A strongly distorted octahedron coordinates copper in malachite. Thus, it stands at the beginning of a series of coordination polyhedra, ending with square planar coordination in Egyptian blue. A strong absorption band influences the color of malachite in the UV so that it varies between blue and green, ►Figure 3.3. ►Section 2.3.6 has discussed details about the origin of the color of azurite and malachite.

Verdigris, Spanish green (PG20, CI 77408)

Historic verdigris is a collective term for blue to green basic copper acetates of variable composition, which were already known since antiquity [48, Volume 2], [47, 49, 109, 409, 410, 416]. The various basic acetates can occur individually or in combinations 1/2 and 2/4 (with Ac=$CH_3COO^{\ominus}$ acetyl radical):

1: $(CuAc_2)_2 \cdot Cu(OH)_2 \cdot 5\,H_2O$ (blue)
2: $CuAc_2 \cdot Cu(OH)_2 \cdot 5\,H_2O$ (pale blue)
3: $CuAc_2 \cdot 2\,Cu(OH)_2$ (blue)
4: $CuAc_2 \cdot 3\,Cu(OH)_2 \cdot 2\,H_2O$ (green)
5: $CuAc_2 \cdot 4\,Cu(OH)_2 \cdot 3\,H_2O$ (blue-green)

Today we understand verdigris as neutral copper acetate dihydrate $CuAc_2 \cdot H_2O$, which occurs as the dimer.

6: $(CuAc_2)_2 \cdot 2\,H_2O$ = dimer of $CuAc_2 \cdot H_2O$ (blue-green)

Verdigris should not be confused with the patina that develops on copper roofs after long weathering periods and consists of basic carbonates, chlorides, and sulfates. As verdigris combines with many resinous binders to form transparent, deep green copper salts, it was often used to produce other colorants; cf. ►copper resinate.

Verdigris has an intensive blue-green color that exceeds that of malachite and green earth, ►Figure 3.3. Therefore, it was artificially produced since antiquity, and the green pigment was used mainly from the thirteenth to the nineteenth century. Transparent copper resinate glazes were also very popular. After discovering emerald green and chromium oxide green, they replaced verdigris entirely.

Contemporary literature considered verdigris a very unstable pigment that changed its color from green to brown due to its reaction with sulfidic pigments such as orpiment, yielding brown copper sulfide. The Old Masters were well aware of this problem and isolated the pigment in layers of oil. Keeping air and moisture away also has a stabilizing effect. However, we cannot be sure that sources always spoke about the same material since, in former times, "verdigris" often implied each artificial or natural green copper compound.

The production was carried out in several ways. One variant involved dissolving copper chips in the acetic acid vapors of fermenting grape marc, and another involved soaking copper strips in vinegar and manure. Afterward, the blue-green coating could be scratched from the copper plates and cleaned by soaking in concentrated acetic acid. During this process, basic acetates form in varying compositions and colors; it was therefore sometimes recommended to convert the product into the blue-green neutral acetate 6 by grinding it in strong acetic acid. This compound can also be obtained directly by reacting acetic acid with copper oxide or carbonate.

From the nineteenth century onward, the production started from copper sulfate and a heavy metal (lead) or alkaline earth acetate (calcium, barium), exploiting the low solubility of the resulting sulfate:

$$CuSO_4\ aq. + Ba(OOCCH_3)_2\ aq. \longrightarrow Cu(OOCCH_3)_2 + BaSO_4 \downarrow$$

Artificial copper blue and copper green

In the context of medieval paint production, we find numerous recipes for the production of green or blue copper pigments [49, 108, 109]. Due to unclear or incomplete recipes and the lack of pure starting materials, the resulting compounds are often difficult to define or detect analytically.

Bartl et al. [108] contains the preparation of *lime blue* from copper or verdigris, lime and vinegar, and NH_4Cl. The resulting blue product was called artificial azurite (blue verditer) or interpreted as a copper-ammonium-lime compound. In more recent times, Krekel assumes the following products:

Calcium copper acetate, turquoise green, $CaCuAc_4 \cdot 6\,H_2O$
Calcium copper hydroxide, deep blue, $CaCu(OH)_4 \cdot H_2O$
Calumetite, greenish-blue, $Cu(Cl, OH)_2 \cdot 2\,H_2O$

A *copper green* is often mentioned as an artificial green pigment. It represents mixtures of acetates, carbonates, chlorides, or tartrates. The "Spangrün" (verbose: chip green), e. g., involves the minerals atacamite and paratacamite $Cu_2(OH)_3Cl$. Copper acetates are commonly described; the basic acetates $CuAc_2 \cdot n\,Cu(OH)_2$ are blue-green and turn green after painting, while the neutral acetate $CuAc_2$ has a stable blue-green color. Grinding verdigris in vinegar is, therefore, useful to obtain neutral acetate.

The required copper is added directly as a copper sheet, through containers made of copper, or many times as verdigris. However, this term refers to any natural or synthetic copper green. We must assume that copper was often also introduced unintentionally as an alloy's component; otherwise, recipes for pigments such as "silver blue" made from silver would be inconceivable. In such cases, it is possible that the silver used was impure and contained traces of copper.

Green dyes were also frequently copper-based, e. g., copper acetates (verdigris) dissolved in water, vinegar, or soluble copper chlorides.

Copper resinate, transparent copper green, copper oleat

Verdigris can form copper salts of oleic and resin acids with oil- and resin-containing binders, the so-called copper oleates and resinates [48, Volume 2], [47, 49, 409, 410]. Carboxylic acids in the binder are oleic acid, linoleic acid, linolenic acid, or abietic acid, ►Sections 7.4.2 and 7.4.8. Copper also reacts with proteinaceous binders to form copper proteinate.

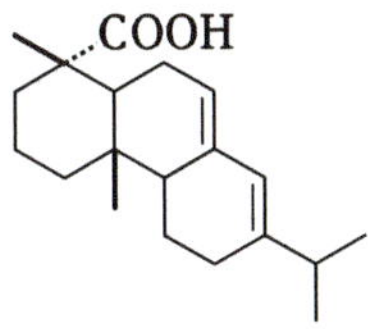

COOH

abietic acid

The copper compounds are glazing colors of a dark green that resemble aged verdigris. They were intentionally obtained by heating verdigris in Venetian turpentine with resins or linseed oil. However, since the same reactions also occur with verdigris and a binder, it is not always clear whether these salts were in fact intentionally produced. According to [109], substances such as these may also cause the greening of azurite in oil binders.

Scheele's green, mineral green, Swedish green (PG22, CI 77412)

In 1775, Scheele discovered the green copper arsenite $n\,CuO{\cdot}As_2O_3{\cdot}m\,H_2O$ ($n = 2, 3, m = 2$ or $n = 1, m = 0$) or $CuHAsO_3$. Under the name "Scheele's Green," it was the first synthetic copper-arsenic pigment to be marketed [48, Volume 3], [49, 109, 416]. Unfortunately, it is known under many less common names, often intermixed with emerald green.

Scheele's green is an unstable pigment showing a not very pure hue. Its exact composition depends on manufacturing conditions and Cu/As ratio: whereas mixtures of 1:1 show a rather yellow-green, 1:3 possesses a dark green color.

Since there were no convincing green pigments in the eighteenth century yet, Scheele's green was a welcome addition to the palette and appreciated in the wallpaper industry. However, soon the brilliant emerald green replaced it.

Emerald green, Schweinfurt green, veronese green, Paris green, Vienna green, Mitis green (PG21, CI 77410)

Copper acetate arsenite $Cu(CH_3COO)_2 \cdot 3Cu(AsO_2)_2$ was independently discovered by von Mitis and Sattler around 1800 and was soon produced on a larger scale, as it is a brilliant green pigment of an intensive color [48, Volume 3], [47, 49, 109, 411–413, 416, 585, 586]. As one of the first pigments, it provided a luminous green that soon found its way onto artists' palettes, fabrics, articles of daily use, and, above all, wallpaper. Caution was advised regarding early cadmium sulfide pigments and other sulfide sources such as air pollution because of the risk of blackening.

The disadvantage of the beautiful color was that the pigment proved highly toxic, which is expressed in the LD_{50} value of 22 mg kg^{-1} and the term "poisonous green." A considerable discussion in the media at the time took place about which arsenic compound results. Today, we know that the organic arsenic compounds that formed only in traces such as e. g., trimethylarsine are nevertheless highly poisonous. The toxicity originated in the housing conditions at that time. For this reason, the pigment disappeared as soon as coloristically comparable alternatives were offered, which took time until the twentieth century. Long after that, it was still traded as rat poison.

The color of the pigment is caused by LF transitions of the $Cu^{2\oplus}$ ion, octahedrally coordinated by oxygen atoms of arsenite anions:

$$\left[\begin{array}{c} O{=}As\diagdown\ Cu\ \diagup As{=}O \\ O\quad O \\ O{=}As{-}O\rightarrow Cu \leftarrow O{-}As{=}O \\ Cu{-}O\quad O{-}Cu \\ O{=}As\diagup\qquad \diagdown As{=}O \end{array}\right]^{2\oplus} \quad 2\,CH_3COO^{\ominus}$$

Starting products for manufacturing are sodium arsenite, copper acetate, and copper sulfate. There are two ways of production:

$$6\,NaAsO_2\,aq. + 4\,Cu(OAc)_2\,aq. \longrightarrow Cu(OAc)_2{\cdot}3\,Cu(AsO_2)_2\downarrow + 6\,NaOAc$$

$$6\,NaAsO_2\,aq. + 4\,CuSO_4\,aq. + 2\,HOAc + Na_2CO_3\,aq. \longrightarrow$$
$$Cu(OAc)_2{\cdot}3\,Cu(AsO_2)_2\downarrow + 4\,Na_2SO_4 + H_2O + CO_2\uparrow$$

Particle size and shape decisively influence the color, so different production conditions (temperature, stirring) yield a variety of light or dark colors. The more uniform the crystals are, the more brilliant and darker the green shade is.

3.3 Ultramarine pigments

Ultramarine [48, Volume 2], [11, 16, 47, 49] is a pigment with a romantic history. The name means "the blue from beyond the sea" and indicates that until the nineteenth century Lapis Lazuli, from which genuine ultramarine derives, was brought from faraway Afghanistan. It was a popular, though expensive blue. After 1806, the structure was elucidated; from 1820 onwards, attempts were made to produce it artificially due to its price and desirable coloristic properties. They culminated in 1828 in the synthesis by Guimet and Gmelin. Seel et al. [459] trace the exciting history of ultramarine blue research, which was complicated by the high stability of the pigment: the colored body, a trisulfide anion, cannot be isolated undestroyed from its sodalite matrix. The situation was even made more perplexing by the fact that attempts of the nineteenth century to describe ultramarine with Lewis or valence line formulas and newly emerged organic color theories had to fail since understanding the color is only possible with the MO theory.

Today, synthetic ultramarine is produced in many shades (yellow, green, reddish-purple, blue). However, only bluish-purple and pink variants (ultramarine violet and ultramarine pink) have practical significance apart from the warm red-tinted ultramarine blue. The designation "ultramarine yellow" does not stand for yellow ultramarine pigments but barite yellow or lead chromate, ►p. 283.

Crystal structure

Ultramarine blue is a sulfurous aluminosilicate. We can imagine it as being derived from sodalite $Na_8[Al_6Si_6O_{24}](Cl, OH)_2$ [16, 458, 459]. In ultramarine, the counterions are replaced by polysulfide radical anions, and the typical composition is about $Na_{6,9}Al_{5,6}Si_{6,4}O_{24}S_{4,2}$.

In sodalite, cage-like structures are strung together. The sulfide anions move quasi-free inside these cages, entirely shielded by it. The inclusion in the sodalite structure and presence of $Na^{\oplus}$ counterions stabilize the otherwise unstable sulfide anions to such an extent that ultramarines are among the exceptionally stable pigments. Tarling et al. [466] offer a wealth of details on the crystal structures of ultramarines.

To understand the complex formula more quickly, we start from SiO_2 or $Si_{12}O_{24}$ and replace six $Si^{4\oplus}$ with six $Al^{3\oplus}$ and six $Na^{\oplus}$, which leads us to the idealized aluminosilicate $Na_6Al_6Si_6O_{24}$. In this way, out of eight gaps for alkali metal cations present in the sodalite structure, six are already occupied by $Na^{\oplus}$ ions, required for charge neutrality of the aluminosilicate. $Na^{\oplus}$ ions from polysulfides can occupy the two remaining gaps. However, since sulfur is present as dianion $S_3^{2\ominus}$, only one molecule

Na_2S_3, and thus only one polysulfide anion can be introduced per sodalite unit cell ($Na_8Al_6Si_6O_{24}S_3$), even if the later oxidation to $S_3^{\ominus}$ introduces a $Na^{\oplus}$ ion that is lost again, yielding the composition $Na_7Al_6Si_6O_{24}S_3$.

If sulfur content increase, the color intensifies. The increase is possible by replacing $Al^{3\oplus}$ with $Si^{4\oplus}$ so that fewer $Na^{\oplus}$ ions are required for charge balance, and more Na_2S_3 can be introduced. By adding a silicon-rich feldspar, the composition $Na_{6,9}Al_{5,6}Si_{6,4}O_{24}S_{4,2}$ can be achieved.

The color

Just as sulfur already offers a rich play of colors in elemental form, it also shows a broad color spectrum in the pigments of the ultramarine type [458, 459]. The color of the blue ultramarine is dominated by the deep blue colored trisulfide radical anion $S_3^{\ominus}$, which causes intense absorption around 600 nm, ►Section 2.5.6 and ►Figure 3.4. The higher the proportion of this anion is in the sodalite framework, the more intense the color becomes.

In steel-blue or even green ultramarine, the disulfide anion $S_2^{\ominus}$ occurs besides the blue trisulfide anion. The disulfide anion absorbs at 400 nm in the blue range, appearing yellow so that a high proportion of this anion in ultramarine shifts the perceived color impression to green, ►Figure 3.4.

The purple and pink ultramarines vary in sulfur oxidation state and have lower Na and S contents. The sulfur chromophore is oxidized to, e. g., $S_3Cl^{\ominus}$ or $S_4^{\ominus}$. The tetrasulfide ion absorbs around 500 nm and, if predominant, is responsible for the reddish color of ultramarine pink. If $S_3^{\ominus}$ is present in addition to $S_4^{\ominus}$, the purple color of ultramarine violet emerges, ►Figure 3.4.

All sulfide anions are intensively colored since electronic transitions between MOs (►Section 2.5.6) are not subject to spectroscopic selection rules. In contrast, the parity prohibition for LF transitions applies, e. g., in the blue copper pigments Egyptian blue and azurite, reducing their color intensity considerably.

Ultramarine analogs containing the homologous selenium instead of sulfur could also be produced [458, 462–464]. They represent brilliantly red pigments of an approximate composition $Na_{6,4}[Al_{5,9}Si_{6,1}O_{24}]Se_{2,0}$; however, they are not used further for ecological reasons. $Se_2^{\ominus}$ acts as the color center, also stabilized by sodalite cages. Even heavier chalcogens such as tellurium were also successfully incorporated into the sodalite structure; the obtained $Na_{9,4}[Al_{6,1}Si_{5,9}O_{24}]Cl_{1,2}Te_{1,1}$ is blue-green and contains $Te_2^{\ominus}$ as the color center [465]. Finally, the sodalite framework can include other colored anions, such as the chromate ion, which leads to yellow pigments [467].

Ultramarine blue, French ultramarine, Guimet's blue, permanent blue (PB29, CI 77007)

Genuine ultramarine blue, one of the noblest and most beautiful blue pigments, shows a light, intensive reddish blue, ►Figure 3.4. Since the refractive index of 1.5

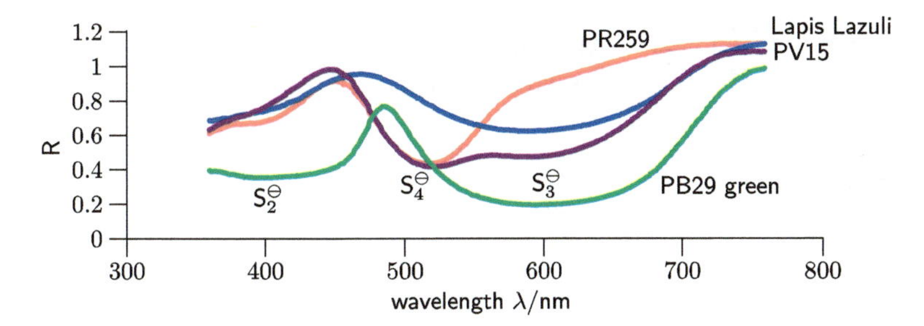

Figure 3.4: Reflectance spectrum of watercolor paints based on ultramarine, normalized to any unit: PB29 (Lapis Lazuli, Kremer no. 10530, about $Na_{6,3}[Al_{4,79)}Si_{7,21}O_{24}] \cdot S_{3,74}$), PB29 (ultramarine green, Kremer No. 447008, about $Na_{7,74}[Al_{5,7}Si_{6,3}O_{24}] \cdot S_{1,88}$), PR259 (ultramarine pink, Kremer No. 42600, about $Na_{3,6}(NH_4)_{0,25}(H_3O)_{1,94}[Al_{4,8}Si_{7,2}O_{24}] \cdot S_{3,01}$), and PV15 (ultramarine violet, Winsor and Newton Professional Watercolor No. 672, about $Na_{6,08}(NH_4)_{0,17}(H_3O)_{1,28}[Al_{5,36}Si_{6,64}O_{24}] \cdot S_{3,83}$). Distinctly visible is the intense absorption by $S_3^{\ominus}$ in the region around 600 nm, which leads to the formation of the typical blue color. An additional absorption by $S_2^{\ominus}$ around 400 nm leads to the greenish to green ultramarine green. A combined absorption by $S_4^{\ominus}$ around 500 nm and by $S_3^{\ominus}$ results in ultramarine violet; absorption by $S_4^{\ominus}$ alone delivers ultramarine pink.

is similar to that of painting agents and plastics, we perceive ultramarine as a deep transparent blue. An admixture of white achieves opacity.

The extremely lightfast pigment is stable to bases; acids decompose it thoroughly to form S, H_2S, and SiO_2. Coating the pigments with SiO_2 achieves a short-term acid stabilization. It is insoluble in water and organic solvents therefore very migration-resistant, and approved for contact with food and use in cosmetics. The thermal stability is high; ultramarine blue is stable up to 400 °C.

Due to its coloristic properties, antiquity has favored ultramarine blue. However, the exceptionally high price of ultramarine blue limited its widespread use, which became only possible after its successful synthesis in 1828. Ultramarine blue is used today in artists' paints, lacquers, coatings, printing inks, and coloring plastics, paper, and generally all kinds of consumer goods, cosmetics, and toys. In detergents and paper, it is added for blue fining, i. e., to correct a yellow tint with blue. According to the widespread application, there are many types. Fine particles have high surface energy and tend to cohesion. Fine-grained types are therefore difficult to disperse. Special surface treatment has led to the development of types with reduced surface energy. There are types with high coloring power for printing inks and hydrophobic types for lithographic inks.

The synthesis is not simple. The construction of the sodalite framework and the introduction of sulfur takes place in a complicated multistep process. Kaolinite, anhydrous soda, sulfur, and a reducing agent such as carbon participate, ►Figure 3.5 [16, 201]. Kaolinite is first thermally dehydrated to metakaolinite and then transformed with sodium polysulfide to pre-ultramarine. Polysulfide forms of soda ash, sulfur, and the reducing agent, e. g., carbon. Pre-ultramarine is converted to ul-

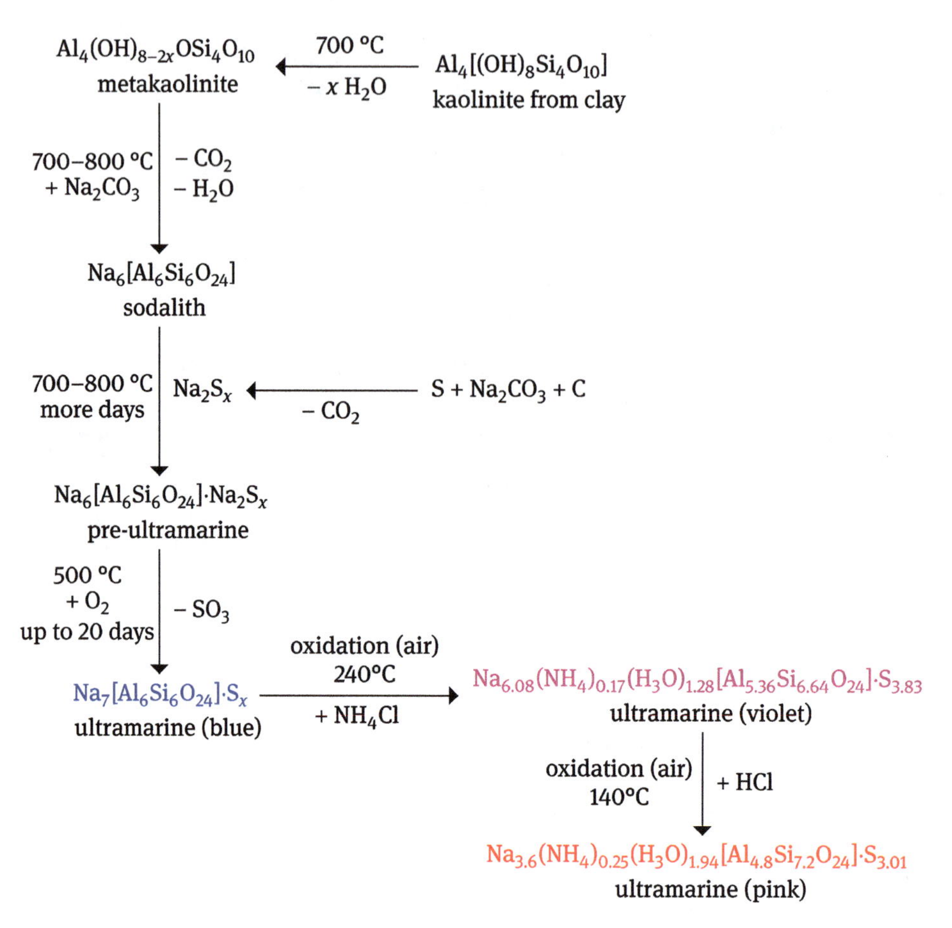

Figure 3.5: Synthesis of French ultramarine from kaolin and sulfur [16], [201, p. 344ff], [200, keyword "Pigmente"].

tramarine by adding air, oxidizing polysulfide dianions to polysulfide anion radicals. The entire process can take 20 days. The final color shade depends on the particle size and chemical composition; fine powders are lighter and greener than coarser ones, so the degree of grinding is commercially an important target parameter for hue and coloring power. Common particle sizes are 0.7–5 µm.

Ultramarine blue is not acutely toxic or irritating the skin or mucous membranes. There have been no known toxicological findings of concern during its very long and intensive history of use. Only toxic H_2S escapes on contact with acids.

Ultramarine pink (PR259, CI 77007), ultramarine violet (PV15, CI 77007)

By varying the proportions of the starting products in the production process, further ultramarine pigments can be produced, namely ultramarine pink and ultramarine violet [16]. They are less intensively colored than ultramarine blue and are hardly used for artistic purposes on a large scale. In industrial applications, the lower thermal stability must be taken into account; ultramarine violet is stable up to 280 °C, ultramarine pink up to 220 °C. Ultramarine pink takes on a purple color under the influence of bases.

All ultramarine variants contain the same aluminosilicate lattice, but the sulfide anions $S_3Cl^{\ominus}$ or $S_4^{\ominus}$ appear as a chromophore. In addition, some $Na^{\oplus}$ is replaced by $NH_4^{\oplus}$ and $H^{\oplus}$, resulting in the following approximate compositions [200, keyword "Pigmente"]:

Ultramarine blue	$Na_{6,3}[Al_{4,79)}Si_{7,21}O_{24}] \cdot S_{3,74}$
Ultramarine green	$Na_{7,74}[Al_{5,7}Si_{6,3}O_{24}] \cdot S_{1,88}$
Ultramarine violet	$Na_{6,08}(NH_4)_{0,17}(H_3O)_{1,28}[Al_{5,36}Si_{6,64}O_{24}] \cdot S_{3,83}$
Ultramarine pink	$Na_{3,6}(NH_4)_{0,25}(H_3O)_{1,94}[Al_{4,8}Si_{7,2}O_{24}] \cdot S_{3,01}$

3.4 Oxide and sulfide pigments

Due to their high chemical stability, oxides and sulfides are among the most common natural compounds. Many of them are colored, as shown in ►Figures 3.6(a) and 3.7(a). Based on the localization of the color mechanisms in the PTE, we can expect that their chromaticity is determined by LF transitions (*d* block metals), SC transitions (heavy metals), and CT transitions (higher *d* block metals).

Artists are limited in their choice of colored oxides and sulfides to natural and synthetic minerals. Unfortunately, rare or nonmineral-forming compounds are not suitable; thus remain the chalcogenides presented in ►Figures 3.6(b) and 3.7(b), some of which were already used as pigments in antiquity. Recently, certain inorganic metal oxides moved again into the focus of the pigment industry as nontoxic substitute pigments.

3.4.1 Classical heavy metal oxides and sulfides

Numerous pigments contain oxides and sulfides of heavy metals such as arsenic, lead, and mercury from antiquity to modern times. They originate from oxidic and sulfidic minerals and have a coloring effect due to their semiconducting nature. For toxicological or economic reasons, most of the heavy metal pigments mentioned here were often phased out already in the past, but at the latest in the twentieth century by new developments. These replacements are, in general, oxides of *d* block elements (►Section 3.4.3) or highly stable organic pigments.

1	2	3	4	5	6	7	8	9	10	11	12	13	14	15	16	17	18
H																	He
Li	Be											B	C	N	O	F	Ne
Na	Mg	Sc	Ti	V	Cr	Mn	Fe	Co	Ni	Cu	Zn	Al	Si	P	S	Cl	Ar
K	Ca	Y	Zr	Nb	Mo	Tc	Ru	Rh	Pd	Ag	Cd	Ga	Ge	As	Se	Br	Kr
Rb	Sr	La	Hf	Ta	W	Re	Os	Ir	Pt	Au	Hg	In	Sn	Sb	Te	I	Xe
Cs	Ba	Ac	Rf	Db	Sg	Bh	Hs	Mt	Ds	Rg	Cn	Tl	Pb	Bi	Po	At	Rn
Fr	Ra											Nh	Fl	Mc	Lv	Ts	Og

(a) Coloration of oxides.

1	2	3	4	5	6	7	8	9	10	11	12	13	14	15	16	17	18
H																	He
Li	Be											B	C	N	O	F	Ne
Na	Mg	Sc	Ti	V	Cr	Mn	Fe	Co	Ni	Cu	Zn	Al	Si	P	S	Cl	Ar
K	Ca	Y	Zr	Nb	Mo	Tc	Ru	Rh	Pd	Ag	Cd	Ga	Ge	As	Se	Br	Kr
Rb	Sr	La	Hf	Ta	W	Re	Os	Ir	Pt	Au	Hg	In	Sn	Sb	Te	I	Xe
Cs	Ba	Ac	Rf	Db	Sg	Bh	Hs	Mt	Ds	Rg	Cn	Tl	Pb	Bi	Po	At	Rn
Fr	Ra											Nh	Fl	Mc	Lv	Ts	Og

(b) Oxides actually used in pigments for painting purposes.

Figure 3.6: Localization of colored oxides in PTE [195, 196, 198]. Included are the colors of different oxidation states and modifications. Group designation according to [194].

Massicot, litharge (PY46, CI 77577)

Lead(II) oxide PbO occurs as a pigment and as a starting product for the manufacturing of chrome yellow [47, 49]. Of the two modifications, the ordinary (tetragonal) one is red, and the metastable rhombic one is delicately yellow. This yellow was already applied 400 BC on wall paintings in ancient Thrace [771] and used until the Middle Ages. In both cases, a band gap of the semiconducting pigment (which is narrower in red lead oxide) causes the color.

The production takes place by burning lead in the air, whereby first the red lead(II) oxide is formed, which changes to the yellow modification at 488 °C:

$$\underset{}{2\,\mathrm{Pb} + \mathrm{O_2}} \longrightarrow \underset{\text{tetragonal}}{2\,\mathrm{PbO}\ (\text{red})} \xrightleftharpoons{488\,^\circ\mathrm{C}} \underset{\text{hexagonal}}{2\,\mathrm{PbO}\ (\text{yellow})}$$

Massicot disappeared from the palette toward the end of the Middle Ages with the advent of lead-tin yellow. Its conversion rate into the red modification at low temperatures is slow; however, being of unstable nature, it discolors into black lead(IV) oxide in light and air and dissolves rapidly in acids and alkaline solutions. Nevertheless, it

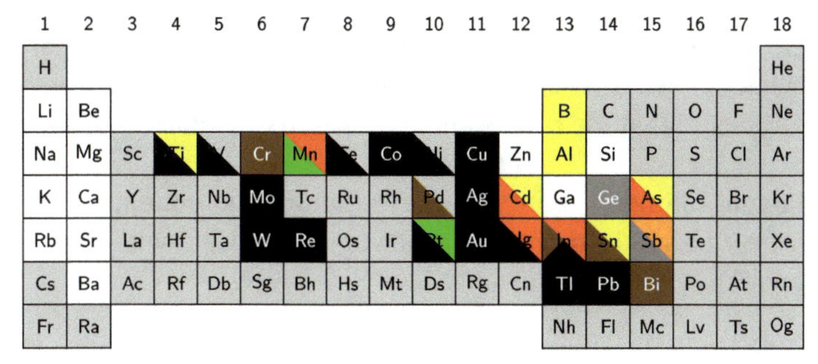

(a) Coloration of sulfides.

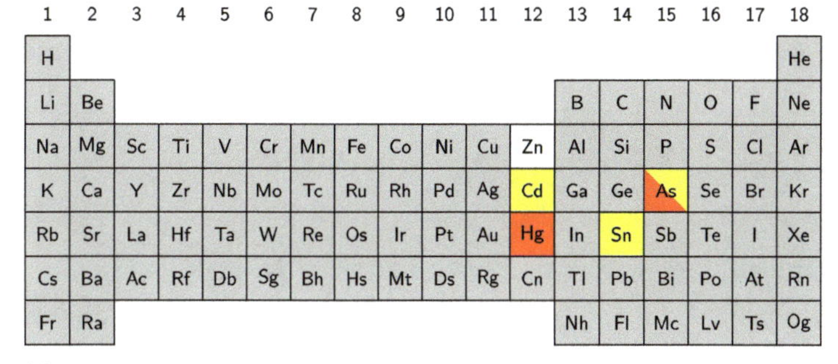

(b) Sulfides actually used in pigments for painting purposes.

Figure 3.7: Localization of colored sulfides in PTE [195, 196, 198]. Included are the colors of different oxidation states and modifications. Group designation according to [194].

provided the advantage that it had a siccative effect as a lead salt in oil painting. Incidentally, the designation Massicot confusingly has often been applied to all yellow lead pigments.

Orpiment, King's yellow (PY39, CI 77086 and 77085)

The bright yellow arsenic(III) sulfide As_2S_3 was already known in Egypt around 1500 BC. We find it in paintings of antiquity, in medieval book illuminations, and as a vital artists' pigment until the nineteenth century [48, Volume 3] [47, 49, 401, 402] since its color was not achievable with any other early pigment. It disappeared from the palettes with the rise of Naples yellow and especially chrome yellow around 1818 with a similar hue and artificial green pigments.

It owes its color to its semiconductor character and a moderately large band gap like all sulfide pigments. The sharp band edge contributes significantly to the color purity of the pigment, and its high refractive index of 2.4–3 helps it achieve high opacity.

As a sulfide, it generally enjoyed the reputation of being incompatible with copper and lead pigments. Since there are numerous examples in which orpiment has remained stable by being embedded in a linseed oil film, this reputation is only partially justified. Studies have shown that orpiment decomposes primarily with traces of acetic acid and moisture and that the sulfide anions released can attack copper or lead pigments.

There are also different opinions about its toxicity. Mining the mineral in hydrothermal deposits is very harmful. In the eighteenth century, it was synthetically produced by the sublimation of sulfur and arsenic As_2O_3:

$$2\,As_2O_3 + 9\,S \longrightarrow 2\,As_2S_3 + 3\,SO_2\uparrow$$

This synthetic pigment was highly toxic due to amounts of soluble arsenic. In the nineteenth century, a wet process was developed to produce it from soluble arsenic salts:

$$As^{3\oplus}\,aq. + 3\,H_2S \longrightarrow As_2S_3\downarrow + 3\,H^{\oplus}$$

Naples yellow, lead antimony yellow, antimony yellow, lead antimonate yellow (PY41, CI 77588)

Naples yellow [48, Volume 1], [47, 49] is a beautiful opaque yellow that can be produced in yellow-cool or reddish-warm shades. Unfortunately, it was often mistaken for lead-tin yellow and led to considerable confusion. Naples yellow had displaced the rival by 1750 until, in turn, chrome and cadmium yellow replaced it about 1850.

Naples yellow, however, by no means is a pigment of modern times but belongs with Egyptian blue to the oldest synthetic pigments: as ceramic paint, we can find it already around 1500 BC on clay tiles, in yellow glasses, and glazes of the ancient cultures in Mesopotamia [23].

The theoretical composition $Pb_2Sb_2O_7$ (corresponding to the mineral bindhemite) is rarely encountered. Instead, we find the formulas $Pb_3(SbO_4)_2$, $Pb(SbO_3)_2$, or $Pb(SbO_4)_2$. Since the old recipes are not precise regarding the exact preparation method, it can be assumed that they achieved a more or less good approximation to the true Naples yellow depending on the production quality. Lead may have even sublimed prematurely at high temperatures and was thus absent.

For preparation, a lead and an antimony compound were roasted: lead, lead(II) oxide (litharge), lead(II, IV) oxide (red lead), as well as antimony, antimony(III) oxide (antimony bloom), or antimony(III) sulfide (gray antimony) served as reactants. The shades of color achieved depend on temperature: a lemon to sulfur yellow one at high heat (above 800 °C) and reaction times of several hours, and a rather orange one at low temperature. Experiments show that in this case, however, hardly the true Naples yellow is formed, but rather diverse mixtures of yellow lead and antimony oxides, chlorides, and oxychlorides emerge, such as the *Kassler yellow* $PbCl_2 \cdot 7PbO$.

Modern paint manufacturers used lead carbonate, lead nitrate, lead hydroxide, and antimony(III) oxide, or antimonates with common salt, lead chloride, or ammonium chloride additives.

With a refractive index of 2.01–2.23, true Naples yellow is one of the pigments with good hiding power. While it is resistant to alkali and light, it reacts with acids and sulfur compounds. Thus, it was preferably bound in oil. If we write the composition $Pb_2Sb_2O_7$ as $2\,PbO \cdot Sb_2O_5$, we recognize an SC transition in the yellow semiconducting lead(II) oxide as cause of color, ►Section 2.2.4.3. The absorption edge typical for semiconductors is visible in the reflectance spectrum, ►Figures 3.8 (idealized) and 3.9.

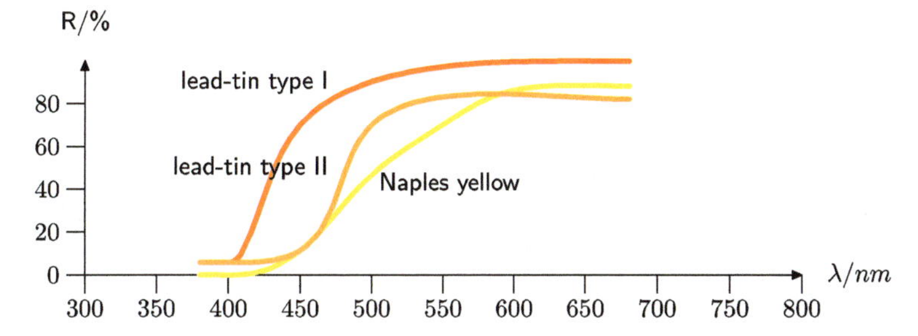

Figure 3.8: Schematic reflectance spectra of Naples yellow, lead-tin yellow type I, and lead-tin yellow type II (spectra drawn after [48, Vol. 1, p. 227]).

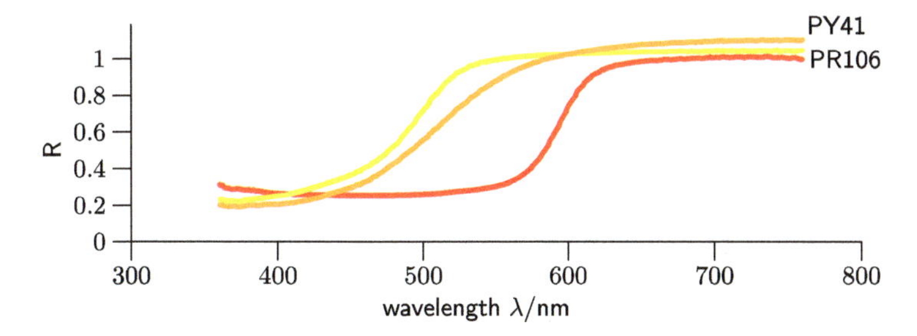

Figure 3.9: Reflectance spectrum of watercolor paints based on lead antimonate and mercury sulfide (semiconductor), normalized to an arbitrary unit: PR106 (vermilion, Kremer no. 42000, HgS), PY41 (Naples yellow light genuine, Wallace Seymour Vintage Watercolour, theor. $Pb_2Sb_2O_7$), and PY41 (Naples yellow deep genuine, Wallace Seymour Vintage Watercolour, theor. $Pb_2Sb_2O_7$).

Lead-tin yellow (CI 77629)

Lead-tin yellow is a medieval pigment [48, Volume 2], [47, 49, 403, 404], whose color ranges from light lemon yellow to warm orange-yellow. Chemically, it is a lead stannate; type I has the composition $Pb_2SnO_4 = 2PbO \cdot SnO_2$, and the glassy type II has the structure $PbO \cdot (Sn, Si)O_2$.

Due to its color and nontoxicity, it was the essential bright yellow pigment from about 1300 to 1700; only orpiment enjoyed higher esteem. In the eighteenth century, it was then displaced by Naples yellow until it was rediscovered in 1940 in the course of painting research.

Having a refractive index of 2.3, it is an opaque pigment that is stable to acids and alkalis. With hydrogen sulfide, it forms brown PbS.

Despite its widespread use, there are few sources for its production. The Bolognese manuscript speaks of roasted lead, tin, and sand, which would correspond to a solid-state reaction of lead and tin oxides. The modern synthesis of 1941 starts from lead(II) oxide and leads from type I to type II:

$$2PbO + SnO_2 \xrightarrow[700\text{–}800\ ^\circ C\ (greenish)]{650\ ^\circ C\ (reddish)} Pb_2SnO_4 \xrightarrow[800\text{–}950\ ^\circ C]{SiO_2} PbO \cdot (Sn, Si)O_2$$

At a low conversion temperature, a warmer red, and at a higher temperature, a cooler lemon-yellow version of the type I yellow emerges, analogous to Naples yellow's preparation. The yellow color is due to a band transition in the lead oxide-tin oxide semiconductor, ►Section 2.2.4.3.

Realgar, sandarach, red orpiment

High toxicity characterizes the orange-red arsenic sulfide As_4S_4. It was used as a red pigment from the second millennium BC to the Middle Ages because its pure orange-red hue could not be achieved by mixtures of alternatives such as red ocher [48, Volume 3], [47, 49]. However, it was employed much less frequently than the extensively used orpiment. Minerally, it occurs as a companion of arsenic and ores in hydrothermal deposits, but reliable indications for a synthetic preparation are missing until now. Problematic is its tendency to turn into an orange-yellow powder when exposed to air, ►Section 7.4.11.

Vermilion, cinnabar, minium (PR106, CI 77766)

“Cinnabar” denotes a red, natural mineral consisting of mercury(II) sulfide α – HgS. Its synthetic form is called vermilion. α – HgS is a red pigment [48, Vol. 2], [47, 49, 57, 389] with a clear, proverbial “vermilion” hue, which varies according to the degree of grinding of the pigment: particles around 2–5 µm diameter are bright, orange-red, and those with a diameter > 5 µm are deep, dark red [58]. This dependence of color on particle size is typical of semiconductor pigments, in which a band gap determines the color, ►Section 2.2.2. Like many sulfides and oxides, HgS is a colored semiconductor, showing the typical absorption edge in its reflectance spectrum, ►Figure 3.9. Besides the red α modification (hexagonal, band gap E_g 2 eV), the polytropic compound also possesses a black, amorphous modification (E_g 1.6 eV) without any significance as a pigment. With a refractive index of 2.8, α – HgS exhibits high scattering, and thus opacity.

Cinnabar was already known in ancient Greece. Thanks to the cinnabar mine in Almaden, it advanced to a crucial red pigment. In China, cinnabar/vermilion is associated with the color of luck, perseverance, and the Heavenly Palace in Beijing, and it plays a vital role in Chinese art history. Cinnabar, and its artificial form, vermilion, surpass red ocher in the purity of color and opacity; not surprisingly, they have been two of the most important red pigments since the Middle Ages. They were known as "minium" in medieval book illumination when mixed with red lead. The term minium (from which the term "miniature" was derived) was later applied to the red lead.

Vermilion was also a crucial substance for alchemy since it represented the unification of the basic principles in an ideal cycle: sulfur (sulfur philosophorum, masculine, hot, dry), mercury (mercurius philosophorum, female, spiritually cool, moist), and lapis philosophorum, together with the separation into the original substances. Thereby the stages of pure substances—gray materia prima, nigredo—solvatio and coagulatio—citrinitas (polysulfides)—rubefactio (cinnabar) could be observed (very nicely depicted in a series of experiments in [59]).

The roasting of cinnabar exposed to air, which again yields mercury, was the primary method of mercury extraction from cinnabar for a long time. Alchemists interpreted it as purification of cinnabar (vermilion), produced from mercury, as the philosopher's stone, or better, as an earthly, apparently still impure approximation since it could neither turn metals into gold nor prolong one's life. (Bucklow [10] provides a lively introduction to medieval thinking. It was written by a chemist working in the field of art, and he managed to explain the somewhat strange-seeming medieval thoughts on matter even to a modern hardcore scientist like the author.)

The ground and washed natural mineral was used as a pigment until about 1900, when alternatives replaced it: around 1880, mercury cinnabar pigments, and later organic pigments such as lithol red. As early as about 800, however, it could also be produced artificially:

$$\mathrm{Hg} + \mathrm{S} \xrightarrow{\Delta} \underset{\text{mineralic Mohr}}{\mathrm{HgS\ (black)}} \xrightarrow{\text{dry (ancient): sublimation, 580 °C}} \mathrm{HgS\ (red)}$$

$$\mathrm{Hg^{2\oplus}\ aq.} \xrightarrow[-\,\mathrm{NH_4^{\oplus}}]{\text{wet (today): }(\mathrm{NH_4})_2\mathrm{S\ aq.},\ \Delta} \mathrm{HgS}\downarrow \mathrm{(red)}$$

The earlier dry preparation yielded only unstoichiometric products until the seventeenth century. Cinnabar (vermilion) was always associated with alchemy, and it took considerable time before the recipes followed precise observations and delivered reliable products, stoichiometrically composed.

Despite its mercury content, vermilion is today a nontoxic pigment because the artificial, pure sulfide is extremely sparingly soluble. Consequently, it practically cannot spread in the body after oral ingestion. It is also inert to other pigments; a lead white and zinc mixture can produce flesh hues without difficulty.

Red lead, Saturn red, minium (PR105, CI 77518)

Mixed-valence lead(II,IV) oxide or lead(II) plumbate(IV) $Pb_3O_4{=}Pb_2[PbO_4]$ is a bright red pigment [48, Volume 1], [49] with the high refractive index of 2.42, which was already used by ancient cultures. It found an important application as a red pigment in book illustrations. First known as “minium,” when mixed with cinnabar, the name was later transferred to the pure red lead. Today it is used as an anticorrosive coating and a colorant in ceramic glazes. It is no longer applied in artists’ paints due to the following problems: it slowly oxidizes to black lead(IV) oxide PbO_2, disturbing the color scheme. Furthermore, it is incompatible with sulfur-containing pigments, yielding PbS and other reaction products, and finally, it is toxic.

Read lead is prepared by oxidation of lead(II) oxide in the presence of air at 500 °C:

$$3\,PbO \xrightarrow[\substack{300\text{–}500\ ^\circ C\ (\text{ancient}) \\ 450\text{–}470\ ^\circ C\ (\text{today})}]{+\frac{1}{2}O_2,\ \Delta} Pb_3O_4$$

This synthesis was already carried out in antiquity, starting from lead white:

$$2\,PbCO_3 \cdot Pb(OH)_2 \xrightarrow[-2\,CO_2\,-\,H_2O]{+\,O_2,\ \Delta\ 430\ ^\circ C} Pb_3O_4$$

The compound gains its color from an IVCT transition between the two valences of lead, ►Section 2.4.2.

3.4.2 Iron oxide pigments, ocher

Since early times, iron oxides and hydroxides in the various oxidation states (II, II/III mixed, and III) have received attention. They cover the broad color spectrum from yellow through orange, red, reddish-purple, and brown to black. Furthermore, many of these colors still have natural, easily accessible deposits. They are also lightfast, weather fast, and alkali resistant. Therefore, they have been used as pigments since the earliest times and remain extremely popular due to their attractive color shades, high durability, nontoxicity, and low price [11, 16], [48, Volume 4], [49, 65, 66], [201, p. 317ff], [634]. At the end of the nineteenth century, the demand and the requirements concerning the quality and properties of these pigments increased to such an extent that the natural sources were no longer sufficient, and the production of synthetic iron oxides started. Today, synthetic iron oxides account for about 80 % of the total quantity of iron oxide pigments.

More than 50 % of all iron oxides are used to color building products, such as concrete roofing tiles, paving stones, or cement. Another 40 % is used to produce paints, lacquers, and coatings. For products to possess the necessary opacity, the pigment particles must be 0.1–1 µm in size. Iron oxides are also a popular pigment for coloring plastics. The low-temperature stability of the yellow iron oxide hydrate prevents the coloring of specific polymer types. However, by encapsulating it in inorganic layers, iron oxide yellow can be stabilized up to approximately 260 °C.

Iron oxide pigments are considered nontoxic. Pigments from natural sources, however, can contain traces of heavy metals. Therefore, only specific synthetic grades are approved for cosmetics, care products, pharmaceuticals, or contact with foodstuffs. They are applied in artists' paints and coloring paper, and also, because of their magnetic properties, in laser toners and coating of data tapes.

3.4.2.1 Mineralogical basis

Iron oxide pigments are based on chemically related iron(III) oxides, ►Table 3.2. They consist of close-packed anions ($O^{2\ominus}$, $OH^{\ominus}$) in a hexagonal close-packed (hcp) or a cubic close-packed (ccp) arrangement. The layer sequence in hcp arrangement is AB AB ..., and in the ccp arrangement ABC ABC .... Both anions are larger than the Fe^{III} (or Fe^{II}) cation; hence they govern the structure. The interstices between the anions are partially filled with $Fe^{3\oplus}$ (or $Fe^{2\oplus}$) in octahedral coordination, thus building blocks are FeO_6 or $FeO_3(OH)_3$ octahedra.

The iron oxides differ in how the $Fe(O, OH)_6$ octahedra are arranged in space. Besides anion packing, the arrangement can be described by how the octahedra are linked together, i. e., do they share faces, edges, or corners. Depending on this linkage type, the Fe-Fe distance varies significantly (0.289 nm for face-sharing octahedra, 0.306 nm for edge-sharing octahedra, and 0.346 nm for corner-sharing octahedra). As a result of the octahedra's arrangement, there are 5 polymorphs of FeOOH and 4 of Fe_2O_3. For those relevant for our topic ►Table 3.2. Also relevant is the mixed-valence compound iron(II,III) oxide Fe_3O_4.

Due to the similarities of the anion frameworks, mutual conversion such as dehydroxylation of α-FeOOH to α-Fe_2O_3 are readily possible, whereby only cation rearrangement and elimination of water is required. ►Figure 3.10 illustrates these pathways:

Table 3.2: Mineralogical properties of iron oxides [65, Chapter 1, 2].

Ligands	α phase, anions in hcp arrangement	γ phase, anions in ccp arrangement
Oxide and hydroxide anions (FeOOH, building blocks are $Fe^{III}O_3(OH)_3$ octahedra)	Goethite α-FeOOH (yellow), diaspore structure (α-AlOOH), acicular crystals	Lepidocrocite γ-FeOOH (orange), boehmite structure (γ-AlOOH)
	Edge- and corner-sharing octahedra	Edge- and corner-sharing octahedra
Oxide anions (Fe_2O_3, building blocks are $Fe^{III}O_6$ octahedra)	Hematite α-Fe_2O_3 (red), corundum structure (α-Al_2O_3), various crystal forms	Maghemite γ-Fe_2O_3 (brown), hematite structure
	Edge-, face-, and corner-sharing octahedra	Edge-, face-, and corner-sharing octahedra

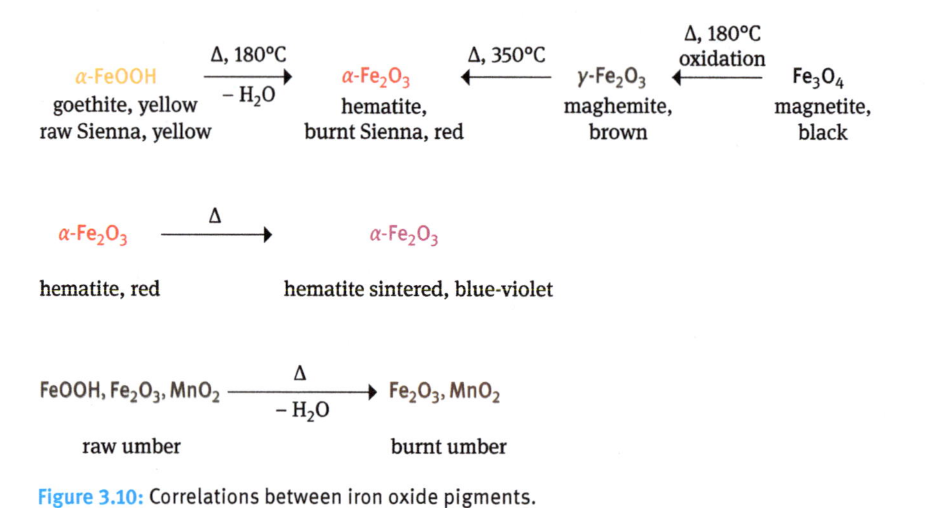

Figure 3.10: Correlations between iron oxide pigments.

- *Calcining* or *burning* leads from yellow goethite to red hematite with water loss. This procedure allows the production of red ocher from yellow. The artist finds the same principle behind the colors "burnt Sienna" and "burnt umber," which both show a distinct red undertone while "raw Sienna" is light yellow and "raw umber" is brown. The transformation goethite → hematite causes the typical change of color to red when firing ceramics.
- Heating (sintering) of (light) red hematite leads to blue, purple, brownish coarse-grained particles, known to the artist as colcothar or Caput Mortuum.
- Heating magnetite under oxidizing conditions also results in hematite, associated with a color change from black to red. The intermediate maghemite results in a brown hue.

The conversions are of great technical importance, as we see in the manufacturing processes below. Already antiquity used two typical conversion reactions for the production of polychrome ceramics (►Section 7.1): the yellow oxides decompose from 180 °C to red α-Fe_2O_3 and deliver pure reds with high coloring power. Above 180 °C, the black oxides convert to brown γ-Fe_2O_3, and from 350 °C also to red α-Fe_2O_3. Only the red oxides are stable up to 1200 °C. Craftspeople could engineer yellow, red, and black ceramics by skillful reactions under reducing and oxidizing conditions. Coating with inorganic materials increases the thermal stability of the yellow iron oxides up to 260 °C.

3.4.2.2 Color of iron oxides

The significantly different colors of the chemically similar compounds (►Figure 3.12) originate in several parallel processes, which were discussed in detail in ►Section 2.3.4, and which depend on crystal phase (α or γ) and nature of the ligands (oxide or hydroxide anions) [65, Chapter 6]:

- strong OMCT transitions $O(2p) \rightarrow Fe^{III}(3d)$ between oxygen and iron atoms in the near-UV (basic yellow color)
- forbidden LF transitions in Fe^{III} d orbitals
- additional transitions of electron pairs in the visual green to yellow spectral range due to magnetic coupling of Fe^{III} cations (EPT, distinguishing yellow and red iron oxides)

These processes are independent of the size and shape of the pigment particles. In addition to these influences, especially in the case of iron oxides, particle size and shape have a visible effect on the color [48, Volume 4], [65, 66] (see Table 3.3).

Table 3.3: Dependence of color on composition and particle size of iron oxide pigments [48, Volume 4], [65, 66].

Mineral	Fine-grained	Coarse-grained
Goethite	Dark yellow, 0.05–0.8 µm darker and more yellow with decreasing needle length, below 0.05 µm brown	Yellow, 0.3–1 µm darker with decreasing needle length, 1 µm yellow
Lepidocrocite	Dark brownish-orange	Bright orange
Hematite	Light red, 0.1–0.2 µm bright red, below 0.1 µm orange	Reddish-brown, 1–5 µm blue-red to purple, above 5 µm purple

There is no apparent cause for these more subtle color changes. Iron oxide is a semiconductor; therefore, *collective electron excitations* or *surface plasmons* play an essential role. These excitations depend strongly on the geometry and spatial expansion of the pigment particles and are explained in ►Section 1.6.4.

Iron oxides have a high refractive index of 2.36 (goethite), 2.87 (hematite), and 2.42 (magnetite) and are therefore opaque. The particle size influences the scattering power (►Section 1.6.8): by selective production of tiny particles with diameter < 0.05 µm, we obtain transparent yellow or yellowish-red glaze pigments. In contrast, opaque ones with diameters of 0.05–1.0 µm are coarse. Hematite particles then show a neutral to cool-red hue [16, 191]. The limit of 0.05 µm results from the variation of scattering as a function of particle diameter, ►Figure 3.11. In this lowest size range, only the scattering Q is practically zero; above it, the scattering increases rapidly to a maximum value before reaching a significantly greater limit than zero.

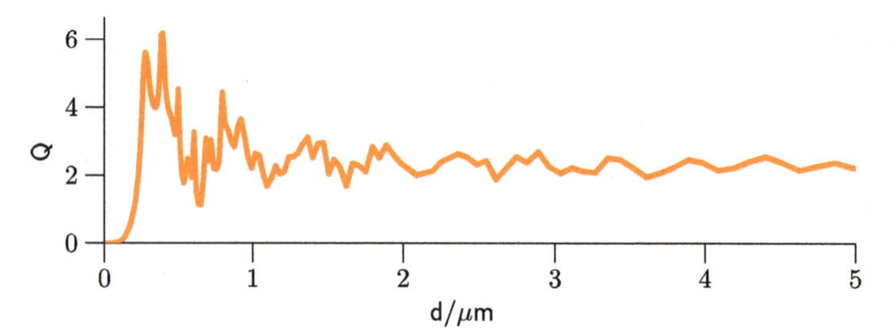

(a) Graph of Q for particles with refractive index $n_1 = 2.3$ (goethite).

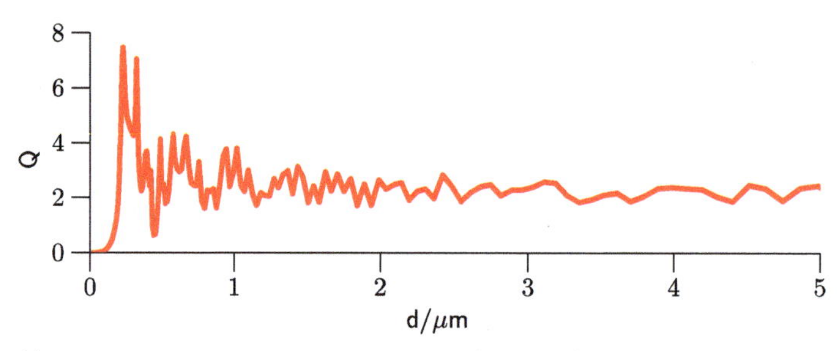

(b) Graph of Q for particles with $n_1 = 2.8$ (hematite).

Figure 3.11: Progress of (Mie) scattering Q as a function of particle diameter d for the red light of wavelength 650 nm (calculated with Mieplot [333]). The refractive index of the medium is $n = 1$ (air). The graphs show that scattering occurs from a particle diameter of about 0.05 µm, so transparent pigment particles must be smaller than this value.

3.4.2.3 Natural iron oxides

The oxides mentioned above are components of various natural pigments. Depending on their proportions and the presence of accompanying substances, they form the basis for "yellow ocher" and "red ocher," "terra di Sienna" and "umber." The iron content depends strongly on the locality of the deposit. The typical Fe_2O_3 content of hematite is 45–95 %; especially large, high-quality deposits are located near Malaga and Hormuz in the Persian Gulf. This "Spanish red" or "Persian red" has a content of > 90 % iron(III) oxide. High-quality deposits of goethite are located in South Africa with the equivalent of 55 % Fe_2O_3 content and southern France with 20 % Fe_2O_3. Accompanying minerals such as clays or silicates, often present in large quantities, are responsible for the residual mass fractions and “dilute” the iron oxides so that natural pigments have a varying and relatively low coloring power.

The raw materials are ground, washed, and ground again for pigment production. Sienna and umber are calcined to remove water; the hue depends on the calcination parameters, temperature, and composition. Higher color purity is achieved by slurrying the iron-rich raw material and then separating it according to particle size, e. g., by sedimentation of the coarser particles and discharging the supernatant water together with the finer particles. The main fields of application of iron oxide pigments are artists’ paints and pastels, inexpensive coatings for shipbuilding, and colorants for building products such as cement, artificial stone, and wallpaper.

The duller, often beige-brown-gray color shades of natural iron oxides result from a low iron content, a broad particle size distribution, and the alternating admixture of various color-active metal cations. Such inhomogeneous distributions occur in poorly crystallized or nodular-amorphous mixtures. They originate primarily as hydrothermally deposited iron oxides or ocher earths as weathering products. As shown in ►Section 1.7.2 at p. 74, highly different crystals vary in all properties, including optical ones. As a result, the features in the absorption spectrum of the ideal iron oxide are broadened, and the color purity decreases. In addition, natural deposits usually contain many (foreign) metals, so the color of pure iron oxide is readily changed by accessory metal cations. In particular, the replacement of iron with manganese strongly affects the color and shifts it in the direction of brown to black.

Yellow ocher, yellow earth (PY43, CI 77492, natural); iron oxide yellow, Mars yellow (PY42, CI 77492, synthetic); Mars orange (PO42, CI 77492)

Yellow ocher is a weathering product of iron-bearing ores. Substantial natural deposits are located in France (Roussillon) and South Africa. Yellow ocher owes its color to goethite FeOOH, which is contained up to 10–60 %; in addition, Fe_2O_3 may occur. The pigment is extracted by slurrying the earth to the desired hue. Typical varieties are “light ocher” and “gold ocher.”

The broad features in the reflectance spectrum of ocher show that the yellow color is more muted and less pure, especially compared to the SC-based or CT-based yellow pigments cadmium sulfide and chrome yellow, ►Figures 3.12 and (idealized) 2.23. Ochers nevertheless exhibit intense yellow hues when carefully washed and sorted by grain size. Furthermore, additions of manganese in umber earths further attenuate the hue, ►Figure 3.13.

Ochers can be opaque or transparent depending on the origin or production conditions. However, especially ochers of the Sienna earth type are characterized by high transparency, which is essentially due to their small particle size.

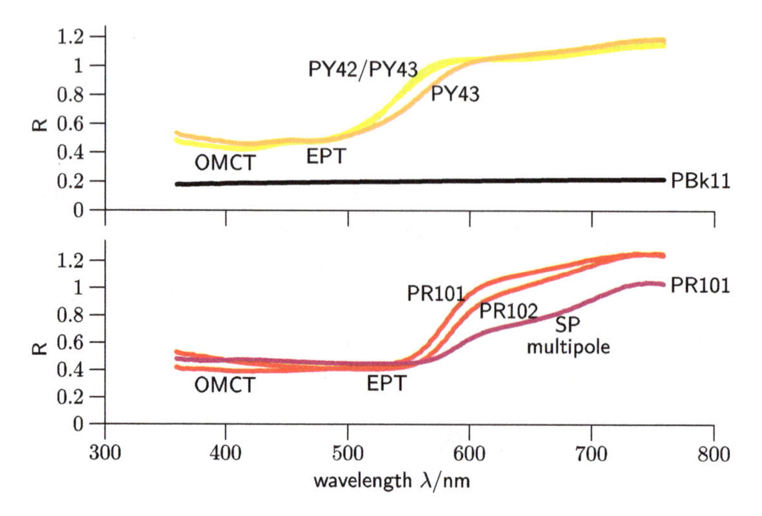

Figure 3.12: Reflectance spectrum of watercolor paints based on iron oxide, normalized to an arbitrary unit: PBk11 (Mars black, Schmincke Horadam No. 791, Fe_3O_4=$FeFe_2O_4$), PR101 (English Venetian red, Schmincke Horadam No. 649, α-Fe_2O_3, synthetic), PR101 (caput mortuum violet, Winsor and Newton Professional Watercolor No. 125, α-Fe_2O_3), PR102 (red ocher RTFLES, Kremer no. 40020, α-Fe_2O_3), PY42 (yellow ochre, Schmincke Horadam No. 655, α-FeOOH, synthetic), PY43 (yellow ocher JTCLES, Kremer No. 40010, α-FeOOH), and PY43 (yellow ocher Havane Orange, Kremer No. 40080, α-FeOH). The complex background of color formation is explained in detail in ►Section 2.3.4 at p. 118. The figure illustrates how EPT leads to different colors: the EPT around 530 nm in red ochers absorbs all yellow components and induces the red color, while the EPT in yellow ochers is around 480 nm and absorbs red and also yellow spectral components leading to a yellow color. Increasing particle size and shifting the EPT toward higher wavelengths allows the development of higher multipole resonances of surface plasmons (SP), which cause additional attenuation in the red region and enhance the development of a blue-purple cast. In addition, an intense OMCT transition in the blue spectral range establishes yellow or red as the fundamental color. Furthermore, the synthetic iron oxide PR101, with its high iron oxide content, has a steeper absorption edge and is therefore purer in color than the natural red ocher PR102.

Terra di Sienna, raw Sienna (PBr6, PBr7, PY43, CI 77491, CI 77492, CI 77499); Burnt Sienna (PBr7, PR101, CI77491, CI 77492)

This pigment originates from Tuscany, consists of 20–60 % iron oxide (as goethite), and may contain small amounts of manganese (< 1 % MnO_2). Compared to other yellow ochers, we find a high content of goethite in Sienna earths. Unfortunately, the CI designation for Sienna is complex and not unambiguous. Since Sienna comprises many materials, it may be described as a yellow iron oxide hydrate (PY43) or brownish iron oxide (PBr6, PBr7). Burnt Sienna may be described as red-brown iron oxide (PBr7) or synthetic iron oxide PR101.

In contrast to other iron oxides, Sienna has been explicitly mentioned only since the eighteenth century, the transition to yellow ochers is fluid and unclear. Chemically, there is no significant difference between yellow ocher and Sienna earths; a possible

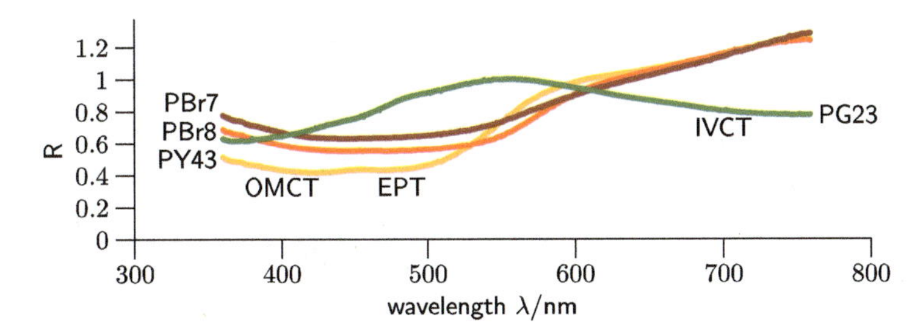

Figure 3.13: Reflectance spectrum of watercolors based on iron oxide $x\mathrm{Fe_2O_3} \cdot y\mathrm{FeOOH} \cdot z\mathrm{MnO_2}$ (yellow-brown) or iron silicate (green), normalized to an arbitrary unit: PBr7 (Cypriote burnt umber, Kremer No. 407208, y large), PBr8 (Italian burnt umber, Kremer No. 40700, y and z large), PG23 (Bohemian green earth, Kremer No. 408108, $\mathrm{(K, Na)(Al, Fe^{III}, Mg)_2(Si, Al)_4O_{10}(OH)_2}$), and PY43 (terra di Sienna, Kremer No. 40400, y large). In (manganese-containing) iron oxides, OMCT and EPT transitions dominate in the blue spectral range and induce a yellow-brown color. In contrast, in green earth, the IVCT transition between $\mathrm{Fe^{II}}$ and $\mathrm{Fe^{III}}$ occurs, absorbing the yellow spectral range. With LF transitions of the yellow $\mathrm{Fe^{III}}$ cation, this results in a yellow-green to blue-green color.

manganese content is not a decisive criterion. The warmer and more transparent color of Sienna compared to yellow ocher is attributed by some authors to the manganese content, by others to low organic content. The transparency could come from the fact that iron is bound in the silicate or aluminate matrix of the material. The considerably small particle size of the Sienna earths (around 5 nm) contributes decisively to the transparency. ►Figure 3.13 shows the spectrum of Sienna earth close to the yellow ocher.

The color of burnt Sienna changes similar to the yellow ocher into reddish-brown shades since a more or less complete transformation of the goethite into hematite occurs. Burnt Sienna is also a good glazing pigment.

Red ocher, red earth, sinoper, Spanish red, Persian red, hematite (PR102, CI 77491, natural); iron oxide red, English red, Indian red, Mars red (PR101, CI 77491, synthetic); sanguine, red chalk, read earth

This pigment contains up to 90 % red hematite $\mathrm{Fe_2O_3}$. The ocher's red color varies with hematite content; the best types with high concentration bear names like "Spanish red" or "Persian red" after their deposits in Spain and Hormuz. *Sanguine* is red ocher mixed with clays, which painters rarely use as a pigment but as red chalk crayons, sticks, or drawing pencils.

Red ocher can be mined naturally or obtained by firing yellow ocher, during which goethite transforms into hematite. Commerce denominates these synthetic red ochers as "Venetian red" or "English red."

►Figure 3.12 displays that the synthetic iron oxide PR101 is purer than the natural red ocher PR102 (containing more iron(III) oxide or hematite), expressed in a steeper absorption edge and a purer color.

Umber, raw umber, burnt umber (PBr6, PBr7, PBr8, PR101, PR102, CI 77491, CI 77492, CI 77499, CI 77730, CI 77728)

The CI designation for umber and its chemical composition is complex and not unambiguous. Since umber comprises many materials, it can be described as brown iron oxide or iron oxide hydrate (PBr6, PBr7). For umber with high Mn content, also PBr8 (manganese dioxide) may be suitable. Burnt umber might be denoted as red-brown iron oxide (PR101, PR102).

The most important deposit of umber in Cyprus is clay. Umber contains 45–70 % iron oxide (goethite) besides 5–20 % MnO_2, which is why the hue of "raw umber" is yellowish-brown, ►Figure 3.13. The two components do not form a mixed oxide but exist in separate crystal phases side-by-side, as could be confirmed experimentally. "Burnt umber" is dark brown with a red undertone caused by calcination of goethite to hematite.

Mars brown (PBr6, CI 77499)

This pigment consists of synthetic brown iron oxide. However, the majority of iron oxide brown, although theoretically possible, is not specific iron oxide but a mixture of the yellow, red, and black iron oxides $FeOOH/Fe_2O_3/Fe_3O_4$ [200, keyword "Pigmente"]. The chemically pure Mars brown is precipitated maghemite or goethite annealed with manganese.

Colcothar, Caput Mortuum, burnt vitriol (PR101, CI 77491)

Colcothar is a brown to brown-purple pigment and consists of burnt ferrous sulfate or strongly heated red ocher. Chemically, it is hematite; its red color is changed into blue-purple red by sintering and coarsening of the particles in the heat, ►Section 1.6.4 at p. 50 and ►Figure 3.12. According to [11, 16], the particle diameter has a decisive influence on the particle color: the oxide exhibits a yellow tinge for diameters around $d = 0.1\,\mu m$, and a purple one for diameters of $d = 1\,\mu m$. ►Figure 3.14 shows the reflectance spectra of some iron oxide red pigments, which depicts this color change from warm red to cold red to colcothar's purple-red with increasing particle size.

In the past, the distillation residue (iron oxide) obtained during the extraction of SO_3 and sulfuric acid (vitriol oil) from iron sulfate (vitriol) was employed:

$$2\,FeSO_4 \xrightarrow{\Delta} Fe_2O_3 + SO_2\uparrow + SO_3\uparrow$$

During early copper recovery from sulfide copper ores with iron (scrap) by the redox reaction, large quantities of ferrous sulfate were obtained in addition to "cement copper":

$$Cu^{2\oplus} + Fe \longrightarrow Cu + Fe^{2\oplus}$$

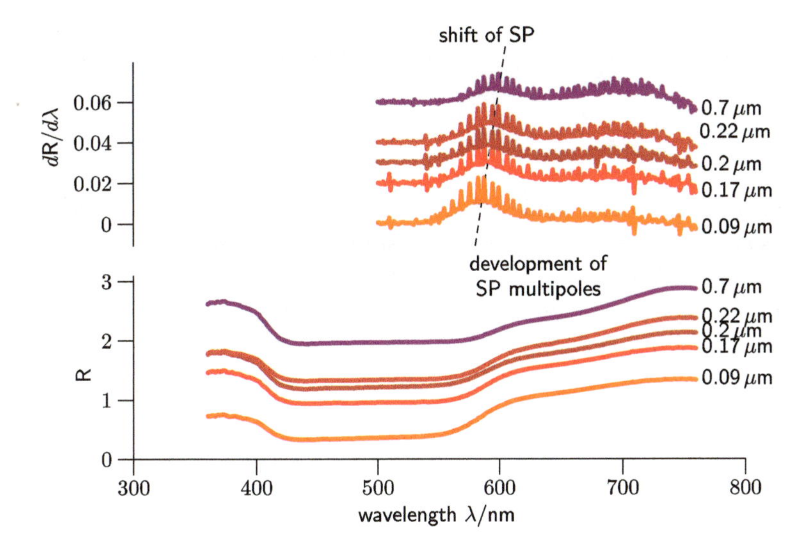

Figure 3.14: Bottom: Reflectance spectrum of watercolor paints based on PR101 (α-Fe_2O_3) with different particle size d, normalized to an arbitrary unit and shifted on the y axis to highlight the individual curves: iron oxide red 110 M light (Kremer no. 48100, d = 0.09 μm), iron oxide red 130 M medium (Kremer No. 48200, d = 0.17 μm), iron oxide red 222 dark (Kremer No. 48250, d = 0.2 μm), iron oxide red 130 B medium (Kremer No. 48150, d = 0.22 μm) and Caput Mortuum 180 M (Kremer No. 48220, d = 0.7 μm). Small particles exhibit an orange-red hue. The figure shows distinctly how higher multipole resonances of surface plasmons (SP) develop in the red spectral range with increasing particle size, changing the hue from warm red to cold red to purple-red. Top: The derivative of reflectivity by wavelength $dR/d\lambda$ shows the absorption in the red spectral region more clearly (broad minimum around 640 nm). Likewise, the shift of SP absorption to higher wavelengths is distinctly visible (shift of the peak in the range 580–600 nm), which indicates an increased absorption of red components and thus causes the growing cool-bluish color impression.

In Falun, Sweden, Europe's best-known copper deposit, the product was contaminated mainly by heavy metals, so the "Falun red" or "Sweden red" was effective against fungal attacks and had high durability. Even today, the waste mountains of the colcothar still exist there.

Magnetite, iron oxide black, Mars black (PBk11, CI 77499)

Magnetite is black mixed-valence iron(II,III) oxide Fe_3O_4 with an inverse spinel structure, ►Section 3.4.3. It is not used as an artist's pigment because it has little coloring power. However, it is a component of black ceramic paints and was of considerable importance, especially in antiquity.

The black color is due to an intense IVCT transition in IR, extending into the whole visible spectral range, ►Section 2.4.2. The origin of this transition becomes clearer if we write Fe_3O_4 as $(Fe^{III})_t(Fe^{II}Fe^{III})_oO_4$, indicating iron in tetrahedral $()_t$ and octahedral $()_o$ sites of the spinel. CT does not occur between t and o sites due to a too large change in geometry, but between Fe^{II} and Fe^{III} in octahedral sites *A* and *B*:

$$(Fe^{II})_{o,A} + (Fe^{III})_{o,B} \xrightarrow{\text{IVCT}} (Fe^{III})_{o,A} + (Fe^{II})_{o,B}$$

Due to the high concentration of Fe^{II} and Fe^{III}, the band broadens and absorbs all visible light. Electrons that have just completed the transition to the neighboring Fe^{III} can immediately move on to the next Fe^{III} and absorb further light. The result is the almost complete absorption of light and a deep, uniform black color. ►Figure 3.12 shows this consistently high absorption over the entire frequency range.

Green earth, terre verte (PG23, CI 77009)

This weathering product of iron silicates occurs universally and has been used as a pigment by every culture since antiquity. Since green earth provides low hiding and coloring power and dull color, it was not predominantly used as a green pigment. Nevertheless, it is well employed as a glaze or water-based color due to its refractive index of about 1.62, imparting a transparent quality. Despite its shortcomings, its transparency and price make it an attractive option for the artist. Astonishingly, its first official record exists only since about 1780 [48, Volume 1].

The color of green earth depends on the place of discovery because it has no fixed composition but is instead defined morphologically. Thus, we often speak of it in plural as "green earths." It can occur as pure material (celadonite) or greenish sand or soil (glauconite). The approximate composition $(K, Na)(Al, Fe^{III}, Mg)_2(Si, Al)_4O_{10}(OH)_2$ is that of a layered silicate with Al, Fe, and Mg in octahedral coordination. By partial reduction of Fe^{III}, the mineral always contains more or less Fe^{II}, which enables an IVCT transition between the two oxidation states of iron. It is responsible for the green color, ►Section 2.4.2 [632]. The transition is displayed in the yellow spectral part, ►Figure 3.13.

Depending on the ratio Fe^{II}/Fe^{III}, the intensity of the transition changes; at high Fe^{II} content, the IVCT transition is distinctly visible, and we obtain blue-colored green earths. At low Fe^{II} content, the IVCT transition is only weak, and the green earth is yellow-tinged due to LF transitions in yellow-colored Fe^{III}.

3.4.2.4 Synthetic iron oxides

Today, natural iron oxide pigments are receding in favor of synthetic iron oxides, called *Mars pigments*, when they appeared after 1800 [48, Volume 4], [11, 16, 49]. Like their natural counterparts, they are composed of the minerals goethite (yellow), hematite (red), magnetite (black), and lepidocrocite (orange).

The great advantage of synthetic oxides over natural pigments is their very high iron content, as they do not contain clay and silicate minerals like natural earths. The iron content of synthetic yellow oxides is typically 96–97 % by weight α-FeOOH; that of synthetic red oxides is 92–96 % by weight α-Fe_2O_3. They are thus stronger in color than natural iron oxides and can be manufactured with reproducible color shades. The control over color shades results from precisely specified reaction conditions and their observance during production, which is hardly possible during the natural formation of iron oxide deposits. Furthermore, due to constant particle sizes, variation in the color shade of synthetic oxides is considerably smaller than that of natural earths; synthetic pigments are therefore much purer in color than natural oxides.

Manufacturing conditions can control the properties of iron oxide pigments, i. e., color hue, particle size, opacity or transparency, and coloring power according to the application purpose. Therefore, the demand for synthetic iron oxides is very high.

The starting materials range from iron scrap to iron sulfate residues from titanium white production. The various iron oxides of pigment quality are obtained in solid-state reactions or wet-chemically. In addition to opaque types, transparent ones can also be manufactured.

Yellow pigment-grade iron oxides have an acicular crystal shape. They are 0.3–0.8 µm long and have a diameter of 0.05–0.2 µm. Typical particles of red iron oxides are 0.1–1.0 µm in size and show a yellowish to purplish-red in this range. Black iron oxides are cubic with a diameter of 0.1–0.6 µm. Iron oxide brown usually consists of yellow, red, and black iron oxides. A less demanded, independent brown iron pigment is γ-Fe_2O_3 (maghemite) or $(Fe, Mn)_2O_3$.

Industry applies several main synthesis routes to obtain controlled iron oxide pigments of defined particle size, distribution, shape, and composition:
- solid-state reactions with solid adducts
- precipitation reactions with aqueous solutions of Fe^{II} salts with the crucial Penniman process
- Laux process

The starting products are usually byproducts of other industries, such as steel and metal production.

Solid-state reactions

Several solid-state reactions are available for the preparation of iron oxide pigments, as shown in ►Figure 3.15 [16]:

1. Red iron oxide pigments can be produced by thermal decomposition of iron compounds, especially iron(II) sulfate $FeSO_4$ and goethite α-FeOH. Oxidation of Fe_3O_4 also delivers red pigments. The pigments are ground to the desired particle size after synthesis. The problem with this manufacturing method is the effluent contamination with iron(II) sulfate. The resulting SO_3 can be converted to sulfuric acid.
2. The direct decomposition of iron(II) sulfate leads to poorly colored blue-tinged pigments of inferior quality. Iron(III) chloride, which is available in large quanti-

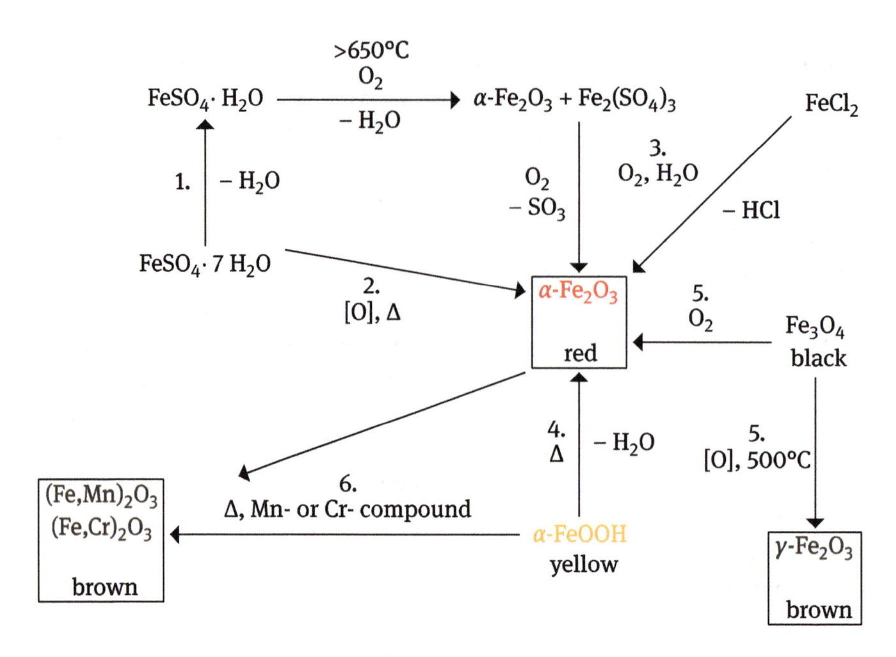

Figure 3.15: Preparation of opaque synthetic iron oxides by solid-state reactions [16]. The numbers 1–6 refer to the enumeration in the text.

ties as a by-product from other industries, delivers Fe_2O_3 during a similar decomposition, which is of no pigment quality.

3. Iron(II) chloride $FeCl_2$ can be converted into red pigment in a spray-drying calcination process.
4. Calcination of α-FeOOH provides red pigments of high purity and coloring power.
5. Fe_3O_4 from the Laux process or other processes can be oxidized to pigments with a wide color range. The color depends on starting material and reaction conditions. Preparing the brown γ phase requires careful control of reaction conditions.
6. Solid-state reactions of iron oxides and hydroxides with small amounts of an Mn or Cr compound deliver brown mixed oxides of the hematite type.

Precipitation reactions

Yellow α-FeOH pigments, red α-Fe_2O_3 pigments, and black Fe_3O_4 pigments can be produced by precipitation reactions of iron sulfate $FeSO_4 \cdot 7\,H_2O$ with alkali hydroxides and air as oxidant [16]. The reagents are identical; reaction conditions such as temperature, pH value, and their progress in time determine the resulting oxide, its hue, color purity, and pigment quality. The required Fe^{II} solution can be obtained from other industries' iron residues. However, in this case, the presence of other metal cations must be avoided to achieve pure colors. ►Figure 3.16 depicts a series of precipitation reactions for the production of iron oxide pigments:

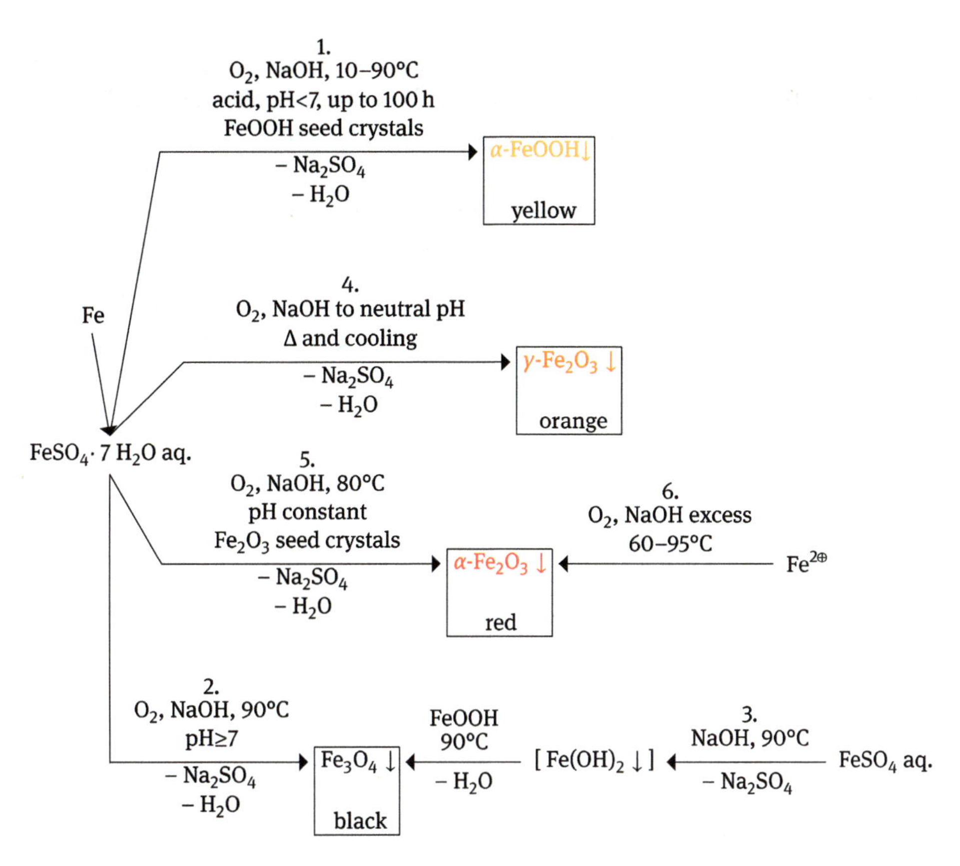

Figure 3.16: Preparation of opaque synthetic iron oxides by precipitation reactions from aqueous solutions of Fe^{II} salts [16]. The numbers 1–6 refer to the enumeration in the text.

1. Yellow, evenly shaped iron oxide pigments of high color purity are constituted by separately generating seed crystals of α-FeOOH, thus triggering uniform and controlled crystal growth of the actual pigment particles.
2. Black iron oxide in the magnetite structure and high coloring power forms by precipitation in an alkaline environment, terminating the reaction at the calculated stoichiometry. Even better qualities are obtained at 150 °C under pressure.
3. Rapid heating of a suspension of FeOOH with $FeSO_4$ and NaOH also produces high-quality black pigments, forming iron(II) hydroxide as an intermediate.
4. Orange pigments of lepidocrocite structure are precipitated from dilute solutions of iron(II) sulfate by adding alkali hydroxides to a neutral pH value. The mixture is briefly heated, rapidly cooled down, and oxidized with air.
5. Pure red pigments are obtained at a constant pH value using seed crystals of α-Fe_2O_3.
6. Red pigments can also be obtained from aqueous Fe^{II} solutions with an excess of NaOH and oxidation with air.

The *Penniman process* is most commonly used to produce yellow, red, and black iron oxide pigments. Starting materials are ferrous sulfate, sodium hydroxide, and ferrous scrap. Foreign metals must be removed so as not to change the color of the pigments. The process comprises two steps:

1. First, α-FeOOH seed crystals are prepared from iron(II) sulfate and sodium hydroxide under air admission. The raw materials for this step are the only required reagents.
2. The suspension of seed crystals is pumped into a reaction tank where ferrous scrap has been placed. In this step, the seed crystals grow into pigment particles. The required FeOOH form when iron scrap reacts with sulfuric acid to iron(II) sulfate and further with air to oxide hydrate. During the oxidation step, sulfuric acid is released, which can react with other ferrous scraps to form sulfate so that a raw material cycle exists.
   The result of the reaction is a yellow iron oxide pigment.
3. Red iron(III) oxide can be obtained by calcination of yellow iron oxide hydroxide.
4. The red iron(III) oxide treatment with a reducing atmosphere containing hydrogen produces black iron(II,III) oxide.

►Figure 3.17 depicts the steps of the Penniman process.

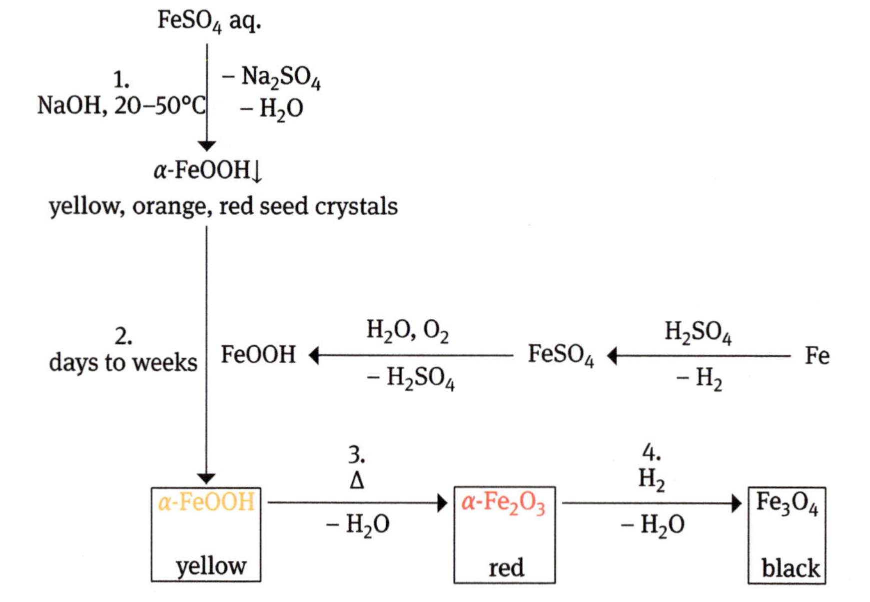

Figure 3.17: Reaction steps of the Penniman process to prepare opaque synthetic iron oxides [16]. The numbers 1–4 refer to the enumeration in the text.

Laux process

The *Laux process* is based on the oxidation of iron by aromatic nitro compounds [16]. This reaction, known for over 150 years, produces iron oxide sludge without any pigment quality. Laux, however, was able to modify it by adding metal chlorides, sulfuric acid, and phosphoric acid to produce high-quality iron oxides that are excellent pigments. Depending on the reaction conditions and additives, the process delivers yellow, red, or black pigments. Calcination of the primary yellow and black pigments produces light red to purple ones, while the mixture of yellow, red, and black pigments is brown. ►Figure 3.18 shows the steps of the Laux process.

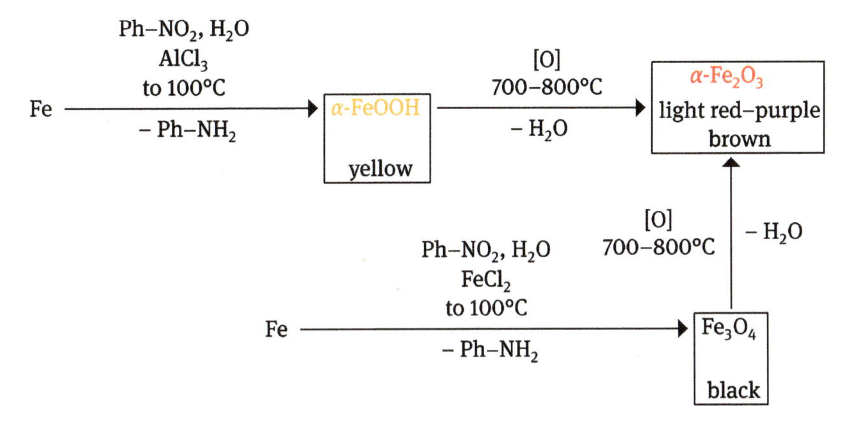

Figure 3.18: Reaction steps of the Laux process to prepare opaque synthetic iron oxides [16].

3.4.2.5 Transparent iron oxides

Transparent iron oxide pigments are always synthetic since precisely defined particle sizes of 1–50 nm must be produced [16]. We already discussed the influence of size at the beginning of the section on iron oxide pigments. There ►Figure 3.11 shows the origin of the upper limit for particle size.

Iron oxides are transparent at the particle size mentioned, show pure yellow and red colors, and have higher coloring power than opaque iron oxides. The manufacturing process influences the size and shape of the primary particles. For example, precipitation reactions produce acicular crystals of FeOOH and Fe_2O_3, whereas decomposition of iron carbonyls produces spherical particles of Fe_2O_3. ►Figure 3.19 illustrates the main preparation pathways:

1. Transparent yellow iron oxide forms by precipitation of very fine Fe^{II} crystallites from dilute solutions of a Fe^{II} salt and air oxidation. The quality depends on concentration, temperature, oxidation time, pH value, and ripening time. Needles 50–150 nm long and 2–5 nm thick are obtained.
2. Transparent red iron oxide is formed from yellow one by calcination. The product is ground until the desired particle size is reached.

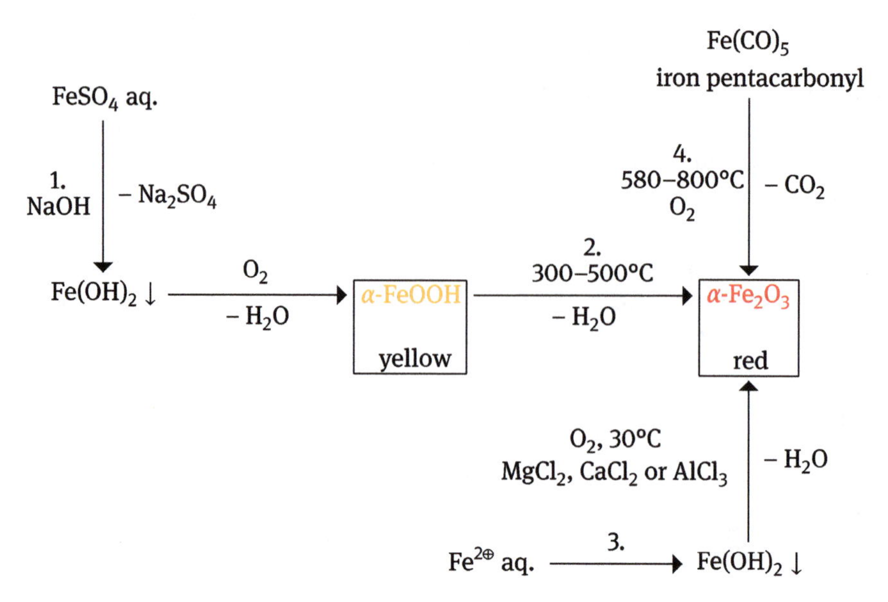

Figure 3.19: Preparation of transparent synthetic iron oxides from aqueous solutions of Fe^{II} salts or by decomposition of iron pentacarbonyl [16]. The numbers 1–4 refer to the enumeration in the text.

3. Transparent red iron oxide can also be produced directly by precipitation of iron(II) hydroxide or iron(II) carbonate from Fe^{II} solutions and oxidation with air, adding chlorides.
4. Very pure semitransparent red iron oxide is obtained by thermal decomposition of iron pentacarbonyl. The color varies from orange to red, and the particles have a size of 10–20 nm.

3.4.3 Complex inorganic color pigments (CICP), mixed metal oxides (MMO)

The discovery that colorless minerals gain color by incorporating metal cations led to the development of artificial colored oxidic minerals. These so-called *complex inorganic color pigments* (CICP) are crystalline, mixed oxides possessing the structures of natural minerals, in which colorless constituents A are replaced by colorizing metals A′ [11, 14, 16], [201, p. 330ff.], [470], [202, keyword "Colorants for Ceramics"], [40, 638, 996]. CICP are solid solutions of metal oxides that are homogeneously distributed in the host lattice, forming new chemical compounds, i. e., they do not represent mere mixtures.

It depends on the metal-oxygen ratio and the radii of the metal cations, which lattice can incorporate which metal. For example, metal cations with a similar radius as Ti^{IV} can often form a rutile structure with a metal-oxygen ratio of 2; in this way, they would also fit into an original rutile lattice. Spinels have a metal-to-oxygen ratio of 1.33, and hematite and corundum have one of 1.5.

The following types are significant in terms of color:
- Spinel and inverse spinel $(A, A')_3O_4$ are based on magnesium aluminum oxides.
- Rutile $(A, A')O_2$ is based on titanium dioxide (rutile).
- Hematite, cassiterite, and corundum $(A, A')_2O_3$ are based on iron, tin, and aluminum oxides.
- To a lesser extent, mixed oxides are based on minerals such as priderite, sphene, and others.

Rutile-based colorants have been known as high-temperature stable ceramic colorants since 1934. Some spinels have enriched our painting palette for a considerably more extended time, such as Thenard's blue (cobalt aluminate, the Impressionists' cobalt blue), cerulean blue (a cobalt stannate of the Realists), and lead-tin yellow (a lead stannate of the Old Masters). However, it was not until the last century that synthetic minerals as color bodies were subjected to a more detailed study, as they offer significant advantages:
- A wide color range includes blue, green, yellow, brown, and black. Only shades of red are rare.
- Due to the choice of metals and the possibility of partially replacing one metal with another, numerous compositions with finely tunable hues are possible.
- High opacity, heat stability up to several hundred degrees Celsius, lightfastness and weather fastness, and high chemical resistance.
- Nontoxicity.

These properties predestine many representatives of CICP for exterior coatings and colorations in the plastics sector.

A disadvantage is that size and shape of mixed oxide particles are decisive for their applicability as pigment and their quality. Reliable color reproduction is particularly problematic since color strongly depends on composition and manufacturing conditions. These also include, e. g., duration, type, and intensity of grinding.

Chromophores for CICP are *d* block metals such as V, Cr, Mn, Fe, Co, Ni, and Cu, which are highlighted in ►Figure 3.20 by a blue background. Their ability for coloration relies on LF transitions in a ligand field created by the lattice, ►Section 2.3. However, since these transitions are often forbidden (►Section 2.3.3), the resulting pigments are relatively weak in color. Improvements in processing have led to enhanced colors. The oxidic solid exhibits semiconductor properties in some cases, and we observe coloration due to the SC mechanism. Examples of this are mixed oxides that do not contain *d* block elements, e. g., lead stannate.

Since the color-bearing metal ions do not necessarily possess the same oxidation state as the original lattice constituents, differences in the electrical charge can emerge. The charge can be balanced, and the crystal structure is stabilized by adding metals shown in ►Figure 3.20 as an equalizing measure. Charge balancing is performed mathematically according to the following scheme: if, e. g., a rutile colorant

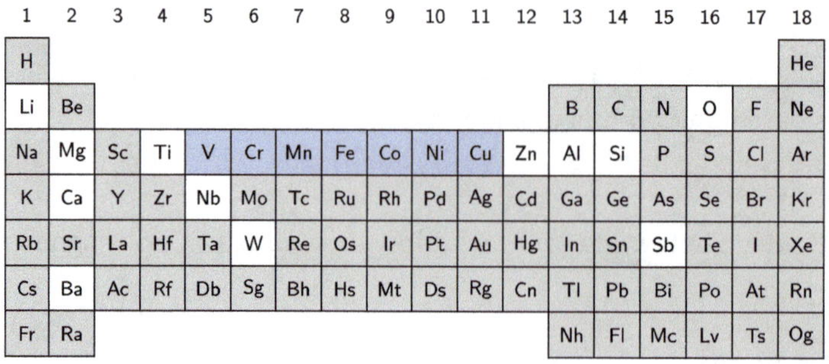

1	2	3	4	5	6	7	8	9	10	11	12	13	14	15	16	17	18
H																	He
Li	Be											B	C	N	O	F	Ne
Na	Mg	Sc	Ti	V	Cr	Mn	Fe	Co	Ni	Cu	Zn	Al	Si	P	S	Cl	Ar
K	Ca	Y	Zr	Nb	Mo	Tc	Ru	Rh	Pd	Ag	Cd	Ga	Ge	As	Se	Br	Kr
Rb	Sr	La	Hf	Ta	W	Re	Os	Ir	Pt	Au	Hg	In	Sn	Sb	Te	I	Xe
Cs	Ba	Ac	Rf	Db	Sg	Bh	Hs	Mt	Ds	Rg	Cn	Tl	Pb	Bi	Po	At	Rn
Fr	Ra											Nh	Fl	Mc	Lv	Ts	Og

Figure 3.20: Localization of elements used for mixed oxides in PTE (group designation according to [194]). A blue background highlights color-active elements, while a white background highlights elements serving for color modification and charge balancing [14].

$Ti^{IV}O_2$ with the divalent metal cation A^{II} should be built, the average oxidation number +IV is obtained by balancing with two pentavalent metal cations B_2^V, since $(2+2\times5) = 4\times3$, i. e., 1 mol A-oxide $A^{II}O$ and 2 mol B-oxide $B_2^VO_5$ are equivalent to 3 mol rutile and form an electrically neutral lattice in these ratios:

$$Ti^{IV}O_2 \longrightarrow A^{II}O \cdot B_2^VO_5 \cdot Ti^{IV}O_2$$

The structure of mixed oxides depends on the size of metal cations and their concentration concerning oxygen. Each type of structure (spinel, rutile, hematite, corundum, and others) possesses a typical oxygen/metal ratio. Therefore, one of these structures can form readily depending on the composition.

3.4.3.1 Manufacturing

A solid-state reaction produces oxide pigments. Starting materials are usually metal oxides or salts. They are mixed in the right proportions and react at high temperatures (several hundred degrees Celsius), as a few examples will show:

$$CoO + Al_2O_3 \longrightarrow CoAl_2O_4$$

$$2\,CoO + TiO_2 \longrightarrow Co_2TiO_4$$

$$0.9\,TiO_2 + 0.025\,Cr_2O_3 + 0.025\,Sb_2O_5 \longrightarrow (Ti_{0,9},Cr_{0,05},Sb_{0,05})O_2$$

Metal salts decompose to oxides during the reaction. Above a specific temperature, metal cations diffuse into the host lattice of the main mineral, corresponding to the given composition. This host lattice partially melts; complicated and constantly changing mineral phases form up to a phase equilibrium, characteristic of composition and production conditions.

The process is complicated in detail and requires strict control in practice since already tiny deviations prevent reproducible results, such as a desired color shade or brilliance.

3.4.3.2 The spinel type

The spinel structure AB_2O_4 is very stable when A is a main group metal and B is a *d* block metal from the titanium to zinc group [224, p. 247], [16, 225]. The metals are usually divalent and trivalent. ►Table 3.4 shows spinels relevant for pigments, an overview of all possible combinations is given in [470].

Table 3.4: The spinel types most important for pigment technology [14, 224, 236, 638].

Type	A	B	Example
Normal (A tetrahedral, B octahedral coordinated)			
$A^{II}B_2^{III}O_4 =$ $(A^{II})_t\ (B_2^{III})_o\ O_4$	Mg, Cr, Mn, Fe, Co, Ni, Cu, Zn, Cd, Sn	Al, Ga, In, V, Ti, Cr, Mn, Fe, Co, Ni	$CoAl_2O_4$ (blue), $CoCr_2O_4$ (blue-green)
Inverse (A octahedral, B tetrahedral and octahedral coordinated)			
$B^{III}[A^{II}B^{III}]O_4 =$ $(B^{III})_t\ ([A^{II}B^{III}])_o\ O_4$	Fe, Co, Ni, Cu, Mg	Fe, In, Cr, Al	$FeFe_2O_4$ (black), $ZnFe_2O_4$ brown
$B^{II}[A^{IV}B^{II}]O_4 =$ $(B^{II})_t\ ([A^{IV}B^{II}])_o\ O_4$	Ti, Sn, V	Mg, Mn, Co, Zn, Ni, Fe	$(Co, Ni, Zn)_2TiO_4$ (blue-green)

A and B can form normal spinels AB_2O_4, in which oxygen surrounds A tetrahedrally and B octahedrally. In the $B[AB]O_4$ arrangement, a so-called inverse spinel, A is coordinated octahedrally, and B is coordinated both tetrahedrally and octahedrally. Oxygen assumes the cubic close-packed lattice (ccp) in all types of spinels.

Spinels form a group that is important for painting if A or B or both are color-imparting or color-modifying metal cations. Depending on whether color-bearing cations replace A or B, octahedral or tetrahedral LF splitting of the metal orbitals occurs. If we consider, e. g., the basic spinel $MgAl_2O_4$, the following possibilities are relevant for pigments, ►Figure 3.21:

- If another metal entirely or partially replaces Mg (A), then the series of *aluminates* with spinel structure and tetrahedral coordination of the color carrier is obtained. If A is cobalt, we gain blue to blue-green *cobalt blue pigments*. They contain cobalt up to 25 % by weight and chromium up to 27 % by weight.
- If a coloring metal such as chromium or iron replaces Al (B) of the basic spinel, the series of *chromites* and *ferrites* with octahedral coordination of the color carrier is obtained. Furthermore, Mg (A) can be replaced by a (same or different) metal, varying the color of chromites and ferrites. For ferrites and iron-chromium spinels

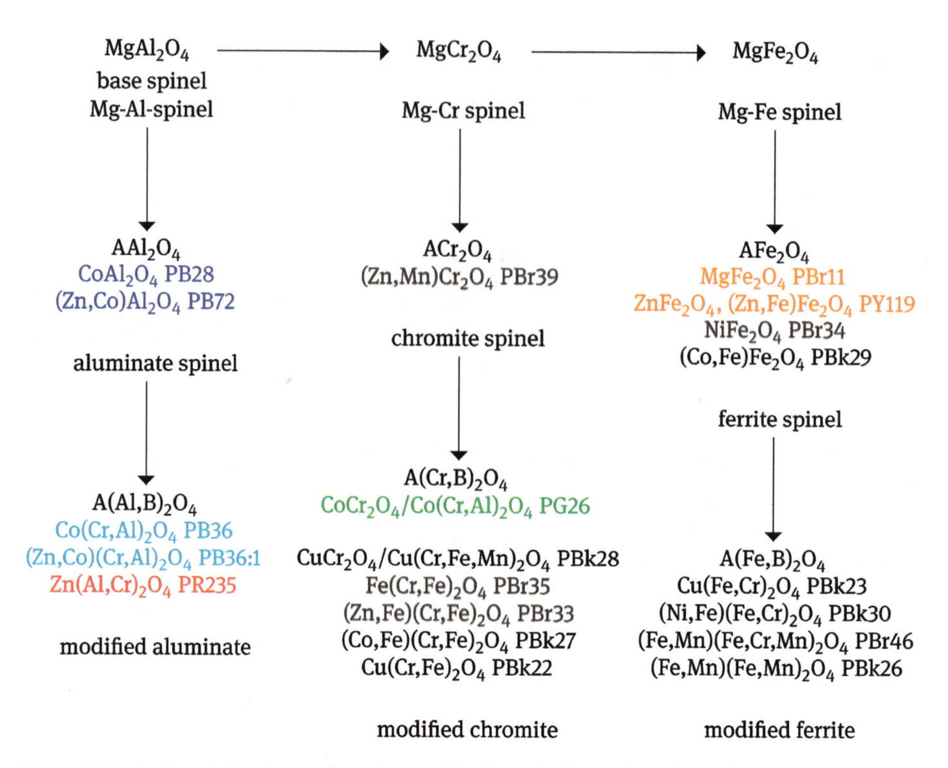

Figure 3.21: Series of aluminate, chromite, and ferrite spinels resulting from the base spinel. Due to fluid transition compositions, often aluminates, chromites, and ferrites cannot be distinguished from each other.

$(A, Fe)(Cr, Fe)_2O_4$, the color ranges mostly from light brown to dark blackish-brown. Here, the structural element $A(B, Fe)_2O_4{=}AO \cdot B_2O_3 \cdot Fe_2O_3$ becomes red in Fe_2O_3 through CT and EPT transitions. On the other hand, structures like $(A, Fe)(Fe, B)_2O_4$ are dark to black if Fe^{II} (A element) and Fe^{III} (B element) are simultaneously present, and IVCT transitions occur between them. An extreme case is black magnetite $FeFe_2O_4$.
The bright blue-green cobalt chromites combine the typical blue of tetrahedrally coordinated cobalt (A) with the green of octahedrally coordinated chromium (B).

Titanates are derived from the inverse spinel Mg_2TiO_4, in which Co, Zn, and Ni are introduced to obtain *cobalt green pigments* $(Co, Ni, Zn)_2TiO_4$ such as PG50. They show a similar color range as chromium oxide green but exhibit brighter and purer colors due to the intense cobalt absorption in the yellow-red spectral range [449]. Iron titanates show yellow-brown to dark red-brown hues and crystallize as inverse spinels.

Any metal replacement can occur entirely or partially, allowing the above-mentioned pure structures to vary in color. ►Table 3.5 lists a small subset of spinels used today, covering mainly the green, blue, brown, and black color range. ►Figure 3.22

Table 3.5: Color of (synthetic) spinel pigments [11, 14], [202, keyword "Colorants for Ceramics"], [40, 638]. Fluid transitions between aluminates, chromites, and ferrites are common, making a simple classification difficult.

Spinel type	Name	Coloring
$A^{II}Al_2^{III}O_4$, aluminate		
$Zn(Al, Cr)_2O_4$	Zinc aluminum pink, aluminum chrome red, chrome aluminum pink, PR235	Pink
$(Co, Zn)Al_2O_4$	Cobalt zinc aluminate blue, PB72	Blue
$CoAl_2O_4$	Thénard's blue, cobalt blue, cobalt aluminum blue, PB28	Reddish-blue
$CoAl_2O_4 \cdot Co_2SnO_4$	Cobalt tin aluminum blue, PB81	Blue
$A^{II}Cr_2^{III}O_4$, chromite		
$(Zn, Co)(Cr, Al)_2O_4$	Zinc chromium cobalt aluminum blue, PB36:1	Greenish-blue
$Co(Cr, Al)_2O_4$	Cobalt chrome blue-green, cobalt blue turquoise, PB36	Greenish-blue
$CoCr_2O_4$, $Co(Cr, Al)_2O_4$	Cobalt chromite green, PG26	Green
$CuCr_2O_4$, $Cu(Cr, Fe, Mn)_2O_4$	Copper chromite black, PBk28	Black
$(Zn, Mn)Cr_2O_4$	Chrome manganese zinc brown, manganese zinc chromite brown, PBr39	Brown
$Fe(Cr, Fe)_2O_4$	Iron chromite brown, PBr35	Brown
$(Zn, Fe)(Cr, Fe)_2O_4$	Zinc iron chromite, zinc iron chromite brown, PBr33	Brown
$(Co, Fe)(Cr, Fe)_2O_4$	Iron cobalt chromite black, PBk27	Black
$Cu(Cr, Fe)_2O_4$	Copper chromite black, PBk22	black
$A^{II}Fe_2^{III}O_4$, ferrite		
$(Co, Fe)Fe_2O_4$	Cobalt ferrite black, iron cobalt black, PBk29	Black
$NiFe_2O_4$	Nickel ferrite brown, PBr34	Brown
$ZnFe_2O_4$, $(Zn, Fe)Fe_2O_4$	Zinc ferrite brown, zinc iron brown, PY119	Deer brown
$(Fe, Mn)(Fe, Mn)_2O_4$	Manganese ferrite black (spinel), PBk26	Black
$(Fe, Mn)(Fe, Cr, Mn)_2O_4$	Chrome iron manganese brown, PBr46	Black
$(Ni, Fe)(Fe, Cr)_2O_4$	Chrome iron nickel black, chrome nickel ferrite black, PBk30	Black
$MgFe_2O_4$	PBr11	Red-brown
$Cu(Fe, Cr)_2O_4$	Spinel black, PBk23	Black
$A_2^{II}Ti^{IV}O_4$, titanate		
Co_2TiO_4, $(Co, Ni, Zn)_2TiO_4$	Cobalt titanium green, cobalt green, cobalt turquoise, PG50	Green
Fe_2TiO_4	Iron titanium brown, PBk12	Yellow-brown

shows typical reflectance spectra for blue to green spinel pigments containing cobalt, chromium and nickel, and ►Figure 3.23 shows those for yellow to brown spinels containing iron.

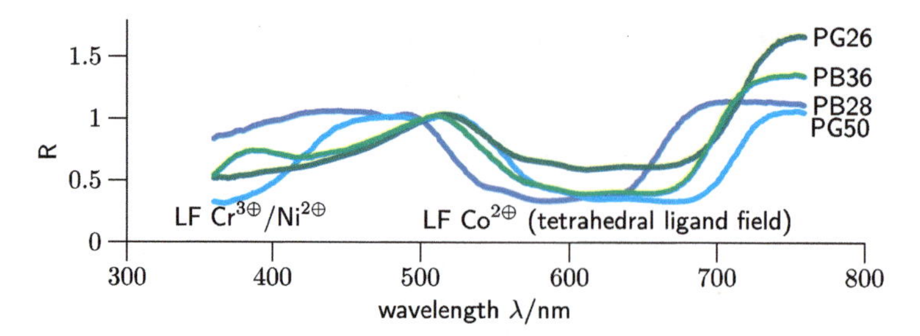

Figure 3.22: Reflectance spectrum of spinel-based watercolors (aluminates, titanates, chromites with cobalt in tetrahedral sites), normalized to an arbitrary unit: PB28 (cobalt blue light, Schmincke Horadam No. 487, $CoAl_2O_4$), PB36 (cobalt green turquoise, Schmincke Horadam No. 510, $Co(Cr, Al)_2O_4$), PG26 (cobalt green dark, Schmincke Horadam No. 533, $Co(Cr, Al)_2O_4$), and PG50 (cobalt turquoise, Schmincke Horadam No. 509, $(Co, Ni, Zn)_2TiO_4$). Cobalt mixed oxide pigments show the intensive LF transition of tetrahedrally coordinated cobalt in the yellow to the red spectral range, yielding blue color. The chromium-containing pigments PG26 and PB36 also show a LF transition in the blue spectral range, modifying the color to green.

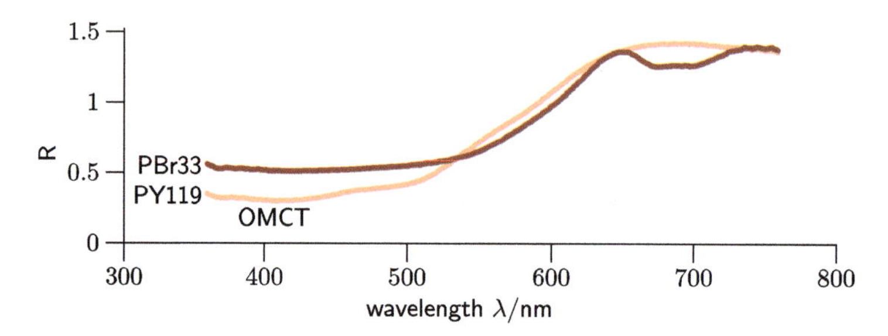

Figure 3.23: Reflectance spectrum of spinel-based watercolors (ferrites), normalized to an arbitrary unit: PBr33 (mahogany brown, Schmincke Horadam No. 672, $(Zn, Fe)(Cr, Fe)_2O_4$) and PY119 (spinel brown, Schmincke Horadam No. 650, $(Zn, Fe)Fe_2O_4$). Iron-containing mixed oxide pigments show the intensive OMCT transition in the blue spectral range, yielding yellow to red colors (iron oxides, ochers).

Cobalt blue, Thenard's blue (PB28, CI 77346)

The deep blue spinel $CoAl_2O_4$ of cobalt oxide and aluminum oxide, cobalt blue, is one of the earliest examples of mixed oxide pigments and an essential one [48, Volume 4], [16, 47, 49, 468]. Around 1800, Thenard developed a process that allowed the pigment to be industrially produced. We find it already in ancient Egyptian glazes, but as a pigment, it seems to have appeared only in use in the nineteenth century. It then revolutionized the possibilities of landscape painting by its unique blue hue, which is particularly suitable for depicting the sky. Because of its high price, only wealthier painters, e. g., of the Barbizon school or the Impressionists [585, 586] initially could afford it. Even today, pure cobalt blue is still an expensive pigment, keeping its unique position due to its high color saturation for which synthetic ultramarine is hardly a match. The coarse-grained pigment is suitable for oil, acrylic, and watercolor painting, but it is not very opaque due to a relatively low refractive index of 1.7. Moreover, due to its oxidic composition, it is very resistant to light and environmental influences.

A transparent grade consists of hexagonal flakes of 20–100 nm in diameter and 5 nm thickness [16]. It is produced by precipitating cobalt hydroxide and aluminum hydroxide from solutions of their salts and annealing them in a solid-state reaction at high temperature:

$$Co^{2\oplus}\text{ aq.} + Al^{3\oplus}\text{ aq.} \longrightarrow Co(OH)_2\downarrow + Al(OH)_3\downarrow \xrightarrow[-\,H_2O]{1000°C} CoAl_2O_4$$

The reaction is followed by grinding to the desired particle size. This pigment also shows good fastnesses concerning chemicals, light, and weather, and thus outperforms organic blue pigments. However, it possesses only a low color intensity.

Thenard's blue shows the typical blue color of tetrahedrally coordinated Co^{II} caused by intense LF transitions. These transitions occur in the yellow-red spectral range corresponding to the small tetrahedral splitting parameter Δ_t, ►Section 2.3.7. Spectroscopic selection rules are ineffective in the tetrahedral field, so the color is intensive, ►Section 2.3.3.

The tetrahedral field is also responsible for the intensive blue color of cobalt aluminate pigments $Co(Al,B)_2O_4$ like PB36. Again, we see a distinct LF absorption in the yellow to the red spectral range in ►Figure 3.22.

Cobalt chromite green (PG26, CI 77344)

In the cobalt blue spinel, we can replace aluminum in octahedral sites with chromium $Co(Al,Cr)_2O_4$ until pure cobalt chromite $CoCr_2O_4$ is reached. The color of cobalt blue spinel changes from blue to turquoise to blue-green due to an additional LF transition of chromium, absorbing the blue spectral range, ►Figure 3.22.

Cobalt green, cobalt titanium green (PG50, CI 77377)

If we introduce cobalt into a titanate instead of an aluminate, we obtain cobalt green $(Co,Ni,Zn)_2TiO_4$. Cobalt is incorporated in this inverse spinel lattice at octahedral and tetrahedral positions. It leads to a blue color via its LF transition in the yellow to the red spectral range. Other LF transitions in the blue spectral range from nickel modify this color to blue-green, ►Figure 3.22. The resulting cobalt green, like cobalt blue and cobalt turquoise pigments, is essential for painting in the blue and green spectral range.

3.4.3.3 The rutile type

Analogous to spinel, rutile also serves as the basis for a metal exchange. Hund [470, 472] shows numerous possibilities for this. Rutile mixed oxides consist of 70–90 % by weight TiO_2 in rutile structure and are therefore similar, i. e., they show semiconductor character and an intense OMCT transition $O^{2\ominus} \rightarrow Ti^{IV}$. Oxide ions octahedrally surround each titanium cation, creating an octahedral ligand field. Consequently, we observe a splitting of the metal's *d* orbitals. While Ti^{IV} is not colored due to its lack of *d* electrons, replacement of Ti with *d* block metals such as Ni^{II}, Cr^{III}, Mn^{II}, and Mn^{III} leads to colored titanates, the so-called DR pigments or "doped rutile" pigments, which are very important for painting. Charge differences to $Ti^{4\oplus}$ have to be counterbalanced by adding colorless metals like Sb^{V}, Nb^{V}, or W^{VI}. Today, many rutile-based colorants exist, covering the yellow-brown range, ►Table 3.6. ►Figure 3.24 shows typical reflectance spectra.

Table 3.6: Color of (synthetic) rutile titanate pigments (DR or doped-rutile pigments) [11, 14], [202, keyword "Colorants for Ceramics"], [40, 638].

Rutile type	Name	Coloring
TiO_2	Titanium white, PW6	White
$(Ti, Ni, Sb)O_2$	Nickel titanium yellow, nickel antimony titanium yellow, PY53	Yellow
$(Ti, Cr, Sb)O_2$	Chrome titanium yellow, chrome antimony titanium brown, PBr24	Orange
$(Ti, Mn, Sb)O_2$	Manganese titanium brown, manganese antimony brown, PY164	Brown
$(Ti, V, Sb)O_2$	Titanium vanadium gray, titanium vanadium antimony gray, PBk24	Gray
$(Ti, Ni, W)O_2$	Nickel tungsten titanium yellow, PY189	Yellow
$(Ti, Sn, Zn)O_2$	Tin-zinc rutile, PY216, PO82	Yellow to orange
$(Ti, Cr, W)O_2$	Chrome tungsten titanium brown, PY163	Brown
$(Ti, Mn, W)O_2$	Manganese tungsten titanium brown, PBr45	Brown
$(Ti, Ni, Nb)O_2$	Nickel rutile yellow, nickel niobtite yellow, PY161	Yellow
$(Ti, Cr, Nb)O_2$	Chrome niob titanium yellow, chrome niob titanium brown, PY162	Brown
$(Ti, Mn, Nb)O_2$	Manganese rutile brown, manganese niob titanium brown, PBr37	Brown
$(Ti, Mn, Cr, Sb)O_2$	Manganese chrome antimony titanium brown, PBr40	Brown

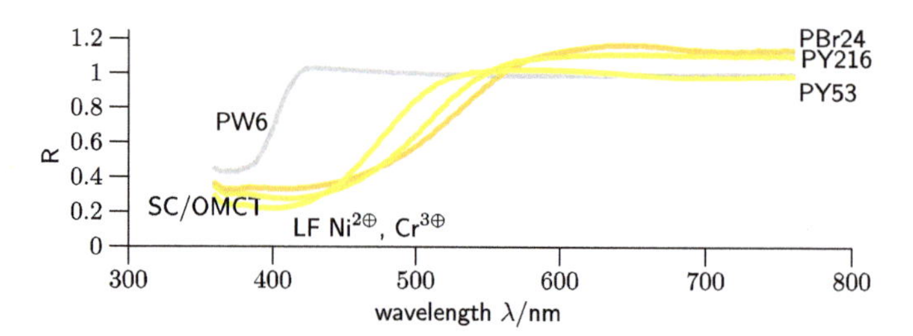

Figure 3.24: Reflectance spectrum of rutile-based watercolors, normalized to an arbitrary unit: PBr24 (titanium gold ochre, Schmincke Horadam No. 659, $(Ti, Cr, Sb)O_2$), PW6 (titanium white, Kremer No. 46200, TiO_2), PY53 (nickel titanium yellow, Kremer No. 432008, $(Ti, Ni, Sb)O_2$), and PY216 (Turner's yellow, Schmincke Horadam No. 219, $(Ti, Sn, Zn)O_2$). The sharp absorption edge conspicuously outlines the pure semiconductor character of TiO_2, which in the titanates appears as an intense OMCT transition in the blue spectral range next to a ligand field transition.

The color impression of rutile pigments is primarily determined by an intense OMCT transition in the near-UV, as electrons can flow from the oxide ligands to the highly charged Ti^{IV} cation [635]. The intense transition explains why DR pigments are white, gray, yellow, or brown. Due to the steep absorption edge, clear, pure colors typical of SC and CT transitions emerge. In the case of pure white TiO_2 (titanium white), this is the SC transition $O(2p) \rightarrow Ti(3d)$, ►Figure 2.9 at p. 90. LF transitions in the added metals determine the exact color of titanates. ►Section 2.2.7 discusses a CT mechanism for the color of the doped rutile pigments. It relies on the admixtured atoms form energy levels in the large band gap of titanium dioxide, reducing it to optical energies.

3.4.3.4 Hematite and corundum type, other types

In addition to spinel and rutile, other minerals are equally suitable as host lattices for color-forming metals. Some examples of such pigments are shown in ►Tables 3.7 and 3.8, spectra of measured pigments in ►Figures 3.25 and 3.26. Hund [470] gives an overview of the formation possibilities of various mixed mineral phases.

Table 3.7: Color of other mixed oxides (I) [11, 14], [202, keyword "Colorants for Ceramics"], [40, 638].

Oxide	Name	Coloration
Corundum Al_2O_3		
$(Al, Mn)_2O_3$	Aluminum manganese pink PR231	Red
$(Al, Cr)_2O_3$	Aluminum chrome pink (synth. ruby), PR230	Red
Cr_2O_3	Chrome green, PG17	Green
Hematite Fe_2O_3		
Fe_2O_3	Hematite PR101, Mars red, iron brown, PR101	Red
$(Fe, Cr)_2O_3$	Chrome iron brown, PBr29	Brown
$(Fe, Mn)_2O_3$	Iron manganese oxide, PBr43	Brown
$MnFe_2O_4 = (Fe, Mn)_2O_3$	Manganese ferrite black (oxide), PBk33	Black
$(Cr, Fe, Al)_2O_3$	Chromium hematite, PG17	Green to green-black
Cassiterite SnO_2		
$(Sn, V)O_2$	Tin vanadium yellow, vanadium yellow, PY158	Yellow
$(Sn, Cr)O_2$	Tin chromium violet, PR236	Violet
$(Sn, Sb)O_2$	Tin antimony gray, PBk23	Gray
$2\,PbO \cdot SnO_2$, $PbO \cdot 2\,SnO_2 \cdot SiO_2$	Lead-tin yellow, CI 77629	Yellow
$n\,CoO \cdot SnO_2$, $n \in [1, 2]$	Cobalt stannate, cerulean blue, PB35	Light blue
Perovskite, see also ►Section 2.2.3 at p. 91		
$PbTiO_3$	Lead titanate, PY47	Yellow
$CaTaO_2N$	Calcium tantalum yellow	Yellow
$Ca_{1-x}La_xTaO_{2-x}N_{1+x}$	Lanthantalum red	Yellow to red

Table 3.8: Color of other mixed oxides (II) [11, 14], [202, keyword "Colorants for Ceramics"], [40, 638].

Oxide	Name	Coloring
Other		
$(Zr, Pr^{IV})SiO_4$	Zircon praseodymium yellow, PY159	Yellow
$ZrSi_{1-x}O_4 : V_x^{4\oplus}$	Zircon vanadium blue, PB71	Blue
$ZrSiO_4 \cdot Fe_2O_3$	Zircon iron pink PR232	Pink
$ZrSiO_4 \cdot Cd(S, Se)$	Zircon cadmium red	Orange to red
$(Co, Zn)_2SiO_4$	Cobalt blue dark, cobalt zinc silicate blue (phenazite), PB74	Blue
Co_2SiO_4	Cobalt silicate blue (olivine), PB73	Blue to purple
Ni_2SiO_4	Nickel silicate green (olivine), PG56	Green
$ZnO \cdot xCoO$	Rinmann's green, cobalt green, zinc green, PG19	Green
$(Zr, (V, In, Y)^{III}, V^V)O_2$	Zircon vanadium yellow (baddeleyt), PY160	Yellow
$3\,CaO \cdot Cr_2O_3 \cdot 3\,SiO_2$	Victoria green, PG51 (garnet)	Green
$CaO \cdot Cr_2O_3 \cdot SiO_2 \cdot SnO$	Chrome tin pink, PR233 (sphen)	Pink
$Sn_2Nb_2O_7$	Tin niobium yellow, PY227 (pyrochlor)	Yellow
$2\,NiO \cdot 3\,BaO \cdot 17TiO_2$	Nickel barium titanium yellow, PY157 (priderite)	Yellow

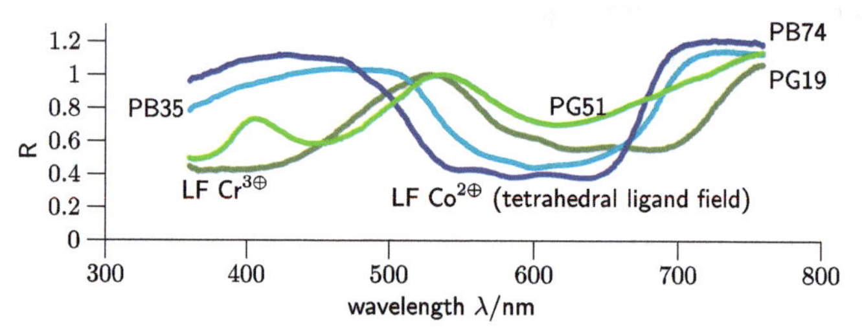

Figure 3.25: Reflectance spectrum of blue and green watercolors based on various mineral lattices, normalized to an arbitrary unit: PB35 (cobalt azure, Schmincke Horadam No. 483, $n\,CoO \cdot SnO_2$, $n \in [1, 2]$), PB74 (cobalt blue deep, Schmincke Horadam No. 488, $(Co, Zn)_2SiO_4$), PG19 (cobalt green pure, Schmincke Horadam No. 535, $ZnO \cdot xCoO$, ►p. 266), and PG51 (lime green, Victoria green, Kremer No. 441908, $3\,CaO \cdot Cr_2O_3 \cdot 3\,SiO_2$). Blue and green cobalt pigments show an intense chromogenic LF transition of the cobalt ion in the yellow to the red spectral range, yielding blue and blueish colors. The color-inducing LF transitions are seen in the red and blue spectral ranges in the green chromium pigment.

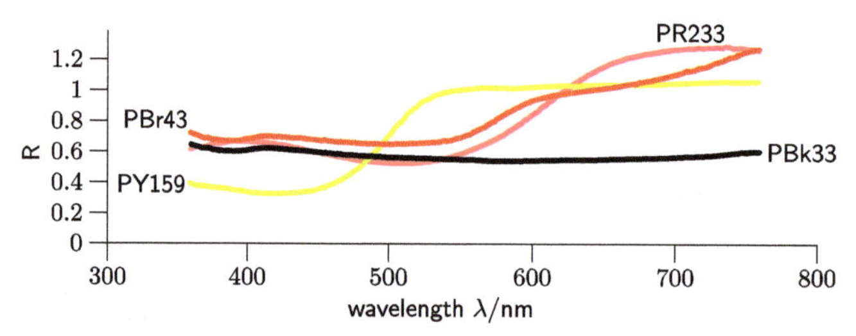

Figure 3.26: Reflectance spectrum of yellow, red, and brown watercolors based on various mineral lattices, normalized to an arbitrary unit: PBk33 (iron manganese mixed oxide 303T, Kremer No. 48445, $MnFe_2O_4 = (Fe, Mn)_2O_3$), PBr43 (iron manganese brown 645T, Kremer No. 48330, $(Fe, Mn)_2O_3$), PR233 (potters pink, Schmincke Horadam No. 370, $CaO \cdot Cr_2O_3 \cdot SiO_2 \cdot SnO$), and PY159 (intense yellow, Kremer No. 438808, $(Zr, Pr^{IV})SiO_4$).

The host lattices of corundum Al_2O_3 and hematite Fe_2O_3 demonstrate the effects of the diverse mechanisms of color formation. Both lattices differ only slightly; oxide ions octahedrally surround the metal cations. The minerals form seamless mixing series with each other, and each of its members displays a typical color:

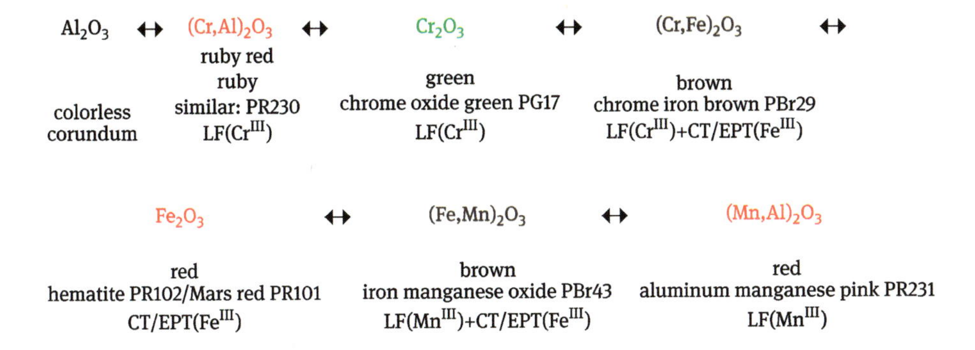

When we add chromium to colorless corundum, it exhibits the typical ruby red color for octahedrally coordinated Cr^{III} in a strong ligand field such as of corundum. On the painter's palette, ruby could be represented by aluminum chrome pink PR230, a Cr^{III} pigment having the sphene host lattice. As we increase the amount of chromium, the field strength decreases; therefore, in chromium oxide green PG17, the typical green color for octahedrally coordinated Cr^{III} becomes visible. Thus, the color change PR230–PG17 is due to the ligand field strength, ►Section 2.3.5. In chrome iron brown PBr29, the green color of chromium oxide overlaps with the red of iron(III) oxide, i. e., green light is absorbed by iron oxide, resulting in a dark brown color impression. This brown resembles the brown shades that the painter can mix by himself from green and red. With increasing iron oxide content, the color shifts to red until pure iron oxide PR101 shows a distinct red color due to CT and EPT transitions, ►Section 2.3.4 at p. 118.

The cassiterite host lattice is vital in artists' paints. Important classical pigments are derived from it: cerulean blue as a mixed oxide with cobalt oxide, lead-tin yellow as a mixed oxide with lead oxide.

For paints applied in ceramics, colored zirconium silicates have become most important, as they show high thermal stability up to approx. 1350 °C and can thus withstand the temperatures of ceramic firing, ►Section 7.1.3. They are based on zirconium silicate $ZrSiO_4$ and are colored by metal cations. After the discovery of zircon vanadium blue with the color center $V^{4\oplus}$, an intensive yellow (►Figure 3.26) soon was put into the market and, after long research, also a red, so that today all intermediate shades are available by mixing these primary colors.

Zirconium iron pink leads to the *inclusion pigments*, in which the color body is protected from external influences by resistant oxides like ZrO_2, SnO_2, and SiO_2. In the case of zirconium iron pink, the protective oxide is $ZrSiO_4$. After the great success of zirconium iron pink, other inclusion pigments followed, comprising cadmium yellow and cadmium red as color bodies to enrich the zircon color palette with brilliant hues of yellow, orange, and red, ►Section 7.1.3.

Lead-tin yellow

Lead-tin yellow (►Section 3.4.1 on ►p. 235) owes its intensive color not to the LF splitting of a coloring metal cation because neither tin nor lead dispose of the necessary *d* electrons. Instead, SC transitions induce the color since the mixture of yellow lead oxide (massicot) with white tin dioxide has semiconductor properties, ►Section 2.2.4.3.

Cerulean blue (PB35, CI 77368)

Another example of a mixed oxide known since about 1860 is cerulean blue, a cobalt stannate of the composition $n\,CoO \cdot SnO_2$ [47, 49]. With typical values of n between 1 and 2, the endpoints are $CoSnO_3$ and Co_2SnO_4. Like Thenard's blue, it owes its color to the LF transitions in the Co^{II} ion, ►Figure 3.25.

Rinmann's green, cobalt green, zinc green (PG19, CI 77335)

The solid-state reaction between a cobalt compound and a zinc compound, each converting to an oxide in the annealing, found its way into the early chemical literature around 1780 as *Rinmann's green* [47, 49]. The bright green powder, whose hue changes to dark green with increasing cobalt content, is hardly used by painters because it is weak in color. In [444], however, considerations are given to its use as a ceramic colorant. It also becomes clear that Rinmann's green is not a uniform spinel-like compound with a fixed composition $ZnCo_2O_4$ but rather a solid solution of colored cobalt(II) oxide in zinc oxide. In the presence of air, traceable amounts of cobalt(III) oxide develop that colorize the product dark green, ►Figure 3.25. Interestingly, solid solutions of MnO and FeO could also be obtained with ZnO, which are yellow-green-olive green and bright yellow-red [471].

3.4.4 Cerium sulfide pigments

Recent developments produced pigments based on cerium sulfide doped with other metals and oxygen [11, 14, 16, 49]. They exhibit very pure red and burgundy hues and high stability and opacity; so far, the color range could be extended to orange. They were developed as nontoxic substitutes for the brilliant but toxic orange and red pigments based on lead, chromate, and cadmium, such as chrome orange, chrome red, cadmium orange, and cadmium red. Artists are not acquainted with them since they are intended primarily for coloring plastics, but other applications are conceivable, too.

Advantageous is their nontoxicity and high opacity (refractive index 2.7), which corresponds to the lead and cadmium-based pigments' opacity and their high light, weather, and temperature resistance. However, the color purity is slightly lower than that of lead and cadmium pigments but considerably higher than the iron oxides.

The high price of the raw materials is problematic, so cerium sulfide pigments currently hold places at the price scale's upper end. On top of that, cerium sulfides have three modifications, of which the desired γ modification forms only above 1100 °C, which makes production difficult. Finally, a drawback is a weak tendency of cerium sulfide to hydrolyze in water and humid atmosphere, producing H_2S. Up to now, treatment of the pigment surface has not solved the problem completely.

Cerium sulfide orange light (PO78, CI 772850); cerium sulfide orange (PO75, CI 77283-1); cerium sulfide red (PR265, CI 77283-2); cerium sulfide burgundy (PR275, CI 772830)

The pigments consist of cerium(III) sulfide Ce_2S_3 (PO78: $Ce_2S_3 \cdot La_2S_3$), a small amount of sulfur is replaced by oxygen. As with virtually all sulfides, the intensive color results from an SC transition, in this case, from Ce($4f$) band into the Ce($5d$) conduction band, ►Figure 3.27. The energy of the $4f$ level depends on the ionicity of the cerium-sulfur bond. We can use the influence of alkali ions and the sulfur-oxygen ratio on the ionicity to change the band gap's size and vary the color between burgundy and light orange. We achieve even brighter shades of orange by partially replacing cerium with lanthanum: $(Ce, La)_2S_3$, PO78. (Patent specifications such as [656] or [657] indicate that replacing Ce and La with rare earth metals RE could create a range of potential rare-earth sulfide and rare-earth oxysulfide pigments of the type $(Ce, La, RE)_2S_{3-x}O_x$).

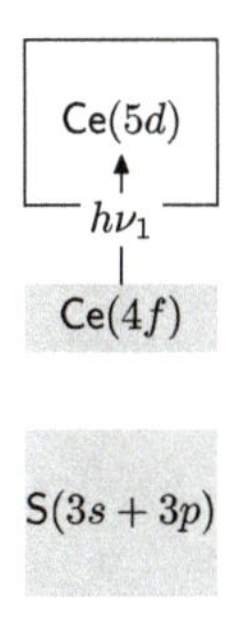

Figure 3.27: Schematic band structure of orange-red cerium sulfide Ce_2S_3 with the Ce($4f$) → Ce($5d$) transition, inducing color [14, 16].

The preparation of cerium sulfide cannot be done in an aqueous solution because the sulfide is sensitive to hydrolysis. Therefore, a solid-state gas reaction is used. In the first step, colorless cerium hydroxide is precipitated from a dissolved cerium salt with hydroxides as cerium-containing precursor for the gas-phase reaction:

$$Ce^{3\oplus} aq. + 3\,OH^{\ominus} \longrightarrow Ce(OH)_3 \downarrow$$

The addition of alkali (earth) salts at this stage affects the color of the finished pigment. The dried cerium hydroxide now reacts between 700 and 1000 °C in a sulfurous atmosphere to the colored sulfide:

$$2\,Ce(OH)_3 + 3\,H_2S \longrightarrow Ce_2S_3 + 6\,H_2O$$

Agglomerates formed in this process must be destroyed to achieve high pigment quality. A surface treatment follows to increase the stability of the pigment.

3.4.5 Chromium oxide pigments

Chromium oxide pigments [11, 16, 49] are structurally related to corundum, in which oxygen anions octahedrally surround aluminum cations. The complete replacement of aluminum by Cr^{III} leads to chromium oxide green, now Cr being octahedrally coordinated by oxygen; it owes its color to *d* orbital splitting in the oxygen's field, i. e., a LF transition, ►Section 2.3.5 at p. 121.

Chromium oxide pigments must not be confused with chromate pigments containing the $Cr^{VI}O_4^{2-}$ anion and hexavalent Cr^{VI} as the color-bearing element. In chromates, the LMCT mechanism induces an intensive color, ►Section 3.6.

In both cases, manufacturers must take care of the toxicity of hexavalent Cr^{VI}, either as final pigment or during production as starting material. At least chromium oxide pigments are nontoxic and approved for toys and contact with food.

Chromium oxide green (PG17, CI 77288)

This pigment consists of chromium(III) oxide Cr_2O_3 and was discovered by Vauquelin in 1809 [48, Volume 3], [11, 16, 47, 49, 201]. The dull olive-green tone is somewhat dependent on the particle size (typically 0.3–0.6 µm): smaller particles are lighter and yellowish; larger ones are darker and bluish but less colorful. Due to the high refractive index of 2.5, its opacity is high.

Chromium(III) oxide is nontoxic and practically inert to light and most chemical influences such as SO_2, acids, and alkali. It is insoluble in water and color stable up to 1000 °C. However, due to the dull color, another chromium oxide, viridian, is usually preferred, ►Figure 3.28; moreover, chrome colors are more expensive than green mixtures or modern organic green pigments. A problem with machine processing of chromium oxide is its very high hardness (9 on the Mohs hardness scale), which leads to high abrasion in printing rollers or spray nozzles.

Since no natural chromium(III) oxide occurrences are known, this pigment is always produced synthetically. Starting materials are alkali metal dichromates, which are reduced as a solid with S or C or thermally decomposed with ammonium sulfate:

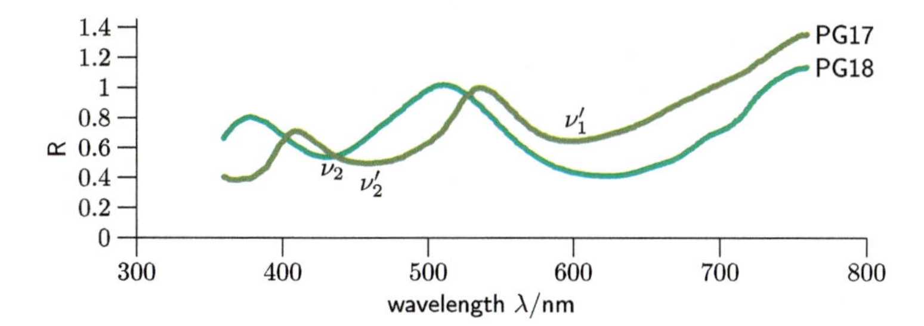

Figure 3.28: Reflectance spectrum of watercolor paints based on chromium oxide, normalized to an arbitrary unit: PG17 (chromium oxide green, Schmincke Horadam No. 512, Cr_2O_3) and PG18 (viridian, Schmincke Horadam No. 513, $Cr_2O_3 \cdot 2\,H_2O$). The graphs illustrate that chromium oxide hydrate has sharper and more intense reflectance peaks with higher blue and green content. Absorbance peaks (approximate): chrome oxide 460 nm wide, 600 nm wide, chromium oxide hydrate 430 nm narrower, $\approx$ 620 nm very broad.

$$Na_2Cr_2O_7 + S \longrightarrow Cr_2O_3 + Na_2SO_4$$

$$Na_2Cr_2O_7 + (NH_4)_2SO_4 \longrightarrow Cr_2O_3 + Na_2SO_4 + 4\,H_2O + N_2\uparrow$$

Ammonium dichromate forms in situ from ammonium sulfate and is then thermally decomposed to nitrogen and chromium oxide. Chromium oxide can also be obtained by wet-chemical reaction; in the first step, chromium hydroxide forms from a solution of sodium dichromate and sodium hydroxide, sulfur being the reducing agent. In a second step, the hydroxide is annealed, yielding chromium oxide:

$$8\,Cr_2O_7^{2\ominus}\text{ aq.} + 12\,S + 8\,OH^{\ominus} + 20\,H_2O \longrightarrow 16\,Cr(OH)_3 + 12\,SO_3^{2\ominus}$$

$$16\,Cr(OH)_3 \xrightarrow{900\text{–}1100\,°C} 8\,Cr_2O_3 + 24\,H_2O$$

Its properties can be changed by precipitating titanium or aluminum hydroxides on the chrome oxide and then calcining it. During the process, titanium or aluminum oxide coatings are formed on the chromium oxide, and the color shifts to a yellowish-green.

Viridian, Pannetier's green, Guignet's green, emeraude green, chromium oxide hydrate green (PG18, CI 77289)

Viridian consists of the hydrate of chromium(III) oxide $Cr_2O_3 \cdot 2H_2O$ [48, Bd. 3], [11, 16, 47, 49, 201]. The altered chemical environment compared to the pure chromium oxide results in a fiery, cool blue-green hue, as shown in the reflectance spectrum, ►Figure 3.28. Chromium oxide hydrate has higher overall reflectivity and is more luminous. Due to the shift of absorption ν_2 by about 30 nm to 430 nm in the blue-violet spectral range and a sharper absorption peak, it has a purer and more blue-green hue.

In addition, the band in the red spectral range is considerably broader and more intense than that of chromium oxide, absorbing more red light so that the color purity in the blue-green range further increases.

The pigment was developed in 1838 by Pannetier and has been produced industrially since 1860. Due to its luminosity and color, it is often used today as artists' paint but was very expensive initially. However, the wealthier painters of Barbizon used it to create a new form of landscape painting with shades of green that were previously unknown [585, 586], and it soon became a popular substitute for the toxic emerald green and an alternative to mixed greens. Due to its low refractive index of 1.6–2, opacity is not high, and it is a good glazing pigment. Chromium oxide hydrate has chemical resistance similar to chromium oxide, but it is sensitive to strong acids. At higher temperatures, as they occur during plastics processing, it loses water of crystallization and is then not color-stable. The pigment is therefore no longer in use for this application.

Chromium oxide hydrate can be produced purely synthetically from alkali metal chromate and boric acid via a borate, hydrolyzing in water to the oxide hydrate:

$$2\,Na_2CrO_4 + 12\,H_3BO_3 \xrightarrow{\text{(at 500 °C)}} Cr_2(B_4O_7)_3 + 4\,NaOH + 16\,H_2O + 3/2\,O_2$$

$$Cr_2(B_4O_7)_3 + 20\,H_2O \longrightarrow Cr_2O(OH)_4 + 12\,H_3BO_3$$

Chrome green

This misleading name conceals green mixtures of yellow and blue pigments, ►Section 3.6.2.

Modern chromium pigments

New developments are chromium-containing mixed oxides, used for harmless, pure, and bright blue-green and turquoise pigments, ►Section 3.4.3.

3.4.6 Titanium oxides and zinc oxides

Titanium dioxide white (PW6)

Today's most important white pigment is titanium white, consisting of titanium dioxide TiO_2 [48, Volume 3], [16, 47, 49, 201]. It was produced for the first time in 1821, but industrially only since 1918 because the natural minerals, rutile, and anatase were not available in sufficient quantity and quality. Since then, the industry has employed this pigment in paints, lacquers, coatings, ceramics, glazes, and colored plastics. In the paper industry, titanium white is favored for producing bright white paper or paper that needs to be very thin yet opaque at the same time. It can be added directly to the pulp or coated superficially. However, one prefers fillers such as kaolin, lime, or talcum in terms of price. Titanium white is the only brightener in printing inks because it has sufficient color strength even for small pigment particles, unlike other white pigments such as lithopone. The requirement for small particle size results from restrictions from the printing industry. There, layer thicknesses below 10 μm for printing layers are commonly used today. Regarding artists' paints, only small quantities of titanium white are sufficient to satisfy the needs of manufacturers.

The reason for the popularity of this semiconducting material is its large band gap and its high refractive index of 2.7 for rutile and 2.55 for anatase. The high refractive index implies an optimal particle size of 0.19 μm for rutile and 0.24 μm for anatase pigments to maximize scattering, ►Section 1.6.6. This very high scattering causes a pure white color, preserved in any usual artists' binder, even in oil. (Other white substances such as limestone, gypsum, or marble powder become transparent in oil because their low refractive index hardly differs from the binders.)

The high uniform reflectance over the whole visual spectral range is distinctly visible in ►Figure 3.29. Following the theory, small particles cause higher scattering of shorter wavelengths, i. e., they show a blue tint while larger particles have a yellowish tint, ►Section 1.6.6.

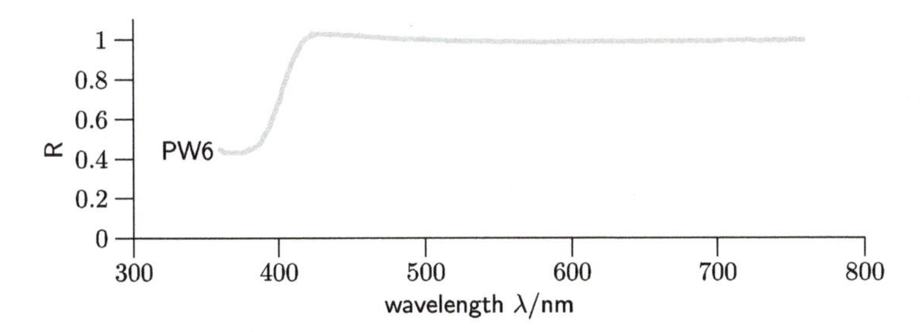

Figure 3.29: Reflectance spectrum of white watercolors normalized to an arbitrary unit: PW6 (titanium white, Kremer no. 46200, TiO_2). The graph depicts the steep absorption edge typical for semiconductors.

Regarding toxicity, titanium white is considered harmless to health.

For the industrial production of titanium(IV) oxide, the *sulfate process* was developed in 1916. Starting materials are ilmenite or titanium slag from the Sorel process (titanium-containing residual slag from ilmenite after separation of iron). The starting materials are converted to oxysulfates by digestion with sulfuric acid. The suspension is again digested with water or diluted sulfuric acid, dissolving titanium into the solution. Interfering metal cations from impurities such as $Fe^{3\oplus}$ must be removed or converted to a more soluble form by reduction. Cooling the solution leads to the crystallization of byproducts such as iron(II) sulfate, which used to be troublesome waste materials but are now, e. g., processed into iron oxide pigments. Hydrolysis of the titanium-rich solution and addition of crystallization nuclei yield titanium oxide hydrate, not yet possessing any pigment qualities. The hydrate still contains metallic impurities, which can be reduced by reaction with zinc or aluminum at 50–90 °C ("bleaching") and removed by washing. In addition, calcination removes residual sulfuric acid trapped in the hydrate:

$$\underset{\text{ilmenite}}{FeTiO_3} \xrightarrow[\substack{-H_2O \\ -FeSO_4}]{\substack{H_2SO_4,\ 80\text{–}98\,\% \\ 170\text{–}220^\circ C,\ 12\,h}} TiOSO_4 \xrightarrow{\substack{H_2SO_4 \text{ dilute} \\ <85^\circ C}} \underset{\text{black liquor}}{Ti^{4\oplus}\ aq.} \xrightarrow{\substack{H_2O \\ 94\text{–}110^\circ C}}$$

$$TiO_2 \cdot nH_2O\downarrow \xrightarrow[-H_2O]{\substack{800\text{–}1100^\circ C,\ 7\text{–}20\,h \\ \text{mineralization aid}}} \underset{\substack{\text{titanium dioxide} \\ \text{rutile or anatase}}}{TiO_2}$$

The addition of mineralization aids (alkalis, phosphoric acid) and ZnO, Al_2O_3, or Sb_2O_3 promotes the formation of the intended crystal structure (rutile or anatase) and stabilizes it. In this process, Zn, Al, or Sb cations are incorporated into the lattice at $Ti^{4\oplus}$ positions, stabilizing it against weather and UV light.

The *chloride process* was developed around 1950. It provides very pure products due to the efficient distillation step. The first step is a carbochlorination of natural or synthetic rutile with chlorine and coke as reducing agents, forming gaseous titanium(IV) chloride and other volatile chlorides of impurity metals. Controlled cooling (freezing out) and sublimation processes remove undesired chlorides. After purification of the remaining $TiCl_4$ by distillation, a reaction with oxygen forms pigment-grade titanium(IV) oxide:

$$\underset{\substack{\text{rutile natural} \\ \text{or synthetic}}}{TiO_2} \xrightarrow[-CO_2]{\substack{Cl_2,\ C \\ 800\text{–}1200^\circ C}} TiCl_4 \xrightarrow[-Cl_2]{\substack{O_2 \\ 900\text{–}1400^\circ C}} \underset{\substack{\text{titanium dioxide} \\ \text{rutile or anatase}}}{TiO_2}$$

The process parameters largely determine the product properties. The addition of water or alkali compounds controls particle growth by forming crystallization nuclei. Similarly, adding $AlCl_3$ promotes the development of the rutile structure, adding PCl_3 and $SiCl_4$ that of the anatase structure.

Most titanium pigments are stabilized after manufacturing by metal oxides or organic substances coating. The formation of the coating takes place by:

– Precipitation of the coating from aqueous solutions containing oxides, hydrated oxides, silicates, or phosphates of Ti, Zr, or Al such as Na_2SiO_3, $NaAlO_2$, $Al_2(SO_4)_3$,

$ZrOSO_4$, or $TiOSO_4$. Precipitation occurs by controlling the pH value, and the precipitate forms a dense coating on the titanium(IV) oxide particles.
- Adsorption of metal oxides or hydroxides onto the particle surface during the milling process.
- Separation of the coating from the gas phase consisting of volatile chlorides or metal organyls.

A surface tailored to the specific application helps to optimally disperse the pigment in the medium, obtaining high-quality layers of color.

According to current findings, titanium white is not acutely or chronically toxic, provided it does not contain heavy metals. Furthermore, it does not irritate the skin or mucous membranes and is approved for food contact and cosmetic applications. So, titanium white is an essential ingredient in makeup and foundations, combined with iron oxides. In sun creams, it provides light.

Zinc white, snow white, Chinese white, permanent white (PW4, CI 77947)

Zinc oxide ZnO, known in medicine as lana philosophorum since the Middle Ages, is generally believed to have been recognized as a pigment only around 1780 and was produced by roasting zinc ores or burning zinc in the air. It is detectable in paintings from 1800 onward and later belongs to the three crucial white pigments (lead white, zinc white, titanium white) [48, Volume 1], [16, 47, 49]. In [400], the assumption is made that zinc oxide was a byproduct of brass production since antiquity so that it could have been used in painting.

Like other white pigments, it owes its cool-white hue to its semiconductor properties (SC mechanism). One advantage is that zinc sulfide is also white, so zinc white does not show any discernible browning or blackening by sulfide formation even in close contact with sulfur-containing pigments. It reversibly turns yellow when heated, which is of no significance for painting.

Since zinc white has a refractive index of 2.0, it has a lower opacity than lead white, so it was traditionally used in watercolors. In oil, it could not entirely replace lead white despite reduced toxicity. Nevertheless, precisely because of its semiopaque properties, it is very suitable for Old Master glazing, as it offers a somewhat finely obscuring white physical quality.

While zinc, from natural sources (ores), and zinc metal were used for production in the past, zinc comes today from industrial residues (galvanizing plants, coating shops, and others). The *direct* or *American process* starts with a high-temperature reduction of zinc-rich raw material to metallic zinc vapor, then oxidized with oxygen to zinc oxide. The raw material typically consists of up to 75 % ZnO and is reduced with coal:

$$ZnO \xrightarrow[1000\text{–}1200\,°C]{C} Zn_{steam} + CO \xrightarrow{O_2} ZnO + CO_2$$

Precise control of process conditions is necessary to obtain defined pigment particles (size, shape). Depending on the intended properties, a surface treatment may be required; thermal treatment at 1000 °C also improves pigment properties.

In the *indirect* or *French process*, zinc is evaporated in muffle furnaces or retorts made of graphite or silicon carbide. After purification of the zinc vapor, oxidation with atmospheric oxygen takes place:

$$\mathrm{Zn} \xrightarrow{\Delta} \mathrm{Zn_{steam}} \xrightarrow{\mathrm{O_2}} \mathrm{ZnO}$$

The resulting zinc oxide can be used directly, but surface treatment improves its properties.

The *precipitation process* starts with aqueous solutions of zinc sulfate or zinc chloride and yields fine particles with high surface area:

$$\mathrm{Zn^{2\oplus}\ aq.} \xrightarrow{\mathrm{Na_2CO_3\ or\ KHCO_3}} \mathrm{ZnCO_3}\downarrow \xrightarrow[-\mathrm{CO_2}]{\Delta} \mathrm{ZnO}$$

A subsequent surface treatment increases weather-fastness and lightfastness, and dispersibility in all cases. The treatment usually involves applying an organic top-coat produced from oil and propionic acid.

3.4.7 Cadmium sulfide pigments

Cadmium sulfide pigments [48, vol. 1], [11, 16, 49] include cadmium sulfide CdS, cadmium sulfoselenides Cd(S, Se), and zinc-cadmium sulfides (Zn, Cd)S. Since the fully synthetic compounds are semiconductors, their color is caused by the SC mechanism, resulting in visual color brilliance. As a result, cadmium sulfide pigments offer the broadest range of brilliant hues of all inorganic pigments, from pale primrose yellow to golden yellow and from light orange to crimson and maroon.

All cadmium sulfide pigments are based on CdS, crystallizing in the wurtzite lattice. Chemically similar elements can replace $Cd^{2\oplus}$ and $S^{2\ominus}$, but only the replacements Cd → Zn, Hg and S → Se have reached practical significance. Zinc-doped cadmium sulfides exhibit greenish-yellow hues, while the selenium-doped cadmium sulfoselenides and the cadmium-mercury sulfides have orange-yellow and red hues. It is possible to replace the elements in almost any quantity to achieve continuous color changes.

Besides the pure compounds, the industry also produces cheaper variants similar to lithopone. These are precipitations of the cadmium compounds on barium sulfate, resulting not in solid solutions but physical mixtures, having similar colorizing properties at a lower price.

Cadmium sulfide pigments have been employed industrially since around 1900, but the raw material and manufacturing prices make them expensive. They are extremely interesting for coloring plastics, glass, and ceramic glazes because of their color brilliance and high thermal stability (up to ≈ 600 °C). The plastics industry consumes about 90 % of the production, and the glass industry about 5 %. Especially the red cadmium sulfoselenides are largely unrivaled in these applications, whereas in the yellow range, inorganic alternatives are available. Cadmium sulfide pigments are almost insoluble in water, bases, and organic solvents but react with acids forming H_2S. They have satisfying migration resistance but not a high weather-fastness, as they decompose under air and light to $CdSO_4$.

Although cadmium sulfide pigments are not acutely toxic or irritating to the mucous membranes, minute amounts may dissolve in acids such as stomach acid and accumulate in internal organs. Processing must observe industrial hygiene regulations to avoid chronic effects detrimental to health. Furthermore, since the disposal of cadmium-containing paints by burning generates cadmium-containing fly ash in outdated plants, the discussion about heavy metals led to the development of alternative, less harmful pigments. As a substitute for the yellow pigments, the semiconductor-based oxynitrides containing calcium and tantalum ($CaTaO_2N$) appear promising. For the red hues, the same applies to oxynitrides with lanthanum and tantalum ($LaTaO_2N$), ►Section 2.2.3 at p. 91.

Cadmium yellow (PY37, CI 77199); cadmium yellow light (PY35, CI 77117)

Pure cadmium sulfides include [48, Volume 1], [11, 16, 47, 49]:

- PY35, zinc-cadmium sulfide (Zn, Cd)S greenish-yellow, with 59–77 % by weight Cd and 13–0,2 % by weight Zn as well as 1–2 % by weight Al_2O_3 for lattice stabilization
- PY37, pure CdS reddish-yellow

The reflectance spectrum of cadmium yellow (►Figure 3.30) shows a typical semiconductor spectrum with a steep absorption edge, explaining the high color purity, ►Section 1.5.3. The color of the pigment depends on its composition and particle size via the size-dependent band gap, ►Section 2.2.4. It can extend into orange and red; ►cadmium orange and cadmium red. Since today's pigment particles have sizes below about 0.5 µm, the band gap also depends on particle size. Small particles have a larger band gap and are brighter and whiter. This size dependence explains why cadmium yellow paints often contain the same pigment PY35 but show a cool, light yellow in one case and a medium yellow in another. The particle shape also plays a role and, thus, the exact manufacturing conditions.

Due to its brilliant yellow color, cadmium yellow has become one of the essential yellow pigments for painters since its discovery around 1820. When mixed with white, yellow ocher, and red ocher, it could also replace the true ►Naples yellow. The high

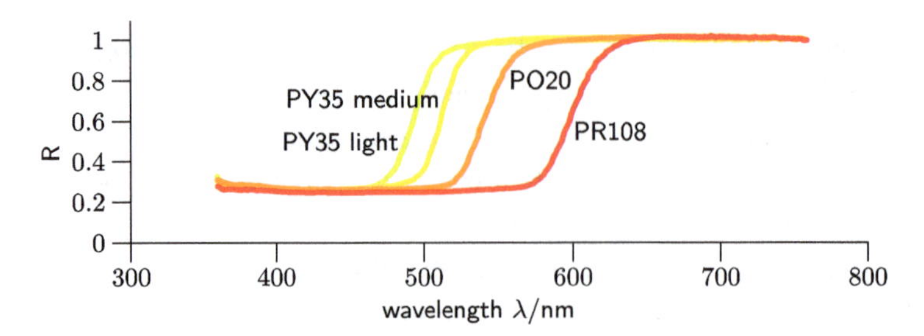

Figure 3.30: Reflectance spectrum of watercolor paints based on cadmium sulfide (semiconductor with the typical, color-defining semiconductor edge), normalized to an arbitrary unit: PO20 (cadmium orange light, Schmincke Horadam No. 227, Cd(S, Se)), PR108 (cadmium red medium, Schmincke Horadam No. 347, Cd(S, Se)), PY35 (cadmium yellow light, Schmincke Horadam No. 224, CdS), and PY35 (cadmium yellow medium, Schmincke Horadam No. 225, CdS).

refractive index and today's small grain sizes result in high opacity. It is sparingly soluble in dilute mineral acids, but it decomposes to form H_2S in strong mineral acids. The early pigments of the nineteenth century were considered unreliable in color hue, which was caused by impurities due to the manufacturing process: They frequently contained sulfur, which through the formation of sulfuric acid, converted into the colorless cadmium sulfate; oxidation of the sulfide could also produce cadmium sulfate. In addition, traces of Cu, Pb, or As changed the hue by forming dark sulfides. Improved manufacturing processes have now solved these problems.

For the preparation of cadmium sulfide, high purity cadmium (> 99,99 %) is a prerequisite; otherwise, precipitation of cadmium sulfides would result in intensively colored sulfides of Cu, Ni, Fe, Co, or Pb. Therefore, the wet-chemical route starts from elemental cadmium, dissolved in mineral acids either via cadmium oxide or directly. From the resulting solution, sodium sulfide precipitates the raw pigment:

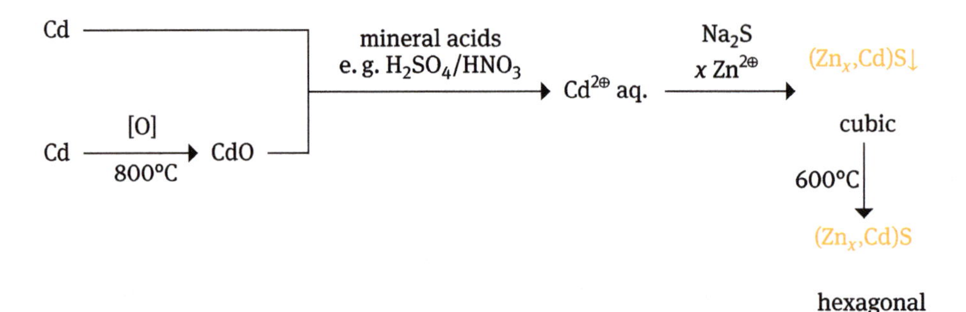

The raw pigment crystallizes in the cubic modification, converted by calcination at high temperature to the desired hexagonal modification. During this process, the pigment particles grow to the size required for pigments (up to ≈ 0.5 µm). Adding zinc salts in the amount necessary leads to the formation of the lighter zinc-cadmium sulfide, whose color depends on the ratios Cd/Zn.

The production conditions (temperature, concentration, pH value, duration, and others) influence particle size and shape. After production, residual cadmium salts must be carefully washed out since they negatively influence pigment properties, and cadmium ions must not enter the product.

A second method is the dry *powder process*. Cadmium carbonate or oxide is reacted with sulfur to form the sulfide. Zinc can also be added here to produce the lighter zinc-cadmium sulfide, followed by a calcination step to create the final pigment:

$$CdO \text{ or } CdCO_3 \xrightarrow{S} \underset{\text{cubic}}{CdS} \xrightarrow{600°C} \underset{\text{hexagonal}}{CdS}$$

The cheaper lithopone variants of the cadmium pigments are prepared by mixing cadmium salt with a soluble barium salt that then precipitates as sulfate:

$$CdSO_4 \text{ aq.} + BaS \text{ aq.} \longrightarrow CdS\downarrow + BaSO_4\downarrow$$

$$CdCl_2 \text{ aq.} + BaS \text{ aq.} \longrightarrow CdS\downarrow + BaCl_2$$

$$BaCl_2 \text{ aq.} + Na_2SO_4 \text{ aq.} \longrightarrow BaSO_4\downarrow + 2\,NaCl$$

Cadmium orange (PO20, CI 77196); cadmium red (PR108, CI 77202)

If we gradually replace sulfur with selenium in cadmium yellow, the resulting cadmium sulfoselenides adapt their color accordingly from orange to red to dark red [48, Volume 1], [11, 47, 49]. ►Figure 3.30 shows the corresponding shift of the absorption edge toward longer (red) wavelengths. The band gap of the semiconducting sulfoselenides becomes smaller with increasing selenium content. High-energy blue radiation and increasingly longer-wavelength radiation from the yellow and red spectral ranges are absorbed, shifting the color impression to orange, red, and finally brown, ►Section 1.5.3. Pure cadmium selenide CdSe is brown-black and has no painting significance.

Cadmium sulfoselenides Cd(S, Se) include PO20 (orange) and PR108 (red) with 76–66 % by weight Cd and 1–14 % by weight Se [11, 16]. The red cadmium pigments have been in great demand since 1820 for their hues and coloristic properties. They possess, like cadmium yellow, high opacity.

Cadmium orange and red are produced by precipitating cadmium sulfoselenide from cadmium-rich solutions with sodium sulfide and sodium selenide. The selenide forms from sodium sulfide, and the calculated amount of selenium is added as a powder. Further processing is the same as for the production of the zinc-cadmium sulfide:

$$Cd^{2\oplus}\text{ aq.} \xrightarrow[]{\text{Na}_2\text{S},\ x\,\text{Na}_2\text{Se}} Cd(S,Se_x)\downarrow \xrightarrow[]{600°C} Cd(S,Se_x)$$

Cadmium cinnabar, cadmium mercury sulfide (PO23, CI 77201; PR113, CI 77201)

Replacing cadmium with divalent mercury (Cd, Hg)S in cadmium yellow yields cadmium cinnabar [48, Volume 1], [49]. The higher the mercury content (typically 10–25 % by weight) is the more the color deepens from orange to dark red or brown. The pigment shows high color brilliance and is equal to cadmium red [406].

Unlike many well-known artists' pigments, it was not discovered by chance, but through planned research, as around 1950, the emerging electronics industry was consuming more and more selenium. At the same time, cadmium orange and red (cadmium selenide-based pigments) had become increasingly important to the plastics industry because of their color and stability. The idea of replacing selenium with substitutes led via the old red pigment cinnabar to a new mixed oxide of cadmium, mercury, and sulfur. This substitution was possible due to the equal lattices of CdS and HgS, as well as the similar ionic radii of the pairs Cd/Hg and Se/S.

For the production, we precipitate soluble cadmium and mercury salts with sulfides. If barium ions are present, they are coprecipitated as white barium sulfate and provide the substrate for the actual pigment, delivering lithophones:

$$x\,CdSO_4\text{ aq.} + (1-x)\,HgSO_4\text{ aq.} + Na_2S\text{ aq.} \longrightarrow (Cd_x,Hg_{1-x})S\downarrow + Na_2SO_4$$

$$x\,CdSO_4\text{ aq.} + (1-x)\,HgSO_4\text{ aq.} + BaS\text{ aq.} \longrightarrow (Cd_x,Hg_{1-x})S\downarrow + BaSO_4\downarrow$$

Cadmium green (PG14, CI 77199)

This is not a pigment, but a mixture of cadmium yellow and ultramarine blue. According to [47, 49], it is a mixture of cadmium yellow and viridian.

3.5 Bismuth pigments

Around 1970, in search of harmless substitutes for the brilliant but toxic pigments chrome yellow and cadmium yellow, pigments based on bismuth orthovanadate $BiVO_4$ were developed [11, 16, 49]. They exhibit brilliant and intense greenish-yellow hues and have no acute toxicity. Standard industrial hygiene prevents risks of damage to health by inhaling dust.

The color is induced by an OMCT transition in the highly charged complex $VO_4^{3\oplus}$ ion, similar to the transitions in chromate ions. Both vanadate and chromate pigments show a steep absorption edge caused by intense OMCT absorption in the blue spectral range, ►Figure 3.31.

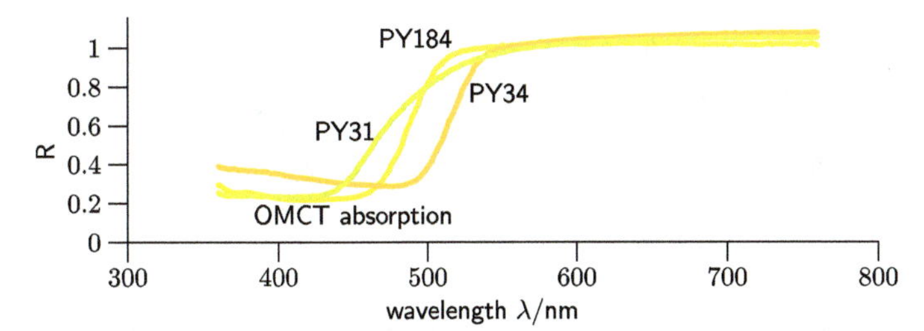

Figure 3.31: Reflectance spectrum of yellow CT-based watercolors normalized to an arbitrary unit: PY31 (barium yellow, own preparation, $BaCrO_4$), PY34 (chrome yellow light, Schmincke, $Pb(Cr,S)O_4$), and PY184 (vanadium yellow, Schmincke Horadam no. 207, $BiVO_4$). The figure depicts the steep absorption edge caused by an intense OMCT absorption in the blue spectral range. In vanadium yellow and barium yellow, this absorption is located at lower wavelengths so that the green spectral range contributes to the cool yellow hue of vanadium yellow. The green spectral range in chrome yellow is partially absorbed due to the OMCT absorption located at higher wavelengths, and the latter shows a warmer yellow.

Bismuth vanadium yellow, bismuth vanadate molybdate (PY184)

The pigment consists of bismuth vanadate $BiVO_4$. It exhibits not only a brilliant greenish-yellow hue but also high coloring power as well as opacity with a refractive index of 2.45 [11, 16, 201]. It is coloristically similar to chrome yellow and cadmium yellow, making it a good base yellow for mixtures in the yellow, orange, red, and green color range. It also has higher coloring power than yellow iron oxides or nickel titanium yellow and has very strong weather-fastness and chemical resistance. Therefore, it is applied in automotive and industrial coatings, paints, and plastic coloring to substitute lead and cadmium colorants.

In addition to the pure vanadate, mixed phases with Mo are also used, up to the composition $4\,BiVO_4 \cdot 3\,Bi_2MoO_6$. Bismuth vanadates occur in several modifications, but only the monoclinic and the tetragonal ones are essential for the production of pigments.

It is prepared in a solid-state reaction from bismuth oxide and vanadium oxide or by precipitation from solutions of bismuth nitrate and sodium vanadate:

$$Bi_2O_3 + V_2O_5 \xrightarrow[500°C]{\text{solid-state reaction}} 2\,BiVO_4$$

$$Bi + 4\,HNO_3 + 3\,H_2O \longrightarrow Bi(NO_3)_3 \cdot 5\,H_2O + NO$$

$$Bi(NO_3)_3\,aq. + NaVO_3\,aq. + 2\,NaOH \xrightarrow{\text{precipitation reaction}} BiVO_4\downarrow + 3\,NaNO_3 + H_2O$$

The reaction solution must be heated under reflux. The bismuth vanadate must be precipitated under precisely controlled conditions (temperature, pH value, concentration) to obtain a fine crystalline product with the desired modification because only monoclinic and tetragonal crystals show the desired color brilliance.

The bismuth vanadate containing Mo is prepared from a solution of bismuth nitrate, sodium vanadate, and ammonium molybdate:

$$10\,Bi(NO_3)_3\ aq. + 4\,NaVO_3\ aq. + 3\,(NH_4)_2MoO_4\ aq. + 20\,NaOH \longrightarrow$$

$$4\,BiVO_4 \cdot 3\,Bi_2MoO_6 \downarrow + 24\,NaNO_3 + 6\,NH_4NO_3 + 10\,H_2O$$

Improved pigment properties are obtained when the precipitate is subjected to an annealing process at about 500 °C. The resulting pigments are then stabilized by coating with additional layers of calcium, aluminum, or zinc phosphate, Al_2O_3, or SiO_2. The coating increases the resistance to weathering, acids, and color changes due to light. It can also increase temperature resistance to approximately 300 °C.

3.6 Chromium pigments

The group of chromium-based pigments includes essential colorants:

- Chromium(III) oxide, which we already discussed in ►Section 3.4.5.
- Chromates and the homologous molybdates and analogous mixed phases.
- Chromates also include chrome green, a mixed pigment of chrome yellow with iron blue pigments, and zinc green, a mixture of zinc chromate with iron blue pigments.

3.6.1 Chromate and molybdate pigments

Thanks to their brilliant colors, yellow to red pigments containing chromate and molybdate have been among the crucial inorganic pigments ever since their discovery in the nineteenth century. Among them are lead chromate (chrome yellow) and lead chromate molybdate (molybdenum orange and red) [11, 16, 49], in which the chromate ion $Cr^{VI}O_4^{2\ominus}$ forms the coloring structure. As a result, their colors range from lemon yellow to bluish-red, which they owe to an OMCT transition in the chromate ion, ►Section 2.4.1.

They exhibit brilliant hues and have high coloring power and opacity. Improved manufacturing methods and stabilizing the pigments with coatings of silicon or aluminum oxides have made it possible to strengthen the lightfastness and resistance to temperature, weather, and chemical influences. Chromate-containing pigments are used primarily in paints, lacquers, and colored plastics. We obtain bright green hues (chrome green, permanent green) in iron blue or phthalocyanine blue mixtures.

Chrome yellow (PY34, CI 77600); primrose yellow, chrome yellow light, chrome yellow lemon (PY34, CI 77603)

Chrome yellow is a yellow pigment [48, Volume 1], [11, 16, 47, 49, 201], based on lead(II) chromate $PbCrO_4$ or lead sulfochromate $Pb(Cr,S)O_4$, in which chromium is replaced to some extent by sulfur. As the sulfur content increases, the hue shifts from pure yellow to a greenish lemon yellow: medium yellow 1 $PbCrO_4 \cdot 0\,PbSO_4$ versus primrose yellow 3.2 $PbCrO_4 \cdot 1\,PbSO_4$ versus lemon yellow 2.5 $PbCrO_4 \cdot 1\,PbSO_4$. The exact color shade is adjusted during precipitation of the mixed-phase by the ratio of chromate/sulfate and the crystal size, which is a consequence of the concentration of the solutions, pH value, reaction temperature, and time. The control of the reaction conditions is crucial for producing the greenish variants, in which lead chromate is present in a metastable orthorhombic crystal modification that readily converts to the ordinary, stable monoclinic modification. Monoclinic lead chromate, also present in the natural mineral crocoite, has a darker yellow hue.

With a refractive index of 2.3–2.7, the pigment has good opacity and is suitable for oil painting; the replacement of chromium with sulfur reduces the coloring power and opacity in the greenish-yellow tones. The intense OMCT transition in the blue spectral range results in a steep absorption edge (►Figures 3.31 and (idealized) 3.32), so that chrome yellow is a pure and very intensive yellow pigment.

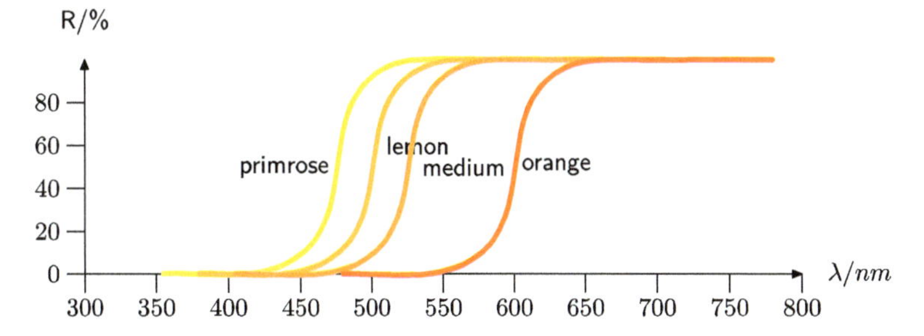

Figure 3.32: Schematic reflectance spectrum of chrome yellow $Pb(Cr, S)O_4$ in different shades (spectra drawn after [48, Volume 1, p. 190]).

Untreated chrome yellow has low lightfastness and darkens due to redox reactions. H_2S or SO_2 also attacks it with browning, i. e., the formation of lead sulfide. Because of this property, it was initially considered an unstable pigment. Today, however, we can buy stabilized chrome yellow for industrial purposes; its pigment particles are coated with oxides of Ti, Ce, Al, or Si. Zinc salts are added as photo inhibitors.

The pigment dates back to the discovery of the mineral crocoite or red lead ore (lead chromate). It was extensively used from 1814 until the twentieth century since its pure yellow color makes it an ideal mixing partner with blue pigments to obtain bright green hues. In addition to its use as an artist's pigment, it was employed in printing inks and paints of all kinds (lacquers, road markings). As the hexavalent Cr^{VI} is toxic, cadmium sulfide replaced it. This one, too, has competed for some time with even safer, heavy metal-free pigments. Alternatives are, e. g., the CICPs, especially PY216 and PY227/PY184. In artists' paints, we do not find chrome yellow anymore. However, due to its low price compared to the cost of other yellow pigments and its high brilliance, it is still of great interest today for industrial paints, coatings, and plastics. Its insolubility supports its continued use. Furthermore, it exhibits neither acute toxicity nor irritates skin or mucous membranes, although care must be taken when handling.

Technically, the pigment is gained from lead and converted into lead nitrate by dissolving it in nitric acid. Lead chromate is precipitated from this solution by adding chromate ions. These come from a solution of sodium dichromate; their pH value is adjusted so that the dichromate-chromate equilibrium is on the side of the chromate:

$$3\,Pb + 8\,HNO_3 \longrightarrow 3\,Pb^{2\oplus} + 6\,NO_3^{\ominus} + 4\,H_2O + 2\,NO$$

$$Cr_2O_7^{2\ominus} + H_2O \rightleftharpoons 2\,CrO_4^{2\ominus} + 2\,H^{\oplus}$$

$$Pb^{2\oplus}\,aq. + CrO_4^{2\ominus}\,aq. \longrightarrow PbCrO_4 \downarrow$$

If the solution also contains sulfate ions from sulfuric acid or alkali sulfates, the result is the lighter lead sulfochromate:

$$Pb^{2\oplus}\,aq. + x\,CrO_4^{2\ominus}\,aq. + (1-x)\,SO_4^{2\ominus}\,aq. \longrightarrow Pb(Cr_x,S_{1-x})O_4 \downarrow$$

In the past, lead acetate was used as a source of lead ions by dissolving lead(II) oxide in acetic acid. Lead oxide, in turn, was obtained by simply burning lead in the air:

$$2\,Pb + O_2 \longrightarrow 2\,PbO$$

This reaction also produced the massicot, an ancient yellow pigment (yellow lead oxide). Massicot is formed from red lead oxide by thermal transformation.

Chrome orange (yellow-orange: PO21, CI 77601; red-orange: PO45, CI 77601), chrome red, Viennese red (PR103, CI 77601)

These pigments are basic lead chromates of the composition $PbCrO_4 \cdot PbO$ or $Pb_2[CrO_4(OH)_2]$, which, in contrast to the neutral lead chromate, extend into the color orange [48, Volume 1], [11, 16, 49]. They form by the precipitation of lead salts from an alkaline solution. Chrome orange precipitates can have different shades from light to dark orange; the exact color depends on the particle size depending on precipitation conditions such as pH value or temperature. Darker and more intensively red variants (chrome red) possess larger particles than lighter ones. The lead oxide content also influences the color shade: the higher the proportion of PbO is the darker the shade becomes.

Chromates with a refractive index of 2.4–2.7 have good opacity and were introduced together with chrome yellow at the beginning of the nineteenth century. Because of their lead content and the toxicity of their chromate ions, they have been replaced today by other harmless, applicable orange and red pigments.

Molybdate red, molybdate orange (PR104, CI 77605)

Molybdate pigments are mixed phases $Pb(Cr, Mo, S)O_4$ with about 10 % $PbMoO_4$, brilliantly colored bright orange to red [11, 16, 49, 201]. In addition to composition and particle size, the crystal form largely determines the exact hue [200, keyword "Pigmente"]. The fastness properties of molybdate red pigments are similar to chrome yellow. Both pigments exhibit high color brilliance. They also show high opacity with a refractive index of 2.3–2.65.

Molybdate red was introduced to the market as a pigment around 1935 and quickly became one of the most important red pigments of the last century. It is used primarily in printing inks, coatings, lacquers, and colored plastics. In addition, molybdate orange can serve as a substitute for chrome orange since the latter's production has been stopped.

Interestingly, pure lead molybdate is colorless, whereas lead chromate is yellow. An intensive orange to red color only occurs when lead molybdate forms a tetragonal mixed-phase with lead sulfochromate. This phase is thermodynamically unstable and quickly changes to a more stable yellow modification, which can already occur when clumsily dispersed. The desired tetragonal phase must therefore be stabilized after preparation.

Lead ions are precipitated in the desired proportions from a lead nitrate solution either with dichromate, ammonium molybdate, and sulfuric acid or with potassium chromate, ammonium molybdate, and sodium sulfate:

$$Pb^{2\oplus}\,aq. + x\,CrO_4^{2\ominus}\,aq. + y\,MoO_4^{2\ominus}\,aq. + (1-x-y)\,SO_4^{2\ominus}\,aq. \longrightarrow Pb(Cr_x,Mo_y,S_{1-x-y})O_4\downarrow$$

Stoichiometry and precipitation conditions determine the exact hue.

Molybdate pigments need to be stabilized to increase lightfastness. Stabilization is achieved by adding sodium silicate or aluminum sulfate and neutralization with NaOH or Na_2CO_3 to coat the pigment with hydrated silicon or aluminum oxides, e. g.,

$$Al^{3\oplus}\,aq. + 3\,OH^{\ominus} \longrightarrow Al(OH)_3\downarrow \longrightarrow Al_2O_3\cdot nH_2O$$

Zinc yellow (PY36, CI 77955, CI 77957, PY36:1, CI 77956); strontium yellow (PY32, CI 77839); ultramarine yellow, barium yellow, lemon yellow (PY31, CI 77103); calcium chromate (PY33, CI 77223)

Since the color of chrome yellow is carried by the chromate ion and not by the metal ion, lead can be replaced by many other metals. From 1850 onward, zinc, strontium, barium, and calcium chromates enjoyed particular popularity, and they show similar brilliance but brighter colors than primrose yellow [47, 49].

While zinc yellow was more often used as artists' pigment, strontium chromate $SrCrO_4$ and barium chromate $BaCrO_4$ (►Figure 3.31) appear only as additives in the color catalogs [49]. The composition of zinc yellow varies with time and author. Modern zinc yellow is regarded to be $K_2O{\cdot}4\,ZnCrO_4{\cdot}3\,H_2O$ (also written as $K_2CrO_4{\cdot}3\,ZnCrO_4{\cdot}Zn(OH)_2{\cdot}2\,H_2O$ (PY36, CI 77955). *Colour Index* [1] lists also $ZnCr_2O_7{\cdot}3\,H_2O$ (as CI 77957). A wide distribution of the $ZnO : Cr_2O_3$ ratio can be observed in older pigments. From 1941, zinc chromate also appears as $ZnCrO_4 \cdot 4Zn(OH)_2$ (PY36:1, CI 77956). These chromates were mainly applied as a yellow component of green mixtures with Prussian blue (►zinc green).

The designation "ultramarine yellow" for barium yellow is misleading since barium chromate has nothing to do with pigments of the genuine type, such as ultramarine blue.

3.6.2 Chrome green (PG15, CI 77510); fast chrome green (PG48, CI 77600); zinc green

The name "chrome green" suggests the presence of a further green compound of chromium in addition to chromium(III) oxide, but this is not the case: Designated as "chrome green," various mixtures of yellow and blue pigments are offered. For example, PG15 is a mixture of chrome yellow and iron blue pigments, PG48 one of chrome yellow and phthalocyanine blue, which replaces the fading Prussian blue and is called fast chrome green [11, 16, 47]. Zinc green is a mixture of zinc chromate and iron blue pigments [47], but this name is also used to describe the green zinc-cobalt compound Rinmann's green. The many possible blue and yellow colorant combinations cause general confusion about how each combination is termed [49]. In addition, sources also use the term " cinnabar green" to express the possibility that a mixed green does not contain chrome at all, e. g., cadmium green.

Chrome green can be produced in shades from light yellowish green to dark blue-green simply by mixing the individual pigments. Of course, a painter could make these himself; however, only industrial production can employ wet-chemical techniques to create particularly brilliant green shades. First, manufacturers deposit one pigment on the particle surfaces of the other. Then, for stabilization, solutions of sodium silicate and aluminum sulfate, or magnesium sulfate, are added to form protective layers of silicon, aluminum, or magnesium oxide. The resulting two-layer pigments exhibit brilliant colors, high color stability, and opacity.

The stabilized chrome greens' fastness and painting properties are good depending on the components used. They generally have the same application areas as chrome yellow and molybdate pigments. Therefore, a transparent variant of the primrose yellow containing lead sulfate is often used for the green mixtures. In contrast to the opaque lead chromate, the low refractive index of 1.89 of lead sulfate causes extensive transparency in oil binders.

3.7 Iron blue pigments (Prussian Blue, Berlin Blue, Milori Blue, Paris Blue, iron blue, PB27, CI 77510, 77520)

This group includes pigments such as Prussian blue, Berlin blue, Paris blue, Milori blue, Turnbull's blue, and many more. They owe their color to a mixed-valence hydrated Fe^{II}-Fe^{III} cyano complex [48, Volume 3], [11, 16, 47, 49, 423, 424]. The traditional names often denote a specific shade of the insoluble iron complex. Today, after systematization and simplification of the varieties, the pigments are generally referred to as *iron blue*.

For a long time, it was unclear how the oxidation states of iron were distributed in the compound, i. e., whether a compound contains $AFe^{III}[Fe^{II}(CN)_6]$ or rather $AFe^{II}[Fe^{III}(CN)_6]$. New analyses show that all variants of iron blue are always the same complex characterized by

$$MFe^{III}[Fe^{II}(CN)_6] \quad \text{with M=Na, K, NH}_4 \quad \text{(CI 77520, so-called “soluble blue”)}$$

$$Fe^{III}_4[Fe^{II}(CN)_6]_3 \cdot n\,H_2O \quad \text{or} \quad Fe^{III}[Fe^{III}Fe^{II}(CN)_6]_3 \cdot n\,H_2O \quad \text{with } n = 14\text{–}16 \text{ (CI 77510)}$$

and contains Fe^{II} in the complex anion. *Soluble* in this context does not entirely refer to the water solubility of the pigment. Instead, it expresses that soluble iron blue forms during synthesis as a colloidal solution whose particles have a size of 0.01–0.1 µm. It also possesses good peptization ability, i. e., the ability to redissolve colloidally. The counterions used in the soluble iron blue are important for hue and dispersibility. Potassium and ammonium industrially are used as counterions, as they provide the most beautiful hues. However, potassium is more expensive than ammonium.

The structure of iron blue pigments is based on a cubic lattice. Fe^{II} and Fe^{III} are alternately arranged; the cyano groups are located between and provide the coloring structural element Fe^{II}–C–N–Fe^{III}. The intensive color of iron blue is a prime example of colorfulness by IVCT transitions: Within the structural part, the irradiation of yellow-red light initiates an electron change from the sixfold C–coordinated Fe^{II} ion to the sixfold N–coordinated Fe^{III} ion, ►Section 2.4.2. This intense broad absorption appears in the yellow-red region and, therefore, the color of intensive blue, ►Figure 3.33.

The water of crystallization and, in the case of soluble iron blue, the monovalent metal cations are distributed within the cubes of the iron lattice. As a result, $\frac{1}{4}$ of the Fe^{II} and the cyano positions are empty in insoluble iron blue. The water of crystallization then replaces nitrogen and is essential for the stability of the lattice.

It is prepared from a potassium hexacyanoferrate(II) solution with iron(II) sulfate or chloride. A white precipitate of iron(II) hexacyanoferrate(II) or Berlin white is cre-

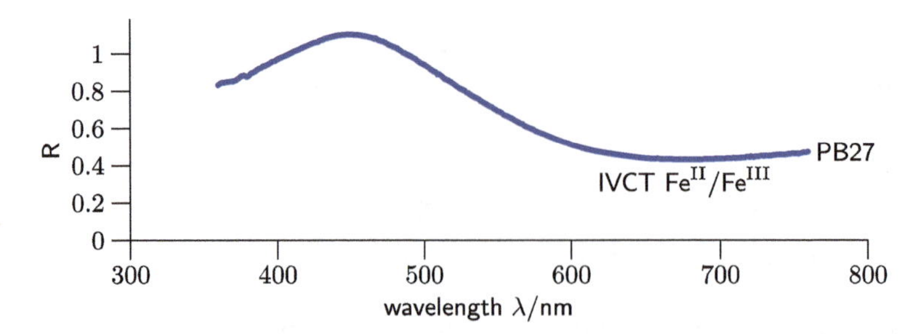

Figure 3.33: Reflectance spectrum of CT-based blue watercolors normalized to an arbitrary unit: PB27 (Prussian blue, Schmincke Horadam No. 492, $(Na, K\,NH_4)Fe^{III}[Fe^{II}(CN)_6] \cdot nH_2O$) with an intensive and broad IVCT absorption in the yellow to the red spectral range.

ated. After heating the suspension with alkali chlorate, H_2O_2, or alkali dichromate, it is oxidized to the blue pigment:

$$K_4[Fe^{II}(CN)_6]\ aq. \xrightarrow[-\ K_2SO_4]{FeSO_4} \underset{\text{Berlin white}}{K_2Fe^{II}[Fe^{II}(CN)_6]\downarrow} \xrightarrow{[O]} \underset{\text{insoluble iron blue}}{Fe^{III}_4[Fe^{II}(CN)_6]_3\downarrow}$$

$$K_4[Fe^{II}(CN)_6]\ aq. \xrightarrow[-\ K_2SO_4]{FeSO_4,\ (NH_4)_2SO_4} (NH_4)_2Fe^{II}[Fe^{II}(CN)_6]\downarrow \xrightarrow{[O]} \underset{\text{soluble iron blue}}{NH_4Fe^{III}[Fe^{II}(CN)_6]\downarrow}$$

The aging and oxidation phases are controlled depending on the target parameters of the pigment. Reaction conditions (temperature, concentration) and the addition of ammonium sulfate determine the size and shape of the precipitated particles and the color of the resulting pigment. Numerous shades of blue can be obtained, which play into the reddish or greenish range. Adding organic compounds improves the pigment's dispersibility in the aging and oxidation phase by preventing agglomeration during drying.

Soluble iron blue is also precipitated by the reaction of potassium hexacyanoferrate(II) with Fe^{III} salts or by the reaction of potassium hexacyanoferrate(III) with Fe^{II} salts:

$$K_4[Fe^{II}(CN)_6]\ aq. + Fe^{3\oplus}\ aq. \longrightarrow \underset{\text{soluble iron blue}}{KFe^{III}[Fe^{II}(CN)_6]\downarrow} + 3\,K^{\oplus}$$

$$K_3[Fe^{III}(CN)_6]\ aq. + Fe^{2\oplus}\ aq. \longrightarrow KFe^{III}[Fe^{II}(CN)_6]\downarrow + 2\,K^{\oplus}$$

Colloidal soluble iron blue form if the reactants are present in a molar 1:1 ratio. Insoluble iron blue can be produced by adding excess Fe^{II} or Fe^{III} ions, but this is not carried out industrially.

Iron blue was observed around 1700 in Berlin by Diesbach as a precipitate in a precipitation reaction, who thus synthesized the oldest coordination compound. Milori introduced the pigment in the early nineteenth century industrially, and pigments of this group rapidly gained in importance. However, from about 1970, their significance declined due to the development of blue phthalocyanine pigments.

Iron blue pigments have high coloring power, lightfastness, and weather-fastness and were a component of blue lacquers and paints. However, white mixtures negatively influence fastness, and a topcoat is necessary depending on the application area. Painters knew about this effect since the eighteenth century, [425–427] confirming the tendency of Prussian blue to fade in white mixtures.

Iron blue pigments are very transparent due to their color strength and low refractive index of 1.56. They are used industrially in pure form, mainly in printing inks and paints, as they have high opacity and allow economical printing. Early large-scale commercial printed products were e. g., wallpapers. For multicolor printing, iron blue pigments were supplemented with phthalocyanine pigments. Iron blue was the most important blue pigment for printing inks until recently. It is still used today for toning black printing inks, i. e., for adjusting the black color. Soluble iron blue is used for low-priced paper coloring, whereas insoluble iron blue is a component of coating colors for surface-colored paper.

For the fine arts, iron blue pigments had great importance as a component of *green colors ready for application*, in which they appear together with a yellow pigment: *chrome green* (iron blue and chrome yellow), *zinc green* (with zinc yellow), *Hooker's green* (with gamboge), ocher, Naples yellow, and other yellows.

Due to their tiny particle size, iron blue pigments rapidly form colloidal solutions; thus, dispersing is problematic. Therefore, alcohol or other wetting agents are required. They are stable to dilute mineral acids and oxidizing agents, but strong acids and alkalis decompose the pigment. Therefore, they cannot be used in media like casein, fresco, or sodium-potassium silicates. However, if the blue pigment is mixed with fillers such as barium sulfate, alumina, or calcium sulfate, we obtain colors such as mineral blue, Antwerp blue, or Brunswick blue.

Despite a cyanide anion, minimal release from the complex occurs. Combined with the low rate of absorption in the animal organism and a high rate of excretion, iron blue pigments show no acute toxicity or proven adverse long-term effects and do not irritate skin or mucous membranes. It is noteworthy that iron blue is used for decontamination of radioactively irradiated persons. The dangerous ^{137}Cs is exchanged for Fe^{II}, and, tightly bound, excreted. Even the high doses of up to 20 g/day used in this process showed no toxicological abnormalities.

3.8 Various metal pigments

3.8.1 Calcium carbonates

White calcium carbonate plays a vital role in art in various forms, be it a pigment, an extender/filler, or a primer [48, Volume 2], [47, 49]. Its primary structures are:

- Calcite is the most common natural form of $CaCO_3$. It builds up sedimentary rocks, e. g., limestone, or when combined with magnesium carbonate, dolomite.
- Marble is the metamorphic form of calcite. It consists of recrystallized calcite with very coarse calcite grains.
- Chalk is a soft, sedimentary rock consisting of fossil remains. They are built from the shells of marine organisms, mainly unicellular algae. Despite its name, blackboard chalk does not contain chalk, but gypsum.
- Precipitated calcium carbonate (PCC). This artificial $CaCO_3$ is made by precipitating $Ca^{2\oplus}$ with various agents and is primarily a byproduct of industrial processes.

$CaCO_3$ is used in massive form, especially for drawing, and in fine form after grinding or precipitating. Nonetheless, its use is limited chiefly by its low refractive index (1.486/1.658 for calcite, or 1.566 for rhombic calcite), causing low opacity in oil binder, so that it is either used in watercolor, fresco, or as pigment for grounds (gesso) or as a filler/extender, e. g., in papermaking.

Lime white, St. John's white, Eggshell white, shell white, bone white, horn white

These white pigments are of artificial origin and described in various manuscripts, among others on producing paints for book illumination [48, Volume 2], [49, 108]. They consist of calcium hydroxide $Ca(OH)_2$, calcium phosphate (apatite) $Ca_3(PO_4)_2$, and hydroxyapatite $Ca_5(OH)(PO_4)_3$, and are obtained by burning and possibly slaking of limestone, marble, eggshells, mother-of-pearl, ivory, or deer horn:

$$\underset{\text{limestone, marble, eggshell, mother-of-pearl}}{CaCO_3} \xrightarrow[800\text{–}900°]{\Delta} \underset{\text{quicklime}}{CaO} \xrightarrow[\text{equimolar}]{H_2O} \underset{\text{lime white, St. John's white, eggshell white, shell white}}{Ca(OH)_2}$$

$$\text{horn, bone, ivory} \xrightarrow[\text{burning}]{\Delta} \underset{\text{bone white, horn white}}{Ca_3(PO_4)_2 + Ca_5(OH)(PO_4)_3}$$

While lime white, eggshell white and shell whites are pure white, apatite is harder and grayer in tone, so bone white and horn white were preferably used in primers. These white pigments have a medium refractive index and are preferred for aqueous techniques such as book illumination but are too transparent for oil. Lime white is used in fresco painting as pure white pigment.

Gofun Shirayuki, Japanese shell white

This warm porcelain white pigment, consisting of calcium carbonate $CaCO_3$, is used in Japan to substitute lead white, which was banned about 400 years ago [48, Volume 2], [62]. It is made from oyster shells, which piled in large heaps, rot in the warm, humid climate of the Japanese coast and eventually break down into tiny mother-of-pearl flakes, sorted into size classes by wet sluicing.

3.8.2 Lead white, flake white, Kremser white, Cremnitz white (PW1)

Since antiquity, the primary lead carbonate $2\,PbCO_3 \cdot x\,Pb(OH)2$ has been known as makeup and pigment. Despite its toxicity, it was the most important white pigment in European panel painting until the nineteenth century [48, Volume 2], [49, 386], as it is a pure white with high opacity even in oil. This property is due to the relatively high refractive index of 1.94. In contrast, other white pigments from this time, such as lime white or gypsum, having almost the same refractive index as oil, appear transparent in this binder, making them suitable only for fresco or watercolor.

Lead white is interesting because it is one of the oldest synthetic pigments and has been produced artificially since antiquity. Lead plates were exposed in closed containers to vinegar vapors (i. e., acetic acid) and warm manure (carbon dioxide). The white coating of lead acetate that initially formed converted into basic lead carbonate and then was dried in the air. Nowadays, the pigment is also made from lead acetate or another soluble lead salt, although it is only available for specific purposes such as professional restoration [16]:

$$Pb \xrightarrow{O_2} PbO \xrightarrow[-\,H_2O]{CH_3COOH} Pb(CH_3COOH)_2 \xrightarrow[-\,CH_3COOH]{+\,CO_2 \; +\,H_2O} PbCO_3{\cdot}Pb(OH)_2$$

The white precipitate can be used as a pigment without further treatment.

One advantage of the pigment was that it accelerated oil drying, just like lead salt. Unfortunately, with oils, it forms transparent lead soaps so that paintings from the time of Rembrandt onward often darken because the formerly opaque layers of lead white increasingly allow dark underpaintings to become visible, ►Section 7.4.11.

Lead white is sensitive to acids as a carbonate, forming the corresponding lead salts. Another disadvantage is its sensitivity to hydrogen sulfide or sulfides, which is particularly noticeable in watercolor. Lead white reacts to black lead sulfide, exhibiting brown to black shades. Therefore, earlier artists sometimes had to apply protective varnishes to lead white layers before they could apply additional layers to separate one from the other. As we can confirm, however, embedding the pigment in an oil film and isolating it in this way has had a protective effect for many centuries up to the present day.

3.8.3 White sulfates

Barite (PW22); blanc fixe, permanent white (PW21, CI 77120)

White barium sulfate $BaSO_4$ occurs naturally as barite and has been used since about 1782 as an alternative to white lead, but mainly in watercolor [48, Volume 1], [16, 47, 49, 201]. Due to a shortage of natural minerals, it was produced synthetically around 1830 and marketed under names like blanc fixe or permanent white. It is still used today as a pigment, but quantities are declining. Due to its low refractive index of 1.64, this very stable, bright white pigment is most suitable for watercolors.

The pigment is prepared simply by precipitating the sparingly soluble barium sulfate with soluble barium salt and sulfate anions:

$$Ba^{2\oplus}\,\text{aq.} + SO_4^{2\ominus} \longrightarrow BaSO_4 \downarrow$$

Precipitation was also used to purify the natural mineral, often delicately colored due to traces of iron. The natural sulfate can be converted into the soluble sulfide by reduction with charcoal or coke:

$$\underset{\text{natural barite}}{BaSO_4} + \underset{\text{charcoal, coke}}{2\,C} \xrightarrow[1000\text{–}1300^\circ]{\Delta} BaS\,\text{aq.} + 2\,CO_2$$

$$BaS\,\text{aq.} + Na_2SO_4 \longrightarrow \underset{\text{blanc fixe}}{BaSO_4 \downarrow} + Na_2S$$

Lithopone (PW5, CI 77115)

Although barium sulfate has little use as a pigment, large quantities are produced to manufacture *lithopone* [16, 47, 49]. This pure white mixture of zinc sulfide with barium sulfate has been known for over 150 years. It was intensively used, especially in the first half of the twentieth century before it was displaced by titanium white [200, keyword “Pigmente”].

Lithopone was never applied as artists’ pigment but for coatings of all kinds (paper, primers) and as a filler and white pigment in the paint and varnish industry (primers, spackles). It is produced by coprecipitation of zinc sulfide and barium sulfate:

$$BaS\,\text{aq.} + ZnSO_4\,\text{aq.} \longrightarrow BaSO_4 \downarrow + ZnS\downarrow$$

Zinc sulfate is obtained by dissolving zinc-rich raw materials in acids such as sulfuric acids. Barium sulfide originates in natural barite, reduced to sulfide as depicted above.

3.8.4 Miscellaneous colored pigments

Cobalt yellow, aureolin (PY40, CI 77357)

Yellow potassium hexanitritocobaltate(III) $K_3[Co(NO_2)_6]$ is known to chemists from qualitative microchemical potassium evidence. It was proposed as a pigment around 1850 and sometimes used in the nineteenth century [48, Volume 1], [47, 49]. Because of its low refractive index of 1.72, the complex was mainly employed as a watercolor, but its fastness has been controversially discussed. In any case, trivalent cobalt is a reactive oxidant, and cheaper and better yellow pigments soon replaced the pigment. Of interest to us is the *yellow* color that the cobalt exhibits here, caused by the octahedral coordination of Co^{III}.

Manganese blue (PB33, CI 77112)

Barium sulfate doped with manganese has been known since about 1920 as barium manganate or manganese blue, showing a beautiful bright hue of cool blue [47, 49, 469]. Unfortunately, it is no longer produced as a pigment, which happened by isomorphous replacement of sulfur with manganese. In the pigment, Mn is tetrahedrally coordinated as complex anion $MnO_4^{3\ominus}$ [469]. The question of which oxidation state (VI or V) Mn assumes has long been unanswered. It cannot be decided by the ion's LF color alone because, e. g., both oxidation states show green to blue colors in an aqueous solution.

In the frequently stated composition $BaMnO_4 \cdot BaSO_4$, Mn^{VI} would be present. According to [469], however, Mn^V is responsible for the color, so we can instead identify the pigment as barium manganate(V) $Ba_3(MnO_4)_2$ or as a mixed crystal $BaSO_4 \cdot BaHMnO_4$. The latter assumption is supported by the existence of a solid solution $BaSO_4 \cdot BaHPO_4$. In a basic host lattice, the following redox reaction would be conceivable:

$$2\,Mn^{VI}O_4^{2\ominus} + H_2O \longrightarrow 2\,HMn^{V}O_4^{2\ominus} + 1/2\,O_2$$

The spectrum, and thus the color is typical for the LF transition $e \rightarrow t_2$ of a $3d^2$ configuration in a tetrahedral ligand field, the dominant transitions being

$$^3A_2 \rightarrow {}^3T_1 \quad (e)^2(t_2)^0 \rightarrow (e)^0(t_2)^2 \quad (14\,500\text{–}16\,500\ \text{cm}^{-1})\ \text{dominant}$$
$$^3A_2 \rightarrow {}^3T_2 \quad (e)^2(t_2)^0 \rightarrow (e)^1(t_2)^1 \quad (11\,500\ \text{cm}^{-1})\ \text{weaker}$$

and correspond to the Mn^{V} ion. Contributions of Mn^{VI} absorption are not detectable in the spectrum. Due to the tetrahedral symmetry, LF transitions are allowed and, therefore, very intense, unlike many other pigments based on LF transitions, ►Figure 3.34.

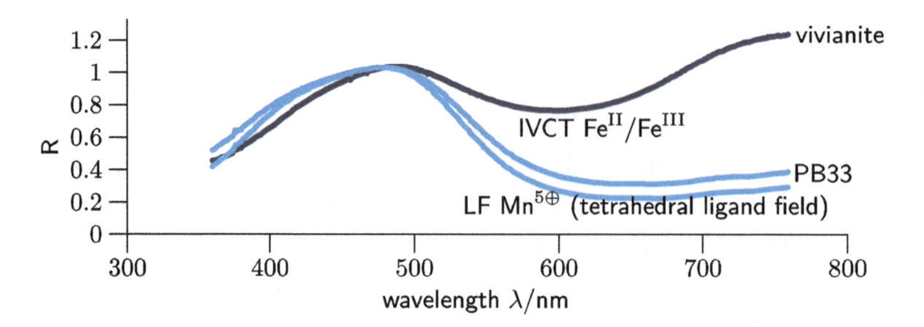

Figure 3.34: Reflectance spectrum of manganese- and iron-based watercolors, normalized to an arbitrary unit: vivianite (Wallace Seymour Watercolour Whole Pans, $Fe_3(PO_4)_2 \cdot 8\,H_2O$ with Fe^{III}) and PB33 (manganese blue genuine, Wallace Seymour Vintage Watercolour and Wallace Seymour Watercolour Whole Pans, "$BaMnO_4 \cdot BaSO_4$").

Iron phosphate blue, vivianite, blue ocher, iron blue, blue iron earth

This blue pigment, used predominantly in the Middle Ages, consists of hydrous iron(II) phosphate of composition $Fe_3(PO_4)_2 \cdot 8\,H_2O$, containing traces of Fe^{III} [49, 56]. Vivianite is a common secondary mineral in crystalline and earthy forms, found chiefly in iron, tin, and copper deposits, lignite mines, or at the bottom of ferruginous bogs. There it is a transformation product of bones and organic material, built from iron and phosphate ions but only turns blue in contact with air through the oxidative formation of Fe^{III}, ►Section 2.4.2 at p. 139. Once formed, the color remains stable until external circumstances change the Fe^{II}/Fe^{III} ratio. ►Figure 3.34 depicts its reflectance spectrum.

Cobalt violet dark (PV14, CI 77360)

Cobalt is often associated with intense blue colors (cobalt glass), but thanks to its LF transition, it can also lend an intense and pure reddish-purple color, ►Figure 3.35 [47, 49, 405]. Cobalt violet dark is cobalt(II) phosphate $Co_3(PO_4)_2$, discovered in 1859; it is still in use but expensive.

A similarly constructed violet cobalt pigment is cobalt arsenate $Co_3(AsO_4)_2$ or cobalt violet light, in which cobalt may be replaced by magnesium: $Mg_3(AsO_4)_2$. In contrast to its frequent mention in literature, it rarely appears in paintings.

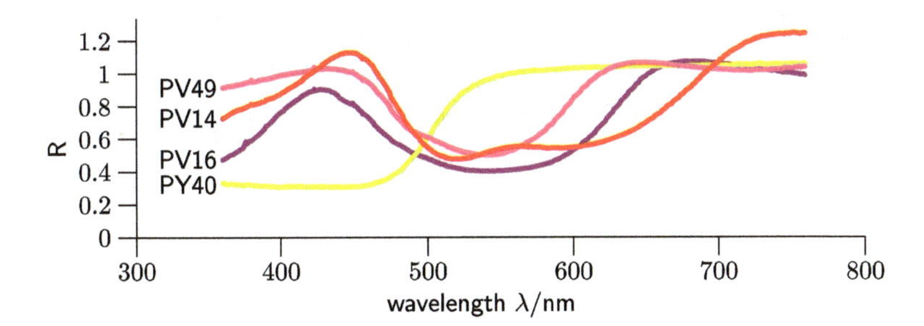

Figure 3.35: Reflectance spectrum of cobalt-based watercolors normalized to an arbitrary unit: PV14 (cobalt violet, Winsor and Newton Professional Watercolor No. 192, $Co_3(PO_4)_2$), PV16 (manganese violet, Schmincke Horadam No. 474, $(NH_4)MnP_2O_7$), PV49 (cobalt violet, D. Smith Extra Fine Watercolor No. 284610030, $NH_4CoPO_4 \cdot H_2O$), and PY40 (cobalt yellow, Kremer No. 435008, $K_3[Co(NO_2)_6]$).

Lithium cobalt phosphate (PV47, CI 77363); ammonium cobalt phosphate, cobalt ammonium violet (PV49, CI 77362)

These are pigments in the red-purple range colored by cobalt, ►Figure 3.35 [405]. They have the composition $LiCoPO_4$ and $NH_4CoPO_4 \cdot H_2O$. In contrast to cobalt arsenate, they are commercially available nowadays.

Manganese violet (PV16, CI 77742)

Manganese is another element that can impart numerous colors to its compounds, thanks to LF transitions. Mn^{III} in manganese ammonium phosphate $(NH_4)MnP_2O_7$ has provided a beautiful bluish-purple since its discovery around 1868, ►Figure 3.35 [47, 49].

3.9 Glasses

We encounter colored materials on the border of painting and decorative arts with glasses. However, they illustrate chromaticity well through LF transitions. Besides, some glass masses were used to a certain extent as pigments. As glass painting also benefits from the knowledge of glasses, we take the liberty to provide a brief outline of glass and its relatives (enamels and glazes). In the case of glazes, we come across one of the oldest synthetic pigments, even if it was not initially used as such. First, we will look at some eminent stages in the history of glass [27, 28].

3000 BC	Egypt, Mesopotamia	First glass of quartz (sand) and halophyte ash (soda, lime), colored green by Cu (halophyte ash glass)
1600 BC	Syria	Lead glass with various color additives, ►Table 3.11
600 BC		Soda-lime glass made from mineral soda
70	Rome	Colorless glass (low iron), flat glass for windows
100		Glassmaker's pipe
2nd–4th century		Gold and enamel decorations
8th century	Carolingian Empire	Early medieval lead flat glass as color windows: green, brown, less frequently blue, red; wood ash glass (potash glass, forest glass)
9th–12th century	Samarra	Islamic glasses, green, blue, colorless
15th century	Central Europe	Wood ash-lime glass
19th century		Soda-lime glass, variety of specialized glass recipes

Glass consists of mixtures that emerge from the melt flow not in crystalline but "glassy" amorphous form. They solidify with near but without far order and consist of [30–32, 605, 606]

- acidic oxides (such as SiO_2, B_2O_3, or P_2O_5),
- amphoteric oxides (such as Al_2O_3 or SnO_2),
- alkaline oxides (like K_2O, NaO, MgO, CaO or PbO and ZnO).

Acidic oxides, the *glass formers*, build extensive three-dimensional networks, while alkaline oxides, the *network converters*, create interfaces and disrupt the networks. They help obtain meltable and workable materials from the hard and brittle glass masses. The amphoteric oxides are not themselves glass formers but can replace them and change the properties of the glass.

The actual glass consists of the network builder SiO_2, which forms a network of corner-linked SiO_4 tetrahedra. Some of the Si–O–Si bridges are split by O^{2-} oxide ions supplied by alkaline oxides (interface former): $Si–O–Si + O^{2\ominus} \rightarrow Si–O^{\ominus} + {}^{\ominus}O–Si$. The more splits occur, the more the softening temperature and melting point decrease.

In the framework formed by SiO_2, $Si^{4\oplus}$ can be replaced by other atoms, for example, $B^{3\oplus}$, $Al^{3\oplus}$, or $P^{5\oplus}$. Glasses such as borosilicate glass are obtained in this way. Since these atoms change the charge of the structure, they require additional network converters to compensate for the differences.

We can find several ancient glass recipes in [27, Chapter 6], [35, 121]. The Recipes reveal a certain temporal and spatial order, ►Figure 3.36. In the Mediterranean and the Orient, the proportion of resistant *calcium soda glasses* is high. Sodium-rich halophytes (salt plants) flourished on the coasts and delivered ash rich in sodium and cal-

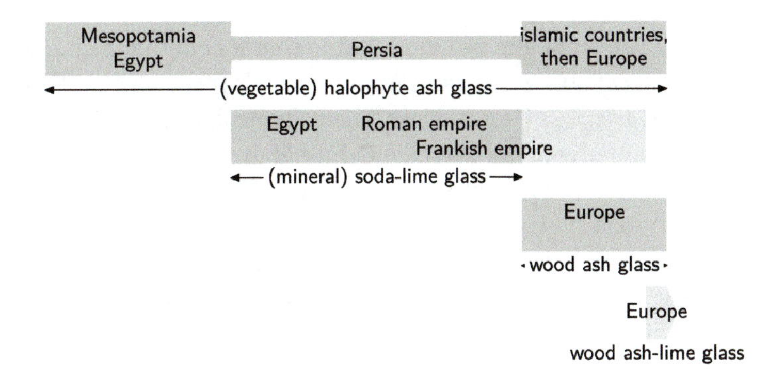

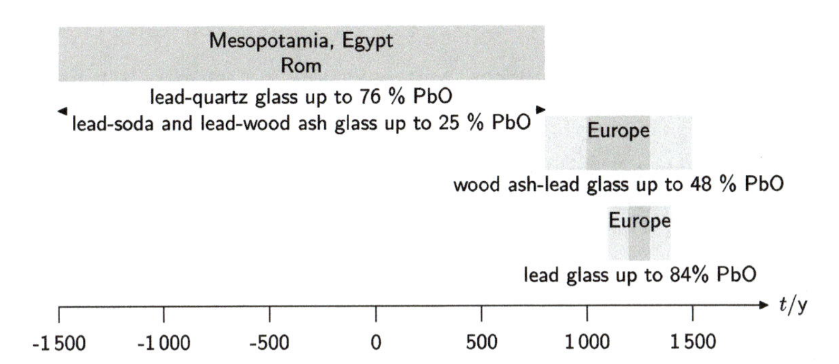

Figure 3.36: Simplified temporal and spatial classification of ancient glass types (drawn after [27, ch. 6], [28, Chapter 13, 14]). The sodium- and calcium-rich glasses made with halophyte or soda ash are also collectively referred to as soda-lime glasses. We also know potassium-rich wood ash glasses as potassium glass.

cium. In addition, natural sodas (sodium carbonate and calcium-rich accompanying substances) occurred in deserts and salt lakes.

Assyrian glass, 7 BC, soda-lime glass [27, Chapter 6] 1 part sand, 3 parts Salicornia ash, 0.1 part saltpeter, 0.04 part lime

Roman glass, first century, soda-lime glass color-compensated [27, Chapter 6] $\frac{9}{12}$ part sand, $\frac{3}{12}$ part natural soda, some manganese dioxide

Syrian glass, seventh to the eleventh century, soda-lime glass [27, Chapter 6] 1 part powder of pebbles, 3 parts ash or 0.8 part alkali of ash

In contrast, Central European glassworks did not have sodium-containing plants that could provide soda; therefore, they had to rely on imported sodium compounds. At the end of the eighth century, there was a shortage of natural soda; consequently, the Mediterranean region had to rely on sodium-rich halophyte ash (which yielded soda-lime glass) or potassium-rich wood ash (which generated potash glass). This shortage also forced the Merovingian empire to use the alkaline component that had to be mainly produced by the ashing of potassium-rich ferns and woods (beeches). This less durable *potash glass* or *wood ash glass* ("forest glass") had a typical, iron-conditioned green coloring and remained predominant from the Carolingian period until the late Middle Ages. Then Venetian glass was produced with sodium-rich halophyte ash from the Levant and converted to soda-lime glass with calcium fractions from the raw materials. Later, a change in furnace design can be observed in Europe. They facilitated higher melting temperatures and the production of calcium-rich and silicon-rich wood ash-lime glass, which was maintained until modern times.

Glass after Theophilus Presbyter, 1123, potash glass [27, Chapter 6] 1 part sand, 2 parts beechwood ash

Green forest glass, northern Europe fourteenth to the sixteenth, potash glass [121] potassium carbonate (wood ash, potash), sand (with iron compounds), or white pebbles

Venetian Cristallo glass after V. Biringuccio 1540, soda-lime glass color-compensated [121], [27, Chapter 6] 2 parts river pebbles or sand (containing lime), 1 part salt from ash (potash plants containing sodium, ferns, lichens), some manganese stone (manganese dioxide)

Glass, according to Agricola, 1556, soda glass color-compensated [27, Chapter 6] 2 parts sand, 1 part soda, rock salt or salt from halophyte ash, small pieces of manganese dioxide. Alternatively: 1 part sand, 2 parts wood ash, a little rock salt, a small piece of brownstone

Venetian Cristallo glass after A. Neri 1612, soda-lime glass color-compensated [121] Rochetta (soda-containing plant) from the Levant or Syria, tartar, tarso (calcareous river pebbles from Ticino with little iron compounds), manganese, common salt

Bohemian chalk glass or lime crystal glass after J. Kunckel, 1689, lime potash glass color-compensated [121] 100 pounds of potash, 150 pounds of sand (with iron compounds), 20 pounds of chalk, 5 ounces of manganese

Craftspeople recognized the possibility of adding lead to the mass of glass to produce an easily meltable and colorable *lead glass* already in antiquity. In Egypt and Rome, we find pure lead-quartz glass with up to 76 % PbO and lead-soda glass with up to 25 % PbO. In Central Europe, lead was produced as a byproduct of the medieval silver mining from lead ores. With the switch to copper ores in the fourteenth century and the decline of the creation of sacred glass windows, lead glass production also declined in the late Middle Ages.

Venetian white mosaic glass, fifteenth century, lead glass [121] lead glass, zinc oxide

The proportion of lead oxide increases the refractive index and dispersion of the glass, lowers both the viscosity of the molten state and the surface tension, and thus improves the wettability of other glass, easily dissolves other metal oxides, and lowers the melting temperature to 700–800 °C. The resulting easily meltable glass provided the basis for transparent or opaque enamel paints, which could be colored by adding metal oxides, ►Section 3.9.3. The above properties can be attributed to the large, easily polarizable $Pb^{2\oplus}$ cations, which are about the same size as the glass-forming constituents $O^{2\ominus}$ and $K^{\oplus}$ [32]. As a result, lead cations do not occupy the vacancies of the silicate network but disrupt the glass order and induce SiO_4 tetrahedra to arrange themselves around the lead ions. Furthermore, due to their deformability, they can adapt to interfaces so that the surface tension is reduced and the solubility for other metal oxides is increased. We have already discussed optical properties caused by polarizability in ►Section 1.6.5.

Today's glasses have the following exemplary compositions:

Silica glass (fused silica) represents almost pure silicon dioxide and contains more than 99 % SiO_2. It is highly temperature resistant, difficult to melt, and very stress-resistant. It is used in apparatus engineering and laboratory.

Lime soda glass is our glass for everyday use and consists of $Na_2O \cdot CaO \cdot 6SiO_2$ or 75 % SiO_2, 13 % Na_2O and 12 % CaO. It is made of silica sand, soda ash, and lime.

Calcareous lime glass is composed approximately according to $K_2O \cdot CaO \cdot 8SiO_2$. It is more difficult to melt and represents the Bohemian crystal or *crown glass*. Crown glasses have low light refraction with low dispersion. Recently lanthanum addition in crown glasses achieves high light refraction at low dispersion.

Thuringian glass is a mixture of the two last-mentioned grades and is to be regarded as a soda-lime glass.

Lead glass replaces CaO and parts of SiO_2 with PbO, BaO, ZnO, and Na_2O with K_2O. In *crystal glasses*, these heavy oxides must account for at least 10 % of the mass; *lead crystal glass* must contain at least 24 % PbO, and the *high lead crystal glass* at least 30 % PbO. A typical lead crystal is composed as follows: 55–65 % SiO_2, 18–38 % PbO, 13–15 % K_2O.

The heavy elements significantly increase the refractive index (►Section 1.6.5 at p. 52), and a lead oxide content of up to 33 % additionally strongly enhances dispersion (flint glasses). Even higher contents increase light scattering. Due to the high dispersion, flint glasses have a strong fire. Recently, lead has been replaced by titanium, which increases the refractive index and dispersion as density decreases (for a discussion of the lead and titanium properties, see [32]).

Borosilicate, Jena, and Duran glass are glasses for particular purposes (laboratory, industry) and contain, besides SiO_2, further glass formers such as B_2O_3 and Al_2O_3. Boron reduces the coefficient of thermal expansion, making the glass more resistant to thermal shock, and aluminum reduces brittleness. The acid-resistant Duran glass used in the laboratory represents a mixture with the typical composition of 70–80 % SiO_2, 7–13 % B_2O_3, 2–7 % Al_2O_3 and 4–8 % Na_2O/K_2O, 5 % BaO/MgO/CaO.

3.9.1 Glass coloring

Similar to crystals, the addition of metals can also color amorphous *glasses* [27, 189]. The coloring proceeds by two mechanisms:
- In *ion coloring*, metal oxides are added to the glass mass, coloring it already in the melt. They dissolve in the glass.
- In *annealing coloring*, metals or II-VI compounds are added and then colloidally distributed in the glass mass. The color is formed only after repeated melting and cooling.

3.9.1.1 Ion coloring

As soon as a metal oxide dissolves in the glass mass in ion coloring, the hue appears. Since this hue depends only on the ion and the glass chemistry, it does not change further during melting or processing. ►Table 3.9 lists the colors that can be achieved with this oxide coloration and typical oxides; colorings of early glass are studied in [612, 613], [27, Chapter 7], [28, Chapter 2.8], [39], ►Table 3.11. It takes between a few grams and a few kilograms of metal oxide per 100 kg of glass (0.1–3 %). Since the color strongly depends on the chemical environment (basicity, coordination number), the given colors are only examples.

The cause of the color is, with a few exceptions, the well-known LF splitting of the *d* orbitals of transition metals. Since, compared to pigments, lower color intensity is sufficient for glass colorations, we also frequently observe the participation of *f* orbitals here. The added metal oxides dissolve when the glass melts, and the metal cations are incorporated into the structure of the glass. The metal acts as a color center via splitting its *d* and *f* orbitals in this crystalline environment.

There are two possible positions for the metal in the glass network. On the one hand, if it takes over the function of the network former (it replaces, e. g., Si), then it is incorporated at tetrahedral positions. On the other hand, if it substitutes a network converter like e. g., Na, it is built-in at octahedron positions. The positions taken depend on the metal, its oxidation number, and the composition of the glass. Cobalt cations, e. g., are tetrahedrally surrounded by oxide anions in silicate glasses and form CoO_4 units with the well-known familiar blue color. In contrast, a phosphate glass contains octahedrally coordinated cations with CoO_6 units, giving it a faint pink color (quite analogous to the cobalt chloride with and without water of crystallization).

It should be noted that, in contrast to crystalline materials with a precisely defined composition, the color of a glass flux varies with the composition of the glass since the chemical environment of the metal cations and thus the crystal field strength changes continuously. Therefore, ►Table 3.9 contains data on glass types and coordination numbers of metals for certain hues.

Glass can react with the added metals to new colored compounds. For example, in lead glass, tin(IV) oxide forms with lead the yellow pigment $PbSnO_3$ (lead-tin yellow),

Table 3.9: Ion colorations of soda-lime glass by metal oxides based on LF transitions in modern glass coloration [42, pp. 72], [32, 33, 39, 43–45].

Metal Oxide	Achieved Coloration
Iron(II) oxide FeO	Blue to green (in reducing environment, also IVCT Fe^{II}/Fe^{III}), light green (in phosphate glass)
Iron(III) oxide Fe_2O_3	Yellow to pink (coordination number 6), red–brown (coordination number 4)
Chromium(III) oxide Cr_2O_3	Green (coordination number 6, with ZnO formation of yellow-brown zinc spinels $ZnCr_2O_4$, with BaO/PbO formation of yellow chromates)
Potassium chromate(VI) K_2CrO_4	Yellow (coordination number 4)
Copper(I) oxide Cu_2O	Colorless to red (coordination number 6)
Copper(II) oxide CuO	Cool blue (coordination number 6; with SnO_2 formation of turquoise cerulean blue) yellow-brown (coordination number 4), green-turquoise blue (coordination number 4, in lead and borate glasses)
Cobalt(II) oxide CoO, also Co_3O_4 (decays to CoO)	Deep blue (coordination number 4), pink (coordination number 6, in phosphate and borate glasses)
Manganese(II) oxide MnO	Colorless to yellow (coordination number 4), orange (coordination number 6), purple (oxidation to $Mn^{3\oplus}$)
Manganese(III) oxide Mn_2O_3	Purple (coordination number 4)
Manganese(IV) oxide MnO_2	Cloudy brown (insoluble) or purple (formation of $Mn^{3\oplus}$)
Nickel(II) oxide NiO, also Ni_2O_3	Yellow (coordination number 6, with Li), purple (coordination number 4, with Na, K), blue (in ZnO-rich glasses)
Vanadium(III) oxide V_2O_3	Green (coordination number 6)
Vanadium(IV) oxide VO_2	Blue-green to blue (coordination number 6)
Vanadium(V) oxide V_2O_5	Colorless to yellow-green (coordination number 4)
Uranium(IV) oxide UO_2, Uranium(IV,VI) oxide U_3O_8, uranium(VI) oxide UO_3	Brown to Black (reducing atmosphere), Yellow to red (oxidizing atmosphere, by $U_2O_7^{2\ominus}$), yellow-green fluorescent (Anna yellow, by $UO_2^{2\oplus}$; Anna green, with copper) [607]
Antimony(V)oxide Sb_2O_5	Naples yellow (in lead and lead-boron glass)
Europium(III) oxide Eu_2O_3	Pink [608, 609]
Neodymium(III) oxide Nd_2O_3	Lavender blue to purple
Praseodymium(III) oxide Pr_2O_3	Green
Erbium(III) oxide Er_2O_3	Pink
Holmium(III) oxide Ho_2O_3	Light yellow
Samarium(III) oxide Sm_2O_3	Yellow (also IR absorber in glass)

which colors the glass yellow. Many of these side reactions do not occur in ordinary silicate glass; tin oxide then merely acts as an opacifier for white glass.

Glassware

At this point, the question may arise as to what causes the coloring of glass articles we use every day. We want to answer it briefly, even if there is little connection to art itself. For the interested reader, [33] offers more details.

Bottle glass

The chromophore of the well-known yellowish-brown bottle glass is the tetrahedral complex $[Fe^{III}O_3S]^{5\ominus}$ ("iron-sulfur amber"). A LMCT transition $S^{2\ominus} \rightarrow Fe^{3\oplus}$ leads to a strong absorption at 400 nm, i. e., in the blue region [610, 611]. We know a similar transition from iron oxide red (Mars red) or hematite. In the bottle glass, this transition is intense compared to LF transitions of other ion coloration.

For the production, sulfur and iron are added to the glass mass in a reducing atmosphere (carbon) so that on the one hand, $S^{2\ominus}$ ions can form, and on the other hand, sufficient $Fe^{3\oplus}$ remains.

Another possibility for brown coloration is the production of colored polysulfides ("carbon amber"). In this process, sulfur is reduced by carbon to different colored polysulfides, which cover the color range yellow, red, and blue [611]. The colored species are $S_2^{\ominus}$ (absorption at 400 nm, red color) and $S_3^{\ominus}$ (absorption at 600 nm, blue color), which we already know from ultramarine blue pigments ►Section 3.3 at p. 227.

Since bottled glass is not produced from high-purity raw materials, the brown hue is modified by the blue-green coloring $Fe^{2\oplus}$ ions present.

Pure blue bottle glass is produced with cobalt oxide Co_3O_4, added in traces (typically 300 ppm).

Green bottles can be produced with the help of 0.2 % by weight chromium oxide Cr_2O_3. A reducing atmosphere is essential; otherwise, oxidation of chromium to chromate induces a yellow-green hue. Depending on the desired green shade, FeO or CoO is added.

Float glass

A transparent float glass must be produced from pure, iron-free raw materials. Depending on the ratio Fe^{II}/Fe^{III}, there remains a blue-green (> 0.5) or yellow-green (< 0.3) color undertone, which we observe in large thickness at the edge of the glass. The green, bronze, or blue-grey float glass, often used in office buildings, is composed as follows:

For *green* float glass, additional iron oxide is added up to about 0.5 % by weight of iron. The ratio Fe^{II}/Fe^{III} is controlled by adding carbon as a reducing agent.

The beautiful *bronze float glass* is a more complicated composition. About 0.002 % by weight of selenium provides a pink color due to molecularly dissolved selenium. This pink color is attenuated by about 0.3 % by weight Fe_2O_3 in the blue, resulting in a red-brown color. Further addition of 0.004 % by weight Co_3O_4 attenuates in the red range to obtain the bronze tone in sum. There is also discussed a tetrahedral color center $[Fe^{III}O_3Se]^{5\ominus}$ with an LMCT transition $Se^{2\ominus} \rightarrow Fe^{3\oplus}$.

Blue-gray float glass contains more cobalt oxide so that the bluish color impression increases.

3.9.1.2 Annealing coloring

In contrast to ion coloring, the coloring in this process is not gained by colored ions dissolved in the glass but by colloidal particles in a size range of a few nanometers, suspended in the glass mass and colored by the development of surface plasmons, ►Section 1.6.4. The resulting colors are pure and characterized by sharp absorption peaks. The position of the peaks and the perceived color impression depends on the size and composition of the particles.

Precious metals (copper, silver, gold) were recognized early on as possible color carriers, so we find yellow, red, and also luster glazes on Abbasid glass from the ninth century, colored with colloidal copper and silver [614]. The required fine distribution of the metals can be achieved by reducing soluble metal salts (chlorides, nitrates) to the element. Sn^{II} can be used as a reducing agent $SnCl_2$:

$$2\,AuCl_3 + 3\,SnCl_2 + 6\,H_2O \longrightarrow 2\,Au + 3\,SnO_2 + 12\,HCl$$

The late medieval genuine solid-colored glass appears yellow by colloidal silver. It was made by applying paint of silver oxide and clay to glass. When heated to about 600 °C, silver cations diffuse into the glass and are reduced by Fe^{II} [39]:

$$Ag^{\oplus} + Fe^{2\oplus} \longrightarrow Ag + Fe^{3\oplus}$$

The temperature profile can control the growth of metal particles to the required size.

In the past, Sb_2S_3, FeS, FeSe, CuS, or Mo_2S_3 were used as color carriers in addition to precious metals. In modern times, cadmium chalcogenides were added, ►Table 3.10. When II-VI compounds are used, a reducing atmosphere must be maintained so that the chalcogenides are not oxidized to the chalcogen oxides (e. g., SeO_2). For an intensive coloration, adding 0.1 % metal or 0.1–1 % of the mentioned compounds is sufficient.

The base glass in which the particles are dispersed is usually a potash-lime glass $K_2O \cdot CaO \cdot SiO_2$, a lead glass $K_2O \cdot PbO \cdot B_2O_3 \cdot SiO_2$ or $K_2O \cdot ZnO \cdot SiO_2$.

Interestingly, the annealing color is often not immediately formed when mixing the components, but only when the initially colorless blown glass is heated again to

Table 3.10: Annealing coloring of glass by colloidally distributed metals and compounds [33, 34, 121].

Colorant	Hue	Colorant	Hue
Gold	Red (gold ruby glass)	Bismuth	Red-brown
Copper	Red (copper ruby glass)	Lead, antimony	Gray
Silver	Yellow	Tin, cobalt	Brown
Selenium, platinum	Pink		
CdS · ZnS	Light yellow	CdS	Yellow
CdS · CdSe	Orange	CdSe	Red
CdSe · CdTe	Dark red		

450–500 °C for about 30 min. Here, the following processes take place [34, Chapter 21, 22]:

1. When heated for the first time, a supersaturated solution of the chromophore forms in the glass melt, which is still colorless.
2. When the melt is cooled, crystallization nuclei are formed in the order of a few hundred atoms. These are too small to be colored.
3. During reheating, the nuclei grow by recrystallization and coagulation up to the size of colloidal particles, and new ones form. When the particles are sufficiently large enough, the color develops. The growth must then be stopped in time by controlling the temperature curve because particles above a specific size are no longer colored but only show the gray-silver metallic luster.

3.9.2 "Decolorization" of glass, color compensation

The glass obtained from most natural raw materials is greenish, as historical glass (forest glass) demonstrates. The color is induced by iron cations contained in the raw materials, up to 1.5 %, especially in the sand used. During weathering of the original rocks, they decompose into iron-rich feldspars and mica, leading to accumulations of iron and heavy metals in sands from rivers or beaches. Iron is predominantly present as Fe^{III}, but during the melting of the glass, an equilibrium between Fe^{III} and Fe^{II} is reached, depending on oxygen content. Since Fe^{II} imparts the glass flow a distinct blue-green color, and Fe^{III} gives a pale yellow shade, the combined result is green glass with a yellow or blue tint.

Manufacturers must use iron-free material to avoid coloring. Therefore, Venetian glass masters used low-iron pebbles from the Ticino, as mentioned before. If this is not possible, adding manganese dioxide MnO_2 can achieve clarification. Its effect relies on the decomposition into manganese(III) oxide and oxygen, oxidizing Fe^{II} ions:

$$2\,Mn^{IV}O_2 \longrightarrow Mn_2^{III}O_3 + O$$

$$2\,Fe^{II}O + O \longrightarrow Fe_2^{III}O_3$$

In this way, not only is the distinct blue-green coloring replaced by the weaker yellow color, the purple Mn^{III} compensates for the yellow tint (complementary color pair). However, it is not possible to use the natural Mn^{II} content of raw materials, as it does not oxidize to the necessary Mn^{III} in the glass flow.

The slight gray coloration resulting from the color compensation is imperceptible due to the lack of a comparison with genuinely transparent glass. Since pure gray only occurs when the concentrations of ions match, manganese dioxide's amount must exactly match the glass. Therefore, raw materials as iron-free as possible are an essential prerequisite for producing transparent glass.

It is possible to carry out the oxidation of Fe^{II} with other oxidizing agents such as arsenic(V) oxide or antimony(V) oxide. Arsenic(V) oxide, e. g., forms at moderate furnace heat from added arsenic(III) oxide (arsenic) and oxygen. Then, at higher temperatures, it decomposes again when oxidating Fe^{II}:

$$As_2^{III}O_3 + O_2 \longrightarrow As_2^{V}O_5$$

$$4\,Fe^{II}O + As_2^{V}O_5 \longrightarrow 2\,Fe_2^{III}O_3 + As_2^{III}O_3$$

Here, the interfering coloration is not compensated to gray as when decolorizing with manganese dioxide but only changed into the weaker yellow color of Fe^{III}.

3.9.3 Ancient glass coloring

Knowledge of metal oxides for glass coloring was already available in the Bronze Age and led to decorative glass jewelry and containers [27, 28, 32]. The colorants known in antiquity are shown in ►Table 3.11 and follow essential principles of today's colorants.

Manganese

Mn^{II} has a faint pink, Mn^{III} an intensive reddish-purple to purple coloring. In medieval enamel colors, manganese appears as *manganese violet* for the incarnate. Mn^{II}/Mn^{III} is always present in an equilibrium whose position depends on the composition of the glass and the melting conditions. Iron can be used to eliminate an undesirable purple color:

$$Fe^{II} + Mn^{III} \longrightarrow Fe^{III} + Mn^{II}$$

Table 3.11: Coloring additives (metals and metal oxides) in ancient glass [28, Chapter 2.8]; t = transparent glass mass, o = opaque glass mass.

Colorant	Hue	Temporal Occurrence
Cu^{II}, $CaCuSi_3O_{10}$ (colloid)	Blue-green (t), blue-green (o)	Egypt, ancient empire
Cu^{II}, Pb^{II}	Emerald green (t)	1350 BC
Cu^{II}, $Ca_2Sb_2O_7$ (colloid), $PbSnO_3$ (colloid)	Turquoise (o)	1427 BC
Cu^{II}, $Pb_2Sb_2O_7$ (colloid)	Green (o)	1550 BC
Cu_2O (colloid), Cu (colloid), reducing furnace	Red (o/t)	1427 BC
Co^{II}	Blue (t)	1390 BC
Co^{II}, $Ca_2Sb_2O_7$ (colloid), $CaSnO_3$ (colloid)	Blue (o)	1479 BC
Fe^{II}	Blue-green (t)	Start of glassmaking
Fe^{III}, Mn^{II}	Yellow-green (t)	Start of glassmaking
Fe_3O_4 (colloid), reducing furnace	Black (o)	1425 BC
Fe_2O_3 (colloid), oxidizing furnace	Red (o)	1450 BC
FeS (colloid)	Yellow (t/o)	1479 BC
Mn^{III}	Purple (t)	1427 BC
Pb^{II} (Sb ?)	Yellow (t)	
$Pb_2Sb_2O_7$ (colloid)	Yellow (o)	1550 BC
$Pb_2Sb_2O_7$ (colloid), $PbSnO_3$ (colloid)	Yellow (o)	1391 BC
$Ca_2Sb_2O_7$ (colloid)	White (o)	1427 BC
SnO_2 (colloid)	White (o)	2nd century BC

Copper

Since the Bronze Age, copper additives have been used to produce blue, green, and red glasses. Cu^{II} creates blue or green, in soda-lime or soda glasses, turquoise, while Cu^{I} creates red by forming the pigment Cu_2O. Metallic copper leads to a deep red color by creating a copper colloid, e. g., from CuO and reduction with Fe^{II}.

Cobalt

Cobalt ores (so-called *zaffer*) were used for blue colorings. In alkali glasses of the early period, only tetrahedrally coordinated cobalt with blue color was present.

Carbon/sulfur

It is interesting to note that iron polysulfides have been known as colorants since early times, in addition to metal oxides. The yellow to amber coloration with carbon and sulfur known as *carbon amber* in bottle glass (►p. 300) was already found in ancient glass. It is caused by the polysulfides $S_2^{\ominus}$ and $S_3^{\ominus}$, which form by reducing sulfur with carbon in the glass mass. They lead to yellow and, in connection with Fe^{II}, to green glasses.

Iron

Iron is introduced into the glass mass by the raw materials and was therefore a common coloring component, ►Section 3.9.2. It colors pale-yellow as Fe^{III} and blue as Fe^{II}, but we always find an equilibrium between both valence states. The equilibrium's position depends on the composition of the glass mass and the manufacturing conditions. Therefore, it usually induces a yellow or blue-tinged green coloring. We can see this particularly well in the greenish forest glass of the Middle Ages (at least from today's point of view).

Lead

High amounts of lead(II) oxide PbO result in lead glass with a low softening temperature of about 750 °C and high brilliance. Lead glass can be yellow, especially when tin oxide or antimony oxide are added. The compounds lead stannate (lead-tin yellow) $PbSnO_3$ and lead antimonate (Naples yellow) $Pb_2Sb_2O_7$ form in this process. When combined with copper oxide in an oxidizing furnace atmosphere, lead glasses assume emerald green coloring and red hues with colloidal copper or Cu_2O. CuO, in combination with Sb_2O_3, induces a turquoise coloring.

Precious metal colloids

Already antiquity practiced coloring with annealing colors. Colomban [614] describes beautiful examples of yellow, red, and luster glazed glasses from the ninth century. They possess a coloration of colloidal copper and silver. The luster is caused by an ordered multilayered structure of colloids and other glass components. The reduction of the metal salts to the metal is achieved by multivalent redox elements such as Fe, Sn, Bi, or Sb, which quickly diffuse into the melting glass and can propagate the effect of the reducing furnace atmosphere.

Colloid glazes, ninth century, Th. Deck [614]
ocher, vinegar, 10 parts CuS, 1 part Ag_2S, 5 parts FeS
or ocher, vinegar, 5 parts CuS, 4 parts $AgNO_3$, 1 part iron oxides
or ocher, vinegar, 8 parts CuO, 6 parts iron oxides
or ocher, vinegar, 5 parts CuS, 2 parts SnO_2

The role of the acetic acid is to dissolve PbO contained in the lead glass, and thus etch the glass. The metal compounds can penetrate the porous structure and become reduced to spherical metal particles during melting.

3.9.4 Frit colors

Colored glasses were mainly made into everyday objects such as cups or jugs. Only a few colored glasses were also of interest as pigments for painters when ground, e. g., smalt, a potassium silicate glass colored blue by cobalt. Any colored glass flow can be ground, but the amounts of chromophores (ions) in the glass mass are low, and the color-generating LF transitions are weak due to spectroscopic selection rules. Accordingly, the coloring power of such a glass powder is too low for use in painting.

Smalt (PB32, CI 77365)

Smalt is coarsely ground blue potassium silicate glass [48, Volume 2], [418, 419] with a composition by weight percentages of about 65–72 % SiO_2, 10–21 % K_2O, 2–18 % CoO, 0–8 % As_2O_3. Blue cobalt glass was used since antiquity (Eridu, Mesopotamia around 2000 BC, Egypt about 1400 BC, and Nineveh around 650 BC) until the Roman seventh century. Between 850 and 1490, there is a gap in the occurrence of blue glass. Nonetheless, it again experienced a significant upswing in the fifteenth century in Venetian glassmaking. Smalt as a cheap pigment is traced back to a Bohemian glassblower around 1550 but was also found earlier in paintings. The main period of use dates to around 1600–1800, but the pigment was quickly displaced as soon as the intensive counterpart, Prussian blue, was discovered.

In potassium glass, the chromophore $Co^{2\oplus}$ or CoO_4 exists. Thus, LF transitions in the tetrahedrally coordinated cobalt cation cause its color, ►Section 2.3 and ►Figure 3.37. Despite the symmetry-allowed and thus intrinsically intense absorption, smalt is a transparent pigment of weak coloring power due to a low cobalt concentration in the glass matrix. The color saturation depends strongly on the particle size. If ground too finely, the color fades; coarse grinding reduces opacity considerably. Further problems are a low refractive index of 1.46, the rapid sedimentation of the coarsely-grained pigment in the oil medium, and grayish-green discoloration caused by air pockets between pigment and binder, e. g., upon decomposition of the binder. On top of that, smalt tends to fade, which has been known for a long time, ►Section 7.4.11.

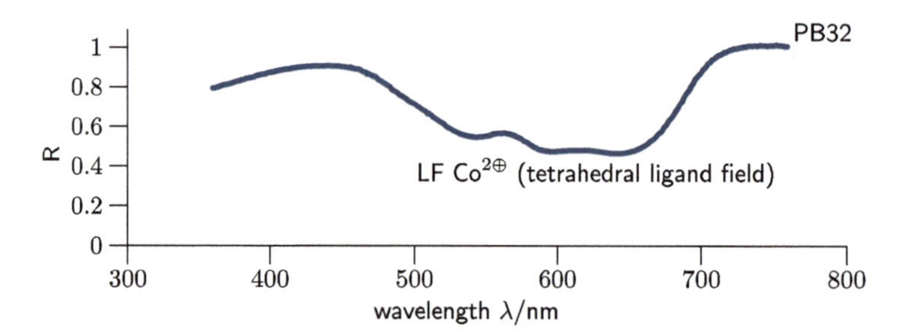

Figure 3.37: Reflectance spectrum of glass-based watercolors, normalized to an arbitrary unit: PB32 (smalt, Wallace Seymour Watercolour Whole Pans, cobalt silicate glass). The blue cobalt glass pigment show an intense chromogenic LF transition of the tetrahedrally coordinated cobalt ion in the yellow to the red spectral range, yielding blue color. Compare to ►Figure 3.25.

3.9.5 Opaque glass

Opaque glass is not used as an (artists') pigment, but we want to address it for complete coverage of colored glasses. Tiny particles are embedded in the glass mass; their refractive index differs from the glass. Typical substances are $Ca_3(PO_4)_2$, CaF_2, NaF, Na_3AlF_6 (cryolite), also white opacifiers like SnO_2, ZrO_2, or TiO_2 are possible. Ancient white glasses were colored with SnO_2 or $Ca_2Sb_2O_7$. In higher concentrations, such opaque glasses lead to the enamel.

3.10 Enamel

The transition from a highly pigmented opaque glass to enamel is fluid. Well-known enamel products are the white milk jugs and pots of the "granny ware." Here, silicate-oxide masses are fused onto a metal body and colored white throughout.

Enamel is glass that is applied and melted onto the metal surface to protect *metal objects*. According to its intended purpose, enamel consists of glass-forming oxides such as SiO_2, B_2O_3, Al_2O_3, and NaF, KF, and AlF_3. However, the added pigments often serve not for opacity but the opaque coloring of the glass mass. Opaqueness is caused by tin oxide, titanium silicate or oxide, and antimony trioxide; colorations come from the oxides of copper, iron, chromium, cadmium, cobalt, nickel, and manganese analogous to the glasses.

A typical composition is 20–35 % borax, 25–50 % feldspar, 5 % fluorite, 5–20 % quartz, 6 % soda ash, 2–5 % sodium nitrate, 0.5–1.5 % oxides of cobalt, nickel or manganese.

4 Organic colorants

Until modern times, artists who wanted to go beyond the dull earth colors had to use either expensive (ultramarine) or toxic (realgar, arsenic sulfide, lead compounds) minerals for pure hues, if they were known or available.

Therefore, artists directed their attention to bright colors in the plant and animal kingdoms [23–25, 108, 303]. They could profit from the knowledge of dyers, because from the earliest times, humankind used a plant basis to impressively dye skin, hair, clothing, and utensils in many colors. Dyes were so important that, e. g., in antiquity, cities and people flourished (and sank back into insignificance) because of their dyeing abilities.

To what extent and when artists adopted knowledge of the dyers and used organic colorants is still unclear. Archeological finds from antiquity are rare. Nevertheless, we know examples of pigments based on purple, madder, and indigo, later yellow flavonoids, red and blue anthocyanins, or brown bark extracts.

The fact that we can only provide scant information about this shows one serious shortcoming of the colorful natural palette: the lightfastness is often poor. Some colors already begin to fade after a few weeks, and most colors are destroyed after a few decades. In many old samples of artwork or clothing, we can only identify remnants of the colorants by using state-of-the-art analysis.

It is not a disadvantage that most natural colorants are primarily suitable for *dyeing* due to their solubility in the painting medium and lack of body; artists found ways and means to produce applicable pigments from dyes already in antiquity. They include precipitation of the dye on chalk or clay and laking.

However, we have to look at natural organic colorants from a historical point of view since practically all of them have disappeared from the palettes in our time. Consequently, we will see in ►Section 4.2 that completely new chemical structures have replaced most of them. As a result, only a few structures have remained. These earlier complex mixtures of natural substances are now synthetically produced and in pure form. Whether the colorants are of natural or artificial origin, they all rely on the MO mechanism. We will explore this vital process in the following sections.

Remarks on commercial artists' colors

There are a plethora of organic pigments on the market, but only a few are relevant for the minute share of artists' paints. To give an impression of how significant a pigment is for artists, we add examples of commercial artists' paints next to the pigment designation, if the pigment is the single chromophore in the paint:

https://doi.org/10.1515/9783110777116-004

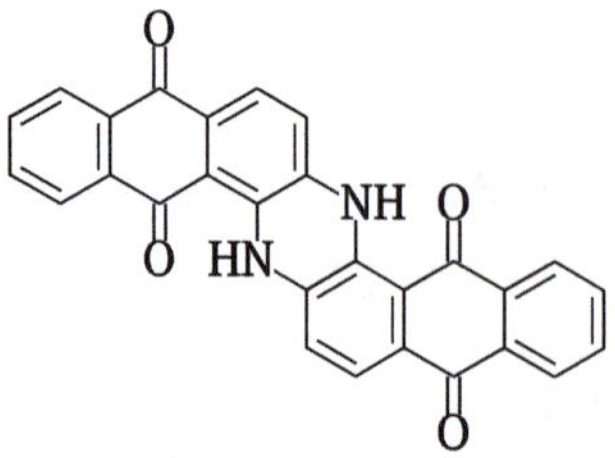

PB60, CI 69800
indanthrone
Delft blue [SC], indanthrene blue [WN]
indanthrone blue [DS]

This list means that PB60 is marketed as "Delft blue" by Schmincke, "indanthrene blue" by Winsor and Newton, and "indanthrone blue" by D. Smith, and these paints are single-pigment paints. Some pigments are widely used by manufacturers, expressed by a long list of names, while others are only rarely used. Considered are oil colors, watercolors, and acrylics; brands are Schmincke in Germany (designation [SC]) [985], Winsor and Newton in England (designation [WN]) [984], D. Smith in the US (designation [DS]) [983], Kremer in Germany (designation [KR]), and Wallace Seymour in England (designation [WS]). The selection of brands is arbitrary and based on availability in regular stores for artist supplies in the author's vicinity. [KR] is a manufacturer of restorer material, selling unprocessed pigments, and [WS] sells pigments no longer in common use in its vintage line.

4.1 Natural organic colorants

MO transitions are the mechanisms underlying most natural organic colorants. We will take a closer look at the following classes of compounds, ►Figure 4.1:

- Polyenes constitute the first class. Cyclic annulenes fulfill biological functions of fundamental importance for life (oxygen transport, energy supply by light). However, they have no significance for painting if we disregard the unintentional coloring of clothes with grass stains and some food colorants. More important are the linear carotenoids, ►Section 4.3. However, they are food colorants rather than pigments.
- Polymethines, the second class, are significant to art in the form of flavonoids (►Section 4.4) and xanthones, ►Section 4.5. These groups and the related betalains are responsible for most flower and fruit colors and have often found application in more or less lightfast dyes and pigments.
- Donor–acceptor systems form the third class. They occur mainly in quinones (►Section 4.6), besides natural indigo and purple, ►Section 4.7.1. The number of colored quinones that exist in nature is significant; they appear, e. g., when fruit

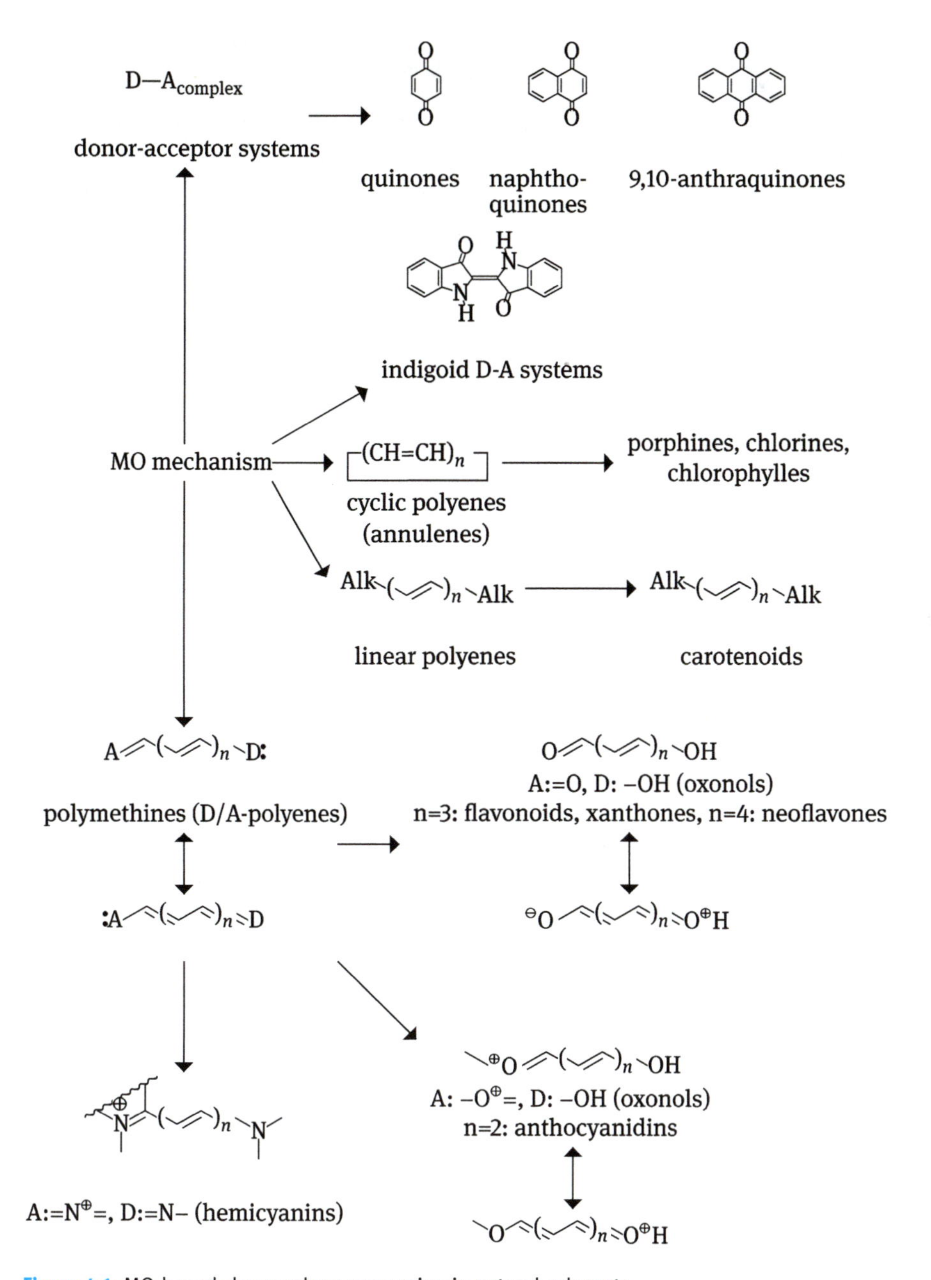

Figure 4.1: MO-based chromophores occurring in natural colorants.

and bark turn brown, organic material decays, or in dark insect carapaces. However, these compounds are unsuitable for artistic purposes. In contrast, simple quinones with one to three benzene rings, such as alizarin or carmine, have been used for dyeing and painting since antiquity.

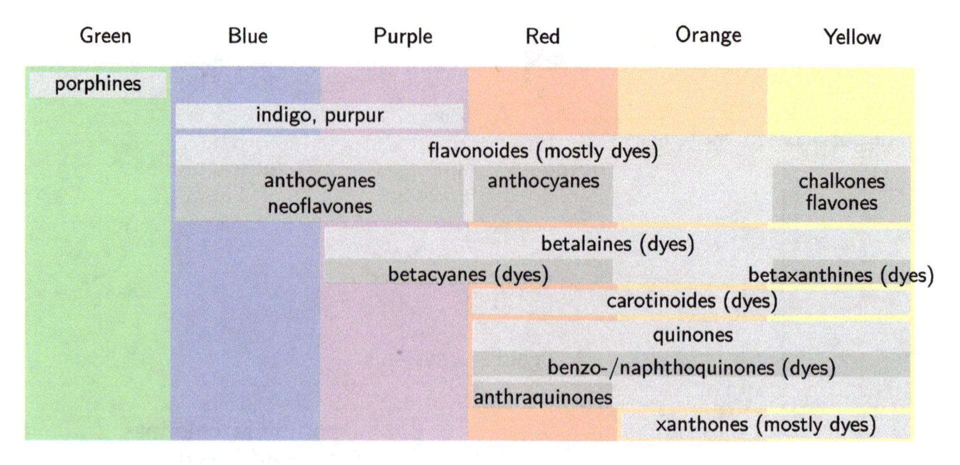

Figure 4.2: Coverage of hues by natural organic colorant classes.

►Figure 4.2 displays the achievable coverage of the spectrum.

4.2 Synthetic organic colorants

This section will look at organic colorants, which were first made possible by the development of organic chemistry. Then we will again meet the structure of some natural colorants, which are used in pure, synthetic form nowadays. However, fully synthetic colorants with fundamentally different structures are employed far more frequently.

With the beginning of scientific chemistry, especially organics, the number of organic colorants grew explosively in the nineteenth century. There are several reasons for this:

- Shades of color can often be changed in the desired direction by a slight variation of the chemical structure.
- New synthetic structures can open up previously unattainable or rare color ranges.
- The properties of synthetic colorants such as insolubility, body, lightfastness, and migration fastness can be optimized through structural variation.
- Synthetic colorants can be adapted to applications such as artists' paints, printing, airbrushing, and coating.
- Synthetic colorants are usually cheaper than natural substances.

In the meantime, organic colorants have succeeded in securing a firm place for themselves among the artists' paints. Many of these pigments have properties not inferior to those of inorganic pigments. However, the proof that they will still be in good condition after 500 years will still take some time.

Interestingly, even at the beginning of the twenty-first century with its highly developed chemistry, not all hues of the same high quality are available at a low price. It is still easier to find good yellow and red organic pigments than blue and especially green ones. Therefore, the color-bearing structures of many organic pigments already discovered in the twentieth century are nowadays further researched for improvement besides searching for new ones.

Hunger and Schmidt [12] give an excellent representation of the whole field of organic pigments. However, we will consider only those classes of colorants that provide contributions to the painter's palette according to ►Section 1.2. Lutzenberger [50] provides a good representation of this development and offers rich reference material on the actual use of organic pigments in painting.

Only a few of the organic chromophores presented in ►Section 2.5 became established industrially. It is instructive if we draw a comparison with ►Section 4.1, which considers natural organic colorants. Few natural structures remain unmodified; the chromophores are mostly introduced with newly developed structures. In detail, we will consider the following classes (►Figure 4.3 provides an overview):

- Polyenes of interest for us are polycyclic hydrocarbons and annulenes. The polycycles are often oxidized to quinones, ►Section 4.6.4. As a result, simple linear conjugated double bond systems such as carotenoids no longer play a role as colorants. Instead, phthalocyanines, which are the most significant blue and green pigments, follow the model of chlorophylls in nature and build on cyclic annulenes, ►Section 4.10.
- Polymethines play an essential role, especially with nitrogen and oxygen as donors and acceptors. Many polymethines are low-molecular-weight colorants from which colored inks can be produced. However, salt formation with suitable counterions also produces pigments, which due to their wide range of colors, are of some interest for fields of application such as printing inks (di- and triarylmethines, ►Section 4.8). The modern diketo-pyrrolo-pyrroles are suitable as pigments from the outset, ►Section 4.14.
- The bathochromic effect of natural donor–acceptor systems could be considerably increased by modern donor and acceptor groups and extended π systems. In this way, modern developments of purple, blue, and blue-green colors have been derived from many yellow and red-colored natural substances.

  Of interest to us are:
  - indigoide colorants with a hetero-substituted indigo parent structure, ►Section 4.7.2
  - quinacridones and azomethines as (phenylogous) donor–acceptor systems, ►Sections 4.12 and 4.15
  - azo colorants, representing modern complex donor–acceptor systems, ►Section 4.11
  - 9,10-anthraquinone and synthetic polycyclic hydrocarbons, ►Section 4.6.4

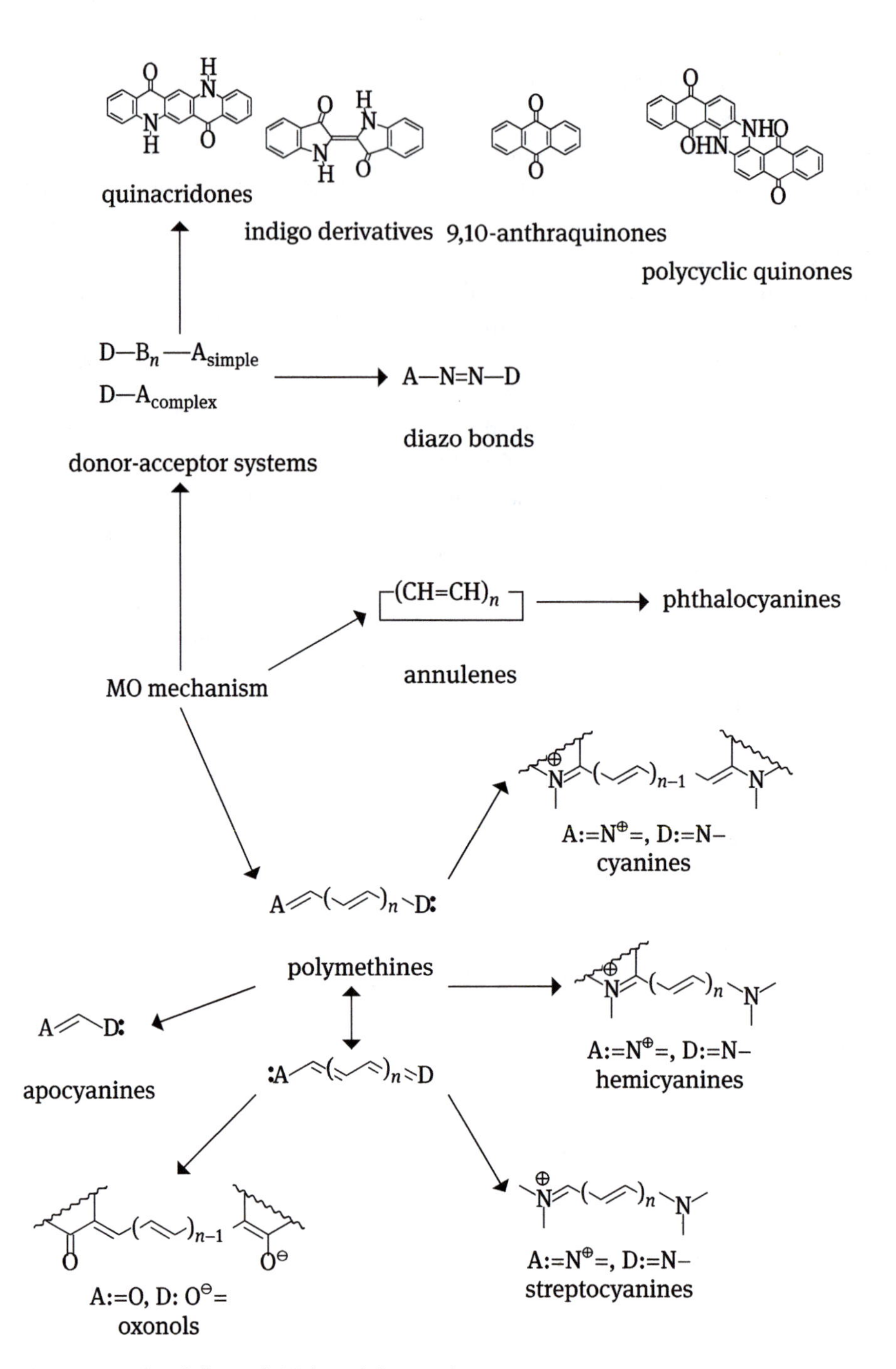

Figure 4.3: Industrially used MO-based chromophores.

Concerning quinones, only 9,10-anthraquinone has persisted as an essential component. Today synthetic polycyclic hydrocarbons are used as basic structures. In addition to bathochromism, the large molecules exhibit improved solvent and migration fastness, which improves their suitability as pigments.

The spectral coverage achievable with these colorants is given in ►Figure 4.4. Furthermore, [9, Chapter 7, Appendix D] contains reflectance spectra for some pigments.

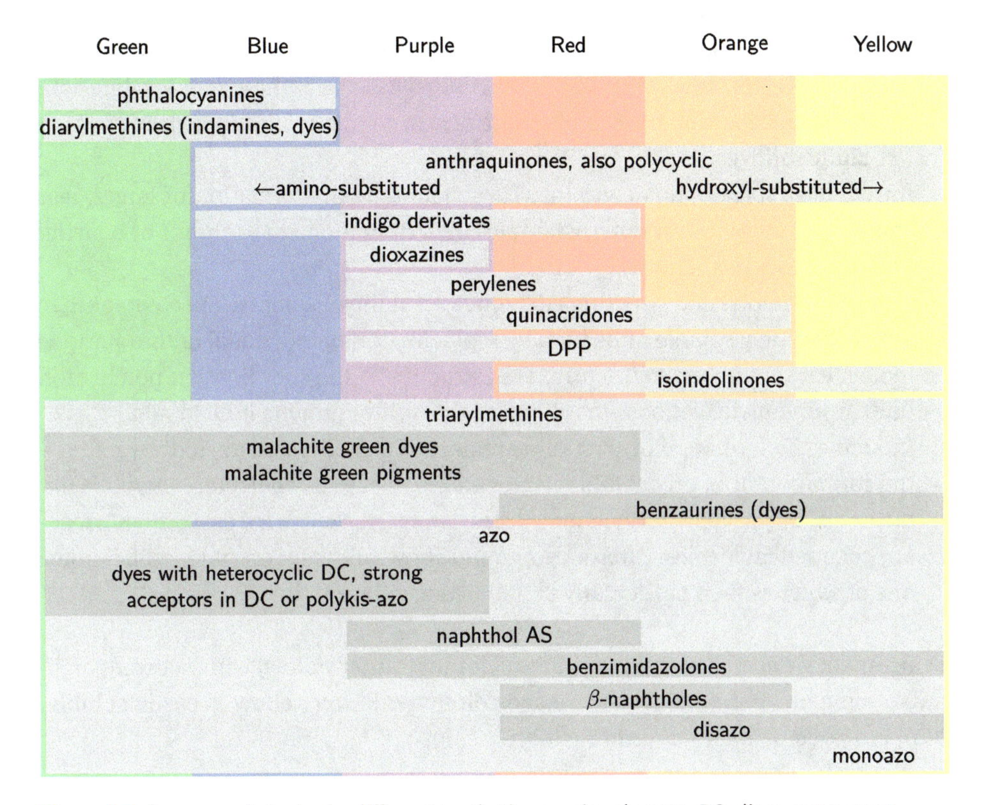

Figure 4.4: Coverage of shades by different synthetic organic colorants. DC: diazo component.

4.2.1 Meaning of molecular structure

In the case of synthetic organic colorants, chemists have tremendous freedom in designing colorant molecules. Once they have found a structure for the desired color, they can try to optimize further application properties by modifying the chromogen.

An essential property of a colorant is that its solubility is as low as possible in various solvents, during the painting process itself, and under the following conditions of use. Many synthetic pigments have disappeared from the market because they were more or less soluble in common solvents. In some cases, a low price can justify poor solubility properties. For mass-produced printed matter, e. g., it is often accepted that the printing ink can dissolve in many solvents. These pigments would be out of the question for art prints, which should have a long shelf life.

There are several possible ways to reduce solubility:

- Using large molecules, which are less soluble than small ones.
  - Replacing benzene systems with naphthalene. If the ring is part of a π system, this often shifts the color to blue.
  - Doubling of the colorant molecule by bridging two individual molecules. The size of the bridge, which itself contributes little or nothing to the coloration, can also be increased. Examples are disazo compounds with benzene or benzidine bridges.
- Introducing solvatophobic groups. These facilitate the formation of larger, less soluble pigment aggregations. Examples include carbonamide groups or benzimidazolone in azo colorants.
- Using supramolecular structures to reduce solubility. Examples are color lakes in which sparingly soluble units form by replacing small, individual anthraquinone molecules with a larger structure. This structure is the coordination polyhedron built around metal cations, coordinating the anthraquinone individuals.
- Inducing the formation of crystalline regions. Crystallinity leads to the preferential formation of poorly soluble aggregations. A regular molecular shape facilitates this, e. g., evenly-formed molecules can be stacked more easily than irregularly shaped molecules. Often a compound shows different crystal modifications, one of which is then particularly suitable as a colorant.

For example, we compare an early yellow colorant (butter yellow) with a current high-quality pigment. Both molecules are azo colorants. Butter yellow is easily soluble; PY126 is a good pigment for printing inks.

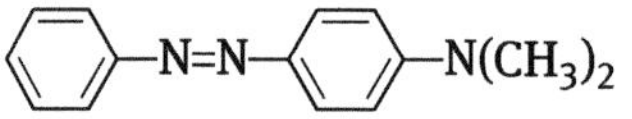

SY2
butter yellow
Winther: A→E

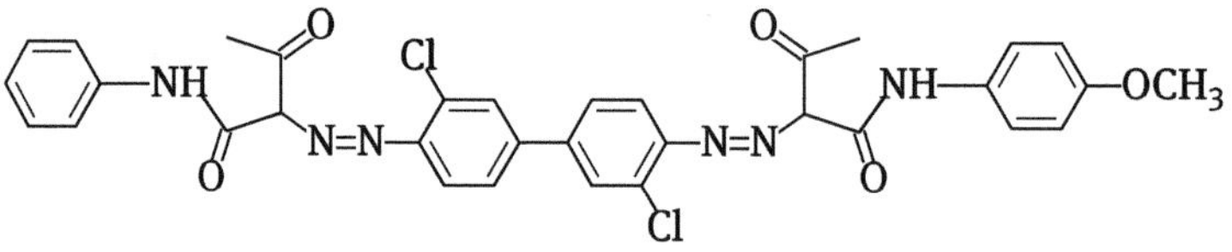

PY126, CI 21101
primary yellow for four-color printing

PY126 is a large pigment due to a bifunctional benzidine bridge and a doubling of the chromogenic structure. Most of the substituents are hydrophobic, and the well-established carbonamide group is used.

If chemists want to design a dye instead of a pigment, they must conversely achieve high solubility and adhesion to the substrate. Both can be achieved, e. g., by acidic or basic groups (see also ►Section 8.5.4). In the best case, the functional groups react with complementary groups of the substrate and form a more or less stable compound (reactive dyes). More commonly, weaker interactions occur, so they merely adhere to the substrate.

4.3 Carotenoids

The polyene chromophore (►Section 2.5.4.1) is perfectly implemented in nature in the *carotenoids*. They are polyenes comprising eight isoprene units (C_{40}-bodies) built up in the course of isoprene metabolism, ►Section 7.4.8 [271]. Carotenoids consist of an isoprene chain, which is almost always in the *all-trans* form, and which is terminated by various end groups, e. g., the β or ψ group:

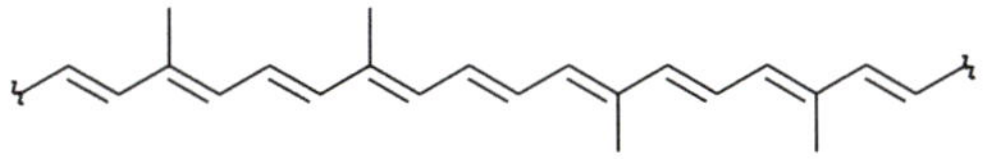

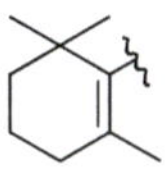

β-group 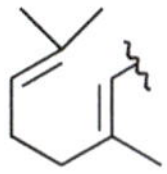 ψ-group

By chemistry, we distinguish two groups of carotenoids [753]:
- *Carotenes* are pure hydrocarbons, as shown. Although they are more or less insoluble in water, they have not gained any importance as pigments since they have insufficient properties for painting. Instead, their fundamental importance comes from the fact that they lend their intensive color not only to vegetables, e. g., yellow (corn, peppers), orange (carrots), and red (peppers, tomatoes) but also to many yellow flowers (examples: roses, tulips, gerberas) [737]. They also appear in green vegetables, but chlorophylls mask their intrinsic color.
- *Xanthophylls* also have oxygen functions such as hydroxy or epoxy groups, in addition to the hydrocarbon structure. For instance, individual oxygen-containing representatives were used as dyes in book illumination.

4.3.1 Xanthophylls

Yellow Orange Red In some branches of carotenoid biosynthesis, hydrocarbons are substituted with oxygen functions [271, 753]. We thus arrive at the *xanthophylls*.

Thanks to the diverse derivatization possibilities, hundreds of them can be synthesized in plants. Oxygen functions are hydroxyl groups, often as fatty acid esters, and carbonyl or ether groups. Especially hydroxylated xanthophylls are considerably more water-soluble than carotenes since they can form soluble glycosides with sugars.

For painters, they are more interesting than carotenes because they used some representatives as dyes because of their water solubility. Nowadays, we find xanthophylls only as food colorants, and every autumn in the bright yellow of the leaves, the latter contain especially violaxanthin, lutein, and neoxanthin besides β-carotene. The yellow compounds are not formed in autumn but constitute together with β-carotene essential chloroplast components. Therefore, they exist in the leaves throughout the year. In summer, the intense green of the chlorophylls masks the color, which gradually degrades toward autumn. However, besides autumn, this yellow color also appears in summer lettuce or cabbage in their inner yellow leaves because they contain etioplasts, stunted, unexposed chloroplasts, developing only the yellow color.

An important representative in the yellow range is *Crocetin*, the yellow dye of saffron *crocus sativus* (Natural Yellow 6). From a small South American tree, *bixa orellana*, we obtain an orange dye, anatto or Orleans (Natural Orange 4), utilized for dyeing since antiquity. In Europe, it has been known since the discovery of America:

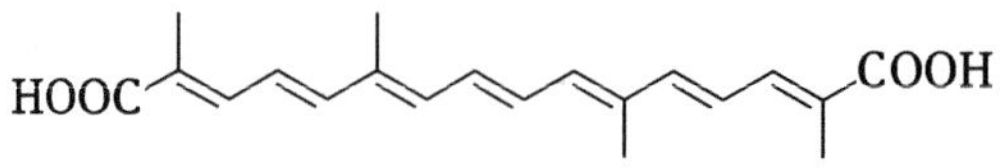

HOOC COOH

Natural Yellow 6
crocetin, historic colored ink

HOOC $COOCH_3$

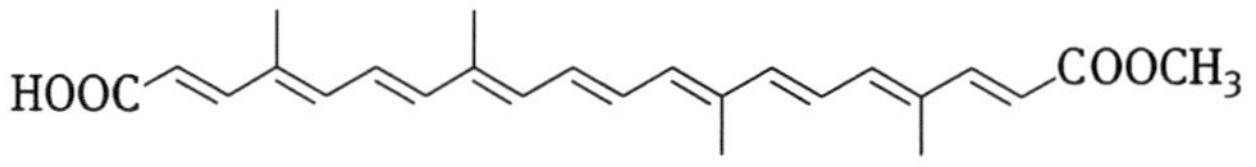

Natural Orange 4, E160 b
bixin, historic colored ink

Both carboxylic acids can be extracted with water from the stamens of the nowadays widespread saffron plant or the seed capsules of the anatto tree. In medieval Europe, illuminators of books and miniatures employed them as direct dyes, ►Sections 8.3.1 and 5.4, e. g., on old painted maps [697]. Since antiquity, they were essential for the yellowing of cotton as direct dye or mordant dye (►Section 5.5); with Al^{III} and Sn^{IV} salts, orange and yellow colors were achieved [70]. The other natural xanthophylls have no significance for painting or art.

4.4 Flavanoids

Flavanoids represent a significant group of plant constituents associated with the metabolism of phenolic bodies. They provide numerous colorants, which are divided into several subclasses according to structural characteristics. One subclass owns an intensely yellow color and is familiar under the name of *flavonoids* (lat. flavus = yellow). Cationic flavonoids, known as anthocyanidins, are intensively red, purple, and blue. For artists, flavanoids are relevant as colored inks (►Section 8.3.1), as mordant dyes (►Section 5.5), or as color lakes, ►Section 2.6.

Over 5000 different flavanoids appear in nature. Flavonoids lend many flowers yellow color, whereas anthocyanidins are responsible for almost all red and blue flowers, fruits, and vegetables [737]. Furthermore, due to their polyphenol structure, flavanoids have other effects on the metabolism, e. g., protection against free radicals).

The basic structure of all flavanoid compounds is flavane I (with the usual numbering of the rings):

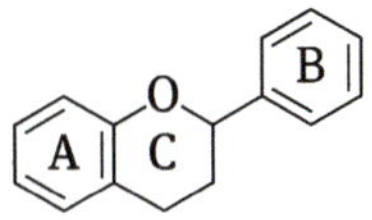

I, flavane

4.4.1 Origin in metabolism

In higher plants, flavanoids are synthesized in secondary metabolism from malonic acid and hydroxylated phenylpropene acids [294]. This exciting part of the metabolic process leads to various phenolic bodies, which form in plants, e. g., flavoring substances such as coumarin (woodruff) or pungent constituents (in pepper or ginger). Due to their importance for nutrition and the economy, they have long been the subject of research [716–732], [295, p. 285], [297, pp. 106, 151].

The primarily formed flavanone III can be transformed in several steps into different basic structures or subclasses, ►Figure 4.5. Interesting for our subject are flavonoids that predominantly occur as 3-mono- or 3,5-diglycosides in nature. As saccharides, predominantly glucose, galactose, glucuronic acid, galacturonic acid, xylose, rhamnose, and arabinose occur.

Oxidation converts flavanoids into o-quinones and further into higher aggregated substances, forming polymers of complicated structures and causing brown staining of overripe or cut fruit, ►Section 8.2 [719]. Brown wood and bark components were highly valued in the Middle Ages to prepare brown inks, ►Section 8.3. Oligomeric flavonoids induce the color of tea and cocoa and are important flavoring substances. They provide the typical astringent taste of tea and cocoa to a lesser extent.

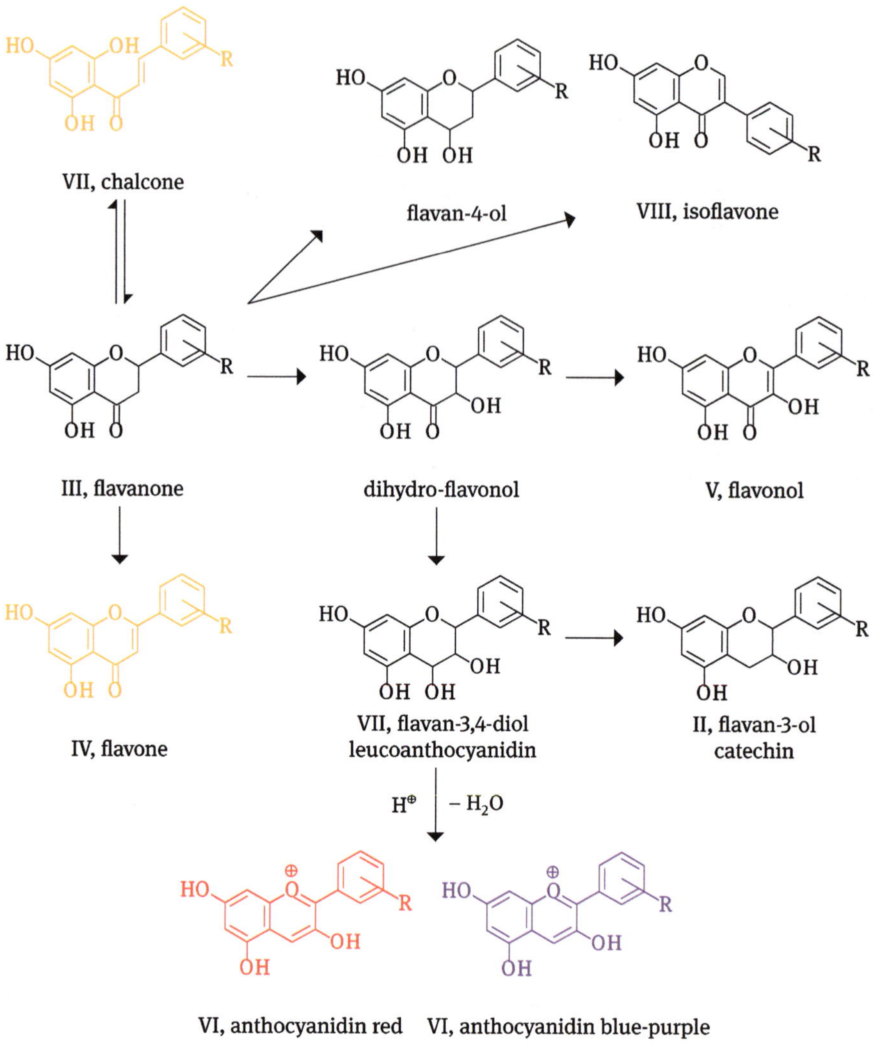

VII, chalcone

flavan-4-ol

VIII, isoflavone

III, flavanone

dihydro-flavonol

V, flavonol

IV, flavone

VII, flavan-3,4-diol
leucoanthocyanidin

II, flavan-3-ol
catechin

$H^{\oplus}$ $- H_2O$

VI, anthocyanidin red

VI, anthocyanidin blue-purple

Figure 4.5: Evolution of the various flavanoid subclasses from chalcone or flavanone [297], [299, pp. 106, 151], [716–718, 732], [295, p. 285]. The hydroxylation patterns shown are exemplary and may vary depending on the underlying phenylpropanoic acid and subsequent modifications.

4.4.2 Classification

As mentioned above, all flavanoids derive from flavane I. The natural substitution patterns of flavane predetermine subclasses. Their representatives possess similar properties and functions (for an overview of structural diversity and classification [295]):

- From flavane I, hydroxylated subclasses such as flavan-3-ols II are derived, the so-called catechins, or flavan-3,4-diol VII. They act as secondary plant constituents in fruits and are colorless. Flavanones III are also colorless, but they are significant substances due to their flavoring contribution to fruits and vegetables.

- The oxo-derivatives IV and V are the parent compounds of the yellow flavones, which are vital as flower, fruit, and artists' colors.
- Anthocyanidins are cations derived from flavane derivative VI. Their glycosides, the anthocyanins, provide the color shades for numerous red and blue flowers and were significant as ink dyes.
- The yellow open-chain chalcones VII are of no importance to the artist, with a few exceptions.

4.4.3 Flavan-3-ols (catechins), flavan-3,4-diols, and flavanones

Flavan-3-ol II is the parent compound of *catechins*, distinguished by the hydroxylation pattern at the benzene rings like other flavanoids. The oxidized compounds of flavanone III type also occur in numerous variants. Both subclasses are colorless and appear together with colored flavonoids and anthocyanins in fruits and vegetables; some typical representatives are

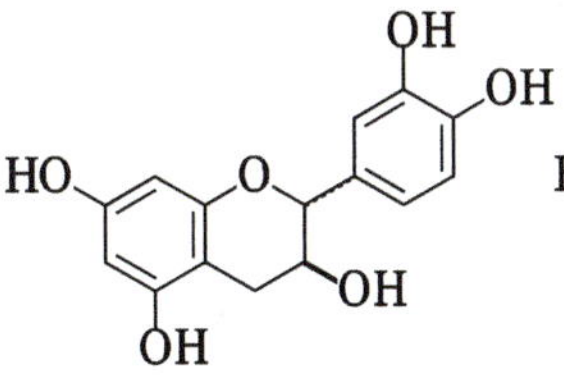

catechin 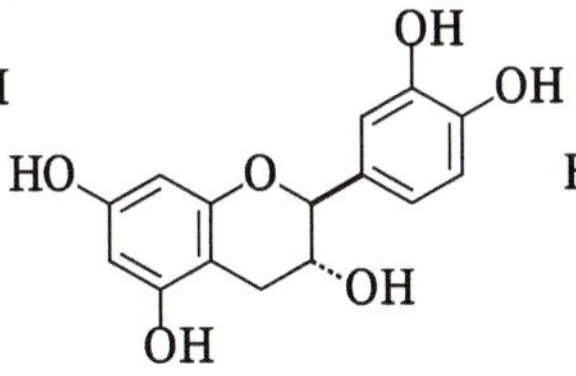 epicatechin 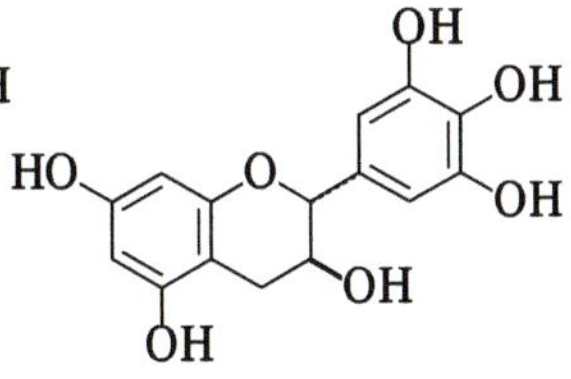 gallocatechin

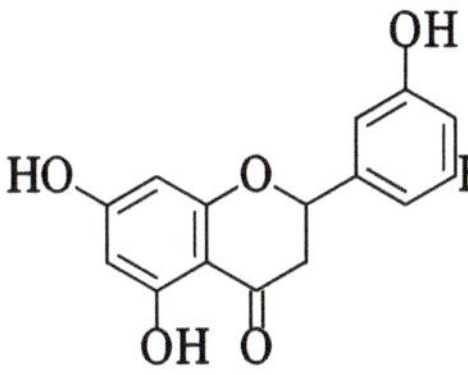

naringenin 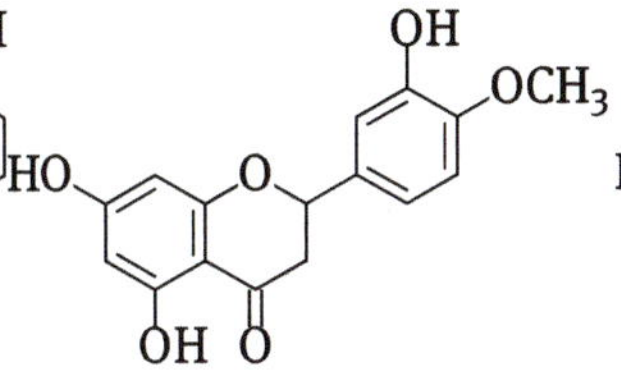 hesperitin 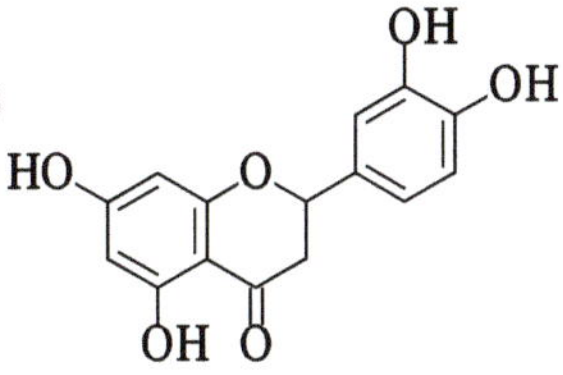 eryodictiol

Catechins build the so-called *condensed tannins* by enzymatic dehydrogenating polymerization. These oligomeric compounds contribute to the astringent taste of fruits and give black tea its red color. They are a part of the tannins in gall apples and similar tree secretions, which we encounter when making inks, ►Sections 8.3 and 8.4.

Flavanones occur as glycosides mainly in citrus and tropical fruits and represent vital flavor substances. The glycosides with neohesperidose, in particular, cause a bitter taste typical for grapefruit and bitter orange. A debittering of the fruits is possible by enzymatic cleavage of the sugar residues.

Black tea

Tea leaves contain large amounts of flavan-3-ols such as catechin, gallocatechin, and their esters with gallic acid [716, 719, 730, 733], [293, p. 203]. After harvesting, the tea leaves are broken for black tea extraction. In this way, an extensive enzymatic activity begins, allowing the colorless catechins partially to oxidize, ►Figure 8.6 in ►Section 8.2.4. The result of the fermentation is colored compounds. Together with the remaining catechins, they provide the main contribution to the color and flavor of black tea infusions:

- Yellow to orange theaflavins (dimeric catechins) are contained in black tea with about 10 % of the flavanols.
- Red to brown thearubigins (polymeric catechins) are retained in black tea with about 75 % of the flavanols. They are poorly known and possess a heterogeneous complex structure.
- Yellow and orange dicatechins, theasinensins, and theanaphthoquinones (complex dimeric catechins) are contained in various fractions.

We will look at these products as components of brown vegetable inks in ►Section 8.3.

In contrast to black tea, the oxidizing enzymes in green tea are inactivated after harvesting so that fermentation does not occur. As a result, the yellow-green color of green tea is preserved due to unchanged chlorophylls, flavones, and flavonols; see below.

4.4.4 Flavones

Yellow Flavone IV or flavonol V is the parent compound of numerous derivatives called flavones, belonging to the most common natural flavonoids. They are distinguished by the hydroxylation pattern at the benzene rings, as can be seen in some frequently occurring compounds:

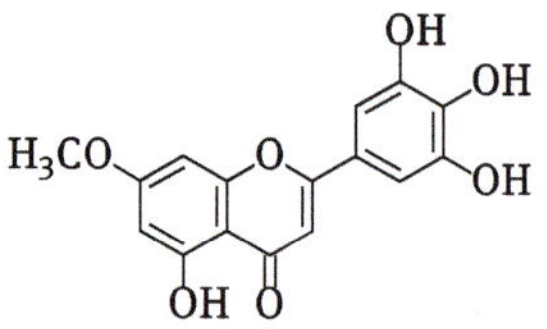

Natural Yellow 13, CI 75690
3,3',4',5-tetrahydroxy-7-methoxy-flavone, rhamnetin

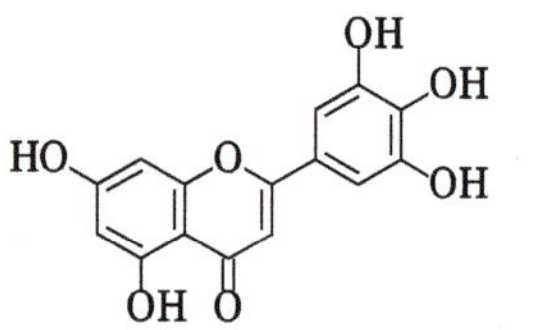

Natural Yellow 10,13, CI 75670
3,3',4',5,7-pentahydroxy-flavone, quercetin

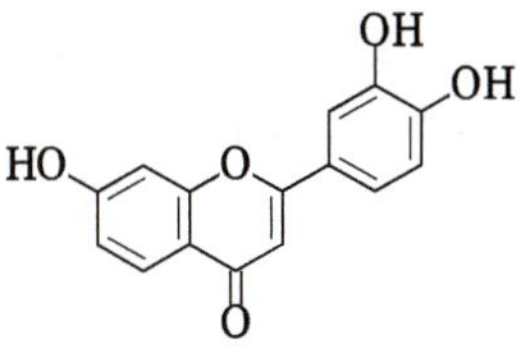

Natural Yellow 2, CI 75590
3',4',7-trihydroxy-flavone, luteolin

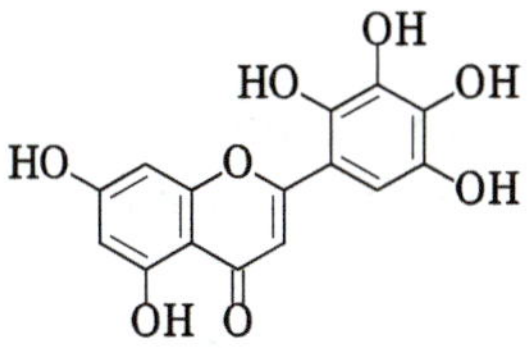

Natural Yellow 8,11, CI 75660
2',3,4',5,7-pentahydroxy-flavone, morin

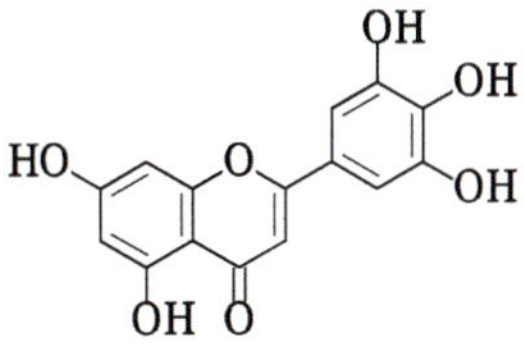

Natural Yellow 13,10, CI 75640
3,4',5,7-tetrahydroxy-flavone, kaempferol

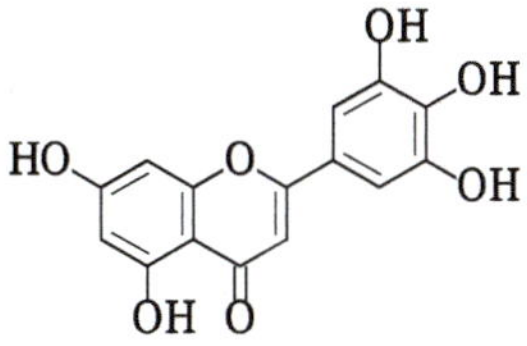

3,3',4',5,5',7-hexahydroxy-flavone,
myricetin

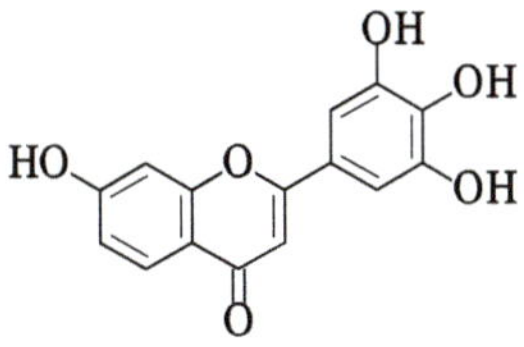

Natural Brown 1, CI 75620
3,3',4',7-tetrahydroxy-flavone, fisetin

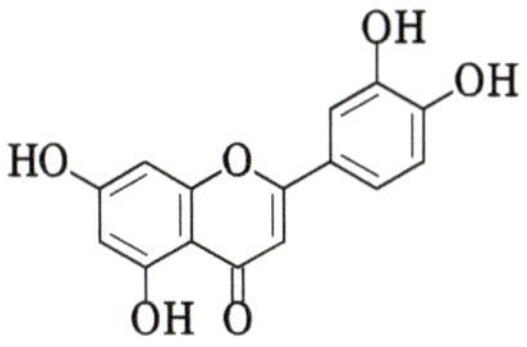

Natural Yellow 1,2, CI 75580
4',5,7-trihydroxy-flavone, apigenin

The number of hydroxyl groups predominantly determines their color. The mentioned flavones are light yellow to yellow; additional hydroxyl groups intensify the yellow color, such as e. g., in 6-hydroxy-quercetin (quercetagenin). In nature, the C^3-hydroxyl groups are present as glucosides, galactosides, and rhamnosides.

Flavones and flavonols occur as yellow dyes in the flowers of many plants; in fruits, leaves, and twigs, they also sometimes appear in appreciable amounts [303], [277, Chapter 18.1.2.5], [23, Chapter III.7]. Together with carotenoids and betaxanthins, they constitute one of the three major groups of natural yellow dyes. Lovers of white wines will appreciate quercetin and kaempferol as these substances play a decisive role in the color of white wines. However, closer to our topic is the use of some flavones for dyeing and book illumination.

Green tea

Flavanols represent up to 80 % of the phenolic compounds responsible for the bulk of tea leaves; the rest are flavones and flavonols. During the fermentation of fresh tea leaves, the flavanols are oxidized. Together with their reddish-yellow follow-on products (theaflavins and thearubigins), they contribute chiefly to the flavor and color of the black tea infusion. In green tea, the oxidizing enzymes are inactivated, so the yellow-green color of green tea is due to flavones and flavonols. In this case, their color is not eclipsed by the more intensive hue of theaflavins and thearubigins.

Yellow color lakes, Stil de Grain, yellowwood lake, Natural Yellow 13, 14

Many different colorants could be obtained from flavones. For example, yellow aqueous extracts were used as *sap dyes* in book illumination, ►Section 8.3.1. Historically, artists created yellow-colored lakes by extracting dyes from berries, barks, stems, or woods with water and precipitating them by adding alum and soda (sodium carbonate) [70]. The complex could also be precipitated on gypsum or chalk and applied in oily or aqueous binders. ►Section 2.6 shows details of laking; [94] is devoted to the in-house production of yellow plant-based color lakes. The following lakes were important for painting:

- Stil de grain was made from the unripe berries of *rhamnus* species containing rhamnetin. The Al lake of the extract of these berries was already used in the first century in dyeing [23]. ►Figure 4.6 depicts reflectance spectra from two batches of Stil de Grain watercolors.
- Yellowwood lake is made from extracts of dyer's mulberry trees containing morin.

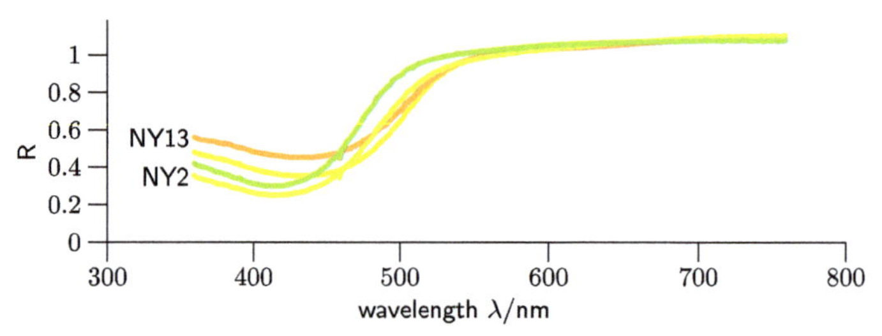

Figure 4.6: Reflectance spectrum of flavone-based watercolors, normalized to an arbitrary unit: NY13 (Stil de Grain, Wallace Seymour Watercolour Whole Pans, rhamnetin/quercetin/kaempferol-based pigment of unknown precise composition), NY13 (Stil de Grain genuine, Wallace Seymour Vintage Watercolour, see above), NY2 (weld reseda, Wallace Seymour Vintage Watercolour, luteolin-based pigment of unknown precise composition), and NY2 (weld yellow, Wallace Seymour Watercolour Whole Pans, see above). NY2 has a cooler, more yellowish color due to its steep absorption edge at slightly higher energies.

Several other yellow lakes are more of interest for dyeing, ►Table 4.1 and ►Figure 4.6 for some reflectance spectra. Bartl et al. [108] list a series of yellow lakes containing flavones, especially for book illumination, and [747] depicts numerous spectra of flavonoid extracts.

One disadvantage of yellow color lakes is their low lightfastness. They can decay within weeks so that yellow plant colorants have only been preserved in works of art that have been shielded from light, e. g., on the inner pages of books. Thick layers of

Table 4.1: Composition and origin of the flavones used for dyeing, and after laking, for painting [70, 108]. For use in book illumination, ►Section 8.3.1..

Origin	Composition	Metals and colors
Natural Brown 1, fisetwood *cotinus coggyna*	Fisetin	Al^{III} (yellow), Cr^{III} (reddish-brown), Sn^{IV} (orange), Fe^{III} (olive)
Natural Yellow 11, yellowwood, dyer's mulberry *chlorophora/morus tinctoria*	Morin, maclurin, kaempferol	Al^{III} (yellow), Cr^{III} (olive yellow), Sn^{IV} (lemon yellow), Fe^{III} (brown)
Natural Yellow 10, black oak *quercus velutina*, onion skin, St. John's wort	Quercetin	Al^{III} (yellow), Cr^{III} (olive yellow), Sn^{IV} (orange), Fe^{III} (olive green)
Natural Yellow 13, berries of a *rhamnus* species (buckthorn)	Rhamnetin, quercetin, kaempferol	Al^{III} (orange-brown), Cr^{III} (red-brown), Sn^{IV} (orange), Fe^{III} (olive)
Natural Yellow 2, weld, yellowweed, dyer's reseda *reseda luteola*	Luteolin, apigenin	Al^{III} (orange-yellow), Cu^{II} (yellow-green), Sn^{IV} (lemon yellow), Fe^{III} (olive green)
Dyer's broom *genista tintoria*	Luteolin, genistein	Al^{III} (pale yellow), Cr^{III} (yellow-green), Cu^{II} (yellow-green), Fe^{III} (chocolate brown)
Dyer's saw-wort *serratula tintoria*	Luteolin	Al^{III} (yellow-green), Fe^{III} (olive green)

paintings offer some protection, too. Saunders and Kirby [395, 396] provide examples of faded yellow lakes in which the fading occurs particularly noticeable in green mixtures. Flemish paintings with strikingly bluish trees and foliage immediately come to mind. As early as 1830, Mérimée noted that "in several Flemish paintings, leaves of trees have become blue, because the yellow lake, mixed with ultramarine, has disappeared." Especially in the seventeenth and eighteenth centuries, yellow lakes (Stil de Grain, Gamboge) were increasingly used instead of stable yellow pigments such as lead-tin yellow or Naples yellow. According to the studies mentioned above, lakes of *rhamnus* berries and black oak are more endangered than lakes of reseda. Today, yellow lakes have fallen into disuse.

Green lakes, sap green, Natural Green 2

During ripening, the berries of *rhamnus catharticus* increasingly develop anthocyanins in addition to flavones. Therefore, if we evaporate fresh fruit juice with alkalis such as potash and alum or lime, we obtain *sap green* [387]. However, it is not a green compound but a mixture of yellow flavones (mainly rhamnetin, as in Stil de Grain) with anthocyanins, which are blue in alkaline environments [490].

According to [108], green plant-based saps were often called sap greens regardless of their exact origin. Thus, *rhamnus* berries (buckthorn), privet, juniper, unripe blueberries, leaves and blossoms of iris, nightshade, rue, violets, and hyssop are named sources. The sap, often brownish at first, acquires a green color only when alum is added. In addition to flavones, other yellow components also contribute to the mixed green color in these cases. Irises, e. g., contain the yellow xanthone dye mangiferin (►Section 4.5), which, together with anthocyanins, leads to the mixed green color. In contrast, rue contains chlorophyll, a true green dye from the beginning, but it fades with time while the yellow flavonoid rutin persists.

Chemistry of color lakes (metal complexes)

The complexation of flavonoids with Al^{III} salts as a mordant for the soluble flavonoids had considerable economic importance for the dyeing industry. Until the discovery of America, dyer's reseda was the predominant yellow dye in Europe which was stained with Al^{III}. (Yellowwood came onto the market as a competitor around 1500, and black oak followed around 1770). However, the resulting complexes also formed insoluble yellow to red color lakes, which could be used as a pigment. Chemically, complexation causes a bathochromic shift of absorption bands from UV to near-UV or even the visible spectral range. Consequently, the lakes are intensely colored. In the wet chemical separation step, e. g., *morin* as a reagent reacts on Al^{III} with strong fluorescence [679].

There is no lack of research on the reaction [301, 677–682, 746]. The complexation of o-dihydroxybenzenes depends on the pH value of the solution. The pH value regulates the deprotonation of acidic phenolic hydrogen atoms, so bonding locations for the metal become free. For example, in the 3',4'-dihydroxyflavane, we find the complex formation in the B ring with stoichiometric ratios of Al^{III} to flavane 1:3, 1:2, and 1:1 (L represents the flavane):

$$\text{in methanol: } LH_2 + Al^{3\oplus} \longrightarrow AlL^{\oplus} + 2\,H^{\oplus}$$

$$\text{in potassium acetate: } 2\,LH^{\ominus} + Al^{3\oplus} \longrightarrow AlL_2^{\ominus} + 2\,H^{\oplus}$$

$$\text{in sodium methanolate } 3\,L^{2\ominus} + Al^{3\oplus} \longrightarrow AlL_3^{3\ominus}$$

If coordination positions on the metal are still vacant, solvent molecules as secondary ligands will occupy them.

If the hydroxyl groups dwell near the 4-oxo group, at C^3 or C^5, the 4-oxo group becomes part of the chelate ring. A hydroxyl group at C^3 is preferred. 1:1 and 1:2 complexes form, i. e., the metal can bridge multiple flavone molecules. In the case of multiple hydroxylations, several chelate rings can occur in the same flavone molecule. Under certain circumstances, complicated multinuclear structures also emerge in which solvent molecules or hydroxyl ligands bridge multiple metal centers. Again, these may be complex with several flavone molecules.

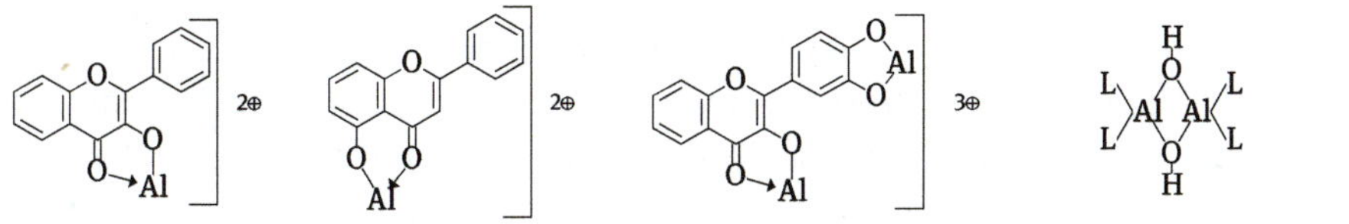

3-hydroxy-flavone 5-hydroxy-flavone 3,3',4'-trihydroxy-flavone multinuclear structure L=flavonoid

While complexation can proceed at all hydroxyl groups, a bathochromic shift of the UV-absorption into the visible spectral range only occurs at 3-hydroxy- and 5-hydroxyflavones since only then the resulting chelate ring influences the chromogenic system (ring B, $\Delta^{2,3}$, and 4-oxo group in ring C), ►Sections 2.6.3 and 4.4.9. The shift in the absorption band associated with metallicity is not pronounced for flavone complexes and does not extend beyond yellow. More decisive color shifts with *d* block metals such as iron and chromium are additionally possible due to LF and CT transitions, which we consider in more detail in the case of iron gall inks, ►Section 8.4.4. It is unclear to what extent the formation of quinoid structures from the flavones (collected in [746]) is involved in inducing colors beyond yellow.

4.4.5 Anthocyanins

Red Purple Blue Type VI flavonoids (2-phenyl-benzopyrylium or flavylium salts) are well known as *anthocyanidins*. In nature, they occur as glycosides, just like flavones. These glycosides are called *anthocyanins*. Their name is derived from the Greek *cyanos* (sky blue) because they are red, purple, or blue. The most common natural anthocyanidins are:

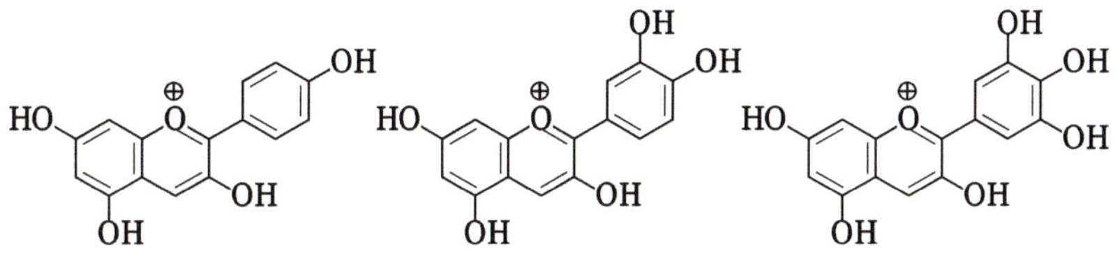

pelargonidin cyanidin delphinidin

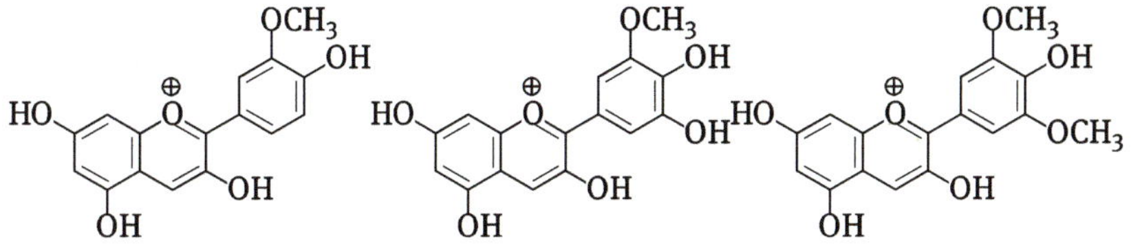

paeonidin petunidin malvidin

Essential for color is a hydroxyl group at one or more of the positions $C^{4'}$, C^5, or C^7 [293, Chapter 6.2]. The hydroxylation pattern influences the color shade considerably. For example, an increasing number of hydroxyl groups on ring B shifts the color from red to blue. Conversely, methylation or glycosylation changes the colors slightly, ►Table 4.2 [277, 293, 303, 734, 737]. Removing a hydroxyl group leads to a yellow shift of the color; at C^3 of the cyanidin, this results in the orange luteolinidin. Taking some fruits from ►Table 4.3 as an example, we recognize typical glycosylation patterns [277], [298, Chapter 3], [734]. Glycosidic linkages develop between the hydroxyl groups on C^3, and C^5, more rarely C^7, and $C^{4'}$, with glucose, but also galactose, rhamnose, xylose, or arabinose. Often the saccharides are esterified with hydroxycinnamic acid or hydroxybenzoic acid. According to [67], [298, Chapter 3], glycosylation at C^3 is important for the stability of the pyrylium cation. It is a prerequisite for plants to accumulate large amounts of anthocyanins. Methylation typically occurs at the hydroxyl groups of $C^{3'}$ and $C^{5'}$.

Table 4.2: Influence of hydroxylation pattern on the color of anthocyanidins [277, 293, 303, 734], [273, p. 446].

Compound	Absorption maximum λ_{max}	Substitution pattern in ring B
Pelargonidin (orange to brick red)	520 nm	–OH
Cyanidin (red)	535 nm	–OH, –OH
Paeonidin (pink)	532 nm	$-OCH_3$, –OH
Delphinidin (bluish-purple)	546 nm	–OH, –OH, –OH
Petunidin (purple)		$-OCH_3$, –OH, –OH
Malvidin (reddish-purple)	542 nm	$-OCH_3$, –OH, $-OCH_3$

Table 4.3: Examples of the occurrence of anthocyanins in fruits [277, 734].

Anthocyanin	Plant
Cyanidin-glucoside (Cy-3-glc)	Peach, plum, sweet cherry, blackberry, raspberry, currant, grape, orange, fig
Cyanidin-galactoside (Cy-3-gal)	Apple, pear
Paeonidin-glucoside (Pa-3-glc)	Plum, grape
Pelargonidin-glucoside (Pg-3-glc)	Strawberry
Petunidin-glucoside (Pt-3-glc)	Blueberry, banana, grape
Delphinidin-glucoside (Del-3-glc)	Blueberry, currant, grape, blood orange, passion fruit
Malvidin-glucoside (Mv-3-glc)	Blueberry, grape

Anthocyanins are the essential group of red and blue pigments in flower and plant colors [67, 303], [23, Chapter III.8]. Surprisingly, the high amount of red to blue colors observable in nature is mainly due to three anthocyanidins (pelargonidin, cyanidin, and delphinidin). Combining anthocyanins with yellow and red carotenoids expands the palette from yellow to orange, as ►Table 4.4 shows. Examples of flowers that owe their color to anthocyanins include cornflowers, roses, and pelargoniums [303, 737]. Crucial for the variety of achievable shades is the dependence of the anthocyanidins' color on environmental conditions. Conditions are pH value, concentration, patterns of esterification and glycosylation, and copigmentation.

Table 4.4: Typical anthocyanidins for certain flower colors [303, 737].

Pelargonidin	Cyanidin	Delphinidin
←pink→	←red–red-purple→←orange (with yellow dyes)→	
←————	red ————→	←blue→
	←———— blue-purple	————→

pH Dependence of color, stabilization and complexation

Anthocyanidins show a pronounced dependence of the color on the pH value and induce shades from red to purple to blue [24, Chapter 7], [67], [277, Chapter 18.1.2.5.3]. While a red flavylium salt is present in an acidic medium (►Figure 4.7), structures of purple quinoid dominate at medium pH values. In the alkaline range, the hydroxylate ion occurs, a strong electron donor, which results in a blue color impression. A well-known example of pH dependence is the color change from blue-violet cabbage (red cabbage) to deep red after adding vinegar.

At pH values of 7–8, anthocyanidins convert to yellow chalcones under ring cleavage; even higher pH values cause decomposition into aldehydes and carboxylic acids. Brown products can originate through polyphenol oxidation within a few minutes, ►Section 8.2.1. Multivalent metal cations counteract these reactions, producing deep blue complexes of the following type, stabilizing the ketoquinone form:

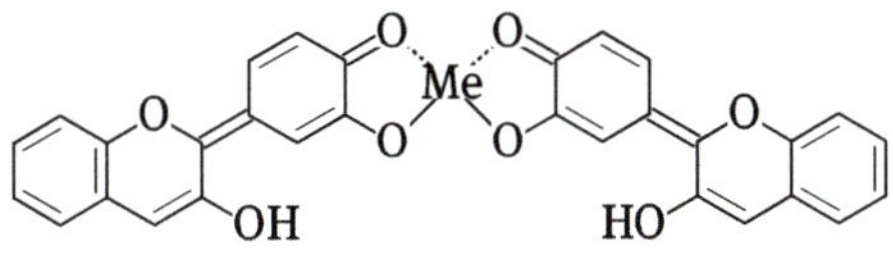

In nature, such complexes occur with Fe^{III} and Al^{III}. In terms of color, red roses and blue cornflowers only differ because cyanidin is a red flavylium salt in roses. In contrast, cornflowers contain cyanidin as a blue complex with aluminum and iron ions. Hydroxyl, chloride, or carboxylate anions saturate the free valences of the metal cation, e. g., from pectin-like polysaccharides [294, p. 410]:

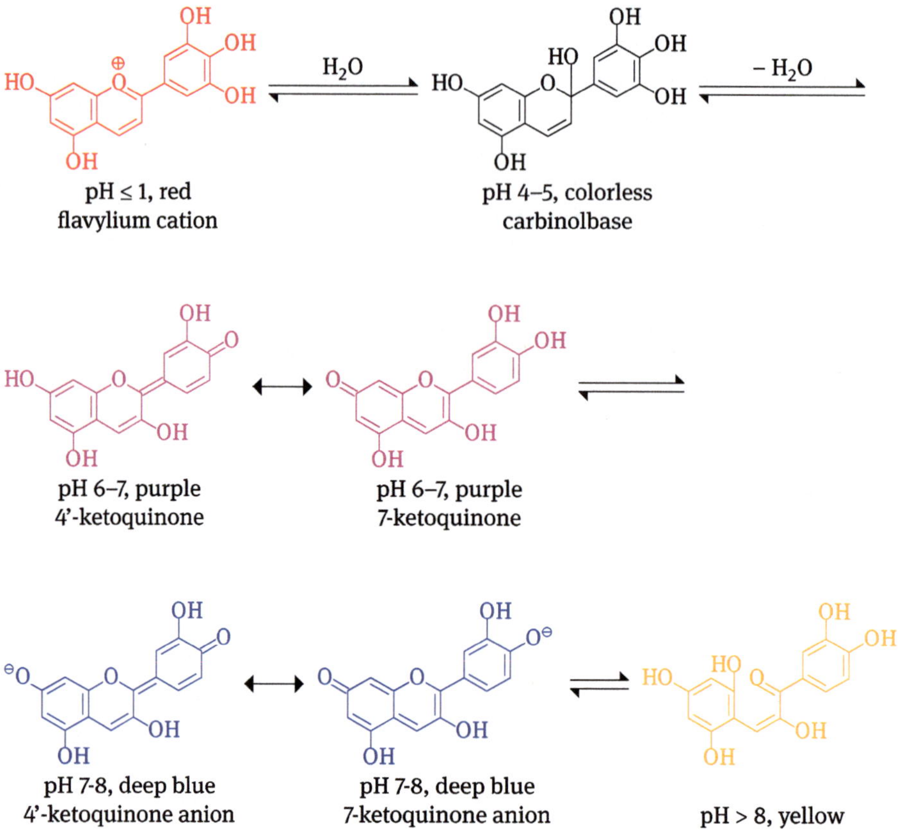

pH ≤ 1, red
flavylium cation

pH 4–5, colorless
carbinolbase

pH 6–7, purple
4'-ketoquinone

pH 6–7, purple
7-ketoquinone

pH 7-8, deep blue
4'-ketoquinone anion

pH 7-8, deep blue
7-ketoquinone anion

pH > 8, yellow

Figure 4.7: Color change of anthocyanidins depending on pH [277]. According to [67], the color change to blue occurs only at pH values greater than 11, and the change to red already occurs at pH values smaller than 3–4. Haslam [293, p. 271] also gives details.

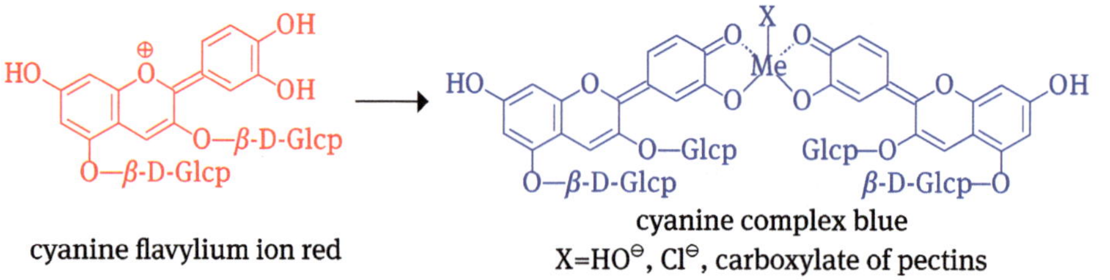

cyanine flavylium ion red

cyanine complex blue
X=HO$^{\ominus}$, Cl$^{\ominus}$, carboxylate of pectins

►Section 2.6.3 discussed the complexation with metals and resulting color shifts in connection with anthraquinone lakes.

Complexes with metal cations are frequently used in nature to assemble anthocyanins and copigments into defined structures.

Copigmentation: stabilization and bathochromy

Due to the low pH value of plant cell saps, most fruits have red to purple colors. Interestingly, according to ►Figure 4.7, the colorless intermediate of carbinol should be a consequence due to hydrolysis in this slightly acidic, aqueous medium and should also decolorize the plant cells. Plants and fruits are nevertheless intensively colored. Thus, the question arose of which mechanism could protect anthocyanidins from carbinol formation and explain the color. The answer was found in *copigmentation*: a (colored or colorless) *copigment* prevents the access of water to the reaction center in the pyrylium ring, which determines the colored form of anthocyanidin [24, Chapter 7], [734], [293, Chapter 6], [754].

An additional anthocyanin molecule or colorless flavonoids that often accompany anthocyanins can serve as a copigment. For example, delphinidin forms a blue structure with azalein [303, p. 179]. Haslam [293, Chapter 6] contains numerous detailed examples from the world of plants. The pyrylium system of the anthocyanin forms molecule stacks with the also flat copigment and thus protects the reaction center from water ingress. Forming esters of the saccharide component of the anthocyanin with cinnamic or caffeic acid can also contribute to copigmentation. Delphinidin again illustrates this. Reacting with Al^{III} and caffeoylquinic acid (chlorogenic acid) form a blue structure. The aromatic ring of chlorogenic acid, a phenylpropene acid, overlays the pyrylium system.

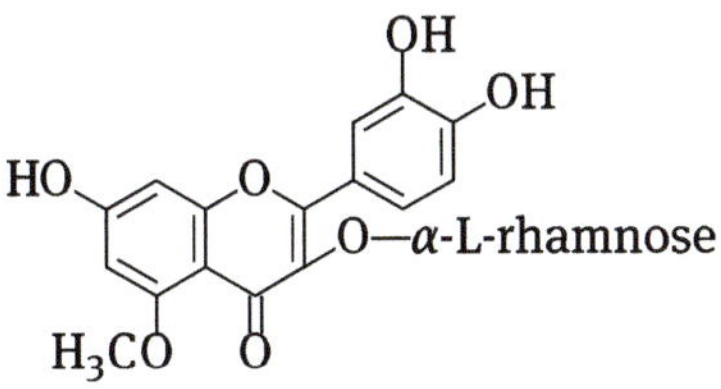

azalein

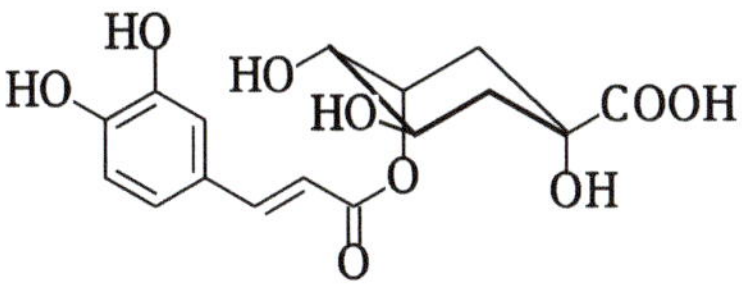

caffeoylquinic acid

Anthocyanins and copigments can be arranged by complexation with metal cations, e. g., Mg^{II}, Fe^{III}, and Al^{III}, into defined supramolecular structures. An illustrative example of this is the complex of cyanidin, apigenin, Fe^{III}, Al^{III}, and Ca^{II} cations, which is responsible for the deep blue color of cornflowers [303, p. 179].

A result of copigmentation is the stabilization mentioned before of the pyrylium ring against water ingress. Another is an intensification of the color or a color shift into the blue range. Since copigments are planar compounds with aromatic rings, intermolecular electronic interactions occur in the stacks of anthocyanins and copigments. The delocalization of electrons extends from the individual anthocyanin to parts of the molecular stack. Similar processes lead to the intensive color of indigo, diketo-pyrrolo-pyrrole pigments, and perylene pigments.

Book illumination, ink, blue and purple color lakes

Although their blue and purple shades cover a spectral range that was previously difficult to achieve with other colorants, anthocyanins have never been of significance to artists, mainly due to their poor lightfastness. Often, we can observe fading and decomposition within days or weeks. As a result, application areas were limited to book illumination and the use of colored ink whenever they could be protected from light, ►Section 8.3.1 and [67, 108]. Commonly used were elderberry, huckleberry, privet, and violet. ►Table 4.5 shows the anthocyanins contained in these plants.

Table 4.5: Anthocyanins in plant extracts often used in book illumination (►Section 8.3.1) [67, 94].

Plant	Coloring anthocyanin
Blueberry, red-purple, blue	Myrtillin, a mixture of delphinidin-3-glycosides, rutin and quercitrin
Elderberry, red, purple, blue, black	Chrysanthemumine = cyanidin-3-glucoside, sambucin = cyanidin-3-rhamnoglucoside
Privet, red, purple, blue	Malvidin-3-glucoside, cyanidin-3-glucoside, delphinidin glycosides
Violet, blue	Violamine = delphinidin-3-rhamnoglucoside

Aqueous plant extracts display red to reddish-purple colors due to the low pH value of the plant cells; the addition of alkalis causes the color change to blue. In this environment, the rapid decomposition into brown oxidation products (alkaline polyphenol oxidation, ►Section 8.2.1) was counteracted by adding aluminum, iron, or tin salts. As a result, stable blue metal complexes appeared whose color shade depended on the metal.

As mentioned above, glycosylation at C^3 is essential for the stability of anthocyanidins. Glycosides were often hydrolytically cleaved during preparation. Therefore, in addition to the metal complexation discussed so far, alum or gum was added to the boiled plant saps. This could have had the purpose of complexing the 3-hydroxyl group with alum or gum, and thus stabilizing the anthocyanidin again.

The plant extracts obtained by pressing or boiling and stabilized could be used directly as inks and watercolors (sap colors). Similar to yellow and red color lakes (►Section 2.6), blue and purple ones could also be obtained by adding alum and potassium carbonate (potash) to the boiled plant saps. They do not show high lightfastness but were applied as cheap wallpaper paints, as a colorant in pastel crayons in the eighteenth century, and until the twentieth century as transparent printing ink. If we want to prepare such lakes ourselves, we can find a rich collection of recipes in [94].

Food colorants

Nowadays, anthocyanins often serve as food colorants for the red, purple, and blue range (E163). They are an obvious choice, being already often contained in foods (e. g., blueberry yogurt). In the case of food products with a limited shelf life and dark storage, those factors that make them unsuitable for painting do not play a role. Furthermore, their lack of body is not a problem since foods only need to be colored.

Wine and fruit juices

Anthocyanins lend various fruit juices their color, e. g., the juice or wine of red grapes. Flavanols (proanthocyanidins, ►Section 4.4.3) are responsible for the typical astringent taste of red wine, and anthocyanins cause the color. Taste and color change during aging: the previously purple-tinged red wine becomes crimson and brown; the taste softens. Here, we observe the transformation of flavonoids into polymeric products and complex pigments [719, 725, 726, 731], [293, p. 245]. Although the process is not yet fully understood, we can imagine copolymers between anthocyanins and flavanols (anthocyan-(4→8)-flavanols, structure I), similar to proanthocyanidins, ►Section 8.2.3. The flavylium structure of the anthocyanins determines the color of these polymers. However, the color is possibly modified after oxidation by quinoid components such as structures II and III [727], [293, p. 338]:

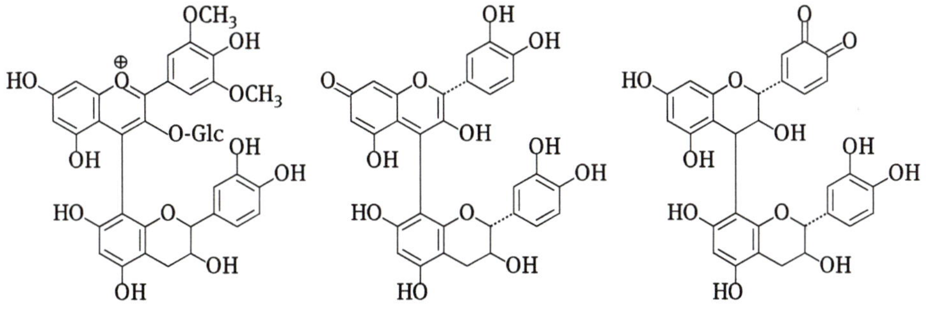

I (red) | II quinoidic dimer | III quinoidic dimer

Dimerizing flavonols with acetaldehyde or glyoxylic acid from alcohol form other colored, complex compounds. The dimers are colorless when formed from flavonols and red-purple from anthocyanins. These products can also further react to a variety of yellow and red compounds, the chromophore of which is a quinone, flavylium system, or xanthylium system, ►Figure 4.8. Unfortunately, the author does not know whether such processes also occur in anthocyanin-containing plant-based inks prepared with wine and whether they form similar dyes.

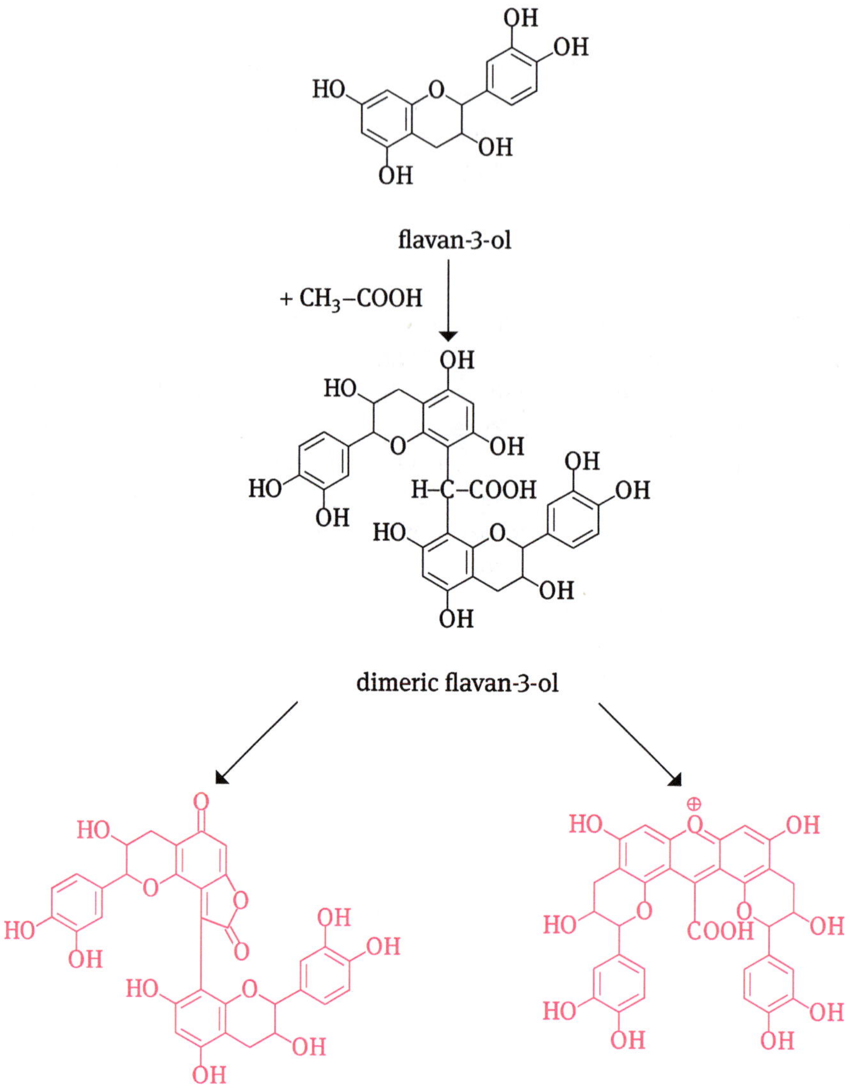

flavan-3-ol

+ CH_3–COOH

dimeric flavan-3-ol

Figure 4.8: Possible formation of red compounds from flavanols in red wine [719]. Precursors are dimers of flavanols with derivatives formed from alcohol, such as acetaldehyde. The dimers can react to form yellow and red compounds with chromophores of quinone, flavylium, or xanthylium.

4.4.6 Neoflavones

Red Purple Blue Neoflavones are the coloring agents in numerous roots and woods. The best known of these color-providing woods are bluewood and redwood. In their countries of origin (Orient, Far East, South America), they have always been known as dyeing materials; in Europe, they also became widely popular after the discovery of

America [70]. Like most flavonoids, they are mordant dyes (►Section 5.5) and provided red and blue-purple colors to medieval inks, ►Section 8.3.

Bluewood, Campeche wood, Natural Black 1

The bloodwood tree *haematoxylum campechianum* native to Central America and the West Indies, has a heartwood that contains the colorless to slightly yellowish dye precursor hematoxylin. This precursor is oxidatively transformed into hematein during prolonged storage and boiling in the dyeing process, exhibiting a brown-blue-purple hue. Due to the laking (►Section 2.6) with various metals, intense colorations emerge: Al^{III} (blue-purple), Sn^{IV} (purple), Cu^{II} (blue-black), Fe^{III} (black), ►Figure 4.9.

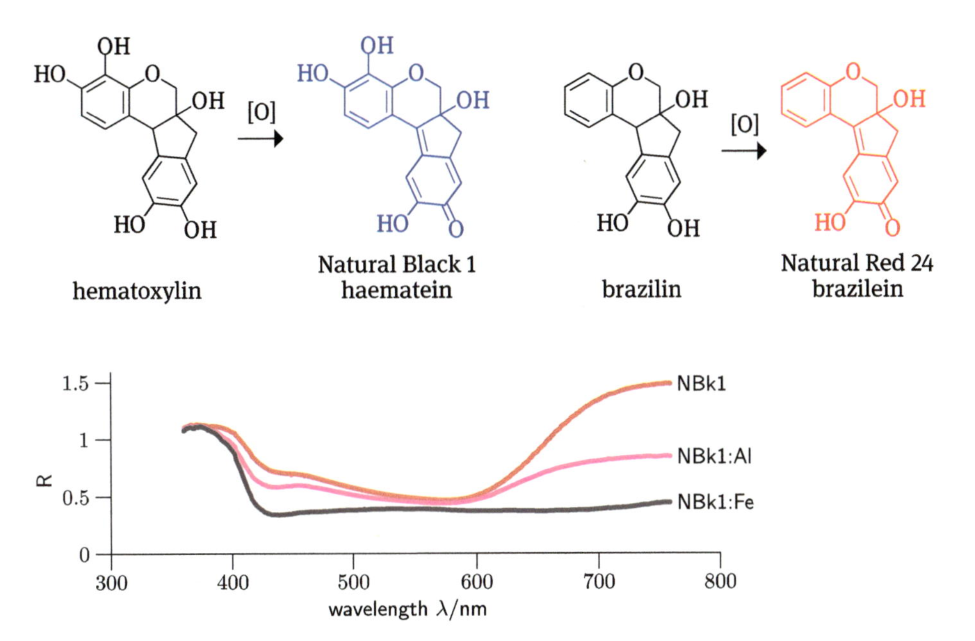

Figure 4.9: Reflectance spectrum of hematoxylin-based colorants normalized to an arbitrary unit: NBk1 (hematoxylin), hematoxylin Al complex (own preparation), hematoxylin Fe complex (own preparation).

An application of the colorant today is cell and tissue dyeing in histology. Formerly, it dyed textiles blue and black. Specifically, in combination with tannin (sumac) and iron vitriol, deep black color was obtained, not unlike black iron gall ink, ►Section 8.4.

Redwood, Natural Red 24

Different tree species (sappan *caesalpinia sappan*, bahia *caesalpinia brasiliensis*, pernambuco *caesalpinia echinata, etc.*) also have a heartwood, which contains the dye precursor brazilein. First, it is extracted by boiling the wood with water, then oxidatively converted into the actual colorant brazilin, similar to bluewood. With metallic mordants, we obtain different color lakes: Al^{III} (red), Fe^{III} (brown), and Sn^{IV} (pink).

In the early Middle Ages, red-colored lakes based on brazilein were very popular as glaze pigments. However, artists were aware of the high light sensitivity of these lakes. They abandoned their use in favor of the more stable ►carmine red from cochineal as soon as this became available in sixteenth century through importation from Latin America. In book illumination, lightfastness was of less importance, so illuminators used economical redwood more often for color production [108].

The oxidation of brazilin to brazilein is facilitated in an alkaline environment but delayed in an acidic one. As a result, the color of the extract depends on the pH value. More yellow extracts are obtained in the acidic range, whereas more red extracts are received in the alkaline range. When an alkaline solution is left to stand, brown coloration occurs. Recipes with alkaline extract and alum addition often resulted in brownish crimson. The extract was used with alum mordant as sap color or precipitated on chalk as a color lake, ►Section 8.3.

4.4.7 Quinone methides

Yellow Orange Red In the section about neoflavones, we have seen that oxidizing neoflavone is responsible for the colored component of the bluewood and redwood. This oxidation produces intensively colored quinoid quinone methides I:

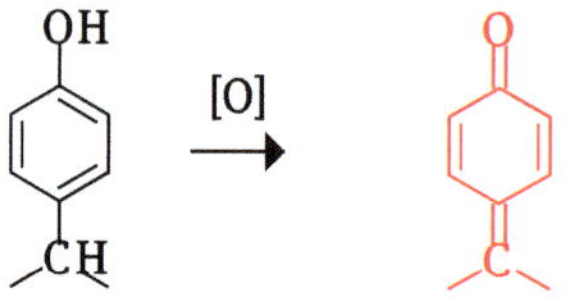

neoflavone I, quinone methide 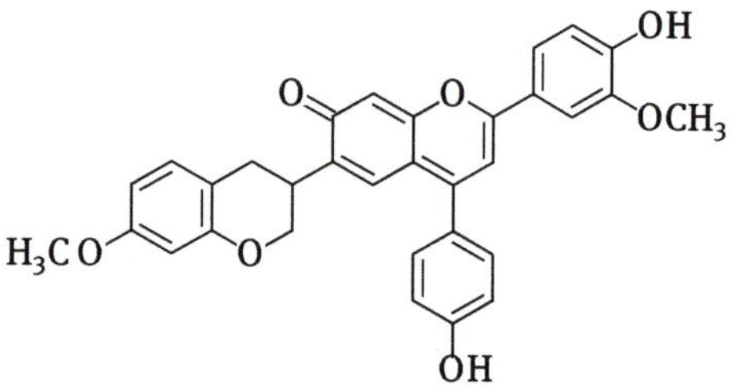 retusapurpurin

The oxidation of most other flavonoids can also create quinone methides. Besides, several phenolic compounds can be oxidatively converted into quinone methides, ►Section 8.2.1 and [766]. We have already encountered an example in the oxidative browning of red wine, ►p. 333; compounds of this type are also involved in the dark coloration of insect larvae during the development of the hard exoskeletons [766].

Yellow or red quinone methides are standard components of the colored resins in the heartwood of shrubs and trees [26], e. g., dracorubin in dragon tree resin, santalin in sandalwood, or retusapurpurin in rosewood (*dalbergia retusa*). For today's artists, they are insignificant, though historically, some red resins (dragon blood) and wood extracts (sandalwood) were employed in book illumination (►Section 8.3) and dyeing.

Dragon blood, Natural Red 31

In the past, red color suitable for book illumination was produced in Socotra or Sumatra from the resin of the dragon tree of the species *dracaena*. It contains dracorubin, which is obtained by oxidative dimerization of a flavone (upper half of the molecule) and a flavane (lower half of the molecule):

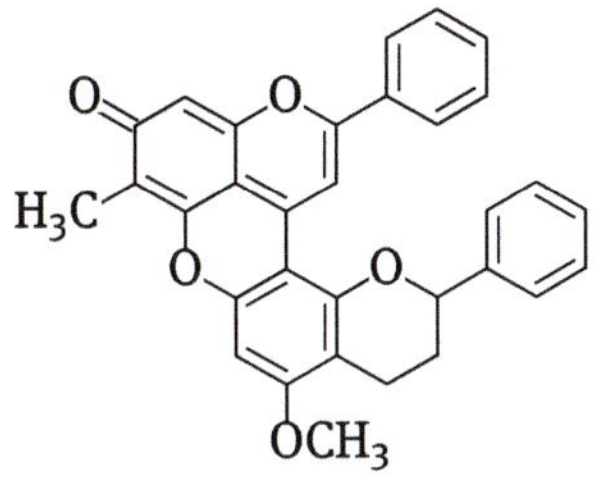

Natural Red 31
dracorubin, historic colored ink

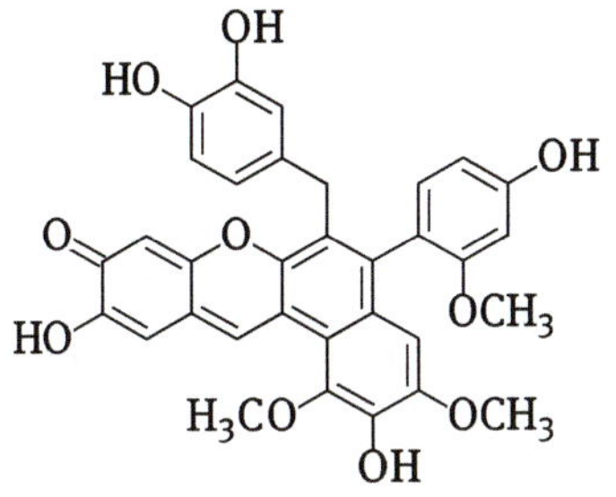

Natural Red 22
santalin A, historic colored ink

Sandalwood, santalin, Natural Red 22

Sandalwood (*pterocarpus santalinus*) is another redwood whose coloring agent is based on santalin, an oxidized isoflavone (left, the lower half of the molecule). Unlike brasilein, santalin must be extracted with alcohol or bases. Depending on the dyeing conditions, a range of colors from red to purple to bronze is possible. Metallic mordants provide various lakes: Al^{III} (orange-red), Fe^{III} (brown), Sn^{IV} (red), and Cr^{III} (brown-red).

4.4.8 Chalcones and quinochalcones

Yellow Red Chalcones, together with flavones, belong to the yellow flavonoids, but in contrast to the latter, they are insignificant for painting. However, studies of safflower have shown that the structural element of chalcone is contained in several colorants derived from this plant, which were applied for dyeing and book illumination.

Yellow root, turmeric, Natural Yellow 3

From the roots of *curcuma domestica*, we obtain a tasty spice but also the well-known yellow colorant curcumin [23, Chapter III.2]:

Natural Yellow 3, Food Yellow 3, E100
curcumin, historic colored ink

Curcumin is a direct colorant and suitable for use in historical color inks, ►Section 8.3.1. However, thanks to the diketo group, curcumin can also be used as a mordant dye, where the metal determines the color: Al^{III} (orange-yellow), Cr^{III} (brown), Sn^{IV} (orange-red), and Fe^{III} (brown). Curcumin remained in use until the nineteenth century, but not as a dye in its own right. Mainly, it supported the shading of cochineal dyes.

Saflorcarmine, carthamine, Natural Red 26

The safflower carmine is obtained from the saffron thistle *carthamus tinctoria* native to the Orient, Africa, and southern Europe. This dye was already known to the Egyptians around 1000 BC [23, Chapter III.3]; we find it on 4500-year-old Egyptian mummy bandages [731]. The actual, valuable colorant of this plant is red carthamine, which in addition to safflor yellow A and B, is contained in the flowers [70, 750, 751]. Good qualities of safflower carmine can be extracted by washing the dried safflower flowers to remove safflor yellow so that red carthamine is left over. Safflor yellow provides unstable yellow dyes and is water-soluble. However, both colorants, i. e., red carthamine, and safflor yellow, were used as inks, ►Section 8.3.

Natural Red 26
carthamin, historic colored ink

safflor yellow A
historic colored ink

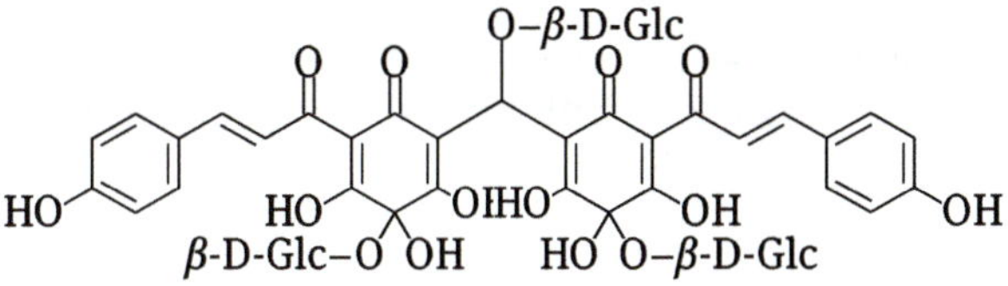

safflor yellow B
historic colored ink

The dyes can be used in a mordant, ►Section 5.5; dyeings on wool with alum with unwashed safflower yield a yellow, without the yellow colorant carmine red (Spanish red). With direct dyes but without mordant, silk and cotton can be stained cold, ►Section 5.4.

Interesting about the structure of carthamine is that we can address it as a dimer of a chalcone containing elements of a quinone simultaneously. Since recent investigations of the safflower plant have found several other colorants of similar structure, they are now named quinochalcones [752].

4.4.9 Cause of color

The color of flavonoids originates from an oxonol system with $n = 3$, extending between the pyran and B phenyl rings. The carbonyl group acts as an acceptor, the hydroxyl group as a donor, ►Section 2.5.5:

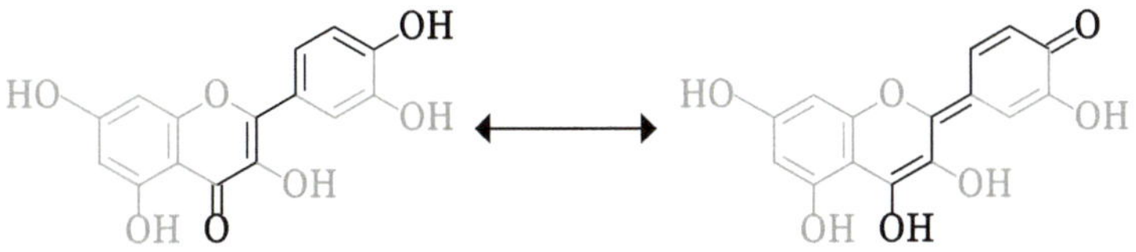

rhamnetin

The hydroxyl groups appear at starred positions, leading to a bathochromic shift, ►p. 186, “Dewar rules.” The structural element of oxonol is also present in chalcones:

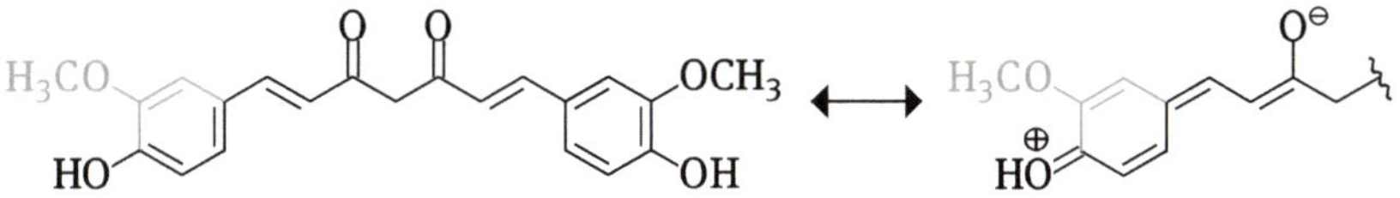

curcumin

Despite the dual presence of the chalcone structure, curcumin is only yellow because both of the oxonol systems are separated by an electronically inactive methane group. Neoflavones contain an oxonol chromophore with $n = 4$:

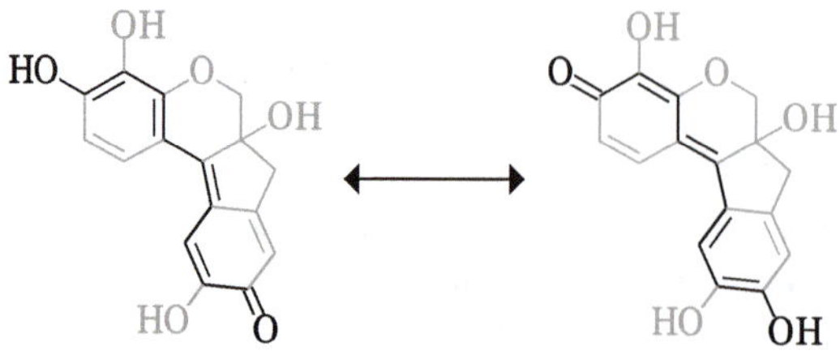

hematein

Anthocyanidins exhibit an oxonol system with $n = 2$ as the chromophore. In contrast to the flavones, the acceptor is formed by the pyrylium oxygen. Hydroxyl groups on ring B act as donors; in addition, hydroxyl groups on ring A can extend the mesomeric system; in this case, $n = 3$:

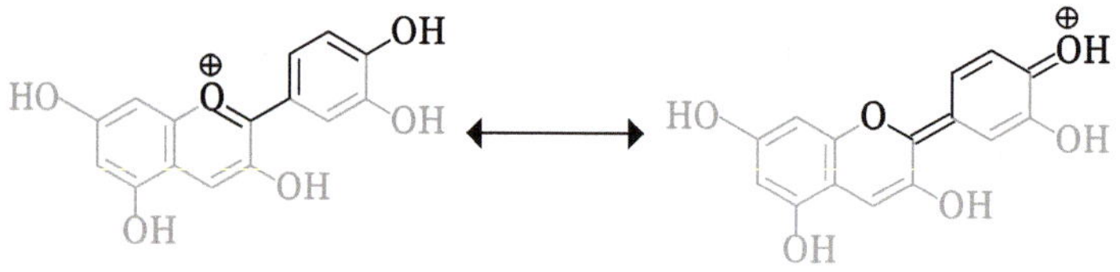

pelargonidin

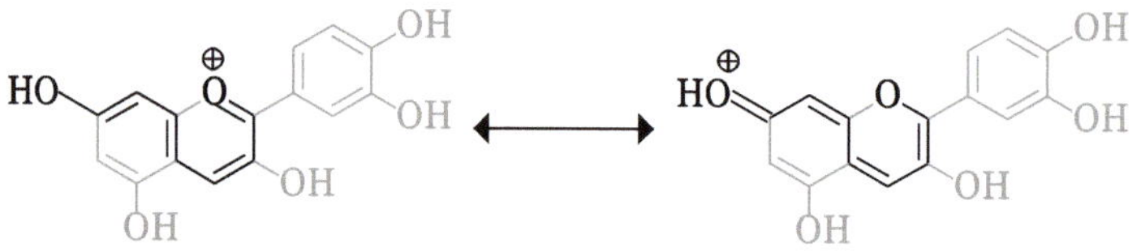

The conversion of a hydroxyl group into a potent electron-donating hydroxylate anion explains the pH dependence and the color change to blue in an alkaline environment:

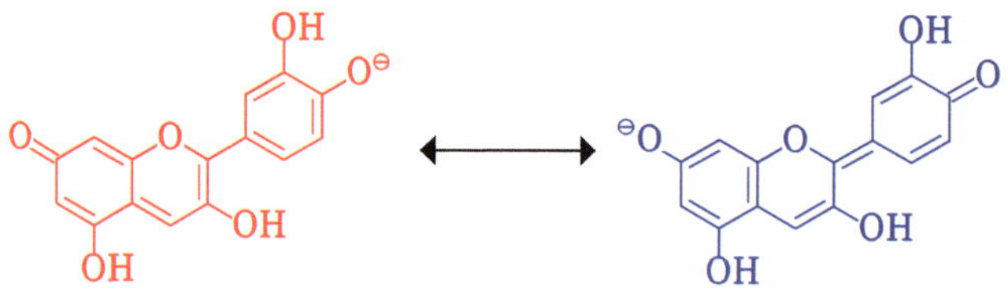

We notice a superposition of the effects of the flavonoid system and the quinoid carbonyl systems in the complex structures of the quinochalcones, ►Section 2.5.3.3. These contribute to the color shift from yellow to red.

Different color phenomena of polyphenols appear when metal cations are added to these compounds. The resulting complexes are often intensively colored, e. g., the blue to black compounds with Fe^{III}. This effect is observable because the chelate rings that emerge between metal cations and hydroxyl groups considerably enhance the electron-accepting and donating abilities of the partners, ►Section 2.6.3. More decisive color shifts with *d* block metals such as iron and chromium are also consequences of LF and CT transitions, as we will see in iron gall inks, ►Section 8.4.4.

Other color phenomena

Colorless phenolic compounds not considered in detail here can also cause color phenomena in fruits and vegetables. They are substrates for polyphenol oxidases, which oxidize diphenols to quinones. The quinones, in turn, undergo numerous complicated further reactions, leading to brown discoloration in the fruit. Inactivation of the enzymes with sulfur dioxide or oxygen deprivation prevents the deterioration and maintains the light color of the fruit. Similar oxidation reactions also play a role in forming brown-colored natural inks, ►Section 8.2.1.

4.5 Xanthones

Yellow Orange Xanthones are aromatic oxo compounds derived from the ring system of xanthene. They represent the coloring substances of some plants that have found their way into the painting. Mangiferin, e. g., is a component of various plants (irises) and provides a yellow ink dye, ►Section 4.4.4:

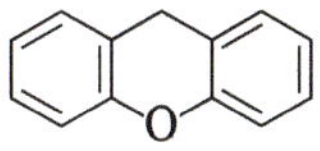

xanthene

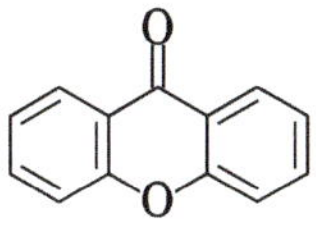

9H-xanthone

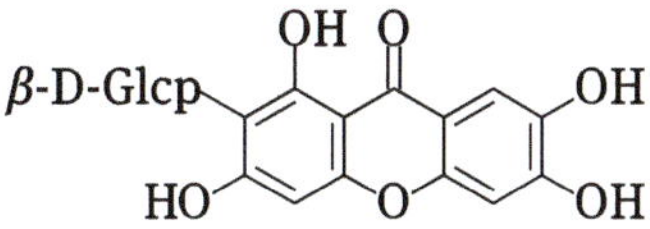

β-D-Glcp, OH, O, OH, HO, O, OH

mangiferin
historic colored ink

Indian yellow, Natural Yellow 20 (CI 75320)

In India, the beautiful yellow pigment *Indian yellow* existed in in the period from the fifteenth century to the nineteenth century. According to vague reports, it was derived from the urine of cows fed with mango leaves [48, Volume 1]. This pigment consists of the magnesium and calcium salt of euxanthic acid. Euxanthone, which does not display pigment or dye properties, may be present and reduce quality. The pressed dye typically contains 65 % euxanthic acid salt, whereas lower qualities hold only 30 %.

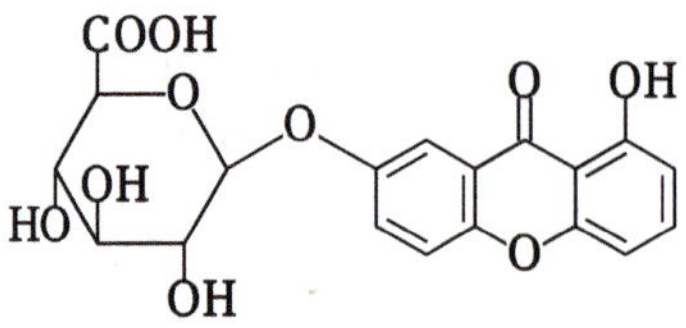

Natural Yellow 20, CI 75320
euxanthic acid

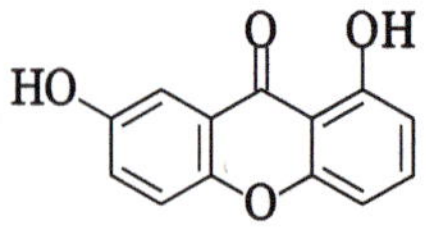

euxanthone

While the pigment was used extensively as a watercolor in India, it appeared late and only rarely in Europe. Since it is in the form of a salt insoluble in oil, it could be well applied as a glazing color. Nowadays, we will hardly ever be able to paint with Indian yellow, as its production already stopped around 1910 for animal welfare reasons. Fortunately, the Acid Yellows 23 (tartrazine) and 63, Acid Orange 1, and the pigments PY40 (cobalt yellow) and PY153 are available as substitutes:

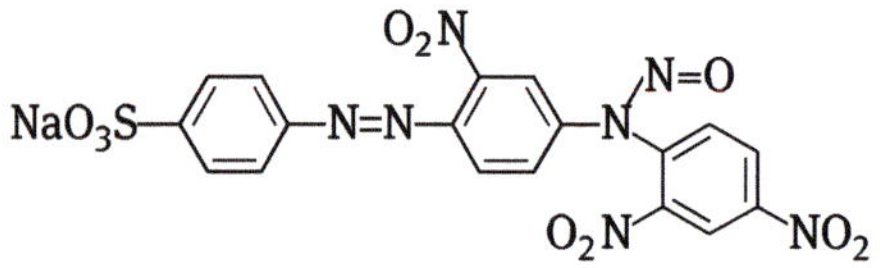

AY63, CI 13095

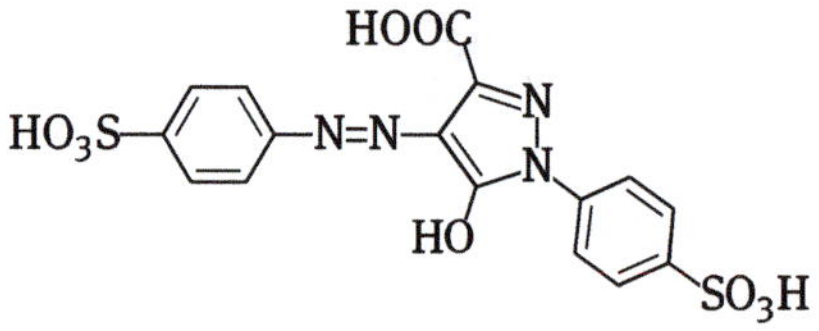

AY23, CI 19140, Food Yellow 4, E102
tartrazine, yellow writing ink
primary yellow for four-color printing
Winther: A→E

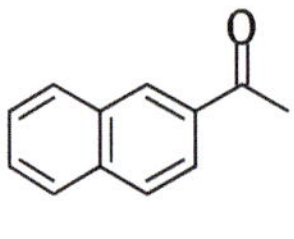

AO1

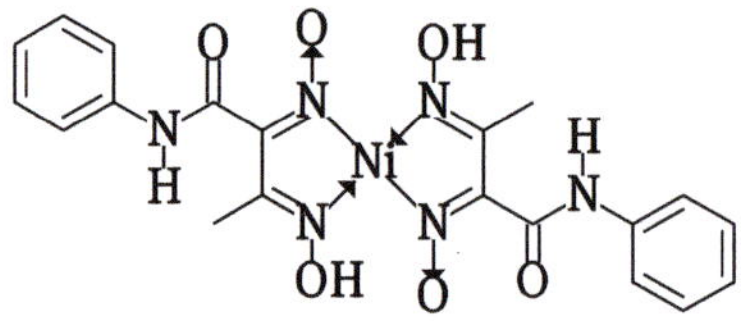

PY153, CI 48545

The substitutes also show a beautiful orange-tinged yellow, which in turn provides warm, sunset shades of yellow when mixed with white.

Gamboge, gumdrop, Natural Yellow 24

Gamboge or gumdrop is a bright yellow powder that used to be popular in book illumination (►Section 8.3) and dyeing. It is made from the colored resin of the Southeast Asian trees *garcinia morella* and *garcinia hanburyi*. Xanthonic acids [269, drug: Garcinia], [691, 695–697] occur as colored components, especially gambolic acid, morellic acid, their iso-forms, as well as their corresponding alcohols, such as isomorellinol:

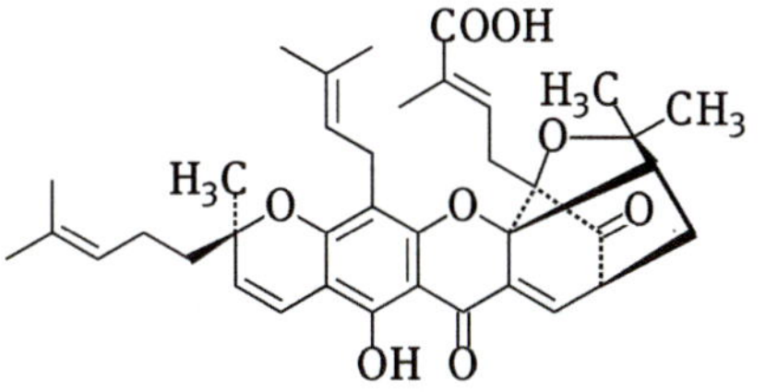

Natural Yellow 24
α-gambolic acid, α-guttic acid, β-guttiferin
historic colored ink

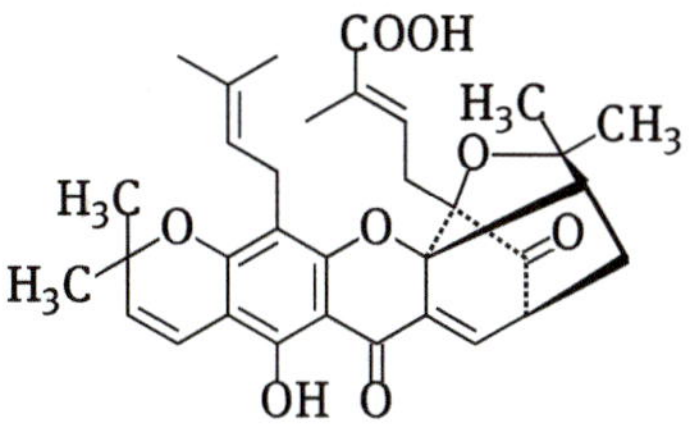

Natural Yellow 24
morellic acid, historic colored ink

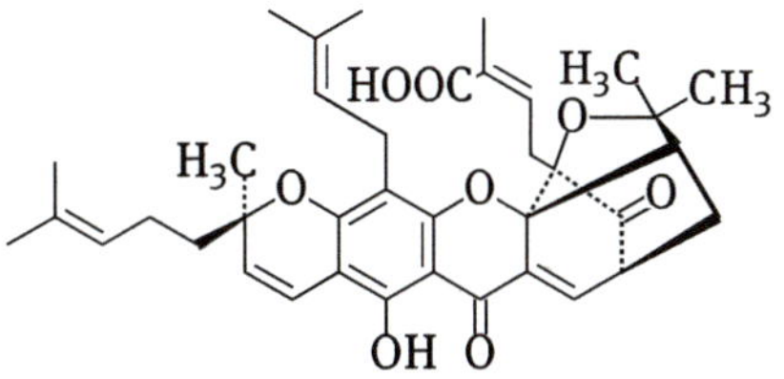

Natural Yellow 24
isogambollic acid, historic colored ink

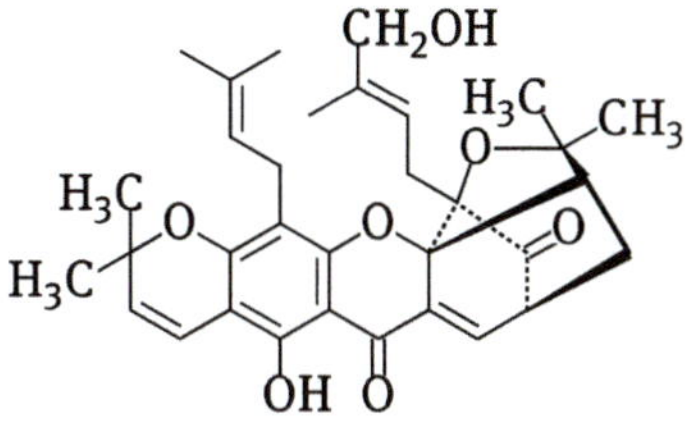

Natural Yellow 24
isomorellinol, historic colored ink

Cause of color

We can consider xanthones as polymethines (►Section 2.5.5) with carbonyl oxygen as an acceptor and pyran oxygen as a donor. In addition, the π systems of benzene rings also act as donors:

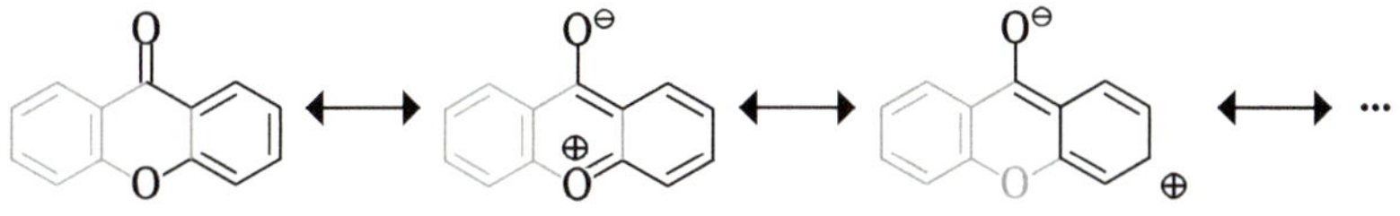

9H-xanthone

4.6 Quinones

In nature, quinones are a component of many red dyes and lakes from woods and roots. Furthermore, lichens, mushrooms, and fruits owe them their color, e. g., butter mushrooms, mangoes, and nuts. Therefore, quinones provided numerous natural colorants from antiquity to modern times. They are derived from the simple aromatics benzene, naphthalene, and anthracene:

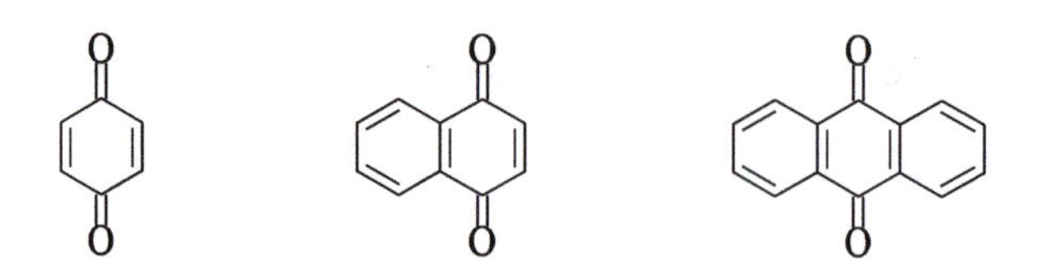

p-quinone naphthoquinone 9,10-anthraquinone

Since modern times, quinones have provided the basis of synthetic pigments inspired by natural models. All artists' paints are based on p-quinones. The o-quinones, which also occur naturally, are found in dark or colored polymerization products. For example, they occur during the browning of fruit or the formation of bark and shells, or in insect carapaces.

4.6.1 Vat dyeing

Natural quinones gained distinction because they were applicable as *vat dyes* for fabrics. Vat dyes have been prominent coloring agents since antiquity. They form as an insoluble pigment only on the fiber, a well-known example being indigo. A mostly colorless, insoluble precursor of the actual dye is put into solution by the action of reducing agents. First, the dyeing material soaks with the dissolved precursor. Then, through an oxidation process, the actual pigment (re)produces from the precursor on and in the substrate's pores [19, Chapter 17].

The dicarbonyl element in quinones is a typical structural element of vat dyes. Therefore, many simple natural quinones were used for vat dyeing. Nowadays, modern quinone structures are produced specifically for this application. The conjugated carbonyl groups allow the coloring molecule to participate in the redox reaction, which is crucial for the vatting of a dye:

reducing agent $HO^{\ominus}$ / oxidizing agent; $H^{\oplus}$ / $HO^{\ominus}$

leuco base leuco base

Sodium dithionite $Na_2S_2O_4$ can act as a reducing agent. As a result, the soluble dinatriumphenolate is often colorless or only slightly colored and is also called leuco base. The reaction must be carried out in a strongly alkaline medium to avoid forming poorly soluble free diphenol, which no longer has any affinity to the substrate.

First, the cellulose-based fabric (cotton) is immersed in the leuco base solution for dyeing. The leuco form bears a high affinity to the substrate and passes from the dyeing solution into the substrate. However, the exact cause of this substantivity is unclear [19, p. 368]. Hydrogen bonds are possible between hydroxyl groups of the substrate, on the one hand, and phenolate and other suitable groups of the dye, on the other hand. However, since the substantivity depends mainly on the surface of the leuco molecule, van der Waals forces must also be powerfully influential, similar to those for direct dyes.

Finally, the leuco form is returned to the pigment by oxidation, which is often already possible with air. Industrially, sodium dichromate, hydrogen peroxide, or sodium perborate are also used.

In contrast to the leuco form, the pigment no longer has any particular affinity to the fabric. It is formed in the cavities of the dyeing material and is held there (as it were, trapped) by a steric hindrance to varying degrees. Dispersion forces and other secondary interactions play a significant role, ►Chapter 5.

4.6.2 Natural quinones and naphthoquinones

Yellow Orange Red Simple benzoquinones are mainly found as colorants in fungi and lichens and have been used to dye fabrics since antiquity. The attainable shades on wool fibers are purple, brown, olive-green, green, or orange-brown, usually with a metal-based mordant (aluminum, copper, iron, or tin salts).

Among the structures, mainly terphenyls such as polyporic acid and atromentin are represented, contained in fungi such as soft sporophytes, paxilli, and boletes. Again, we find structures such as boviquinone and grevillins A–D, the dyes of slippery jacks (*suillus luteus*):

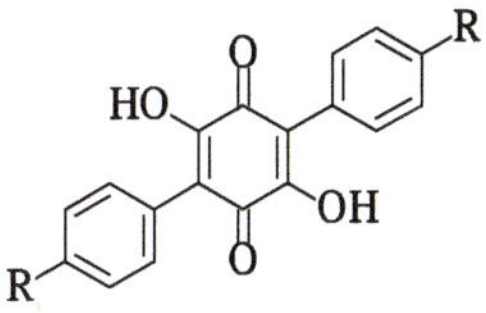

polyporic acid, R = –H
atromentin, R = –OH

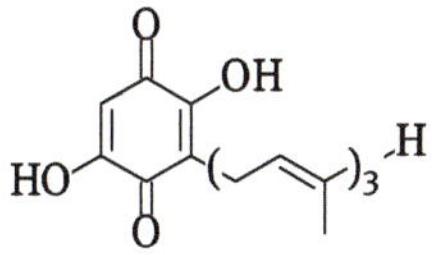

boviquinone-3

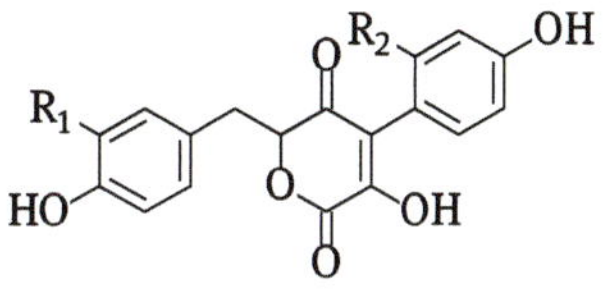

grevillin-A, R_1 = –H, R_2 = –H
grevillin-B, R_1 = –OH, R_2 = –H
grevillin-D, R_1 = –OH, R_2 = –OH

Naphthoquinones are coloring agents in numerous roots and woods; as the following examples demonstrate:

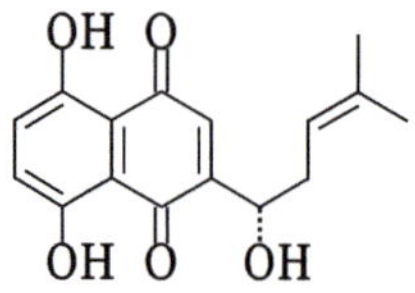

Natural Red 20
alkannin

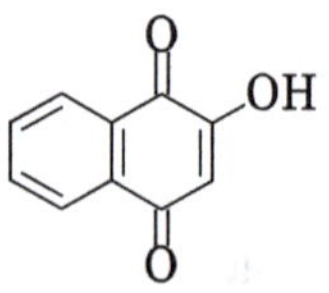

Natural Orange 6
Lawson

Alkannin is present in the alkanna root *alkanna tinctoria*, a dark red colorant for food and cosmetics (Natural Red 20, CI 75530) [70]. Antiquity knew and applied the dye (Egypt, Mesopotamia) until the Middle Ages. With Al^{III} mordants, we obtain purple shades on wool and silk.

Henna (Natural Orange 6) was already popular 3000 years ago; Egyptians dyed mummy bandages with henna [771]. The dye is extracted from the henna tree *lawsonia inermis* native to the Mediterranean, Orient, Africa, and Asia and dyes proteins such as hair, fingernails, or toenails orange-red. Henna consists mainly of Lawson [70], a direct dye for wool, silk, and other proteins. Wool and silk dyings deliver orange-brown colorations directly or with Al^{III} mordant.

4.6.3 Natural anthraquinones

Red Derivatives of 9,10-anthraquinone provide the primary basis of naturally occurring quinone colorants. The colors displayed are pink, deep-red, purple, and brown, ►Figure 4.10. They were significant dyes for the red range until the development of synthetic dyes. Depending on the starting product (plant or insect), mainly alizarin, purpurin, and pseudopurpurin (madder), or carmine and kermic acid (kermes, cochineal) participate in the formation of the color. Natural sources of anthraquinone colorants contain countless other mono and polyhydroxyanthraquinones and their carboxylic acids and methyl derivatives. These usually do not contribute to coloration, do not form stable compounds like the monohydroxyanthraquinones, or are soluble.

Anthraquinones can be converted by laking into a water-insoluble, paintable pigment, ►Section 2.6. Variations in coloring are due to the different compositions of the natural anthraquinones. Historically, Al^{III} and Fe^{III} complexes could be produced using natural alum, $KAl(SO_4)_2$, various natural iron salts such as iron vitriol $FeSO_4$, and iron-containing moor waters. With the advent of tin ash, tin complexes became available later. In some cases, such as on old painted maps [697], anthraquinones were also used as aqueous dye solutions without prior laking.

Red anthraquinone lakes have appeared on European panel paintings since the fifteenth century. However, unfortunately, the analytical detection of small amounts of similar organic materials has only recently been perfected; therefore, we have not

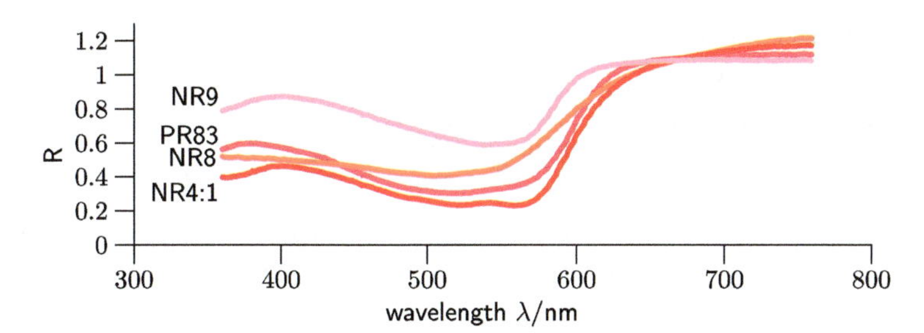

Figure 4.10: Reflectance spectrum of anthraquinone-based watercolors, normalized to an arbitrary unit: NR4:1 (carmine Naccarate, cochineal red, Kremer No. 421008, Al lake of carminic acid), NR8 (madder lake, root madder lake, Kremer No. 3721418), NR9 (rose madder genuine, Winsor and Newton Professional Water Color No. 587), and PR83 (alizarin crimson, Winsor and Newton no. 002, PR83). The purity in the shade of synthetic alizarin crimson PR83 is evident in the steep absorption edge around 600 nm. Similarly, the relatively pure carmine is brilliantly red. In contrast, the natural madder lake from roots has a duller color due to the complex composition of similar anthraquinones, recognizable by the more indistinct features in the spectrum.

been able to gather as much knowledge about the exact nature of these lakes as other colorants. Nevertheless, the frequency of use helps us to gain a deeper understanding [393]:

- Between 1400–1890, red lakes consisted of Al lakes comprising brazilwood, madder (orange-red), kermes carmine (red), cochineal carmine (bluish-red), and lac dye (bluish-red).
- By 1600, the colorants brazilwood and lac dye dominated. In addition, kermes and cochineal carmine were employed in Italy and madder lake in northern Europe. From the fourteenth century to the seventeenth century, it was a common practice to extract kermes, cochineal, and madder dyes from dyed textile remnants.
- After 1600, cochineal carmine increased, and after 1800, lakes consisted mainly of cochineal carmine and madder lake. To the same extent, the direct extraction of the dyes from the raw materials (cochineal louse, madder plant) increased. From the early seventeenth century onward, also American cochineal became available.

A complicating factor for the analysis is the comparatively high light sensitivity of early red lakes; their partial or complete fading in many paintings is sad proof, ►Section 7.4.11.

Madder lake, madder carmine, Natural Red 8, 9, 6, 14, 16, 18, 19 (CI 75330, 75340, 75350, 75370, 75390, 75410, 75420, 75430, 58050)

A source of beautiful red dyes renowned since antiquity (1600 BC) is the madder plant, which grows in varieties from southeastern Europe through the Mediterranean to the Caucasus, Asia, and America. Its color is due to the content of anthraquinones. Depending on the origin of the madder (►Table 4.6), the main colorants are alizarin,

Table 4.6: Composition and origin of anthraquinones used for dyeing and laking from madder and related plants [23], [48, Volume 3], [70].

Origin	Composition
Natural Red 8 Madder *rubia tinctorum* (widely used)	Alizarin, purpurin, pseudopurpurin, purpuroxanthine, rubiadin, munjistin, and many others
Natural Red 8 Levantine madder *rubia peregrina* (Mediterranean, Asia Minor)	Pseudopurpurin, purpurin, alizarin
Natural Red 16 Indian madder *rubia cordifolia* (India)	Alizarin, pseudopurpurin, purpurin, purpuroxanthine, munjistin
Natural Red 6 Chay root *oldenlandia umbellata* (India)	Alizarin
Natural red 14 Galium *galium verum* (widespread) and others	Alizarin, purpurin, pseudopurpurin, purpuroxanthine, rubiadin, lucidin
Relbunium root *relbunium* (South America)	Pseudopurpurin, purpurin, munjistin
Natural red 18, 19 Indian mulberry *morinda citrifolia* and *umbellata* (India, Southeast Asia)	Morindon, soranjidiol, alizarin, rubiadin

pseudopurpurin, rubiadin, and munjistin [48, Volume 3], [70, 394, 711]. The composition may change during processing. While the madder roots dry, e. g., purpurin from pseudopurpurin and purpuroxanthin from munjistin form by decarboxylation.

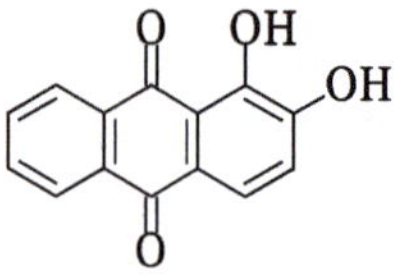

1,2-dihydroxy-anthraquinone
alizarin, PR83, CI 75330

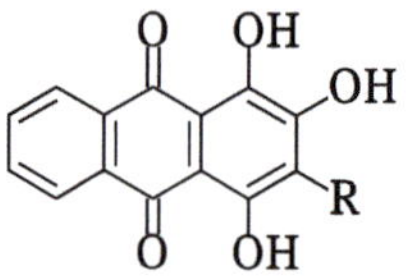

pseudopurpurin, CI 75420, R=COOH
purpurin, CI 75410, R=H

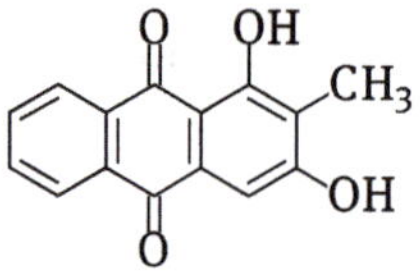

rubiadin, CI 75350

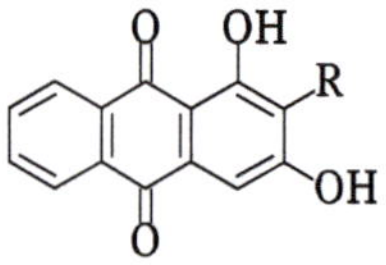

munjistin, CI 75370, R=COOH
purpuroxanthine, CI 75340, R=H

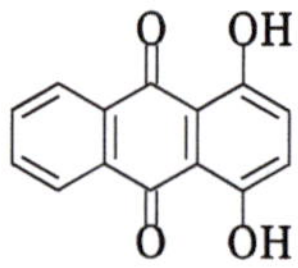

1,4-dihydroxy-anthraquinone
quinizarin, CI 58050

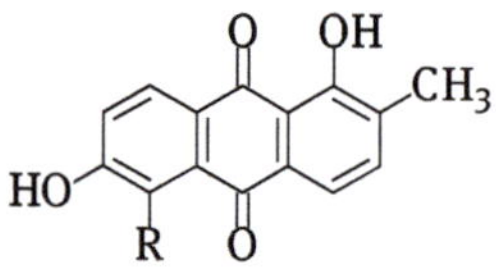

morindon, CI 75430, R=OH
soranjidiol, CI 75390, R=H

Dyes are found primarily in the bark of madder roots (Natural Red 8). Although they can be obtained by simply drying the roots, better qualities are obtained by separating the bark from the woody parts. The exact composition of the anthraquinone mixture varies depending on the origin of the roots and influences the coloring considerably. In the spectrum (►Figure 4.10), the complex composition of a lake derived from natural madder root containing different anthraquinones results in a rather dull red color, visible in an indistinct reflectance spectrum.

Madder has been used since the second pre-Christian millennium in textile dyeing, at the latest since the first century also as a mordant dye [23]. An Al^{III} mordant with natural alum was employed, yielding medium to dark red colorations. Dyers in the Orient achieved the particularly brilliant and thus sought-after color shade called *Turkish red* with a Ca^{II}-Al^{III} mordant. In the sixteenth century, the discovery of the Sn^{IV} mordant led to dyeings in orange, with Fe^{III} mordant in brown, and Fe^{III}-Ca^{II} mordants in purple.

By laking, madder root yields important, paintable lake pigments [48, Volume 3], [70, 391–393]. The lake is precipitated with alum and soda (sodium carbonate) and applied to chalk, clay, or gypsum. The resulting *madder lake* (Natural Red 9) is a red, quite light-resistant Al lake, already known in Greek and Roman antiquity and found in Pompeii. With the end of the first post-Christian millennium, due to the constant availability of native madder plants, red madder lakes were more frequently used as painting pigments. Over time, the production methods were even more refined so that around 1828, Garancine lake, a substance highly enriched with alizarin, became available. Synthetic alizarin (►Section 4.6.4) was already produced in 1869 on an industrial scale. In artists' paints such as "madder lake" or "alizarin crimson," synthetic alizarin is used. Other natural lake variants are:

- Sn lake with the colors scarlet, pink or orange
- purple Fe lake, *madder purple*
- pink madder lake, *rose madder*, made from source materials with high (pseudo) purpurin content
- *madder carmine*, a lake enhanced with cochineal (crimson)
- red-brown Cr lake

Despite its long history and the knowledge of alizarin as chromophore, the structure of the lakes remained elusive for a long time. We will follow the exciting detours concerning the structure in ►Section 2.6 at p. 190.

Turkish red

Fabrics dyed with *Turkish red* were highly sought after in earlier times because they were lightfast and washing resistant. Unfortunately, dyers could only obtain them from special dye workshops or factories at high costs for a long time. This madder red dyeing applied the Ca-Al lake of alizarin to the plant fiber, ►Section 4.6.4.1. In contrast to dyeings with the ordinary Al lake of alizarin, the complicated dyeing process comprised more than ten steps, which could later be somewhat simplified. Nevertheless, and by no means simple sequence of oiling, staining, fixing, and dyeing had to be strictly followed to achieve this unique shade. Many steps like the oiling were necessary to make the vegetable fibers receptive to the aluminum mordant [711, p. 23].

Kiel and Heertjes [709] then established that the fastness of Turkish red is based on the multinuclear metal complex formed directly in the fiber. As a result, even simple dyeing processes achieve this fastness if they observe the conditions for complex formation (pH value, concentration of the reactants) and prevent premature migration of the reactants or the resulting complexes.

A decisive factor that influenced the concentration of calcium ions in the acidic solution was the water hardness. Therefore, if soft water was available, chalk (Europe) or calcium-containing plants (Asia and India) had to be added to increase the calcium content to the correct level. Another step of the dyeing process was removing purpurine and pseudopurpurine in the madder plant to obtain a pure alizarin lake. This step was necessary because the alizarin lake alone provides lightfastness lacking in the purpurin lakes.

Carmine red; Natural Red 3 (CI 75460, kermes); Natural Red 4 (CI 75470, cochineal)

Specific louse species, including the Kermes louse in the Mediterranean region and Spain and the cochineal louse in Central America, have high concentrations of red in their carapaces [394, 955]. They are among the oldest dyes for textile application; around 1300 BC in ancient Egypt, kermes seems to have been used for dyeing [23].

The species originally known initially in Europe and employed for dyeing are the Kermes louse *kermes vermilio* and the Polish and Armenian cochineal louse [23]. In the New World, the Mesotecs of Oaxaca harnessed the native cochineal louse *dactylopius coccus*. The dyes consist mainly of carminic acid and kermic acid, and in a few species, also laccic acid, ►Table 4.7:

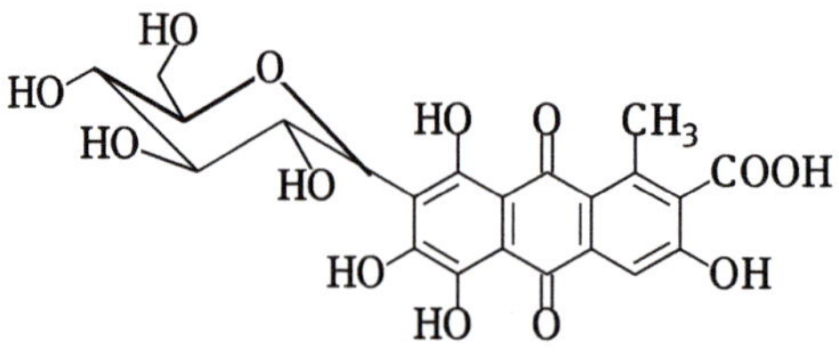

Natural Red 4, CI 75470
carminic acid
carmine Naccarate [KR]

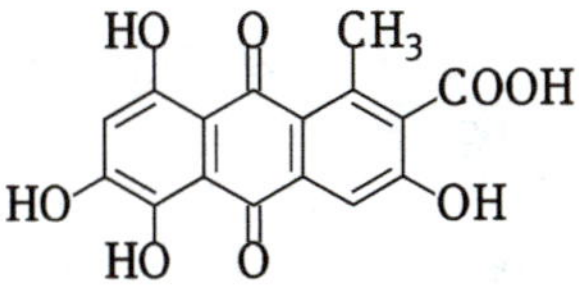

Natural Red 3, CI 75460
kermic acid

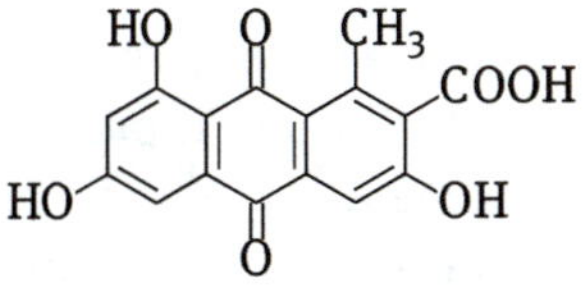

flavoceramic acid, laccaic acid D
yellow

Table 4.7: Composition (in percent) and origin of anthraquinones used for dyeing and laking from lice [70].

Origins	Kermesic acid	Laccaic acid D	Carminic acid	Laccaic acid A–C
Kermes vermilio, Southern Europe, Turkey	75–100	0–25	–	–
Porphyrophora polonica, Eastern Europe	12–38		62–88	–
Porphyrophora hameli, Armenia	1–4		95–99	–
Dactylopius cocca, Mexico, South America	0,4–2		94–98	–
Kerria lacca, India, Southeast Asia	3–9		–	A: 71–96, B: 0–20

Depending on the louse species, the ratio of carminic to kermic acid varies greatly; kermes lice contain almost only kermic acid, and cochineal lice almost only carminic acid. When the Spaniards introduced cochineal lice to Europe in the sixteenth century, they rapidly displaced the native kermes lice since cochineal lice have a dye content of 10 to 14 %, which is about ten times higher than that of the native species. On top of that, the American lice (and the Armenian ones) contained almost exclusively red carminic acid. In contrast, the European Polish cochineal louse contained a high proportion of yellow flavokermic acid.

Since all acids are water-soluble, the pigments needed for painting are obtained by laking, ►Section 2.6. Due to the natural occurrence of Kermes lice, red color paints in Europe were initially based on kermic acid. However, with the import of the color-intensive cochineal from America, a loss of importance of Kermes took place in the sixteenth century. Since then, cochineal carminic acid lakes have dominated. Besides their importance in dyeing [70], carmine lakes were famous in miniature painting and book illumination in the Middle Ages (►Section 8.3) but too costly for panel painting until the seventeenth century. In the eighteenth or nineteenth-century paintings, they often appear next to madder lake [391–393].

Best known is the Al or Ca-Al lake of carminic acid, which provides intensely dark red, transparent *carmine* or *crimson*. (In ►Figure 4.10, the color purity of carmine red is exemplified by the steep absorption edge of carmine Naccarate, an Al lake of carminic acid from cochineal.) The lake is precipitated from the hot, aqueous extract of cochineal lice with iron-free Al^{III}, Ca^{II}, and sodium carbonate. Then, by adding baryte white or kaolin, lighter shades are obtained. Two other lakes are not lightfast and have gained no importance for painting: the scarlet Sn lake (*cochineal scarlet*), discovered around 1640, and the purple Fe-Pb lake *carmine purple*.

The structure of the Al lake is proposed according to formula I [713], for the Ca-Al lake, according to formula II [714]. Like alizarin lakes, carmine lakes possibly have structures similar to Turkish red, ►Section 2.6.1.

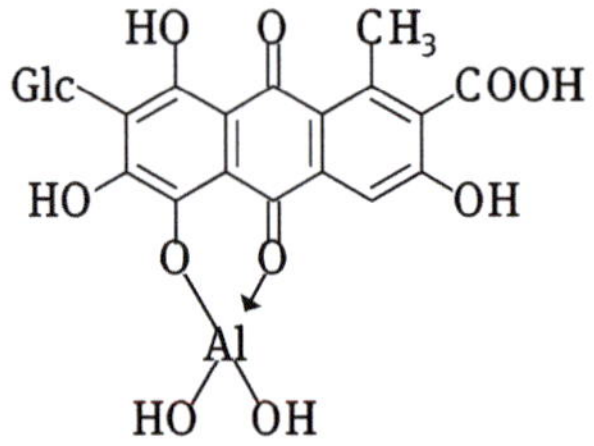

I, carminic acid Al lake

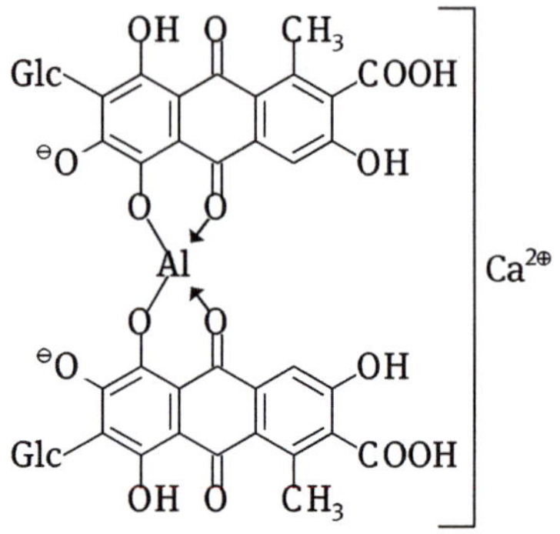

II, carminic acid Ca-Al lake

Lac dye, Natural Red 25

The louse species *kerria lacca* produces with the yellow *erythrolaccin* a related compound [955] that gives shellac its color, ►Section 8.8.2. Reddish variants of shellac contain a mixture of laccic acids that act as red dyes forming *lac dye* [394]:

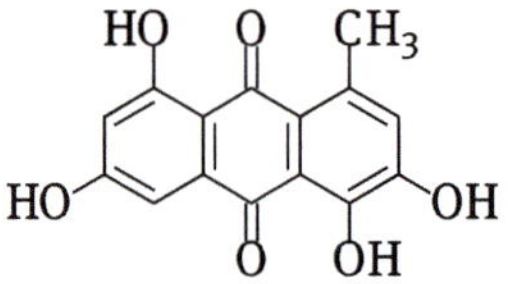

part of Natural Red 25
erythrolaccin

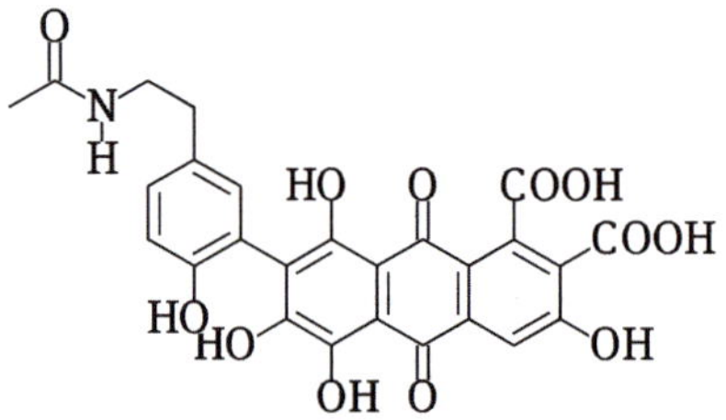

part of Natural Red 25
laccaic acid A

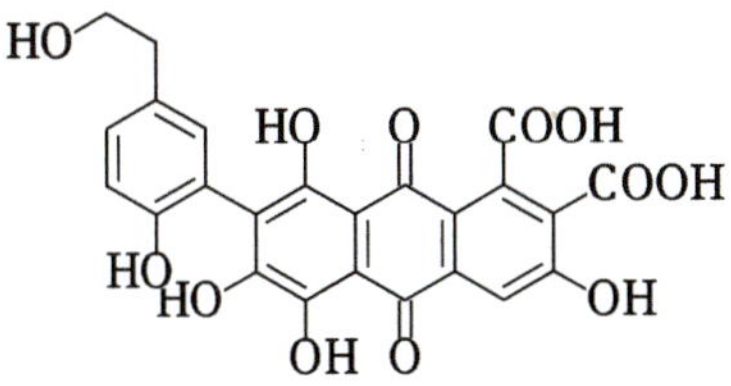

part of Natural Red 25
laccaic acid B

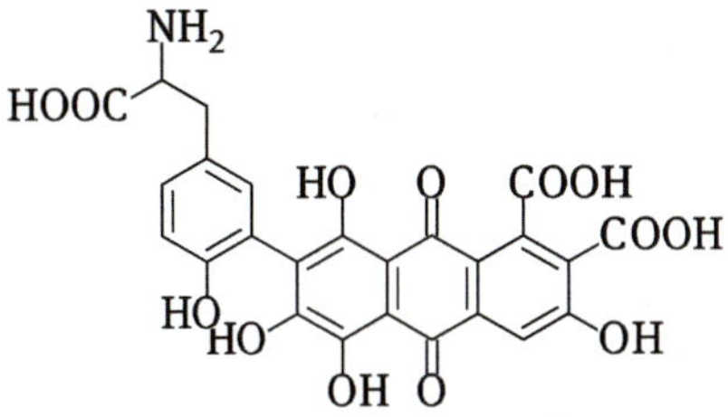

part of Natural Red 25
laccaic acid C

Colored shellac is extracted with water or alkali carbonates before the crimson lake is precipitated with Al^{III} salts or lime [70]. Sn^{IV} gives a scarlet lake, Fe^{III} a purple one.

Since antiquity, the lake has been known in Asia and was applied in the Mediterranean region for dyeing [23]. In European painting, lac dye played a role from the Middle Ages until the nineteenth century, but from 1600 onward, it began to be replaced by carmine red from cochineal lice. Lac dye also possessed some importance for medieval book illumination.

4.6.4 Synthetic quinones

Synthetic colorants based on quinones are derived exclusively from anthraquinone, which frequently carries amino and other functional groups next to hydroxyl groups. Frequently, the quinone is part of a polycyclic structure. The reasons for this are:

- Besides red, many other colors are obtained by choice of the functional groups.
- More highly condensed ring systems increasingly exhibit properties of a pigment and are traded as commercial, high-quality pigments.

4.6.4.1 Hydroxy-anthraquinone

Red Purple Nowadays, the naturally occurring hydroxylated anthraquinones, especially alizarin and purpurin, are only used as starting materials for synthetic pigments [12, Chapter 3.7.2]. According to the structure-color discussion from ►Section 2.5.3.3, they show red, purple, and brown colors. The soluble anthraquinone colorants can be converted into sparingly soluble compounds in two ways:

- by formation of color lakes with metal salts, ►Section 2.6
- by salt formation of anthraquinone sulfonic acids with metal cations

Both groups have lost their importance as pigments in industry today. However, PR83 is an example of a color lake still used in artists' paints. It corresponds to the early known and much sought-after Turkish red, ►p. 349; its complicated structure has only recently been revised, ►Section 2.6.

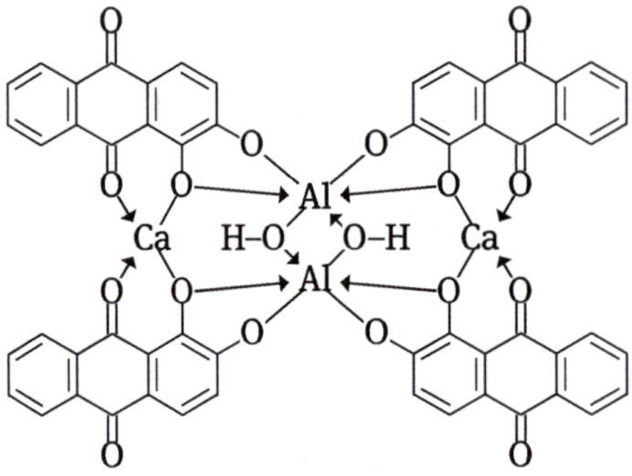

PR83, CI 58000:1
Turkish red
alizarin crimson [WN][DS]

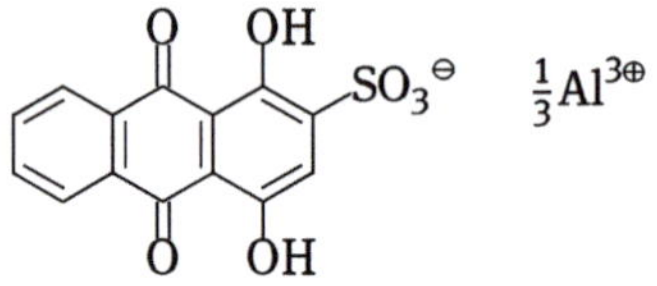

PV5:1, CI 58055
alizarin purple

The purity of alizarin can be regarded as an advantage over the natural product since the natural byproducts of madder root (purpurin and pseudopurpurin) strongly reduce the lightfastness of the lake. This purity is reflected in the lake's hue, as we can see from the steep absorption edge of PR83 in the spectrum, ►Figure 4.10.

An example of a metal salt pigment is the Al^{III} salt of quinizarin sulfonic acid.

4.6.4.2 Amino-anthraquinone colorants

Red Purple Blue ►Section 2.5.3.3 showed that the color range of an anthraquinone could be extended from red to blue. Hydroxyl groups would be replaced with more powerful electron donors such as amino or alkylamino groups.

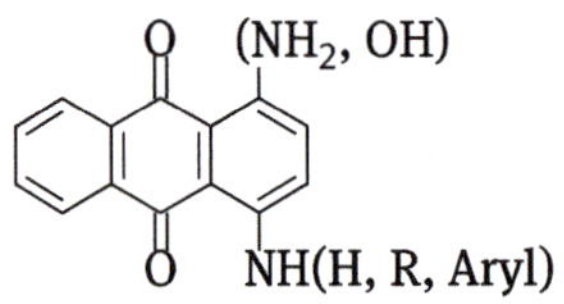

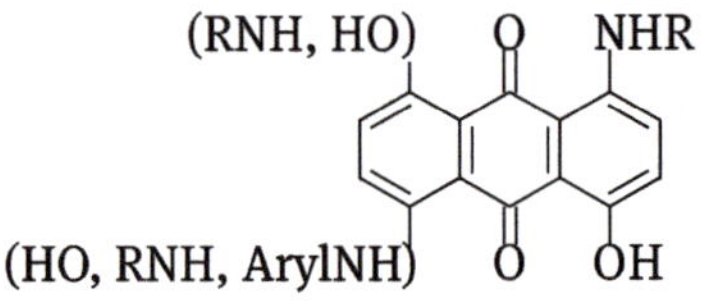

While such compounds are rarely utilized as pigments (PR177 is an exception here) [12, Chapter 3.7.1.3], they are eminently significant as pure red to blue dyes, as two examples may show:

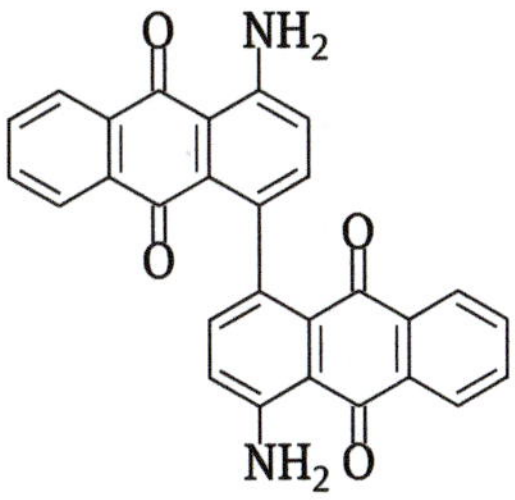

PR177, CI 65300
anthraquinoid red [DS]
permanent alizarin crimson [WN]

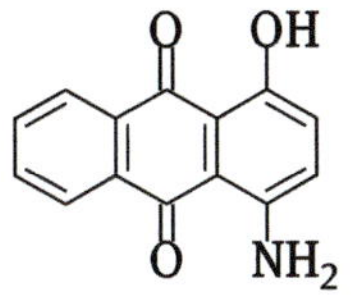

dispersol red A-2B

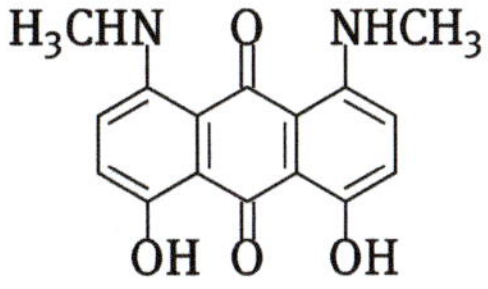

dispersol blue B-G

4.6.4.3 Polycyclic anthraquinone pigments

Yellow Orange Red Purple Blue Condensing several anthraquinones or aminoanthraquinones yields polycyclic compounds applicable as colorants [12, Chapter 3.7.3]. However, in artists' paints, they rarely occur, their main application being in (vat) dyeing and industrial coating technology. Their suitability for vat dyeing is due to their dicarbonyl structure, allowing the colorant to be dissolved as vat and return to its insoluble form, ►Section 2.5.4.4.

We arrive at anthrapyrimidine, indanthrene, and flavanthrone pigments if we start from amino anthraquinone. Anthrapyrimidine pigments [12, Chapter 3.7.3.1] are derived from 1,9-anthrapyrimidine, with PY108 as a commercially successful example, used in artists' paints as a gumdrop substitute:

1,9-anthrapyrimidine

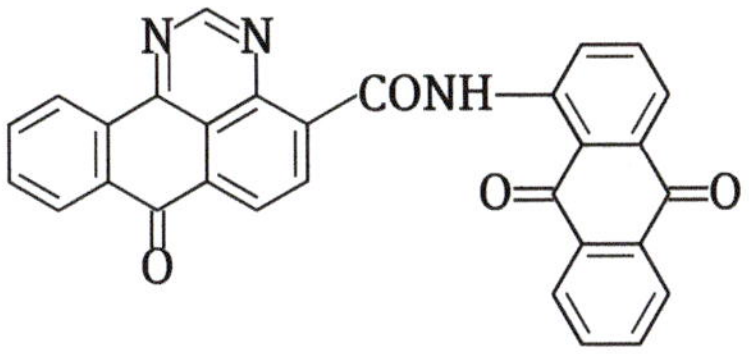

Vat Yellow 20, PY108

Indanthrone and flavanthrone [12, Chapter 3.7.3.2] were synthesized and used as vat dyes as early as 1901. They are among the oldest synthetic vat dyes ever. The basic structure of indanthrone pigments is indanthrone, which acts under the name PB60 predominantly as a vat dye but also frequently in artists' paints, ►Figure 4.11. The former name "indanthrene" became the eponym for this class of extraordinarily high-quality colorants, including halogenated derivatives.

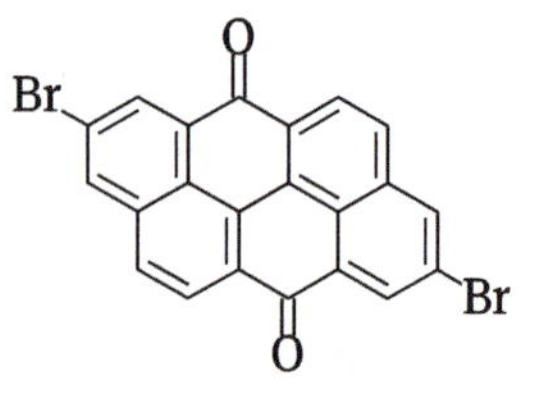

PB60, CI 69800
indanthrone
Delft blue [SC], indanthrene blue [WN]
indanthrone blue [DS]

PR168, CI 59300
4,6-dibromoanthanthrone
primary magenta for four-color printing
anthraquinoid scarlet [DS]

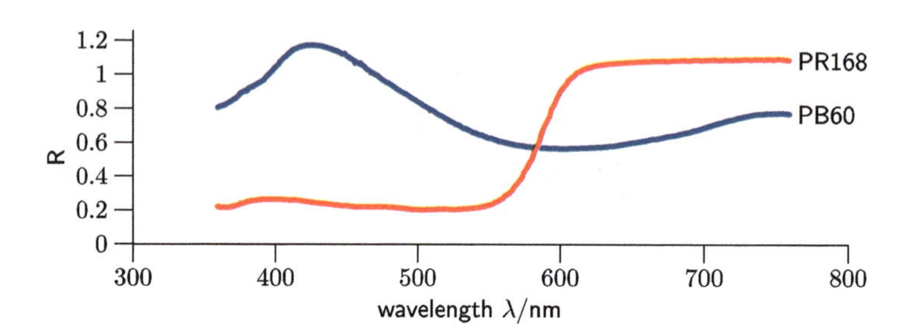

Figure 4.11: Reflectance spectrum of indanthrone- and anthanthrone-based watercolors normalized to an arbitrary unit: PB60 (Delft blue, Schmincke Horadam No. 482) and PR168 (anthraquinoid scarlet, D. Smith Extra Fine Watercolor No. 224).

During the condensation of several anthraquinone structures with no amino groups, we obtain pure carbocyclic compounds, some of which serve as colorants. Although the basic structures are already colored (yellow to red), we can achieve purer color shades and higher fastnesses by halogenation. Only the anthanthrone pigments [12, Chapter 3.7.4.2] derived from the orange type are of interest to artists. These include orange to scarlet red halogenated pigments with their most prominent representative 4,6-dibromo-anthanthrone. Referred to as PR168, it is a possible magenta component in four-color printing. PR168 is also used in artists' paints, ►Figure 4.11.

4.6.5 Cause of color

In simple quinones, a donor-substituted quinone (►Section 2.5.3.3) is present, in which the $n \rightarrow \pi^*$ transition of the carbonyl group is neither critical for the color of natural nor commercial quinone colorants. Instead, the aromatic rings and conjugated carbonyl groups form a complex system of electron acceptors. In natural compounds, the aromatic system and hydroxyl groups act as donors, present in typical substitution patterns. However, in synthetic quinone colorants, powerful donors like amino groups are introduced, which sometimes even replace the hydroxyl groups.

Polycyclic quinones develop color through $\pi \rightarrow \pi^*$ transitions, ►Sections 2.5.4.3 and 2.5.4.4, an example being dibromoanthanthrone. The underlying hydrocarbon anthanthrene is yellow due to its relatively sizeable conjugated π system; the quinone anthanthrenequinone is orange-yellow [307, Volume II, p. 206]:

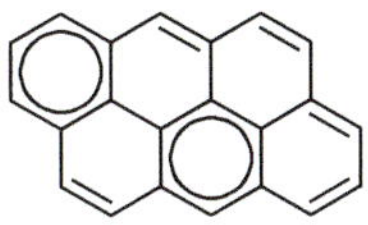

anthanthrene
golden yellow
($p \approx 433$ nm, $\beta \approx 310$ nm)

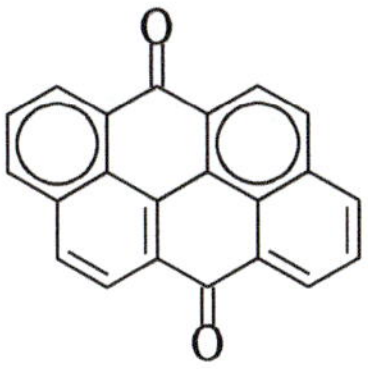

anthanthrone
orange-yellow

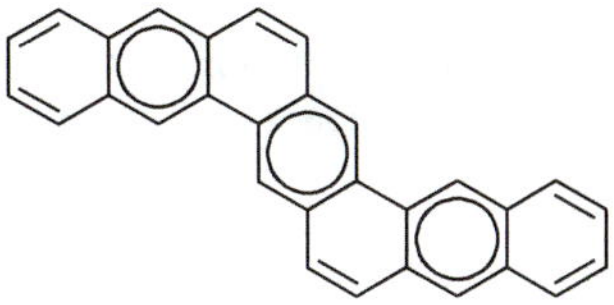

naphtho[2'.3':3.4]pentaphene
yellow
($\alpha \approx 436$ nm, $p \approx 390$ nm, $\beta \approx 330$ nm)

The color of anthanthrone slightly shifts bathochromically because the carbonyl group acts additionally as an acceptor and takes some charge from the two naphthalene subunits. These are only weak donors, but in dibromoanthanthrone, the halogen, an atom capable of mesomerism, participates in this process. As a result, a further bathochromic shift occurs, ►Figure 4.12(a).

In contrast, the deep blue color of the indanthrone does not result from MO transitions in the large molecule. Naphtho[2'.3':3.4]pentaphene, the underlying hydrocarbon, is only yellow [307, Volume I, p. 371]. Due to the presence of three aromatic benzene units, symbolized by the three Robinson circles, the large π system is divided into small fragments. Consequently, only a weak bathochromic shift of the HOMO-LUMO transition into the visible spectral range ensues, ►Section 2.5.4.3. The intense color of indanthrone originates from the fact that indanthrone is a donor-substituted quinone, more specifically, a dimeric amino anthraquinone. When excited with light, electrons flow from the amine nitrogen of the middle rings to the carbonyl oxygen and into the anthraquinone system, ►Figure 4.12(b).

4.7 Indigoid colorants

The indigo molecule presents a colored structure that has been of eminent consequence since antiquity. It provided blue and purple colors, and it was readily available since plants and some animals were natural sources. Early on (around 4000 BC), the Central American cultures knew how to produce an attractive coloring. Moreover, the indigo derivative purple was associated with luxury for a considerable time.

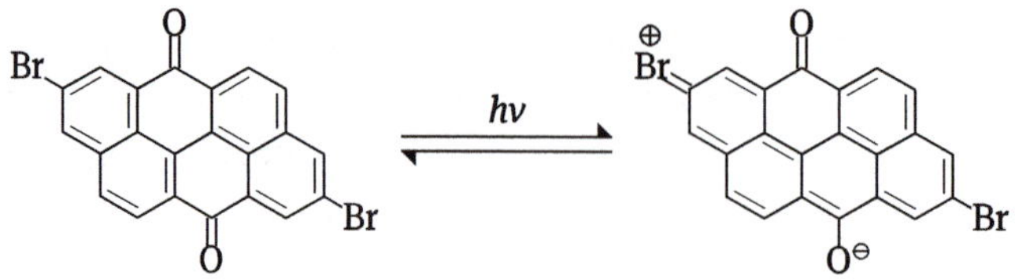

4,6-dibromoanthanthrone
red

(a) Halogens are capable of mesomerism and can provide charge, which can flow into the quinone system. A bathochromic shift results.

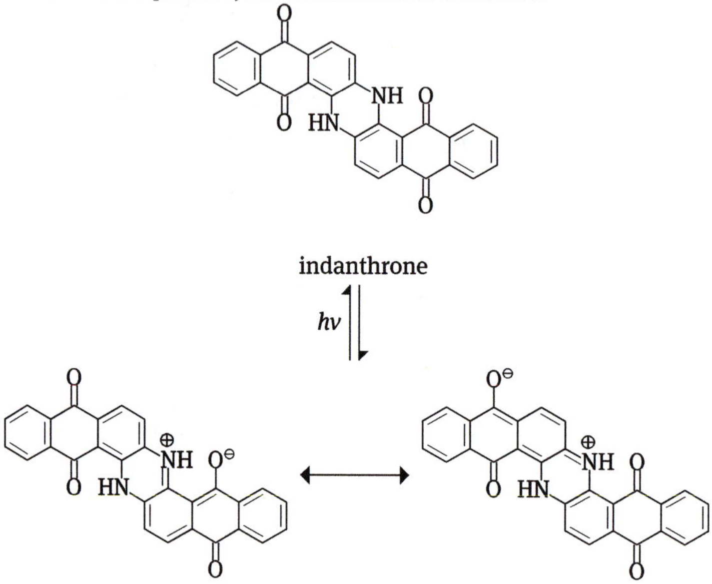

indanthrone

(b) In indanthrone, a dimeric amino anthraquinone, a charge can be transferred from the amine nitrogen to both quinone systems.

Figure 4.12: Some possible processes in polycyclic colorants when excited by light.

4.7.1 Natural indigoid colorants

Blue Purple Since the earliest times, two representatives of the natural indigoid colorants, namely indigo and purple, have been economically crucial because of their attractiveness and applicability as vat dyes. Indigo was also of some interest for painting due to its insolubility in oily and aqueous binders.

Indigo, Natural Blue 1 (CI 75780); synthetic indigo, Vat Blue 1 (PB66, CI 73000)

Indigo [12, Chapter 3.6], [48, Volume 3], [23] is a plant extract and one of the oldest colorants at all: Egyptian mummies from around 1500 BC were found with cloth already dyed with indigo [23]. Caesar mentioned that the Britannians painted their bodies blue using woad, a plant yielding indigo. Although it was mainly used for dyeing in ancient times, it was also imported as a pigment and appeared in European panel painting in the twelfth century. To this day, natural and synthetic indigos defend their place in artists' paints, especially in watercolors. Along with indigo and purple, we will look at how the Mayans combined indigo with clays to produce a bright turquoise pigment, ►Figure 4.13.

Natural Blue 1, CI 75780, Vat Blue 1, PB66, CI 73000
indigo
indigo [SC]
PB84, Mayan blue genuine [DS]

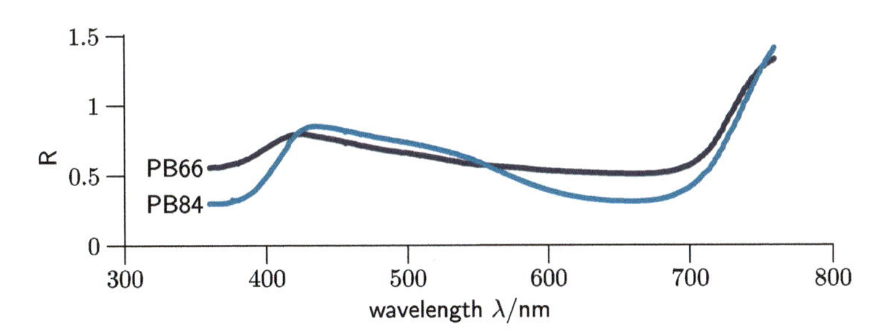

Figure 4.13: Reflectance spectrum of indigo-based watercolors normalized to an arbitrary unit: PB66 (indigo) and PB84 (Mayan blue genuine, D. Smith Extra Fine Watercolor No. 211).

Unlike other plant extracts, indigo is insoluble in oil and water and can be used directly as a pigment. However, since it is present in the plants as a colorless enol form (indoxyl glucoside), it must first be converted into its colored type by fermentation and subsequent air oxidation.

Natural indigo is readily obtainable from various dye plants, so indigo, as mentioned above, was already known long before the discovery of America:

- dyer's woad, *isatis tinctoria*, in Asia (Caucasus), Europe
- indigo plant, *indigofera tinctoria*, in Asia (India), Africa, America

The plants, especially their leaves, or sometimes as a whole, were pounded and left to ferment for 1 to 2 weeks. After that, small pellets were formed from the pulp, then treated with urine and left again for another long fermentation time. Finally, a vat was prepared for use by mixing the pellets with urine and potassium carbonate.

During all of these steps, no blue coloring appears since the plants contain only the colorless precursor indican, a glucoside: The glycosidic bond is cleaved by acid (HCl) or enzymes yielding the likewise colorless indoxyl. Eventually, indoxyl passes by oxidative condensation into the actual pigment indigo. By beating the vat with rods, air provided the necessary oxygen:

O–β-D-Glcp; N H — HCl, enzymes / – D-Glc → O; N H — × 2, + O_2 / – 2 H_2O → O, H N; N H, O

indican indoxyl indigo

Maya blue, nowadays Pigment Blue 84

Indigo further appears in a surprising shade: Since the year 800, for some time, *Maya blue* materialized in the mural paintings, ceramics, and books of the Mayan people.

In contrast to the matt, dark blue indigo, Maya blue is a bright turquoise pigment that is exceptionally resistant to light, chemicals, and temperature. Many shades of the color can be achieved: depending on the concentration of indigo, light cerulean-blue and sea-green colors emerge; depending on the pH value during production, turquoise to pure blue colors arise.

The pigment is composed of indigo molecules embedded in the structure of sepiolite or palygorskite clay. It is prepared simply by mixing the components and heating them to at least 100 °C. While it is heated for one day, the dark indigo color fades rapidly and lightens in the further course.

The exact structure is not yet evident; Hunger and Schmidt [12, Chapter 4.5.11] speak of a hybrid pigment made from clay and natural plant and dye extracts. Polette-Niewold [430] and Chiari [431] represent a detailed overview of the topic. According to these sources, the structure forms through the intercalation of indigo in the substrate, efficiently protecting the pigment against color loss or change. Electronic interactions occur during the intercalation process:

- The carbonyl oxygen of indigo binds to aluminum ions of the structure.
- The geometry of the indigo molecule alters.
- Dehydroindigo forms.

How dramatic the color change is due to the formation of such complexes illustrates [430] using the example of red thioindigo, which develops a whole range of colors from purple to blue upon intercalation.

Overall, Maya blue demonstrates how the color of a pigment is not solely determined by the color-active substance alone but can also be altered to a large extent by the emergence of a superordinate structure.

Purple, Natural Violet 1

Another colorant based on indigo is *purple* [23, 70, 457]. In antiquity, purple was already produced by the Phoenicians in Tyre on the Mediterranean coasts to a considerable extent. However, its production originates probably in Minoan Crete [23] as early as 1600 BC. Unlike indigo, purple was not extracted from plants but animals: areas of the Mediterranean provide the habitat of some species of purple snails. The hypobranchial glands of the snails contain only tiny quantities of the colorless precursors of purple; therefore, numerous glands are needed to produce one gram of purple.

Due to its rare shade (a reddish-purple) and scarce occurrence, purple was a precious, coveted color, fabulously expensive, and reserved only for high-ranking personages. Therefore, until the second millennium AD, this color demonstrated the reification of power and wealth in the form of senatorial, imperial, and ecclesiastical purple and established the wealth of entire cities and empires. Finally, in 1464, purple lost its significance when Pope Paul II decreed that church robes had to be dyed with domestic kermes dyes.

Natural purple consists of a series of similar compounds based on the structures of isatin, indigotine, and indirubine. Their essential difference is the degree of bromination [454, 455]. The significant component is 6,6'-dibromo-indigotine:

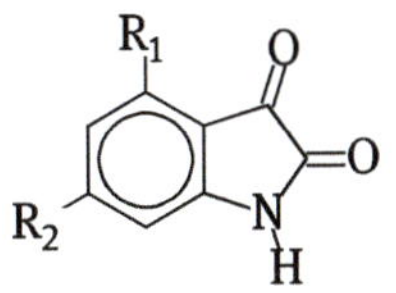

isatinoids
isatin (IS), R_1=R_2=H
4-bromo-isatin (4BIS), R_1=Br, R_2=H
6-bromo-isatin (6BIS), R_1=H, R_2=Br

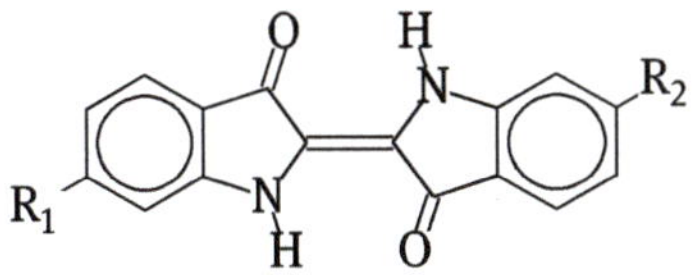

indigoide
indigo (IND, indigo), R_1=R_2=H
6-bromo-indigo (6MBI), R_1=Br, R_2=H
6,6'-dibromo-indigo (66'-DBI, purple, Natural Violet 1), R_1=R_2=Br

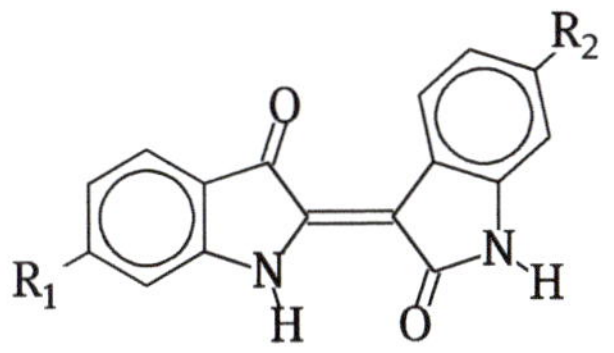

indirubinoids
indirubin (IR), R_1=R_2=H
6-bromo-indirubin (6MBIR), R_1=Br, R_2=H
6,6'-dibromo-indirubin (66'-DBIR), R_1=R_2=Br

The gland of snails does not contain the finished purple but various sulfated precursors, which are converted into the components of purple by exposure to oxygen and light [455, 456].

A peculiarity of purple is that its leuco form loses the halogens under solar radiation and turns increasingly blue. At first, due to the loss of bromine, blue 6-bromo-indigo emerges, then finally indigo. The Phoenicians were aware of this phenomenon, and they used it to produce blue cult robes dyed with purely animal material for ideological reasons.

4.7.2 Synthetic indigoide colorants

Red Purple Blue Nowadays, many derivates of natural indigoid colorants are available [12, Chapter 3.6.2]. The development of synthetic pathways to artificial indigo substituted its natural form. Moreover, numerous derivatives could be prepared, which have the structure I; they are bridged via various heteroatoms instead of nitrogen: –NH– (indigo), –NR–, –S– (thioindigo), –O– or –Se–:

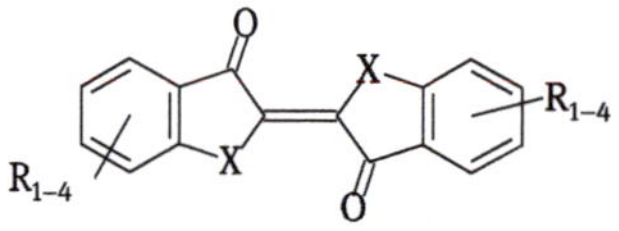

I, indigo pigments, yellow to purple
X = NH, R = H: indigo, PB66, CI 73000
X = S, R = H: thioindigo

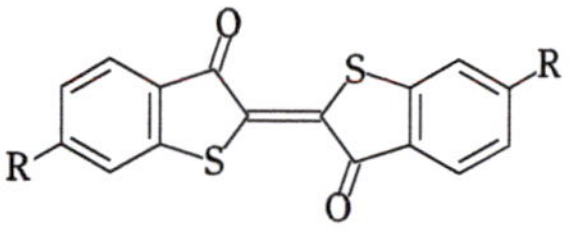

6,6'-diethoxy-thioindigo, yellow-orange, R = O–CH_2–CH_3
6,6'-dichloro-thioindigo, orange-red, R = Cl
6,6'-dinitro-thioindigo, red-purple, R = NO_2

Substituents have little effect on the blue color in indigo. Instead, they significantly vary the hue, covering the range from yellow to purple in the heterosubstituted compounds. Industry interest has therefore shifted from indigo to thioindigo.

Despite their flexibility in color, only a few thioindigo derivatives are important today. Their fastness properties are not high since they are strongly dependent on the position of the substituents. In artists' colors, indigo is often already displaced by mixtures of ultramarine blue and copper phthalocyanine.

Pigments based on thioindigo span the color range from red to purple to brown and possess structure IV, essentially substituted with chlorine and methyl (PR88 red-purple, PR181 bluish-red). They exhibit excellent light, weather, migration, and solvent fastnesses and are used in the coatings sector, while indigo itself and thioindigo are partly applied in spinning and plastics coloring. ►Figure 4.14 displays the reflectance spectrum of thioindigo as an example for purple indigoid compounds:

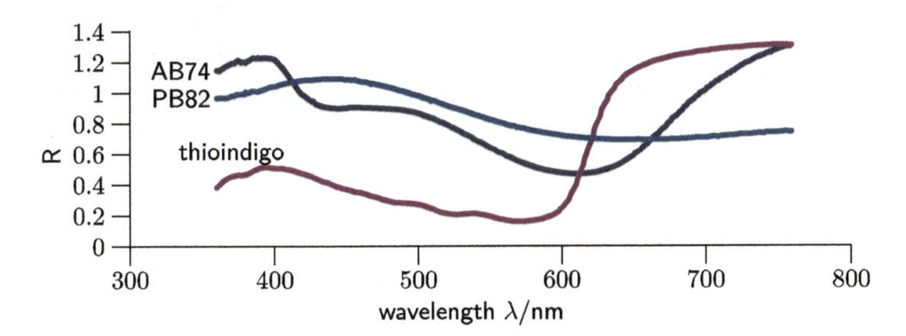

Figure 4.14: Reflectance spectrum of synthetic indigo-based dyes and watercolors normalized to an arbitrary unit: AB74 (indigocarmine, Natural Blue 2, E132), PB82 (Mayan dark blue, D. Smith Extra Fine Watercolor No. 213), and thioindigo (Wallace Seymour Vintage Watercolour).

Vat Red 41, CI 73300
thioindigo, red
thioindigo [WS]

IV, thioindigo pigments

PR88, CI 73312
4,4',7,7'-tetrachloro-thioindigo, blue

PR181, CI 73360
red

In addition to the substitution pattern, the position of the methylene bridge can also be changed from 2,2' to 2,3' or 3,3' (structure VII):

VII, thioindigo scarlet

VIII, AB74, CI 73015, Natural Blue 2, E132
indigocarmine, indigotine, red
PB82 (precipitated on clay), Mayan dark blue [DS]

Indigocarmine, Natural Blue 2, Acid Blue 74, E132

Around 1740, Barth discovered indigo carmine, a water-soluble direct dye with structure VIII, ►Figure 4.14. It facilitated dyeing and spread rapidly. Unfortunately, indigo carmine remains water-soluble even when aluminum mordant is applied, so dyeings fade after a certain amount of time. Chemically, indigo disulfonic acid forms when indigo is dissolved in concentrated sulfuric acid.

Pigment blue 82 (PB82)

PB82 denotes indigocarmine precipitated on clay, ►Figure 4.14.

4.7.3 Dyeing with indigo and derivatives

The coloring power of indigo is not based on a chemical bond between the colorant and the fiber. Instead, indigo particles hold in the fabric purely mechanically and wash out relatively quickly (see jeans). The basis of the dyeing process represents the following equilibrium:

$$\text{indigo} \underset{[O]}{\xrightarrow{+ Na_2S_2O_4,\ +4\ NaOH;\ -2\ Na_2S_2O_3,\ -2\ H_2O}} \text{leuco indigo}$$

indigo

leuco indigo

Blue indigo, obtained from plants or synthetically, is finely dispersed and reduced with sodium dithionite $Na_2S_2O_4$. In the past, soaking indigo in a fermenting mash for several days achieved the reduction effect.

Now, the material to be dyed is immersed in the leuco base solution. Subsequently, it is reformed into a pigment by final oxidation. In the case of indigo, oxidation to the blue form can already take place in the air. In the past, the dyeing material was spread out simply in the open: the fabric took on a blue color within a few hours. (From the sphere of this old craftmanship originated the proverb “to make blue”: with their hands in their pockets, so to speak, the dyers could watch their fabric turn blue, and thus had “made blue.”)

4.7.4 Cause of color

Indigo derivatives owe their color to a special carbonyl chromophore, called *H chromophore* I, which we have already discussed in ►Section 2.5.3.2. It is a donor–acceptor system with oxygen as acceptor Y and two amine or sulfur bridges as electron donors X (II):

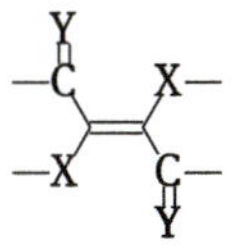

I, H chromophore
e. g., X = –S–, –NH–, Y = O

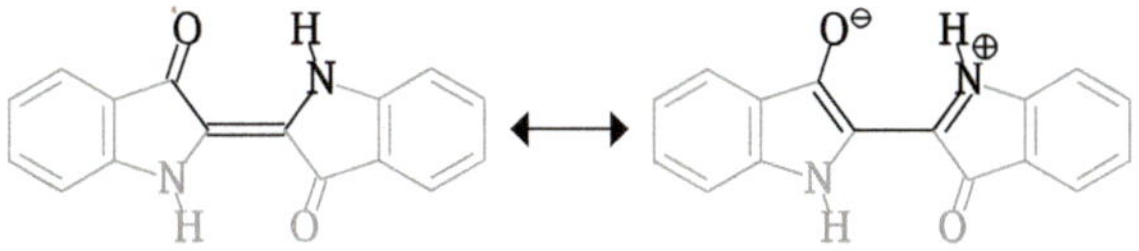

II, carbonyl chromophore

The two phenyl rings have no significance for the color of the compounds and are either retained in technical derivatives for technical reasons or replaced by naphthalene rings. Compound I is already colored with no benzene rings, and the carbonyl groups can be simplified by acceptors such as boron in compound II.

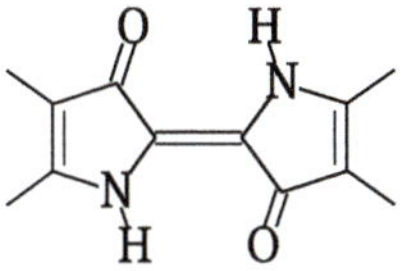

I

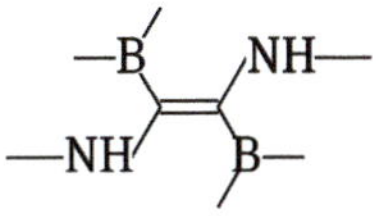

II

4.8 Polymethine colorants: di- and triarylmethines, quinone imines

As we outlined in ►Section 2.5.5, polyenes with an odd number of carbon atoms between an electron donor D and an electron acceptor A, the so-called polymethines, are strong chromophores. By starting from the shortest polymethine I, we obtain a large number of colorants by structural variation of D, A, and the central atom X in several steps, ►Figure 4.15:

- In the first step, we set X = C and apply the phenylogy principle to structure I. We obtain larger molecules II and III, which are electronically equivalent to I. The central atom X = C can carry two or three aryl groups, yielding *diarylmethines* II and *triarylmethines* III, respectively. The first two aryl residues are essential for inducing color; the third one modifies the hue. II and III are parent structures for numerous dyes.
- In a second step, we can replace the central atom X with N to get the heteroanalogous *azomethines* and *quinone imines* IV. However, for our subject, only IV is of limited relevance.
- In a third step, we can bridge the two essential aryl residues in II and III by nitrogen, oxygen, or sulfur to yield complex trinuclear aromatics V and VI. More specifically, we obtain diazines (phenazines), oxazines, and thiazines V and, respectively, acridines, xanthenes, and thioxanthenes VI.

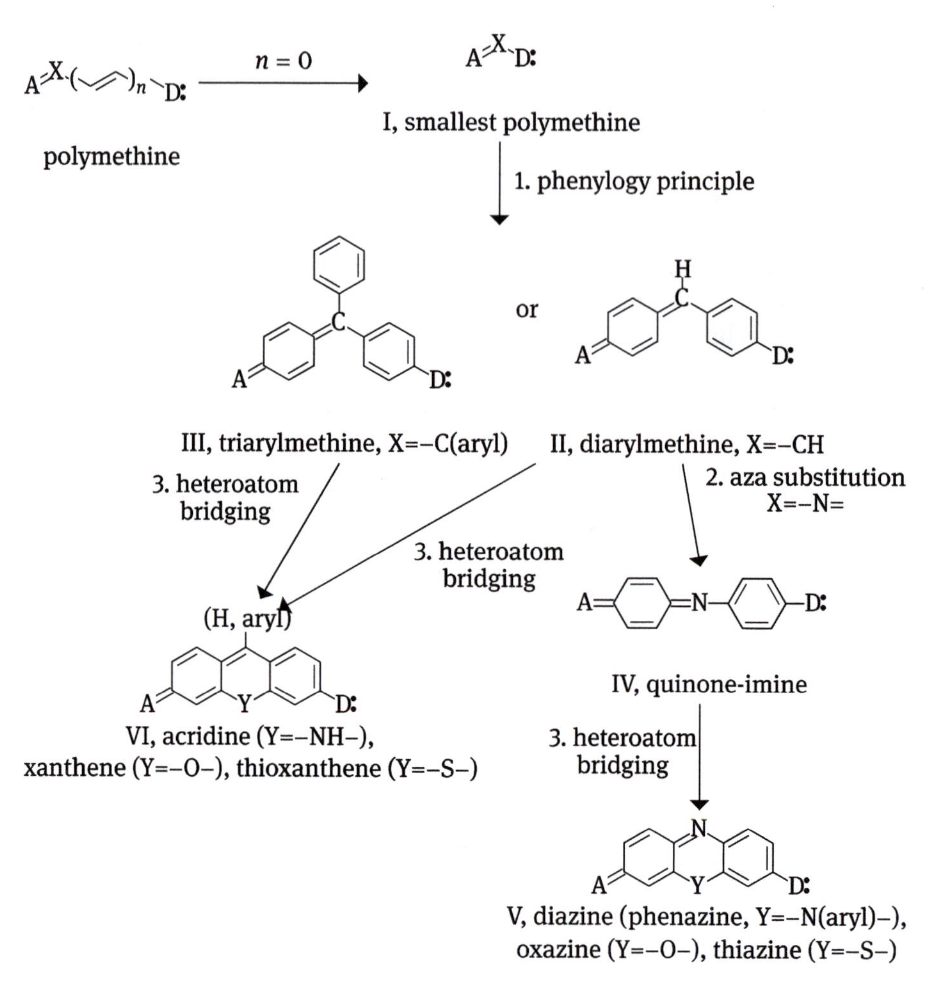

Figure 4.15: Industrially applied types of polymethine colorants. Three modification steps 1 to 3 lead from the smallest polymethine I to several related types II–VI of dyes.

– In the last step, we select A and D as fits our application's needs. Frequently, O or $NR_2^{\oplus}$ act as acceptors, and OH or NH_2 act as donors.

In principle, the intensive coloring of these groups' members can be explained by their derivation from polymethines, which will be discussed in more detail in the following sections. The elementary process is the mesomeric equilibrium:

II, diarylmethine

4.8.1 Triarylmethine colorants

Triarylmethine III in ►Figure 4.15, having X = =C(aryl)– as a typical element, is the parent structure of numerous triarylmethine dyes. By applying the specific substitution X = =C(C_6H_5)–, we obtain the most significant group among them, the so-called *triphenylmethines*, which are particularly influential as colorants [12, Chapter 4.1]. We distinguish two types depending on electron acceptor A and electron donor D, ►Figure 4.16:
- benzaurine type II, A=O, D=OH, X=C–Aryl, e. g., benzaurine
- malachite green type III, A=$NR_2^{\oplus}$, D=NR_2, X=C–Aryl, e. g., malachite green

Colorants of the benzaurine type II are dyes (e. g., indicator dyes such as phenolphthalein). In contrast, the malachite green type III is significant for pigment chemistry: the triphenylcarbonium cation can form soluble alkali salts IV. However, it can also be converted into pigments by forming inner salts V or salts with complex heteropolyacids VI.

Benzaurine type (xanthene group)

Yellow Red No pigments are derived from benzaurine. Instead, by introducing an oxygen bridge, we obtain a xanthene structure parental to several vital dyes for fluorescence and writing inks, the diversity obtained by halogenation, ►Table 4.8 and ►Figure 4.17.

As we will discuss later for malachite green derivatives, two aryl substituents are required at the central carbon atom to induce color. The third aryl group only changes the hue. Nonbridged compounds such as benzaurine itself release their excitation energy by molecular movements after irradiation with light. These movements are impossible for xanthene dyes since the bridge forces both aryl-rings relevant for color into a planar, rigid xanthene structure. This rigid structure provides ideal conditions for a delayed release of energy, and thus for fluorescence; therefore, xanthenes are typically fluorescent dyes, suitable for highlighter inks. (Similarly, many planar aromatic hydrocarbons are capable of fluorescence.)

Benzaurine type (phthalein group)

By carboxylation of benzaurines, we obtain the so-called phthaleins, another class of dyes, phenolphthalein being their best-known representative and also the simplest phthalein. The primary reaction product, a carboxylic acid, is in equilibrium with its lactone that forms through an internal nucleophilic attack of the carboxyl oxygen onto the central carbon atom:

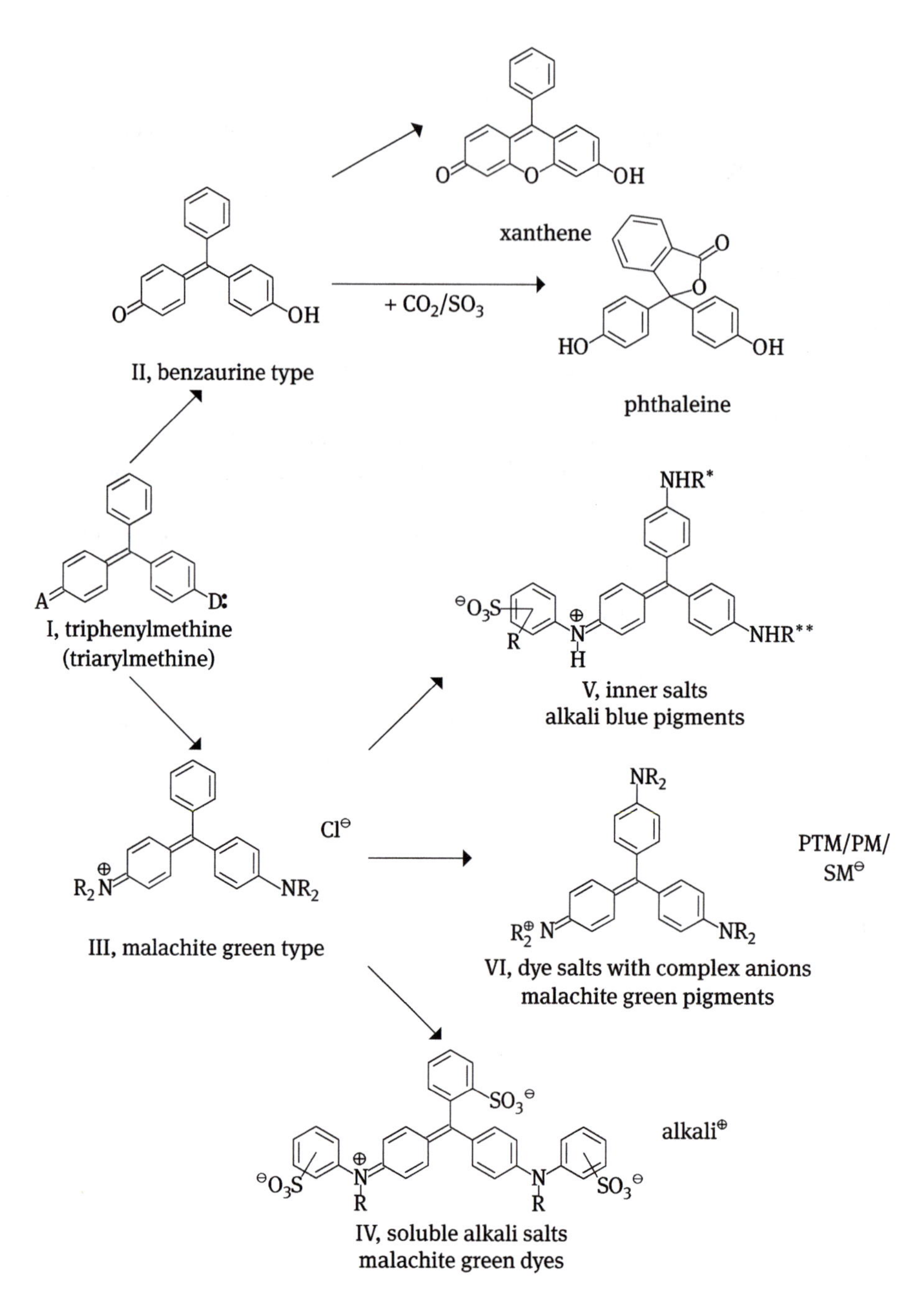

Figure 4.16: Industrially applied types of triphenylmethine (triarylmethine) colorants. The nature of donor and acceptor in I is responsible for the difference between the benzaurine type II (only dyes) and the malachite green type III (dyes and pigments).

Table 4.8: Typical benzaurine dyes with xanthene structure.

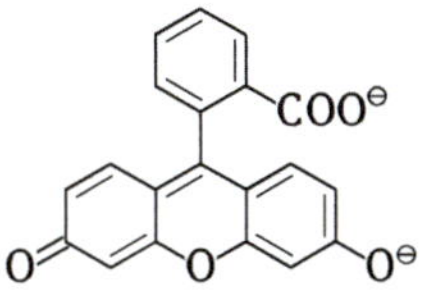 AY73, CI 45350 fluorescein, yellow highligther ink	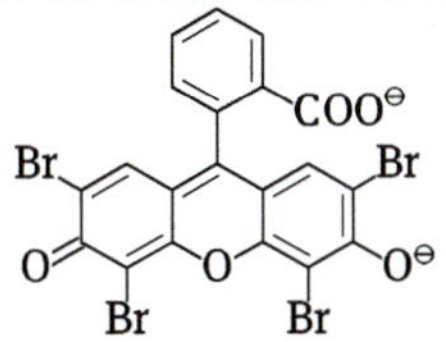 AR87, CI 45380 eosin yellowish, red highligther ink
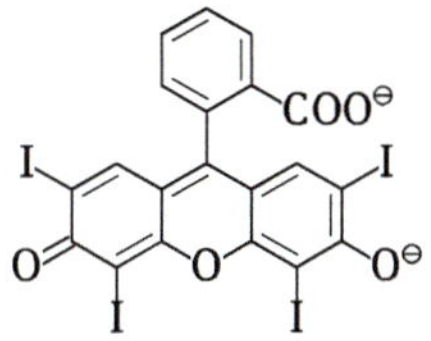 AR51, CI 45430, Food Red 14, E127 erythrosine, eosin J, red writing ink	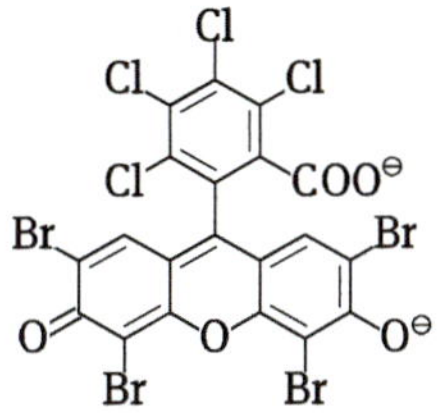 AR92, CI 45410, Food Red 104 eosin B, phloxine, red writing ink

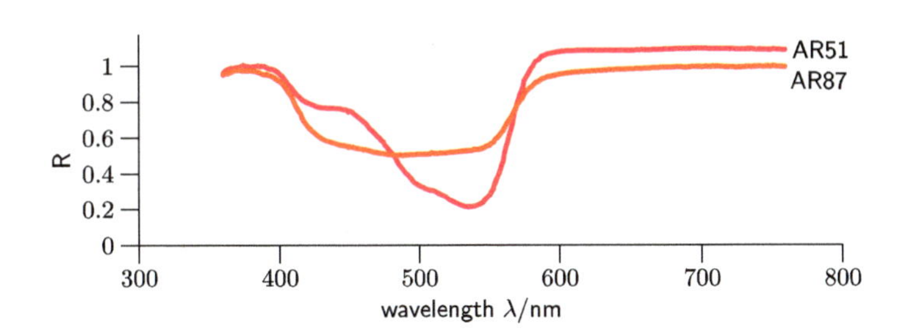

Figure 4.17: Reflectance spectrum of benzaurine-based colorants normalized to an arbitrary unit: AR51 (erythrosine, FR14, E127), AR87 (eosin yellowish).

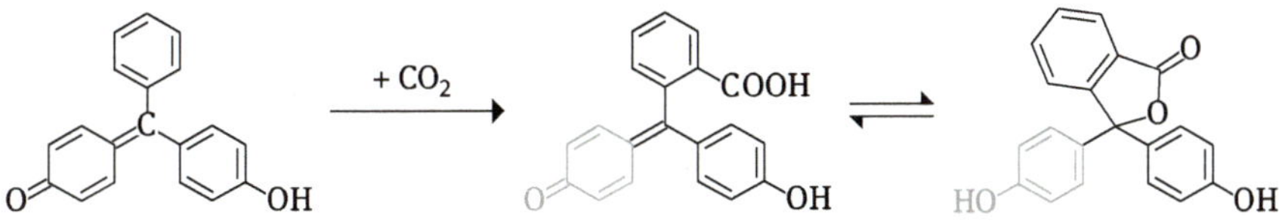

+ CO_2

benzaurine

phenolphthalein

The lower left phenolic structure drawn in gray is the reason for the name: "phenolphthalein" is formed from two phenol molecules (and a benzene ring). Similarly, two resorcinol molecules combine into a phthalein named resorcinolphthalein, better known as fluorescein. In a condensation reaction, one hydroxyl group of each resorcinol molecule forms the oxygen bridge, yielding a xanthene structure:

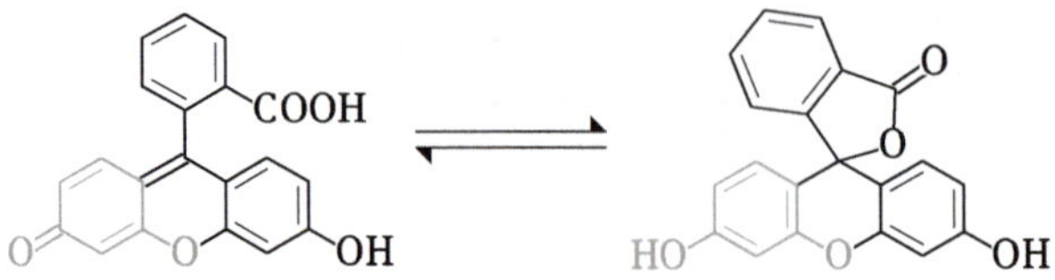

fluorescein
resorcinolphthalein

In the same way, we can use o-cresol, m-cresol, and thymol as building blocks for phthaleins. We can also introduce a sulfonic acid group instead of a carboxyl group to obtain sulfo-lactones, ►Table 4.9. Due to a pH-dependent cleavage of the lactone ring associated with a color change, the table lists many indicator dyes. The pH-dependent equilibrium is the following:

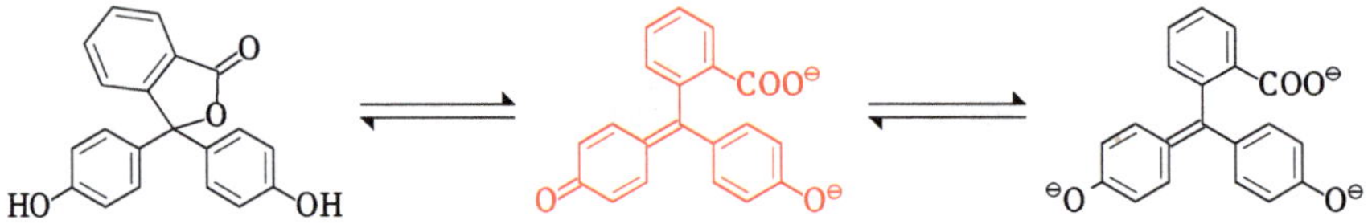

phenolphthalein
acidic or near neutral
colorless

alkaline
pink

strongly alkaline
colorless

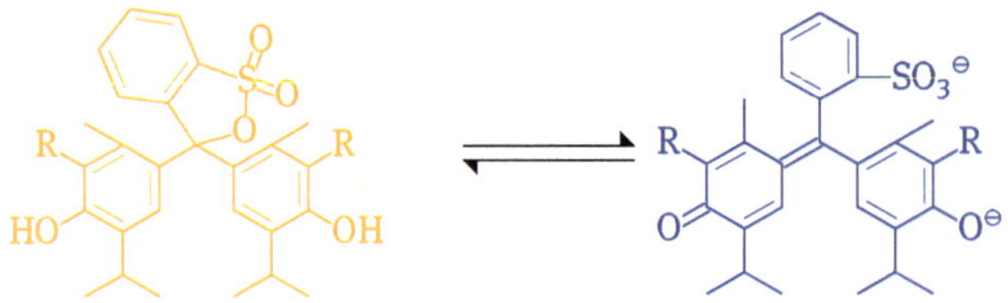

thymol blue, R=H, colorless
bromothymol blue, R=Br, yellow

blue
blue

Table 4.9: Various phenols from which phthaleins and sulfo-phthaleins are derived.

Building block	Carboxylated phthalein	Sulphonated phthalein
Phenol	Phenolphthalein	Phenol red
Resorcinol	Fluorescein, eosine, rose bengal	
o-Cresol	o-Cresol-phthalein	Cresol red
m-Cresol	m-Cresol-phthalein	Cresol purple, bromocresol green
Thymol	Thymol-phthalein	Thymol blue, bromothymol blue

Malachite green dyes

Red Purple Blue Green Malachite green dyes are characterized by nitrogen both in the acceptor and the donor. Their positive charge is delocalized over the molecule and must be balanced by a counterion. ►Table 4.10 lists some malachite green dyes, the counterion being chloride ($X^{\ominus} = Cl^{\ominus}$). Each dye corresponds to an insoluble dye salt if we replace chloride with anions of heteropolyacids such as phosphomolybdic acid or phosphotungstic acid.

Table 4.10: Typical malachite green or triphenylmethine (triarylmethine) dyes and their corresponding pigments. PM represents dodecamolybdate phosphate $P(Mo_3O_{10})_4^{3\ominus}$, PT represents dodecatungstate phosphate $P(W_3O_{10})_4^{3\ominus}$, and PTM is the mixture of PM and PT.

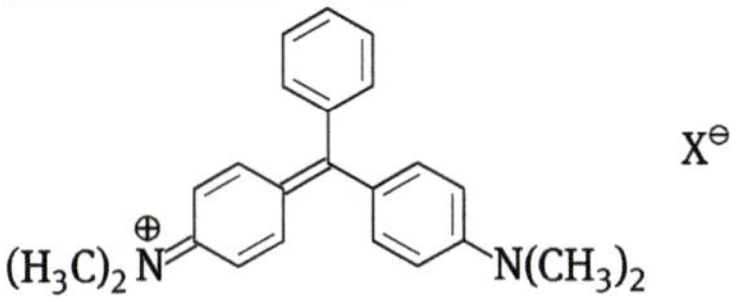

$(H_3C)_2\overset{\oplus}{N}$ … $N(CH_3)_2$ $X^{\ominus}$ BG4, malachite green, X=Cl, writing and inkjet ink PG4, CI 42000:2, X=PTM, PM	$(H_5C_2)_2\overset{\oplus}{N}$ … $N(C_2H_5)_2$ $X^{\ominus}$ BG1, Brilliant Green G, X=Cl PG1, CI 42040:1, X=PTM, PM
$N(CH_3)_2$; $(H_3C)_2\overset{\oplus}{N}$ … $N(CH_3)_2$ $X^{\ominus}$ crystal violet, X=Cl PV39, CI 42555:2, X=PTM, PM	NHC_2H_5; $(H_5C_2)_2\overset{\oplus}{N}$ … $N(C_2H_5)_2$ $X^{\ominus}$ BB7, Victoria Rein Blau B, X=Cl, writing ink PB1, CI 42595:2, X=PTM, PM PB62, $X=Cu_3^I[Fe^{II}(CN)_6]$
$COOCH_2CH_3$; H_3C; CH_3; $H_5C_2H\overset{\oplus}{N}$ … O … NHC_2H_5 $X^{\ominus}$ BR1, rhodamine 6G, X=Cl, writing ink PR81, CI 45160:1, X=PTM	COOH; $(H_5C_2)_2\overset{\oplus}{N}$ … O … $N(C_2H_5)_2$ $X^{\ominus}$ BV10, rhodamine B, X=Cl, inkjet ink PV1, CI 45170:2, X=PTM
NH_2; CH_3; $H_2\overset{\oplus}{N}$ … NH_2 $Cl^{\ominus}$ fuchsin	$N(CH_3)_2$; $(H_3C)_2\overset{\oplus}{N}$ … $N(CH_3)_2$ $X^{\ominus}$ BV1, methyl violet, X=Cl PV3, CI 42535:2, X=PTM, PM PV27, $X=Cu_3^I[Fe^{II}(CN)_6]$

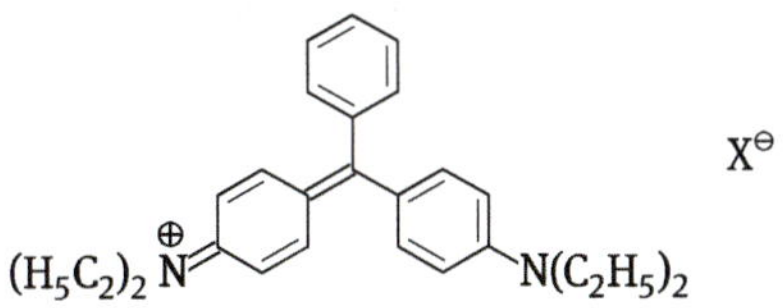

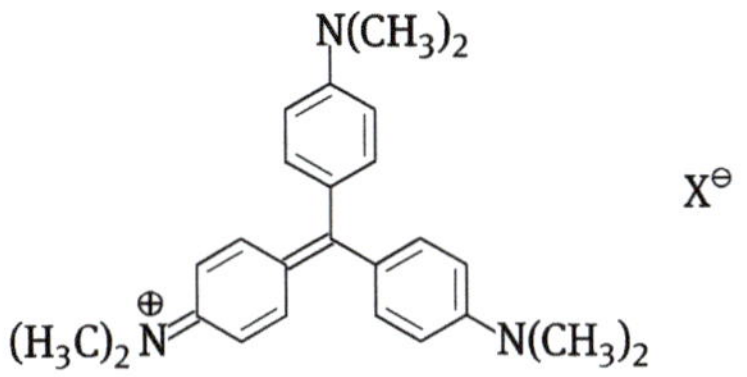

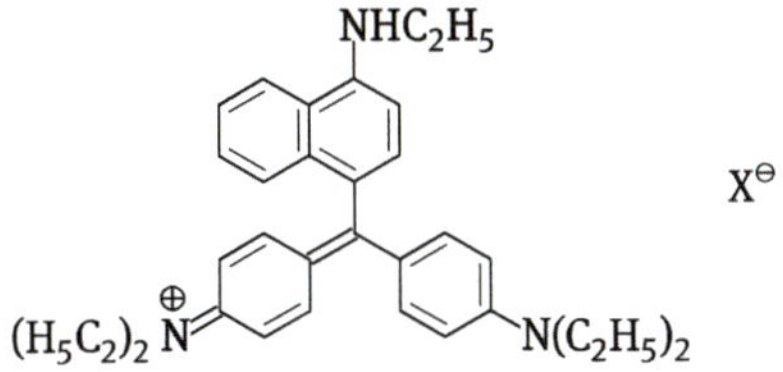

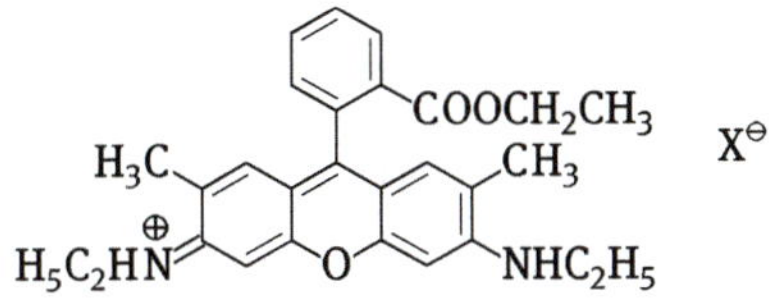

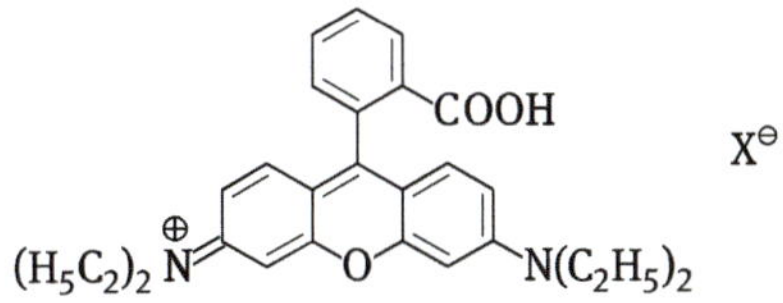

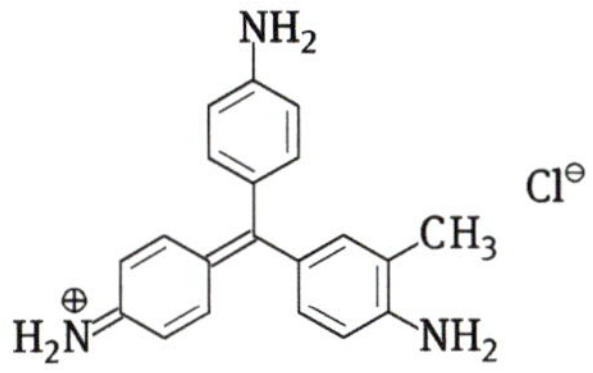

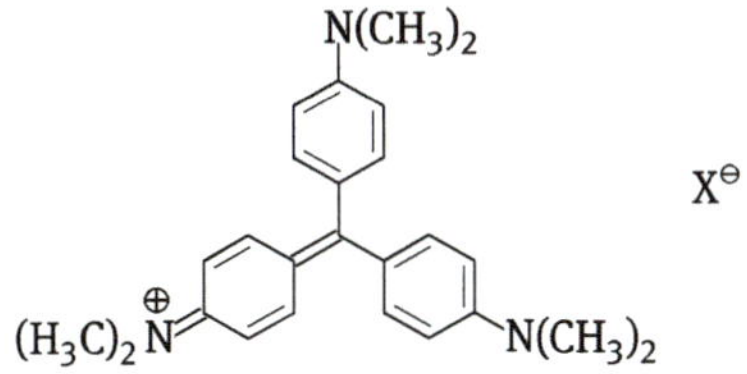

Malachite green dyes exhibit brilliant shades of red, purple, blue, and green, ►Figure 4.18. Unfortunately, their lightfastness is low, and they are also unstable against alcohols and alkalis. ►Figure 4.19 depicts the process during acidification: protonation of amino groups reduces their donor strength, and less electron density is available for the cyanine system, decreasing its size. Thus, the absorption band shifts to shorter wavelengths, and the hue changes to yellow.

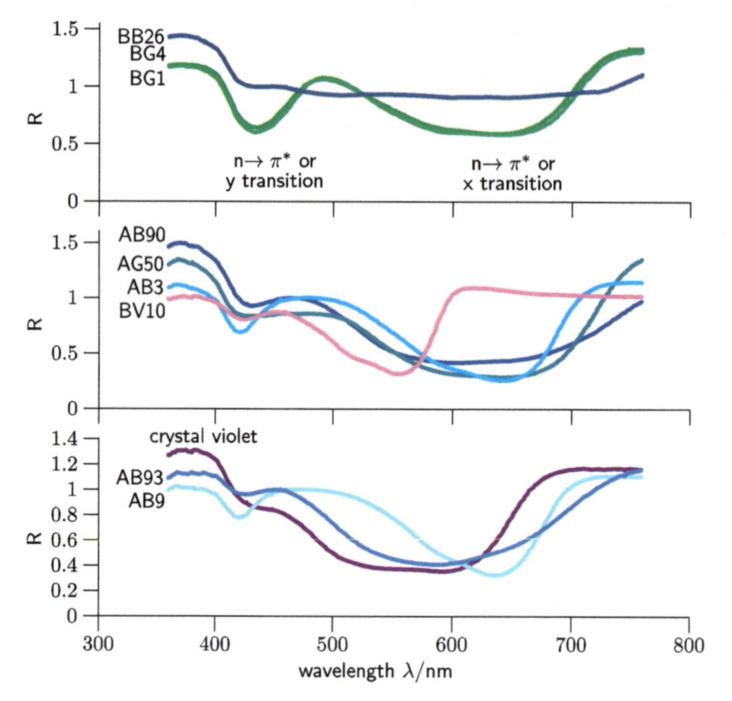

Figure 4.18: Reflectance spectrum of malachite green-based colorants normalized to an arbitrary unit: AB3 (Patent blue V, FB5, E131), AB9 (Brilliant Blue FCF, FB2, E133, primary cyan for four-color printing), AB90 (Brilliant Blue G-250), AB93 (royal blue, Helvetia blue), AG50 (Lissamine Green, E142), BB26 (Victoria Blue B), BV10 (rhodamine B), BG1 (Brilliant Green), BG4 (Diamond Green B), and crystal violet.

Malachite green dyes are versatile; their application ranges from low-cost printing inks and fast fading writing inks for fountain pens (►Section 8.5) to paper colorants. We recognize crystal violet from purple stamping inks (schoolchildren and chemistry students probably remember the seemingly unavoidable smearing when trying to produce crystal violet); fuchsin is an essential dye in microscopy. Some of the dyes, such as Patent Blue, are significant for food coloring.

The solubility of malachite green dyes is considerably enhanced if we introduce one or more sulfonic acid groups, thus obtaining familiar red, blue, blue-green, and purple dyes for inks, food, and paper coloring:

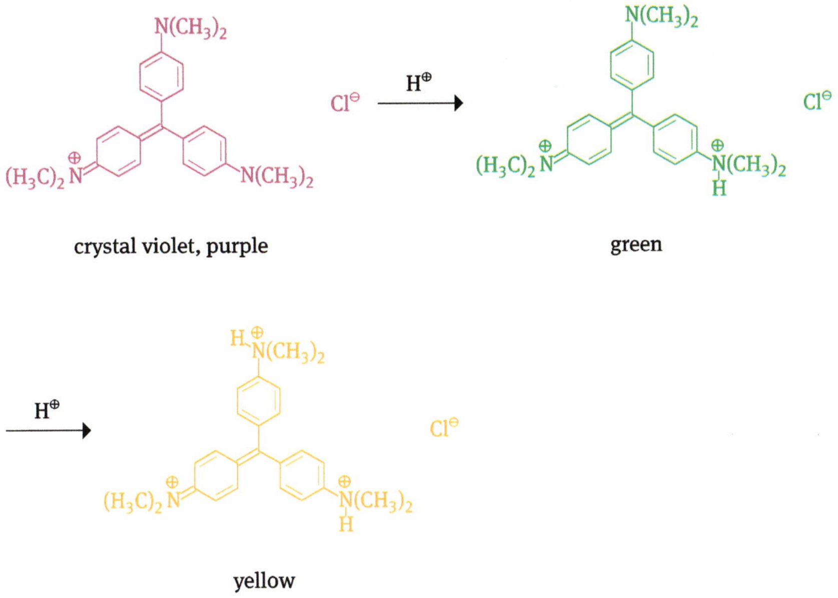

crystal violet, purple

green

yellow

Figure 4.19: Reaction of malachite green colorants with acids. Protonation reduces the donor strength and decreases the size of the mesomeric system, resulting in a hypsochromic color shift to higher wavelengths.

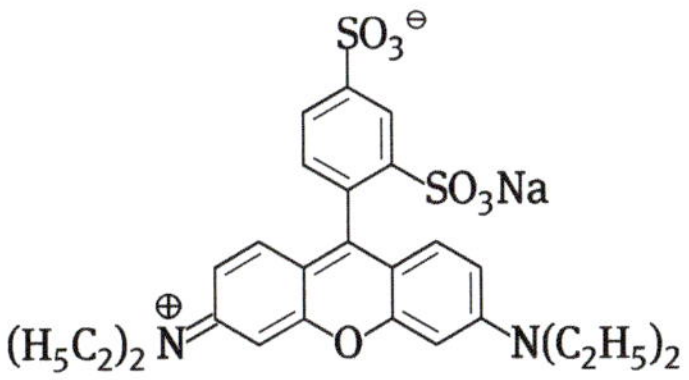

AR52, CI 45100, Food Red 106
red writing ink
primary magenta for inkjet

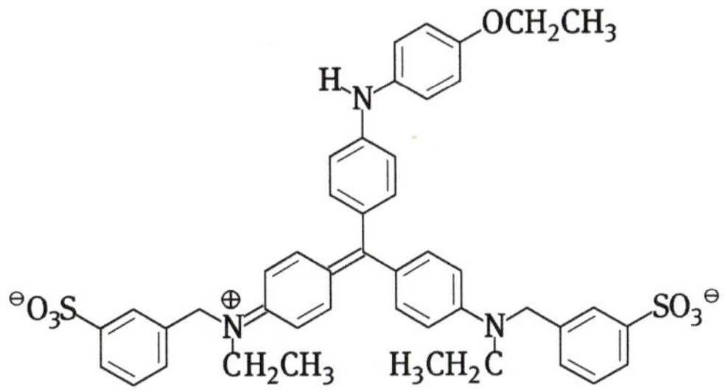

AB90, CI 42655
Brilliant Blue G, blue writing ink

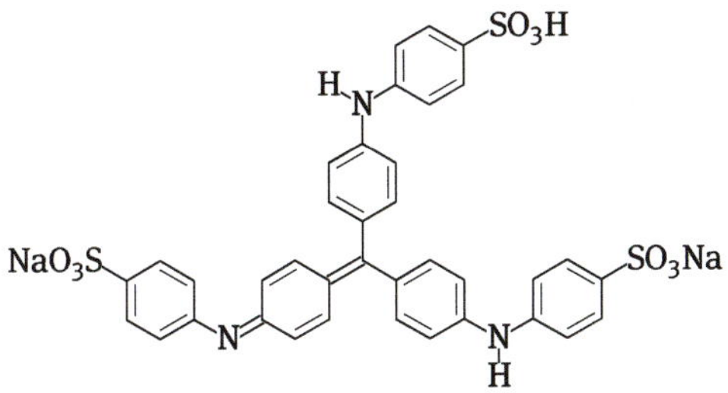

AB93, CI 42780
Helvetia blue, royal blue writing ink

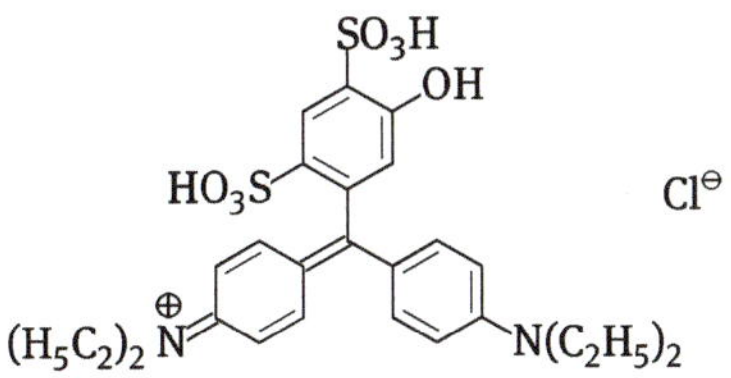

AB3, CI 42051, Food Blue 5, E131
Patent blue

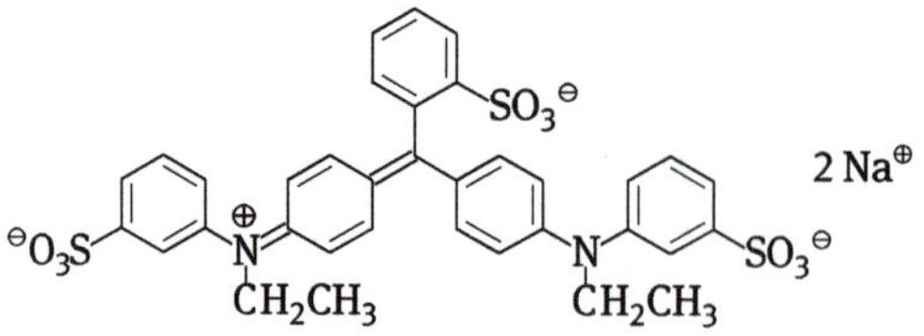

AB9, CI 42090, Food Blue 2, E133
Brilliant Blue FCF, blue writing ink
primary cyan for inkjet

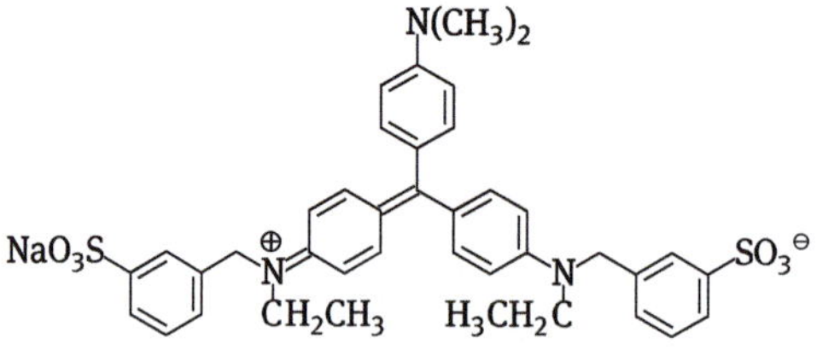

AV49, CI 42640
benzyl violet, purple writing ink, purple inkjet ink

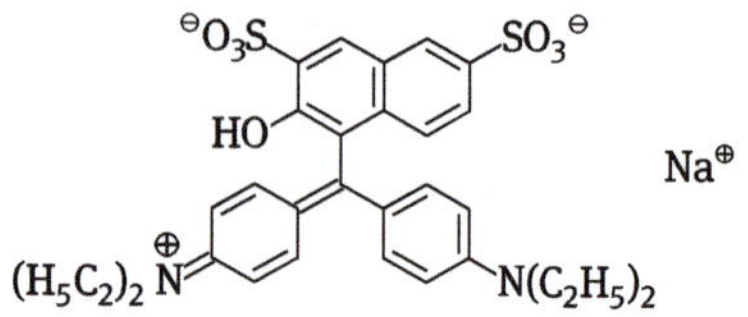

AG50, CI 44090, E142
Lissamine Green, Brilliant Acid Green BS

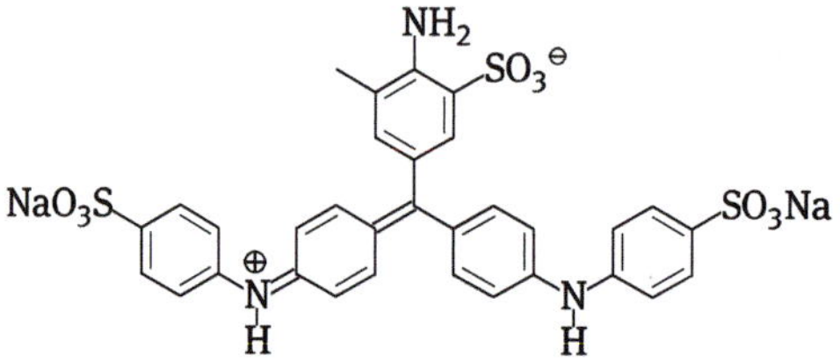

CI 42755
water blue

Malachite green pigments, alkali blue pigments

Red Purple Blue Green The malachite green dyes described before were formerly precipitated with antimony potassium tartrate on aluminum hydroxide. The resulting lakes possess splendid colors but are hardly lightfast. Today, the dye cations (triarylcarbonium cations) are converted into malachite green pigments by salt formation, following two paths:

- formation of an inner salt by introducing sulfonic acid groups into the phenyl rings (alkali blue type) [12, Chapter 4.1.1]
- formation of dye salts with complex anions by precipitating the dye cations with anions of complex heteropolyacids such as phosphotungstic acid or phosphomolybdic acid [12, Chapter 4.1.2]

The dye parafuchsin represents the basic structure of alkali blue pigments. Technically interesting are derivatives with structure I. As a rule, two ($R^* = -H$) or all three amine positions are substituted. Two substituents induce red tints, three substituents green ones. Due to the complexity of reactions occurring during the manufacturing process, usually mixtures of various composition are obtained, with I being the main component, accompanied by compounds of higher or lower degrees of arylation and sulfonation:

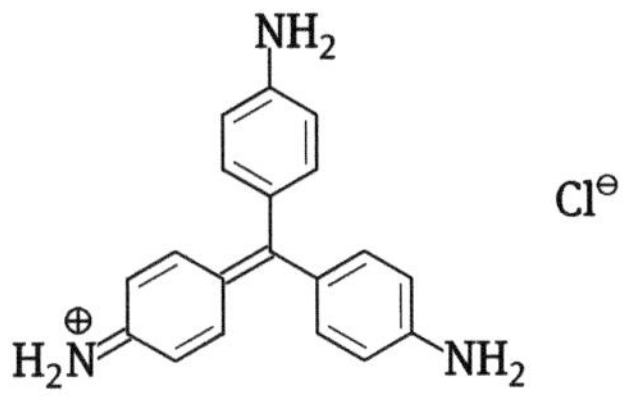

parafuchsin

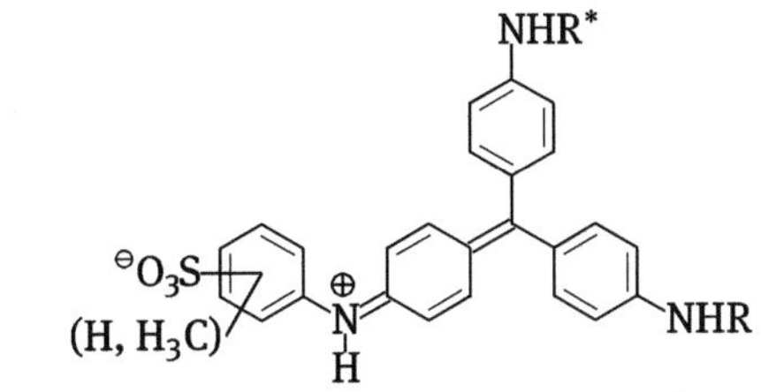

I, R, R* = –H, $-C_6H_5$ and $-C_6H_4CH_3$ (meta)
triarylmethine, alkali blue type

The resulting triphenylmethines are blue and show a more distinct red tint than phthalocyanines. They are very intensive in color but unstable against alcohol, toluene, and moderately lightfast. Alkali blue pigments are used in printing inks for letterpress and offset printing but rarely as a colorant in its own right. Instead, they are added as shading pigments to tone black printing inks. They correct the brown touch of carbon-based black printing inks to achieve a neutral, deep black.

The second group of triarylmethine pigments, the salts with complex anions, are also derived from parafuchsin but carry no sulfonic acid groups and thus form dyes IIa. ►Table 4.10 shows some malachite green dyes compared to their corresponding dye salts or pigments. For the dyes, chloride acts as counterion X, while complex phosphoheteropolyacids play this role for the pigments. Common heteropolyacids for industrial pigments are phosphotungstic acid $P(W_3O_{10})_4^{3\ominus}$ (PT or PTA) and phosphomolybdic acid $P(Mo_3O_{10})_4^{3\ominus}$ (PM or PMA). A combination is the phosphotungstomolybdic acid (PTM).

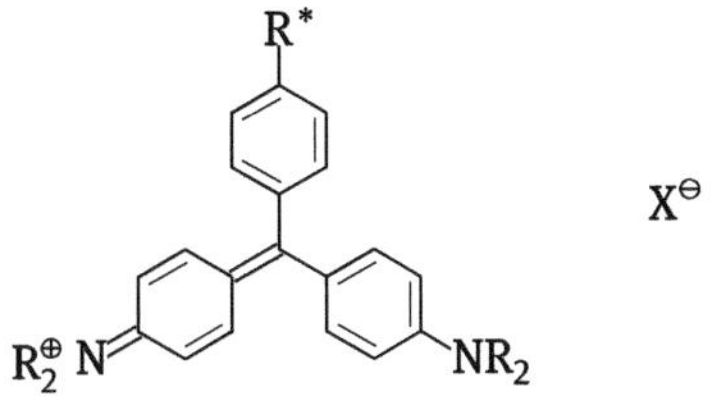

IIa, R=–Me, –Et, R*=–H, $-NMe_2$
X = PM, PT or PTM

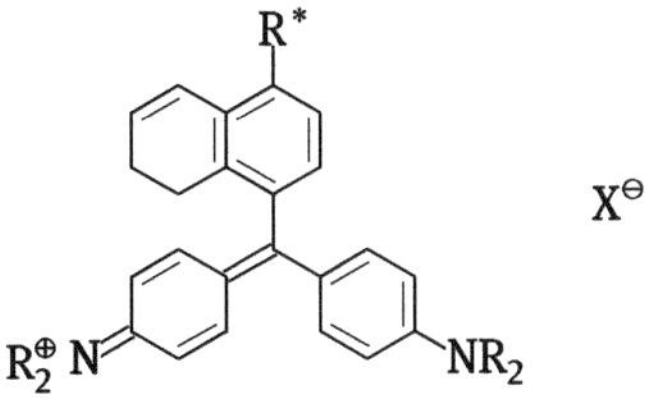

IIb, R=–Me, –Et, R*=$-NEt_2$
X = PM, PT or PTM

Triamino compounds exhibit shades of purple (crystal violet → PV39). The elimination of an amino group leads to shades of green (malachite green → PG4, brilliant green → PG1). The replacement of a phenyl ring by a naphthyl system IIb results in a shift of purple to blue (Victoria blue → PB1).

Another basic structure is phenylxanthene III, derived from triphenylmethines by incorporating an oxygen bridge. ►Table 4.10 displays more examples. Again, PM and PTM occur as counterions (rhodamine 6G → PR81, rhodamine B → PV1).

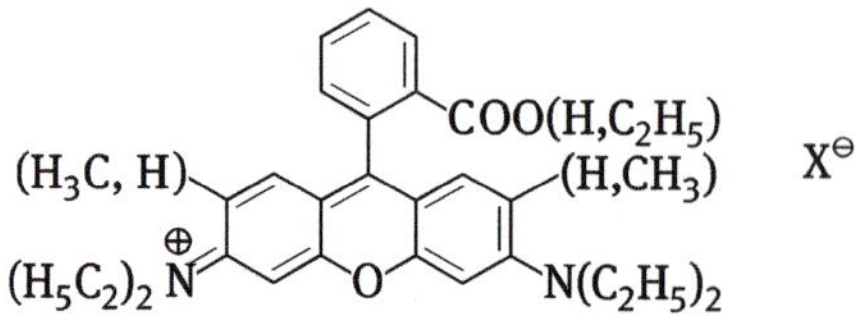

III, phenylxanthene, X = PM, PT or PTM

Complex salts of phenylxanthene cover a wide range of red, purple, blue, and green shades. They exhibit very pure hues but are only moderately lightfast and unstable against alcohols and alkalis. Therefore, they are only employed for inexpensive printing and writing inks or paper coloring. However, some of them, like PV1, PV3, or PR81, are included in artists' inks.

Cause of color

The color of triarylmethine colorants is caused by their derivation from the smallest possible polymethine with $n = 0$, as shown in ►Figure 4.15. For benzaurines, this is oxonol I, and for malachite green derivatives, it is streptocyanin II. Moreover, ►Section 2.5.5 showed that these compounds possess mesomeric structures inducing color upon irradiation with visible light:

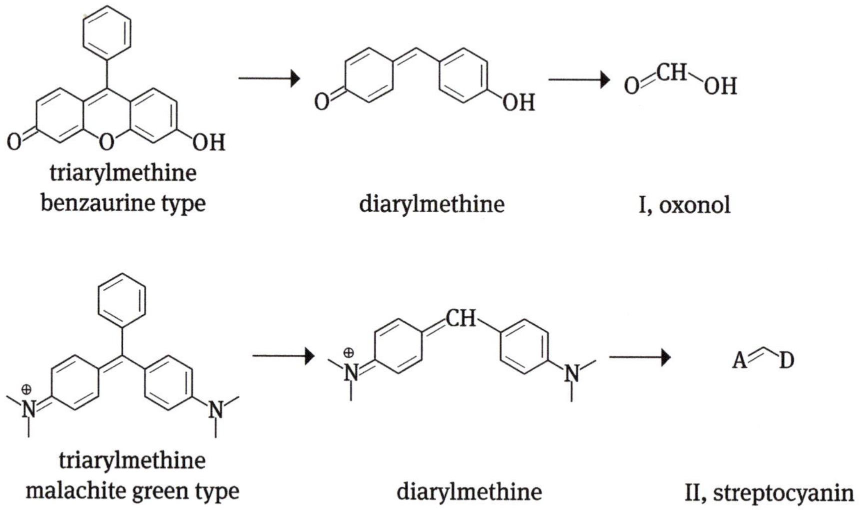

triarylmethine benzaurine type → diarylmethine → I, oxonol

triarylmethine malachite green type → diarylmethine → II, streptocyanin

The existence of an NBMO in methine systems approximately halves the energy difference between HOMO and LUMO. Thus, the absorptions in diarylmethines are strongly bathochromically shifted so that Michler's hydrol is blue ($\lambda = 607$ nm):

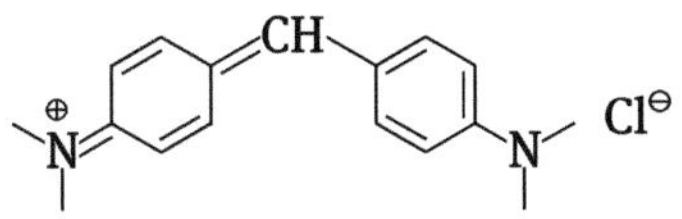

Michler's hydrol

If we introduce a third phenyl ring into this diarylmethine, we lay the foundation for the considerable variety of colors shown by triarylmethines [4, Chapter 9.3]. The increased conjugated π system supports bathochromic shifts; this effect remains small and the absorption still occurs in the orange to red spectral range ($\lambda = 629$ nm), inducing in principle blue colors. Instead, the compound shows a green color caused by a newly formed second absorption band at $\lambda = 430$ nm that effectively absorbs the blue spectral range so that predominantly green light is reflected. Two absorption bands are present in triarylmethines, ►Figure 4.18:

- An x band occurs at long wavelengths, caused by $n \rightarrow \pi^*$ transitions within the cyanine system, inducing blue shades. The third phenyl ring is not directly involved. However, when the ring is substituted with electron acceptors, it can stabilize the central carbonium cation, which is highly charged in the excited state. The observed bathochromic shift of the x band can be attributed to this stabilization. On the other hand, if the third ring is substituted with an electron donor, we observe a hypsochromic shift of the x band.
- The new y band is caused by $\pi \rightarrow \pi^*$ transitions, starting from a π MO below the NBMO, and is thus located at short wavelengths in the blue spectral range. This absorption is polarized transversely to the cyanine system. In contrast to the x transitions, y transitions are hardly influenced by substituents on the third ring, and electron donors in this ring only lead to slight bathochromic shifts.

Both phenomena taken together give rise to x and y bands in the blue and red spectral range, inducing shades of green. Both bands move together with the increasing donor strength of the substituents in the third ring. Finally, they converge to crystal violet ($R = p - N(CH_3)_2$) at $\lambda = 589$ nm, yielding the purple color.

If we replace the third phenyl ring with a naphthalene ring, we observe a bathochromic shift to blue due to the higher conjugation, an example being Victoria blue.

An exciting feature of triarylmethines consists of the three rings that do not lie within the molecule's plane but are twisted propeller-like by about 30° against each other. The influence of this rotation is explained in detail in [4, Chapter 4.6], we will only state that the loss of planarity does not decisively influence the color since the underlying diaryl system is a proper cyanine system. Unlike highly conjugated aromatic compounds, such a system does not have to be planar to develop color. Substituents in the ortho position of the third ring, such as chlorine, increase the twist even more. Therefore, we observe a hypsochromic shift of the *y* transition, causing a color shift toward blue.

4.8.2 Diphenylmethines, diarylmethines, indamine dyes

Blue Green If we replace the central atom X of a polymethine with a CH group (►Figure 4.15), we obtain diarylmethines, or more specifically, diphenylmethines. However, more significant is the replacement by an aza group, yielding quinonimines. By adding electron acceptors A and donors D, we get several types of industrially important quinonimines, ►Figure 4.20:

- indophenoles I, A=O, D=OH, X=N:
- indoanilines II, A=O, D=NR_2, X=N:
- indamines III, A=$NR_2^{\oplus}$, D=NR_2, X=N:

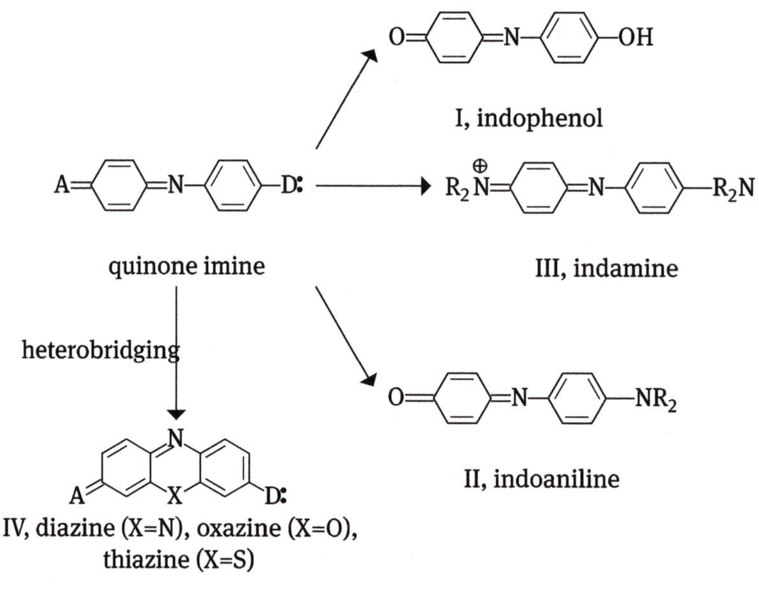

I, indophenol

quinone imine

III, indamine

heterobridging

II, indoaniline

IV, diazine (X=N), oxazine (X=O), thiazine (X=S)

Figure 4.20: Industrially applied types of diarylmethine (quinonimine) dyes. A indicates an electron acceptor, D an electron donor.

We obtain more type IV (diazines, oxazines, and thiazines) by bridging I to III with a hetero atom, usually N, O, or S. Indamines III and the hetero-bridged oxazines and thiazines IV are of particular interest for the chemistry of dyes, diazines (phenazines) being now only of historical interest. Regrettably, however, pigments in this class do not exist.

III and IV represent the structures of numerous dyes. Phenoxazines (example: Capri Blue GN) and phenthiazines (example: methylene blue) were used early in silk dyeing. Due to their lack of lightfastness, they disappeared from the market over time. After observing that their lightfastness significantly increased on acrylic fibers, these dyes experienced a renaissance. Methylene blue is an essential dye in analytics, ►Figure 4.21.

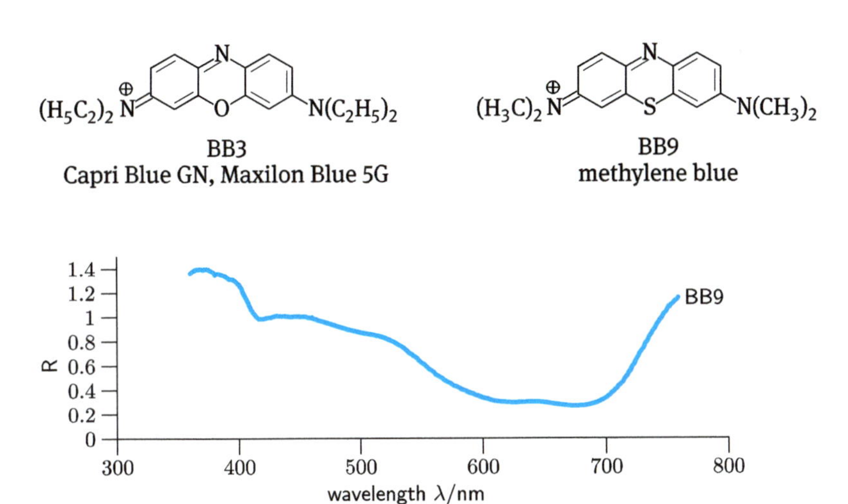

Figure 4.21: Reflectance spectrum of thiazine-based dyes normalized to an arbitrary unit: BB9 (methylene blue).

Orseille, litmus

The inconspicuous gray lichens of the *roccella*, *lecanora*, and *variolaria* species supplied two related dyes: Orseille and litmus. Orseille already surfaced in the second century for dyeing second century for dyeing [23] and later in the Middle Ages for book illumination, ►Section 8.3.1. Litmus is popularly known for the classic pH indicator paper.

The lichens are extracted with aqueous ammonia under air admission. In the case of litmus, calcium salts and potassium carbonate are added [689, 690]. The resulting purple dye is called orcein in purified form. It is complex in composition [685], comprising three series of phenoxazines [687–689]:

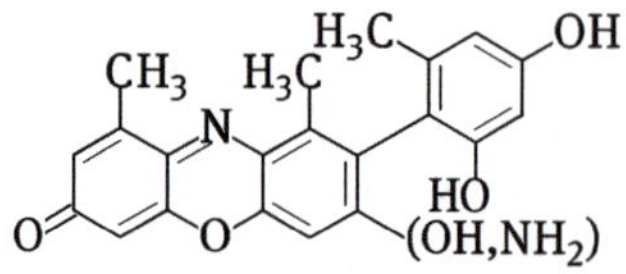

H_3C OH CH_3 H_3C N HO (OH,NH_2) O O

α-hydroxyorcein/α-aminoorcein

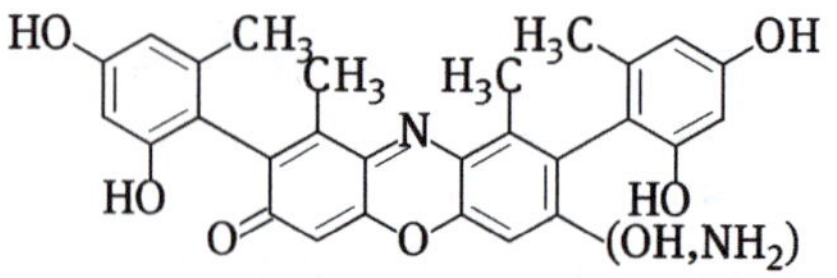

HO CH_3 H_3C OH CH_3 H_3C N HO HO (OH,NH_2) O O

β-hydroxyorcein/β-aminoorcein (*cis*)
γ-hydroxyorcein/γ-aminoorcein (*trans*)

The phenoxazine structure forms a reference plane for the phenyl substituents. It invokes *cis-trans* isomerism: in the β-orceins, the two hydroxyl groups of the phenyl substituents point to the same side of the reference plane (*cis*), and in the γ-orceins, the hydroxyl groups point to different sides of the reference plane (*trans*).

Litmus contains the same constituents as orcein but includes a complex polymeric fraction, probably formed from hydroxyorcein by oxidative linkage. The reason for litmus' pH-dependent color change is the protonation of a phenolate or an amino-nitrogen atom [686].

Cause of color

As already shown for triarylmethines, the color of diarylmethines is caused by their descent from polymethines with $n = 0$, ►Figure 4.15. Indamines have mesomeric structures, enabling a transition and absorption in the visible spectral range:

O= =N– –$O^{\ominus}$ ⟷ $^{\ominus}O$– –N= =O

indophenol

O= =N– –NR_2 ⟷ $^{\ominus}O$– –N= =$\overset{\oplus}{N}R_2$

indoaniline

$R_2\overset{\oplus}{N}$= =N– –R_2N ⟷ R_2N– –N= =$\overset{\oplus}{N}R_2$

indamine

The NBMO, a distinctive feature of methine systems, roughly halves the HOMO-LUMO energy difference compared to polyenes and leads to a distinct bathochromic shift of the $n \rightarrow \pi^*$ transition, located about 600 nm in the red spectral range.

4.9 Dioxazine pigments

Purple The orange triphenodioxazine I is the parent compound of dioxazine pigments [12, Chapter 3.8]. In practice, derivatives II are relevant as colorants:

I, triphenodioxazine, orange

II, X=–Cl, $-NHCOCH_3$,
Y=$-NHCOCH_3$, $-NHCOC_6H_5$

Dioxazine pigments have purple hues and show high lightfastness and weather resistance. Varnishes, coatings, printing inks, and artists' paints contain them. PV23 (►Figure 4.22) constitutes an exception to the general substitution pattern II and illustrates that the residues Y can be part of a ring system. PV37 is equivalent in hue to PV23 but halogen-free. Depending on the source, its structure is depicted as I [12] or II [432]:

PV23
carbazole violet [DS], Schmincke violet [SC]
Winsor violet [WN]

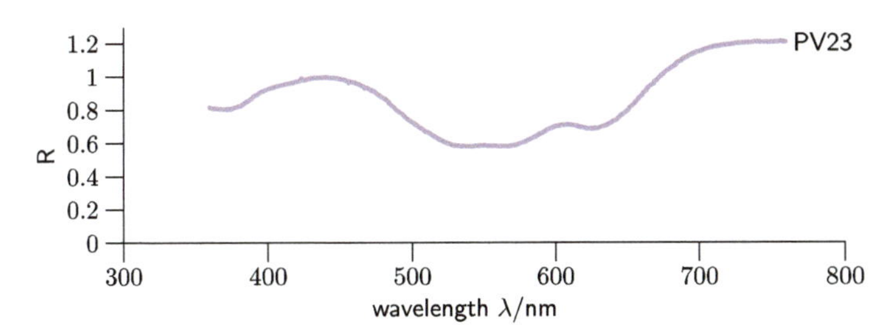

Figure 4.22: Reflectance spectrum of dioxazine-based watercolors normalized to an arbitrary unit: PV23 (Schmincke violet, Schmincke Horadam No. 476).

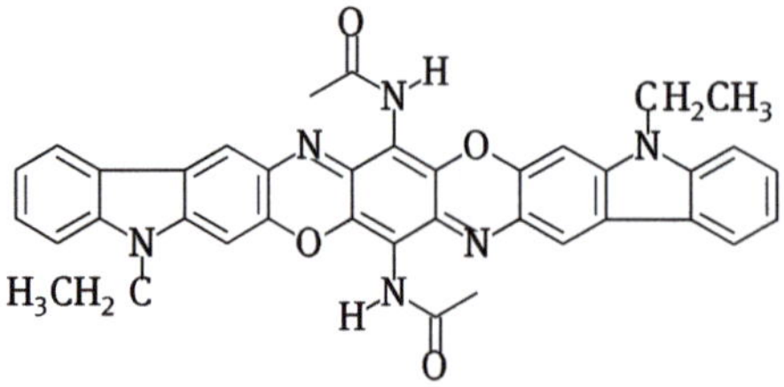

I, PV37
dioxazine violet

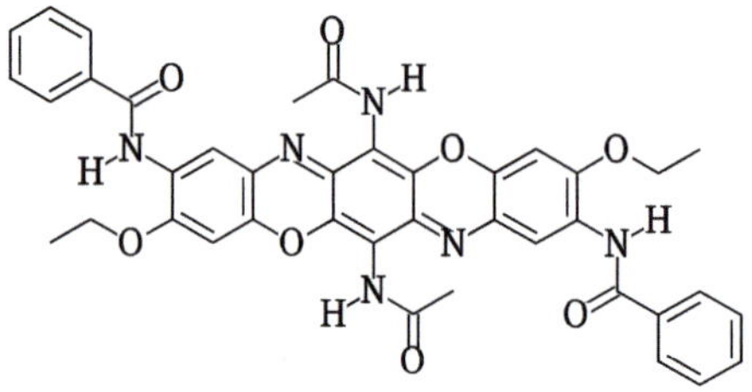

II, PV37
dioxazine violet

4.10 Phthalocyanine pigments

Green Blue Phthalocyanines [12, Chapter 3.1] can be traced back to an accidental discovery made in 1928. More detailed research in 1940 led to the first commercially viable pigments. They have enjoyed appreciation ever since due to a simple production method, high resistance against many detrimental influences, and above all, their colors, ranging from cool to warm blues and greens, a color range where only a few other competitive pigments exist if any.

Phthalocyanine pigments are derived from phthalocyanine, a pigment itself (PB16). They correspond to a polyene, more precisely an [18]annulene. A derivative, tetraaza-[18]annulene, widely occurs in nature. As structures of porphyrins or corrines, natural tetraaza-[18]annulenes play an essential role in the metabolic process: Chlorophyll and its derivatives are green pigments in leaves, whereas myoglobin in muscles and hemoglobin in the blood are red ones. The coloring of chlorophylls captures sunlight as excitation energy, which is then converted into chemical energy by a complicated multistationed system, the so-called photosystems I and II. The color, as such of muscle and blood pigments, is not needed but a side effect of their structure. Porphyrin derivatives participate in redox reactions and have excited states of low energies, making them adequate redox partners and chelating agents for multivalent cations like iron.

The basic structure of phthalocyanine and all derivatives in practical use are vast and symmetric molecules, offering no targets for harmful chemical or environmental influences. Moreover, including 18 π electrons in the ring, they also obey the Hückel rule 18 = $4n+2$, making them aromatic and chemically very stable. Both characteristics in combination lead to high light and solvent fastness and very low solubility. These advantages are especially obtained in derivatives complexed with metals, e. g., copper (PB15) in industrial practice.

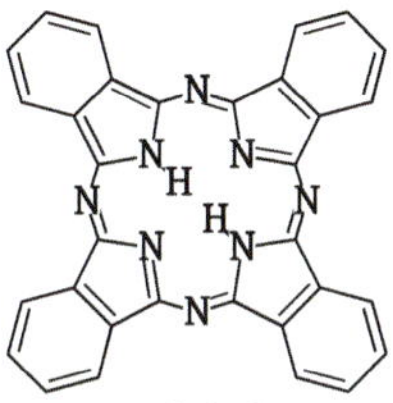

PB16, phthalocyanine
helio turquoise [SC], phthalo turquoise [WN]
transparent turquoise [SC]
phthalo blue turquoise [DS]

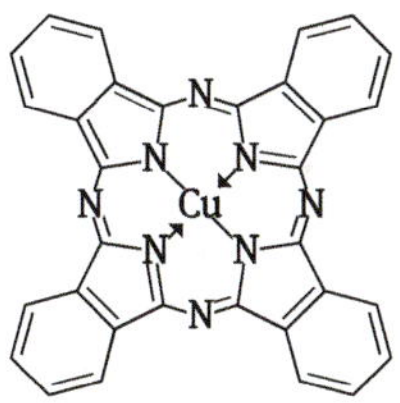

PB15, copper phthalocyanine
primary cyan for four-color printing
PB15 (α phase), manganese blue hue [DS][WN]
Winsor blue (red shade and green shade) [WN]
phthalo blue (red shade and green shade) [WN]
phthalo turquoise [WN]
PB15:1 (α phase), phthalo blue [SC]
phthalo blue (red shade) [WN]
PB15:3 (β phase), helio cerulean [SC]
phthalo blue (green shade) [DS][WN]
transparent cyan [SC]
PB15:6 (ϵ phase), phthalo sapphire blue [SC]
phthalo blue (red shade) [DS], oriental blue [WN]
transparent oriental blue [SC]

Blue pigments, PB16, PB15, and variants

Phthalocyanine itself figures as blue pigment PB16 in industrial and artists' paints, ►Figure 4.23. However, PB16 is less used in industrial practice than its metalized derivatives, predominantly the copper complex PB15, abbreviated as CuPc.

CuPc exists in several crystal phases, the unstabilized α phase PB15 being reddish-blue. The greenish-blue β phase PB15:3 belongs to the essential blue pigments for the printing industry, and it is therefore produced in large quantities; only small fractions of them are consumed for artists' paints. The α phase PB15:1 is stabilized against phase change by addition of 0.5–1 chlorine atoms per molecule; therefore, its shade is slightly greener. The stabilized ϵ phase PB15:6 is even more reddish-blue than PB15.

►Figure 4.23 illustrates the influence of crystal structure on the color of copper phthalocyanine pigments. Phthalocyanines occur in 11 different phases, α and β being crucial. All phases differ in the arrangement of CuPc molecules in the crystal lattice and show distinctive hues, ►Figure 4.24:

greenish blue	β-CuPc	α-CuPc	α-CuPc	ϵ-CuPc	reddish blue
	PB15:3	PB15:1	PB15	PB15:6	

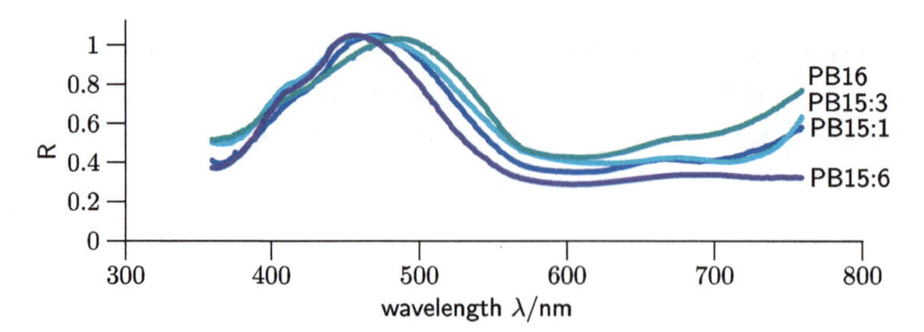

Figure 4.23: Reflectance spectrum of phthalocyanine-based blue watercolors normalized to an arbitrary unit: PB15:1 (phthalo blue, Schmincke Horadam No. 484, α-CuPc, primary cyan for four-color printing), PB15:3 (helio cerulean, Schmincke Horadam No. 479, β-CuPc), PB15:6 (phthalo sapphire blue, Schmincke Horadam No. 477, ϵ-copper phthalocyanine ϵ-CuPc), and PB16 (helio turquoise, Schmincke Horadam No. 475, CuPc). The β phase PB15:3 has the highest green tint among the copper complexes, followed by the slightly chlorinated PB15:1. PB15:6 has the highest red tint among the copper complexes. Overall, the copper complexes are bluer than the free phthalocyanine PB16.

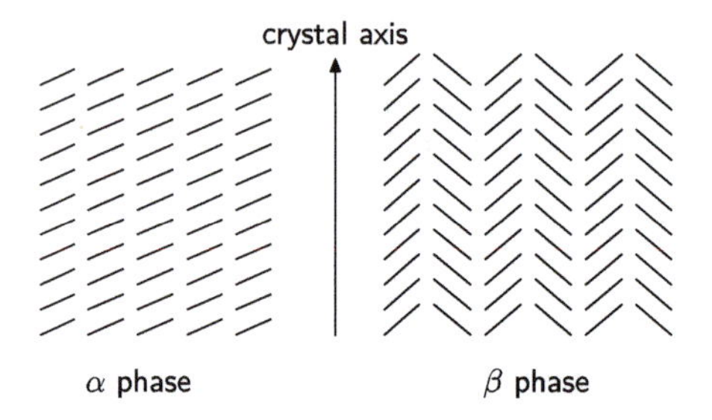

Figure 4.24: The two phases of PB15: In the α phase, the flat phthalocyanine rings are stacked in the same way; in the β phase, they are stacked against each other in a herringbone pattern [776]. The reference axis is the principal crystal axis.

The thermodynamic stability of the phases decreases in the order $\beta > \epsilon > \delta > \alpha \approx \gamma$. They can be mutually converted via various mechanisms that are not yet fully understood:

$$\underset{\text{reddish}}{\alpha\text{-CuPc}} \underset{\substack{H_2SO_4/\text{ice water} \\ \text{milling with NaCl}}}{\overset{\text{organic solvents, } \Delta}{\rightleftharpoons}} \underset{\text{greenish}}{\beta\text{-CuPc}}$$

Therefore, the pigments may need to be protected against unwanted phase change for practical applications. The protection can be achieved by a light halogenation of the phthalocyanine ring. PB15:1, e. g., is shielded against phase changes typically by halogenation, introducing 0.5–1 chlorine atoms per molecule. In addition, coating the pigment with a thin layer of resin can prevent unintended phase change.

The greenish-blue hue of β-CuPC (PB15:3) makes it an ideal primary cyan for four-color printing. PB15:3 is therefore produced in large quantities for the printing industry.

Green pigments, PG7, PG36

β-copper phthalocyanine already possesses a clear green tint, but it is still perceived as a blue pigment. Chlorination of the unsubstituted CuPc leads to bluish-green pigments, while additional bromine atoms shift the hue of the green pigments further to a yellow-tinged green, ►Figure 4.25. The exact ratios of halogens vary: the blue-tinged PG7 has 14 or 15 chlorine atoms, whereas the yellower PG36 includes 2–8 chlorine atoms and 9–4 bromine atoms:

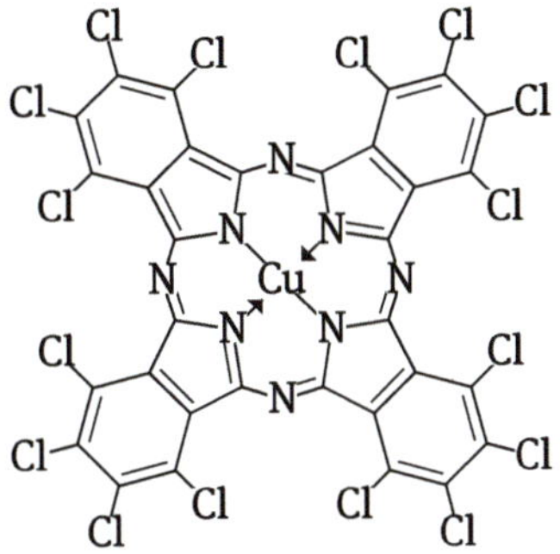

PG7
phthalo green [SC], helio green deep [SC]
Winsor green (blue shade) [WN]
Winsor green (phthalo) [WN]
phthalo green (blue shade) [DS]

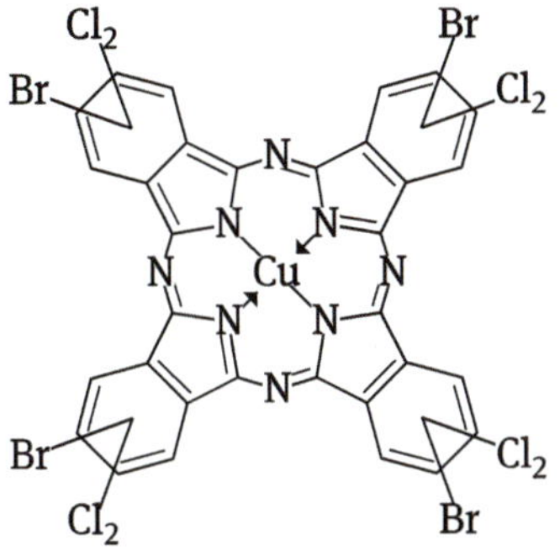

PG36
helio green [SC], helio green light [SC]
Winsor green (yellow shade) [WN]
phthalo green (yellow shade) [WN]
phthalo green (yellow shade) [DS][SC]
cobalt turquoise [WN]

Although interesting due to their green hue, phthalocyanine pigments are considerably more expensive than mixtures of cheap phthalocyanine blue and yellow pigments. Their cost can be attributed to the additional halogenation step. As a result, green CuPc pigments are employed industrially only to meet special requirements, but they enrich the blue and green artists' palette, ►Figure 4.25.

Dyes

The introduction of sulfonic acid groups into the outer rings increases the solubility and thus the applicability of phthalocyanines as dyes. Therefore, DB86 and DB199 are vital dyes for blue writing and inkjet inks, ►Figure 4.26:

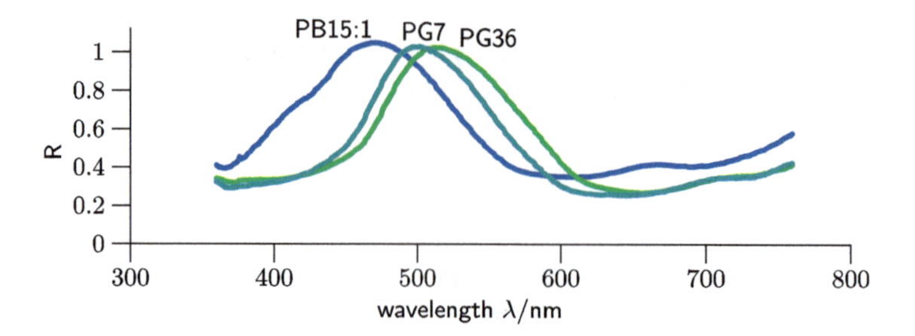

Figure 4.25: Reflectance spectrum of phthalocyanine-based watercolors, normalized to an arbitrary unit: PB15:1 (phthalo blue, Schmincke Horadam No. 484, copper phthalocyanine CuPc, primary cyan for four-color printing), PG7 (phthalo green, Schmincke Horadam No. 519, CuPc chlorinated), and PG36 (helio green, Schmincke Horadam No. 514, CuPc chlorinated/brominated). The shade of green increases with the degree of halogenation, specifically with heavy halogens such as bromine.

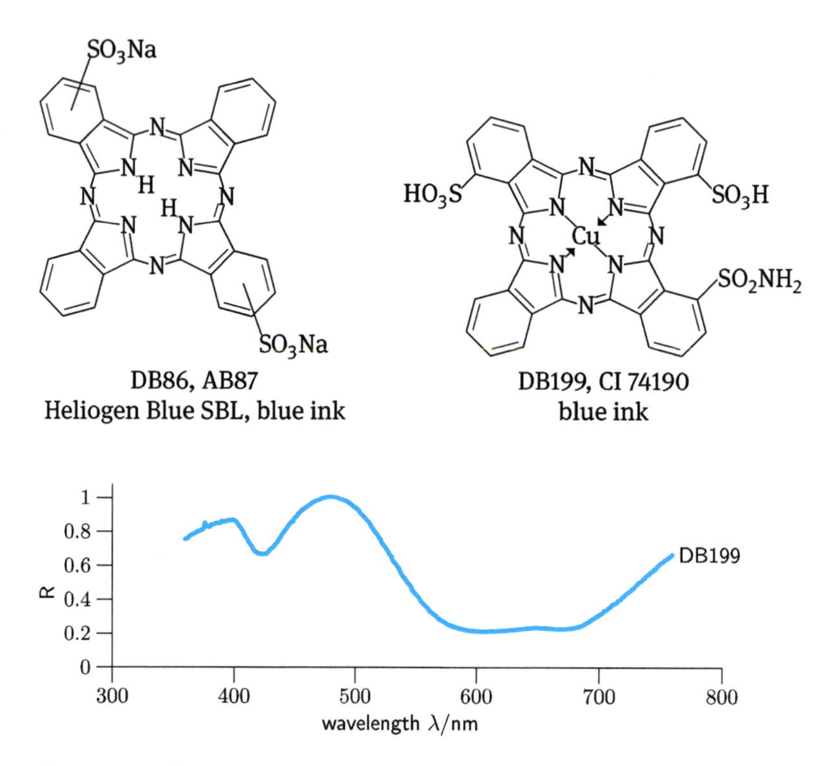

Figure 4.26: Reflectance spectrum of phthalocyanine-based dyes normalized to an arbitrary unit: DB199.

Cause of color

The benzene rings shown are not relevant for the coloring and can be omitted or replaced by other systems such as naphthalene. The intensive coloring of the phthalocyanines is due to their nature as [16]annulenes and is discussed in ►Section 2.5.4.6.

4.11 Azo colorants (hydrazone colorants)

Yellow Orange Red Purple Blue *Azo colorants* impart color to many pigments and dyes in the yellow, red, and purple range [12, Chapter 2]. Their characteristic feature is the diazo group –N=N–, in industrial colorants linked to an aryl system and a substituent R_2:

R_1–C_6H_4–N=N–R_2

The aryl group, substituted with R_1, is called the *diazo component*, and R_2 is the *coupling component*. The name azo colorants is derived from the formation reaction, the so-called *diazotization*, which was discovered in 1858. The reaction starts with the diazo component, carrying a primary amino group where the azo bridge will be generated. The aromatic amine is converted into a diazonium salt and reacts ("couples") subsequently with the coupling component, yielding the diazo compound:

R_1–C_6H_4–NH_2 —($NaNO_2$, HCl)→ R_1–C_6H_4–$N{=}N^{\oplus}$ —(coupling component (CC))→ R_1–C_6H_4–N=N–CC

diazo component | diazonium salt | p-amino-azobenzene

The coupling component comprises phenols, aromatic sulfonic acids, amines, or methylene-active compounds. For a long time, the colors achieved by azo colorants were yellow, orange, and red. As an example, we show the reaction leading to methyl red. Anthranilic acid acts as a diazo component, and the N,N-dimethylaniline acts as a coupling component:

NH_2/COOH-benzene —($NaNO_2$, HCl)→ $N{=}N^{\oplus}$/COOH-benzene —(Ph–NMe_2)→ COOH-benzene–N=N–benzene–$N(CH_3)_2$

anthranilic acid | diazonium salt | methyl red

As we will see in ►Section 4.11.2, the coupling component can be selected from a broad spectrum of aromatic and aliphatic compounds, yielding a large variety of azo colorants. ►Table 4.11 shows a selection of classic azo dyes in the color range yellow to red. They have no pigment character and do not exhibit the qualities needed for

Table 4.11: Examples of early azo dyes. For the use of Winther's symbols, ►Table 4.12.

N=N–N(CH_3)$_2$ SY2 butter yellow Winther: A→E	NaO_3S–N=N–N(CH_3)$_2$ methyl orange, helianthin Winther: A→E
NO_2 HO O_2N–N=N– PO5, CI 12075 dinitraniline orange Winther: A→E	NO_2 HO H_3C–N=N– PR3, CI 12120 toluidine red Winther: A→E

colorants in artists' paints. Azo colorants suitable as pigments for art and industrial coloration, having complex structures, came only later.

For today's azo pigments, the selection of diazo and coupling components aims at optimizing specific properties of the resulting pigments, such as insolubility and migration fastness, ►Section 4.2.1. These pigments cover the entire spectrum, as shown in ►Figure 4.4. Some azo dyes that are built upon benzidine, 2-naphthylamine, or 4-aminobiphenyl have been classified as carcinogenic and are now no longer used. In contrast, azo pigments, in general, are considered harmless due to their insolubility and improved chemical structure [780], ►Section 4.11.6.

To better depict the structure of azo colorants, we will use a shorthand notation, the so-called *Winther symbols*, which is used in [1], but created no more widespread interest. The system comprises five letters, denoting types of diazo component and coupling component, ►Table 4.12. Additionally, arrows between the letters depict the direction of coupling, pointing from the diazo component to the coupling component, e. g., A→E. Freeman [782, p. 148] explains the notation.

Table 4.12: Symbols of the Winther notation [1] [782, p. 148].

Symbol	Meaning	Occurence
A	A diazotizable amin	diazo component
D	A bis-diazotizable diamin	diazo component
E	A coupling component	coupling component
Z	A coupling component, reacting twice	coupling component
M	An amine that couples and can be diazotized	coupling component, then diazo component

4.11.1 Azo-hydrazone tautomerism

Azo compounds are classically notated as possessing the azo group –N=N–, a tradition that we keep throughout the book for its simplicity of presentation. However, in the last few decades, research on azo colorants proved that the so-called azo form is in equilibrium with its tautomeric hydrazone form. In fact, the current view is that azo colorants occur consistently in the hydrazone form, stabilized by hydrogen bonds, ►Figure 4.27 [12, Chapter 2].

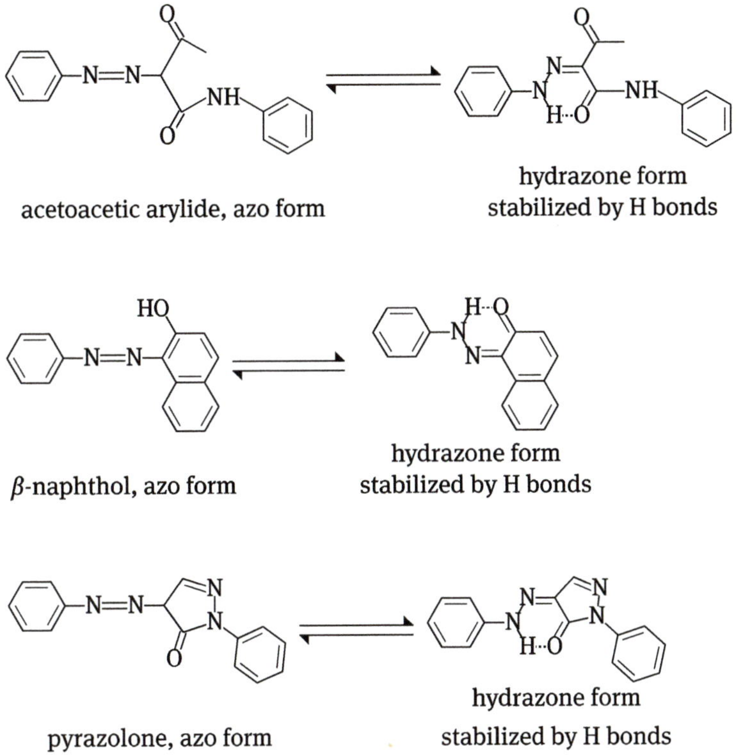

acetoacetic arylide, azo form

hydrazone form
stabilized by H bonds

β-naphthol, azo form

hydrazone form
stabilized by H bonds

pyrazolone, azo form

hydrazone form
stabilized by H bonds

Figure 4.27: Azo-hydrazone tautomerism of azo colorants exemplified by acetarylides, β-naphthols, and pyrazolones [12, Chapter 2]. Hydrogen bonds stabilize the hydrazone form.

4.11.2 The diazo component

The diazo component comprising a single amino group is denoted by letter "A" in Winther's notation, ►Table 4.12. The aryl group of the diazo component can be an aromatic ring or an aromatic heterocycle. The corresponding primary aromatic amines such as I are employed to prepare the diazonium salt. Mono, di, and trisubstituted anilines (II to IV) are of industrial importance. Common groups R are $-Cl$, $-CH_3$, $-NO_2$, $-COOH$, $-COOCH_3$, $-CONH_2$, and $-CONH-C_6H_5$.

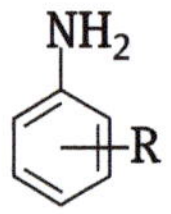

I, aniline

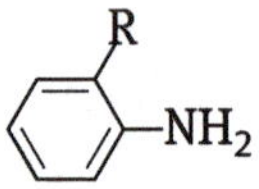

II
Winther: A

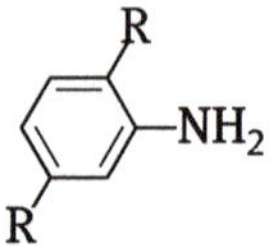

III
Winther: A

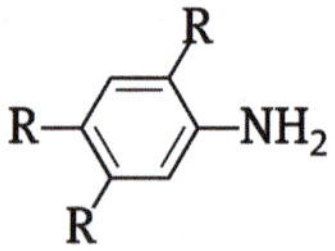

IV
Winther: A

Bifunctional diamines such as the 4,4'-diamino-biphenyls (benzidines, I) are converted into doubly diazotized salts, capable of reacting with two molecules of the coupling component per molecule of a diazonium salt. Such a diazo component is denoted by letter "D" in Winther's notation, ►Table 4.12. Di and tetrasubstituted benzidines II with R, $R' = -H$, $-Cl$, $-CH_3$, and $-OCH_3$ are industrially significant:

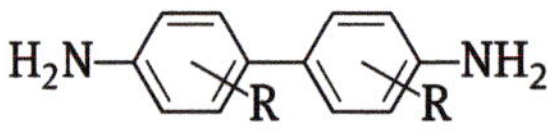

I, benzidine
Winther: D

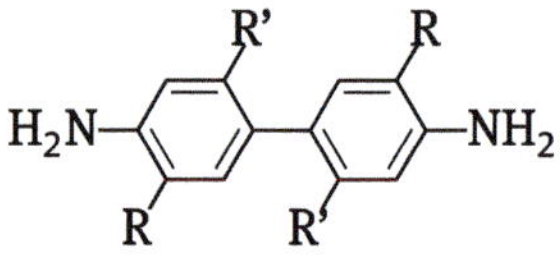

II
Winther: D

►Table 4.13 shows some diazo components of historical and modern times. Historical diazo components were simple monosubstituted anilines. Current aniline-based diazo components carry up to three substituents selected from a specific set of proven structures to obtain the required donor and acceptor properties. The substitution patterns of diamino compounds are designed to reduce the carcinogenicity of the benzidine component [780].

4.11.3 The coupling component

The coupling component comprising a single coupling position is denoted by letter "E" in Winther's notation, ►Table 4.12. For the coupling component, several structural types are used in industrial practice (the arrow marking the coupling position):

- Aryl group type I. Significant are the β-naphthols II and III. Azo *pigments* could be produced only after including naphthalene structures into the colorant.

Table 4.13: Examples of diazo components in azo colorants (letters "A" and "D" in Winther's notation, ►Table 4.12).

Examples of historical diazo components		
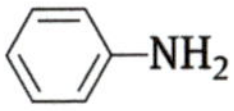	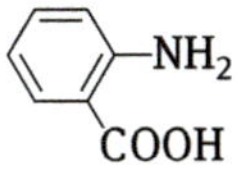	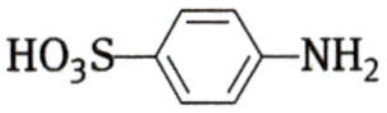
aniline	anthranilic acid	sulfanilic acid
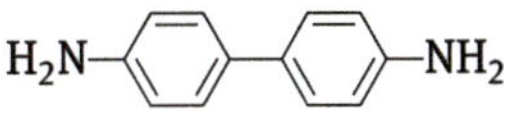	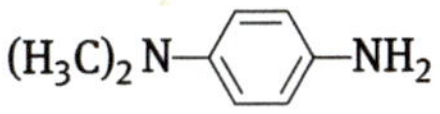	
benzidine	p-amino-N,N-dimethylaniline	p-nitroaniline
Examples of current diazo components		
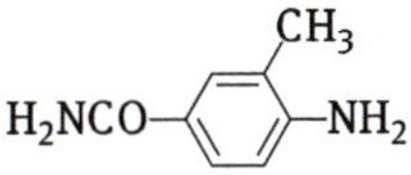	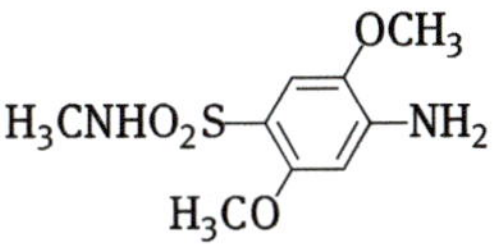	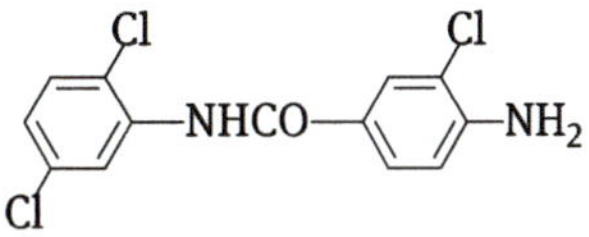
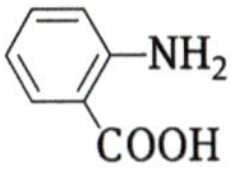	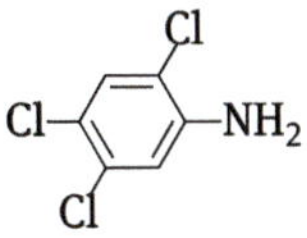	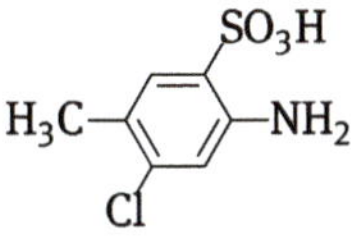
anthranilic acid	2,4,5-trichloroaniline	2B acid
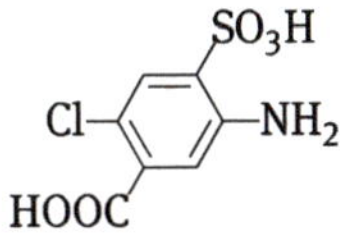	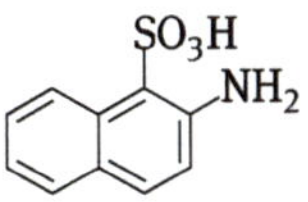	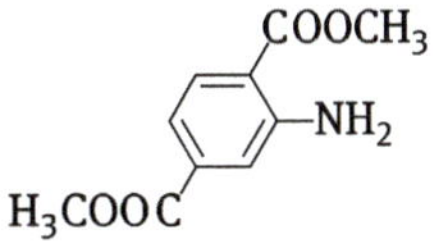
CA acid	Tobias acid, A acid	2,5-anilindicarboxylic acid methyl ester
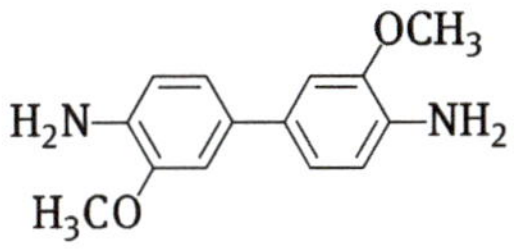	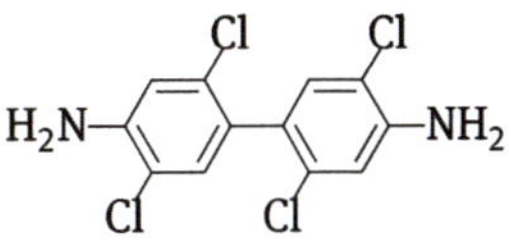	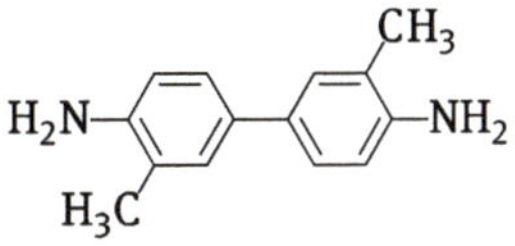
o-dianisidine	2,2',5,5'-tetrachlorobenzidine	o-tolidine

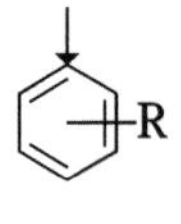

(CH$_3$, OCH$_3$, OC$_2$H$_2$, NO$_2$, Cl)

I, benzene

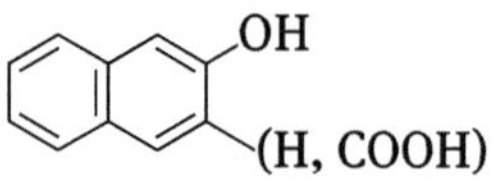

II, β-naphthols
Winther: E

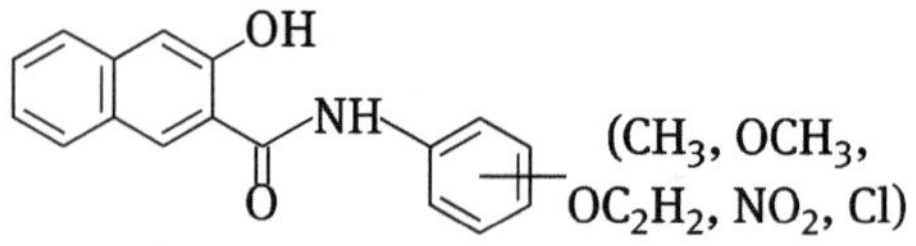

III
Winther: E

– Methylene active type I. Acetoacetic arylides II are of particular importance.

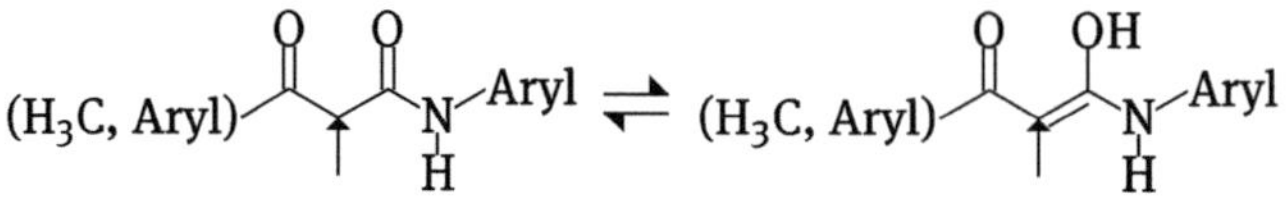

I, active methylene group

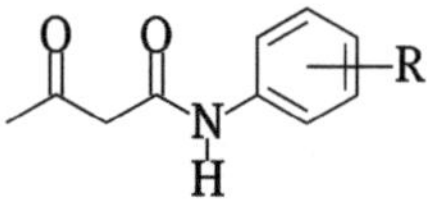

II, acetoacetic-arylide
Winther: E

– Pyrazolone type I. Industrially used are 1-aryl-pyrazolones II.

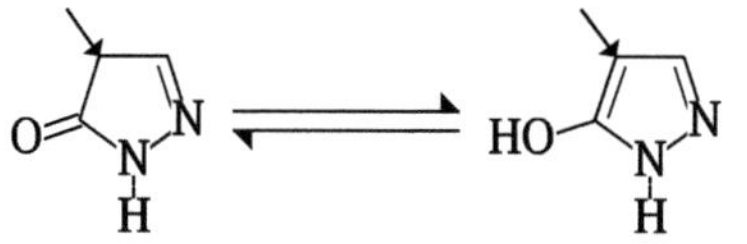

I

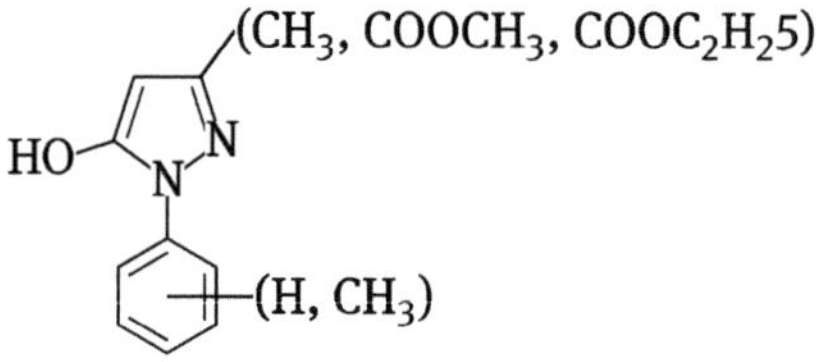

II, 1-aryl-pyrazolone
Winther: E

Colorants with an aryl ring as a coupling component, such as β-naphthol, were already known a century ago. Their disadvantage, however, was that their color could not be shifted beyond yellow or orange, dinitraniline orange being an example of this limitation. Equally early was observed that the methylene-active 1,3-dicarbonyl compounds easily form carbanions and are therefore powerful coupling components, yielding pure yellow diazo compounds (1910, Hansagelb):

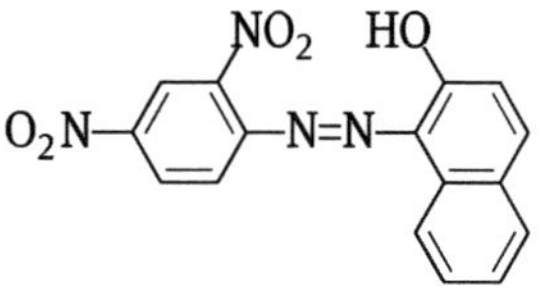

PO5, CI 12075
dinitraniline orange
Winther: A→E

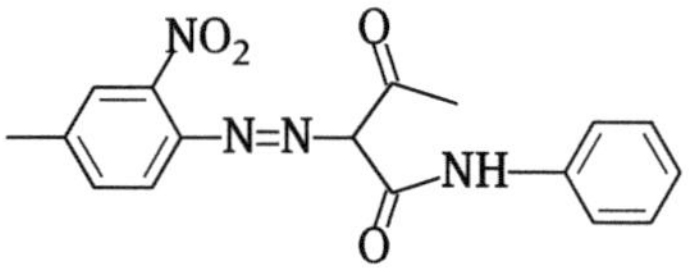

PY1, CI 11680
Hansagelb G
primary yellow for four-color printing
Winther: A→E

The heterocyclic pyrazolones can also be used as coupling components to yield yellow to red colorants. Yellow tartrazine was found as early as 1884. An example of the red range is PR38:

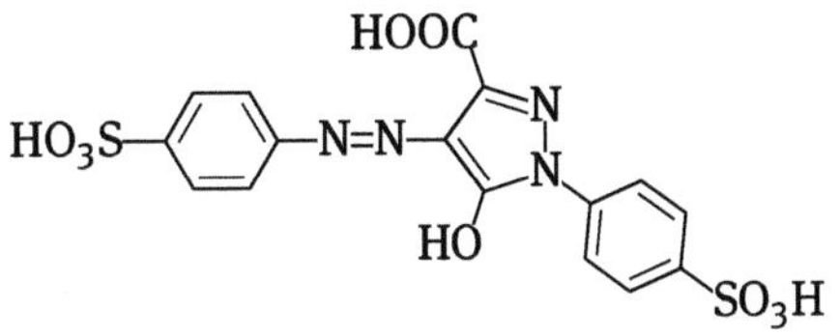

AY23, CI 19140, Food Yellow 4, E102
tartrazine, yellow writing ink
primary yellow for four-color printing
Winther: A→E

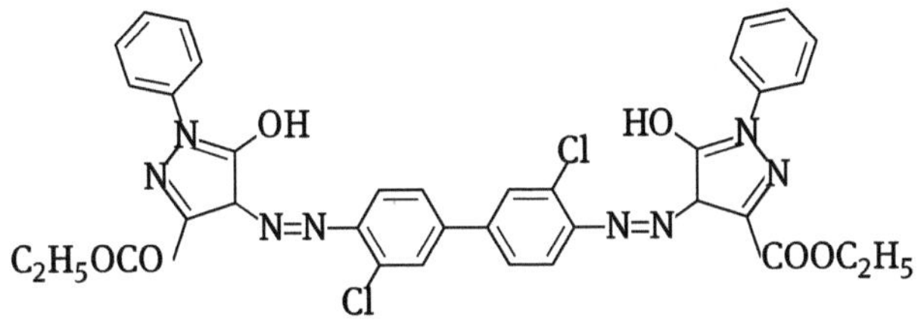

PR38, CI 21120
Winther: E←D→E

►Table 4.14 depicts some historical and modern coupling components. The arrows mark where the diazonium salt will form the diazo bridge. Some coupling components can couple at several positions, and the reactivity of each of these positions is pH-dependent.

Compared with diazo components, coupling components underwent a more pronounced historical change. The first section of the table shows some of the early coupling components, the constituents of the first known azo dyes. Already during the first half of the twentieth century, they were replaced by compounds that imparted better properties to the dyes. β-Naphthols are still in use as coupling components today. However, the large number of similar β-naphthol compounds indicates that systematic substitution experiments were carried out to find optimal coupling components, ever improving the final colorant.

Current coupling components comprise aryl, acetoacetic arylide, and pyrazolone types. While acetoacetic arylides yield pure yellow pigments, carboxylated β-naphthols possess attractive solubility and migration properties due to the introduction of carbonamide groups and heterocycles.

Table 4.14: Typical coupling components of azo dyes (letters "E" and "Z" in Winther's notation, ►Table 4.12).

Examples of historical coupling components		
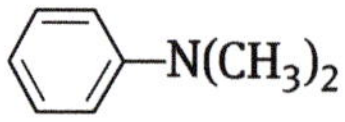 $-N(CH_3)_2$ N,N-dimethylaniline	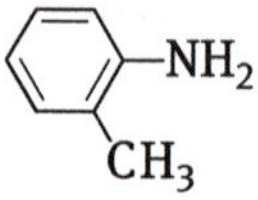 $-NH_2$, CH_3 2-methylaniline	
Examples of recent coupling components		
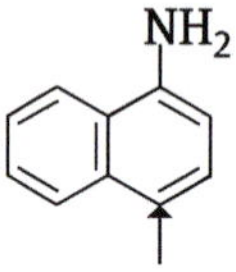 NH_2 α-naphthylamine	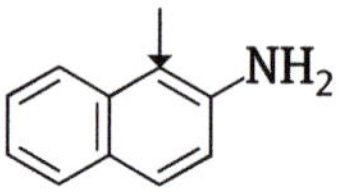 NH_2 β-naphthylamine	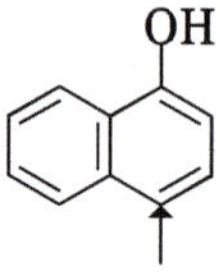 OH α-naphthol
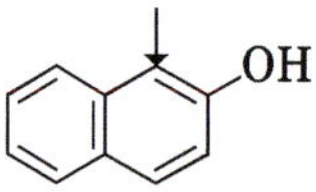 OH β-naphthol	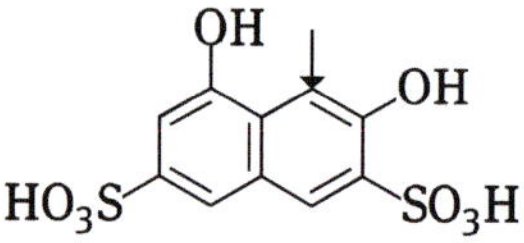 OH, OH, HO_3S, SO_3H chromotropic acid	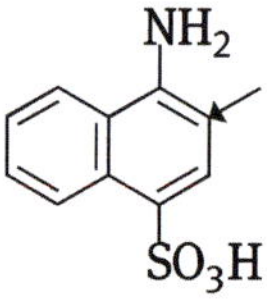 NH_2, SO_3H naphthionic acid
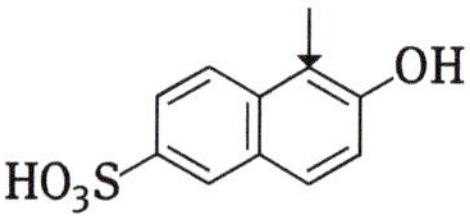 OH, HO_3S Schaeffer acid	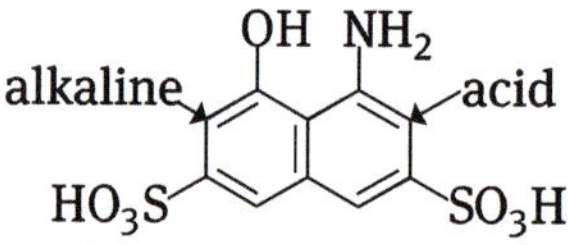 OH, NH_2, alkaline, acid, HO_3S, SO_3H H acid	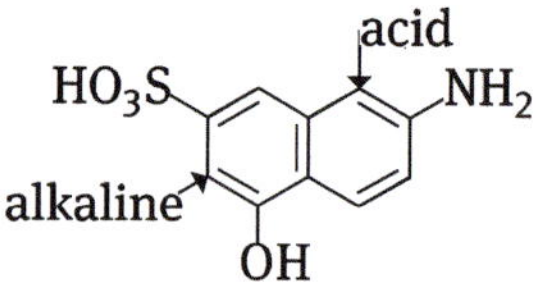 acid, HO_3S, NH_2, alkaline, OH I acid
Examples of current coupling components		
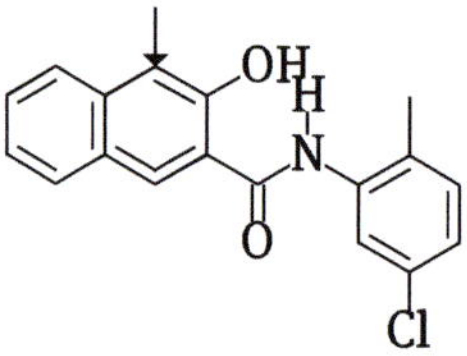 OH, H, N, O, Cl naphthol KB	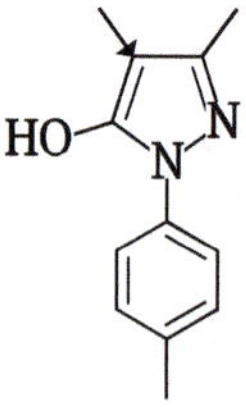 HO, N, N 1-methylphenyl-3-methyl-pyrazolone-5	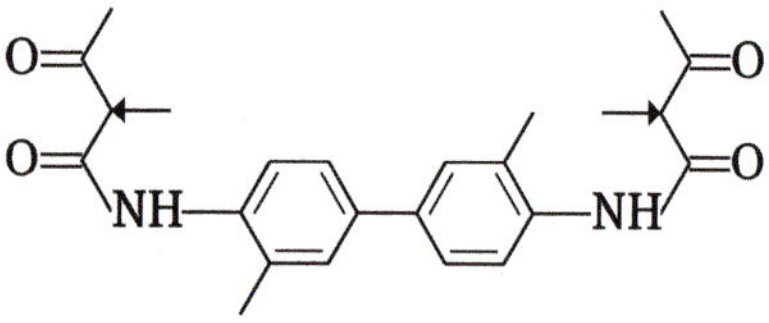 O, O, NH, NH, O, O naphthol AS-G

4.11.4 Classification of azo pigments (hydrazone pigments)

The azo pigments (new: hydrazone pigments) in today's industrial coloration, printing, and manufacturing of artists' paints can be classified according to the coupling component. In contrast to the used diazo components, the coupling components allow a sharper distinction of individual types in fastnesses and solvent resistance. From the abundance of types, we will focus on those that are of interest for artists' paints and printing inks:

- monoazo yellow and monoazo orange pigments (new: monohydrazone yellow and monohydrazone orange pigments)
- disazo pigments (new: dihydrazone pigments)
- *β*-naphthol pigments
- Naphthol AS pigments
- benzimidazolone pigments

4.11.4.1 Monoazo yellow/monoazo orange pigments (monohydrazone yellow/monohydrazone orange pigments)

Yellow Monoazo yellow and monoazo orange pigments [12, Chapter 2.3] are now called monohydrazone yellow and monohydrazone orange pigments, described by Winther as A→E. They consist of a central azo group linked to either acetoacetic arylide I or pyrazolone III as a coupling component (Winther symbol "E"). Acetoacetic arylide pigments can also be laked with metals II:

- acetoacetic arylides R = H, CH_3, OCH_3, OC_2H_5, Cl, Br, NO_2:

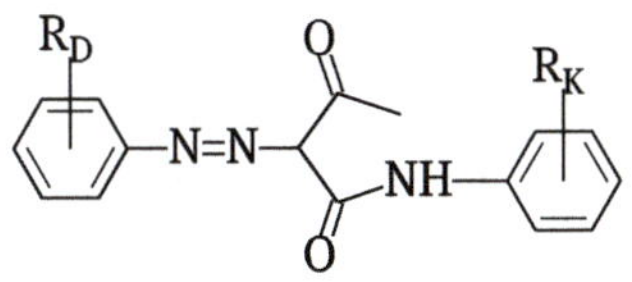

I, acetoacetic arylide pigment
Winther: A→E

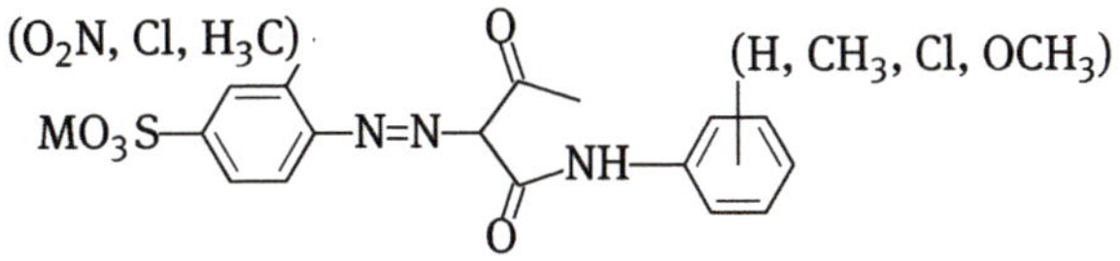

II, acetoacetic arylide pigment, laked, M = 1/2 Ca, Ba
Winther: A→E

- pyrazolones:

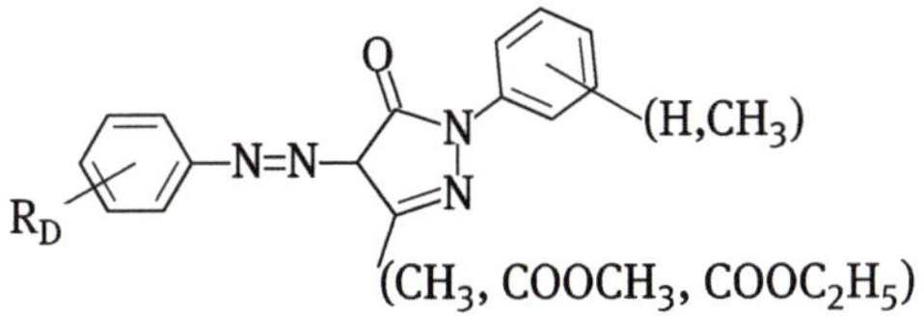

III, pyrazolone pigment
Winther: A→E

Simple acetoacetic anilides (R=H) in both diazo and coupling components were long in use and called Hansagelb pigments (German for "Hansa yellow"). They substituted lead chromate in printing inks and paper coloring. For example, PY1 and PY3 were used as yellow pigments in four-color printing:

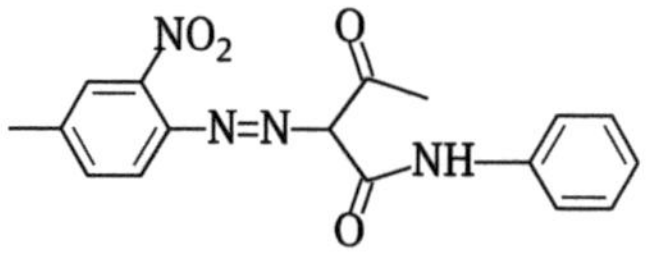

PY1, CI 11680
Hansagelb G
primary yellow for four-color printing
Winther: A→E

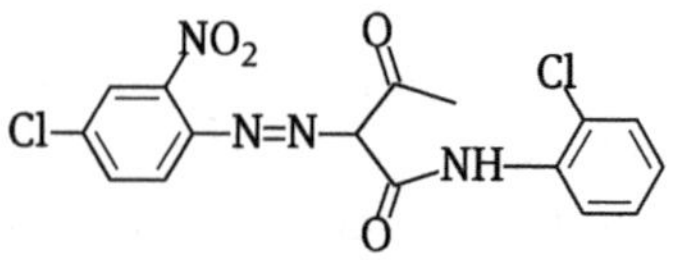

PY3, CI 11710
Hansa yellow light [DS], lemon yellow [SC][WN]
Winsor lemon [WN]
primary yellow for four-color printing
Winther: A→E

Monoazo yellow pigments are opaque and cheap but soluble in organic solvents. Their successors in the printing ink sector are diaryl yellow pigments, e. g., PY12 or PY13, having significantly higher molecular weights and, therefore, reduced solubility. To artists' ink, other requirements apply, so a corresponding range of monoazo yellow pigments is in use, ►Figure 4.28:

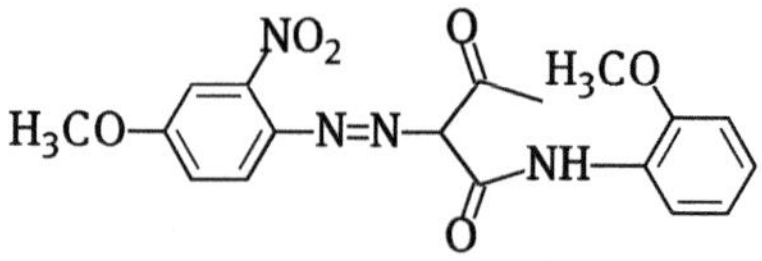

PY65, CI 11740
Hansa yellow deep [DS], Winsor yellow deep [WN]
chromium yellow hue deep [SC]
azo yellow deep [WN]

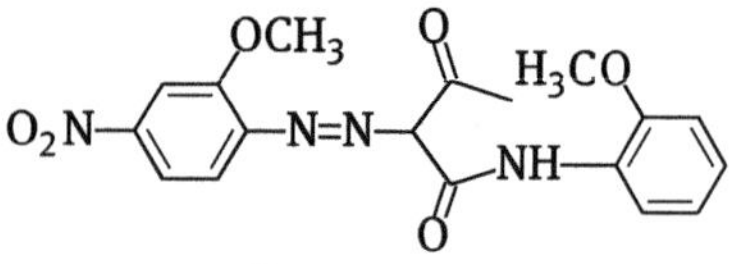

PY74, CI 11741
azo yellow medium [WN]
Hansa yellow medium [DS]
Winsor yellow [WN], yellow ink

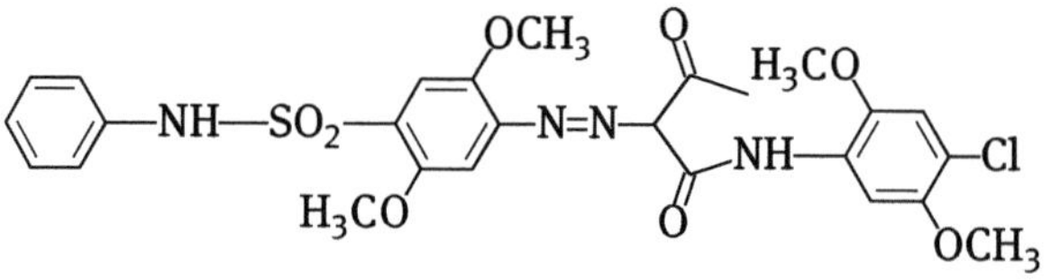

PY97, CI 11767
Hansa yellow medium [DS]

PY65 is suitable as a substitute for the toxic chrome yellow dark.

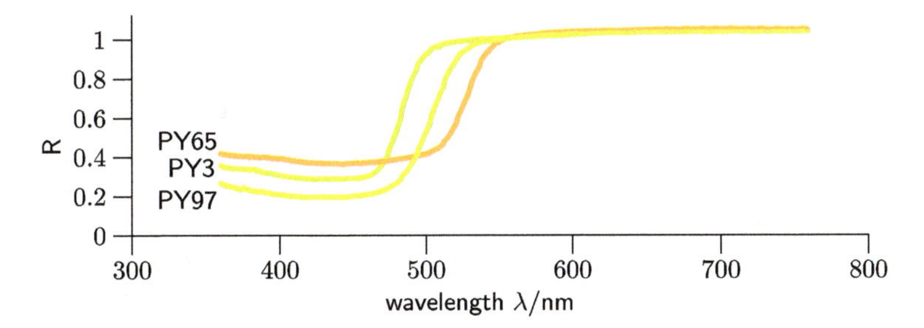

Figure 4.28: Reflectance spectrum of monoazo yellow-based watercolors normalized to an arbitrary unit: PY3 (lemon yellow, Schmincke Horadam No. 215), PY65 (chromium yellow hue deep, Schmincke Horadam No. 213), and PY97 (Hansa yellow medium, D. Smith Extra Fine Watercolor No. 039).

Laked acetoacetic arylides and pyrazolones carry sulfonic acid groups in the diazo or the coupling component's N-phenyl group. They are laked with Al^{III}, Ca^{II}, or Ba^{II}, and the resulting color lakes are more stable in terms of solvent fastness than the simple acetoacetic arylides but are not used for artists' paints. (Note: "lakes" originally denoted dyes, absorbed onto alumina hydrate, such as madder lakes and other anthraquinone lakes, ►Section 2.6. Today, insoluble metal salt pigments, in general, are termed lakes. In Europe, these colorants were also known as "toners.")

4.11.4.2 Disazo pigments (dihydrazone pigments)

Yellow Orange Red Disazo pigments (new: dihydrazone pigments) [12, Chapter 2.4, Chapter 2.9] are derived from a bifunctional diazo component and possess two azo groups (Winther symbol "D"). Since a condensation reaction might form them, some are also called disazo condensation pigments (new: dihydrazone condensation pigments). Standard bifunctional diazo components are p-diaminobenzene I and 4,4'-diamino-diphenyl II:

H_2N–(benzene ring)–NH_2 with substituents (H, Cl, OCH_3, CH_3) and (H, Cl, H_3CO, H_3C)

I, p-diaminobenzene
Winther: D

H_2N–(biphenyl)–NH_2 with substituents (H, Cl), (Cl, CH_3, OCH_3), (Cl, H_3CO, H_3C), (H, Cl)

II, 4,4'-diaminodiphenyle
benzidine
Winther: D

The bifunctional diazo component must be diazotized twice, either in one or in two steps:

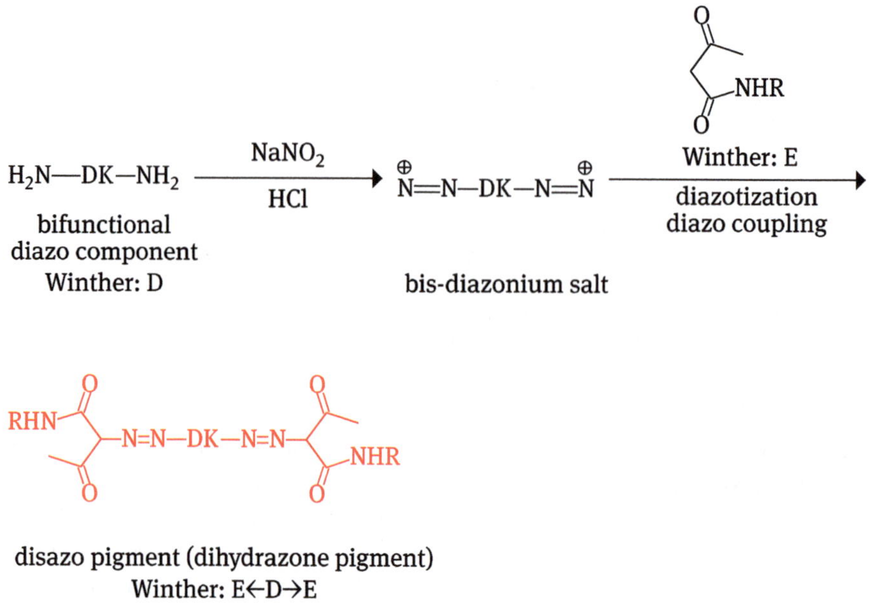

$H_2N{-}DK{-}NH_2$ $\xrightarrow[HCl]{NaNO_2}$ $\overset{\oplus}{N}{\equiv}N{-}DK{-}N{\equiv}\overset{\oplus}{N}$ $\xrightarrow[\text{diazotization, diazo coupling}]{\text{Winther: E}}$

bifunctional diazo component
Winther: D

bis-diazonium salt

NHR (coupling component, Winther: E)

RHN … N=N–DK–N=N … NHR

disazo pigment (dihydrazone pigment)
Winther: E←D→E

This reaction sequence can be difficult or even impossible to achieve. Consequently, an alternative process was developed as "disazo condensation." The condensation is the reverse of the above reaction sequence: the bifunctional diazo component (Winther symbol "D") is replaced by a bifunctional coupling component (Winther symbol "Z"), consisting of two simple coupling components (Winther symbol "E," e. g., acetoacetic arylides), connected via a bridge B (in Winther's symbols Z=E–B–E). The bifunctional coupling component Z reacts with two molecules of the diazo component A after its diazotization:

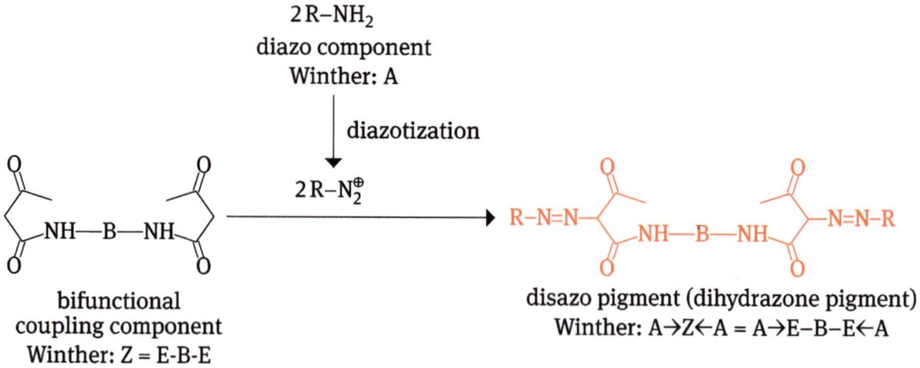

$2\,R{-}NH_2$
diazo component
Winther: A

diazotization

$2\,R{-}N_2^{\oplus}$

NH—B—NH

bifunctional
coupling component
Winther: Z = E-B-E

R–N=N– … NH—B—NH … –N=N–R

disazo pigment (dihydrazone pigment)
Winther: A→Z←A = A→E–B–E←A

Alternatively, the diazo coupling happens first, yielding a monoazo pigment. Two molecules of them are bridged in a subsequent step to the final disazo pigment:

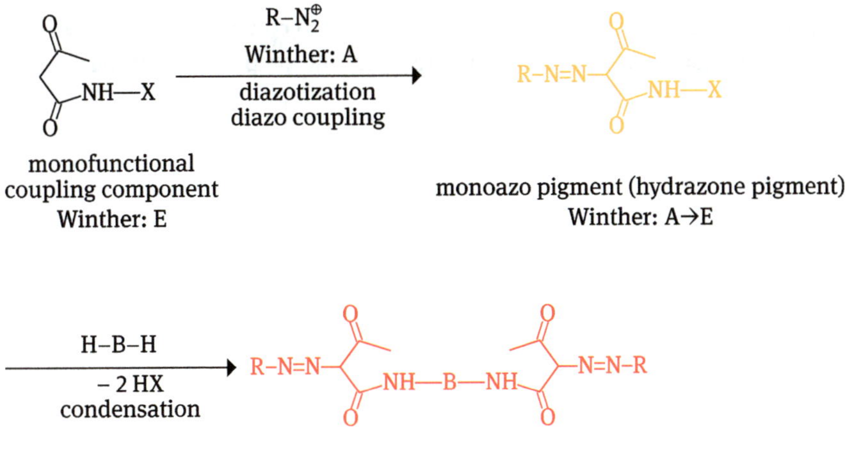

R–$N_2^{\oplus}$
Winther: A
diazotization
diazo coupling

monofunctional coupling component
Winther: E

monoazo pigment (hydrazone pigment)
Winther: A→E

H–B–H
– 2 HX
condensation

disazo pigment (dihydrazone pigment)
Winther: A→Z←A = A→E–B–E←A

Bridging occurs through a condensation reaction with H–B–H, which always involves the elimination of small molecules (in the example, HX). The name disazo condensation pigment (new: dihydrazone condensation pigment) is derived from this. Possible bridges (bifunctional diazo components) are diaminobenzenes I and benzidines II, and coupling components are acetoacetic arylides III, IV, or *β*-naphthols V (in Winther's notation: Z = E–B–E):

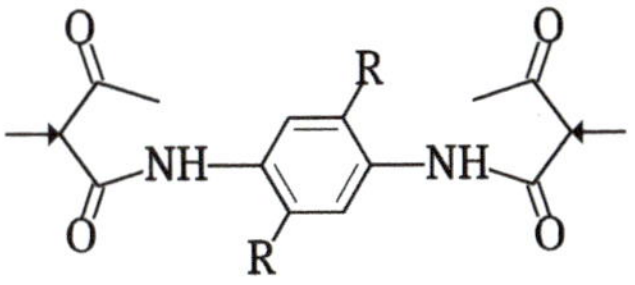

III, R = H, CH_3, OCH_3, Cl
Winther: Z = E–B–E

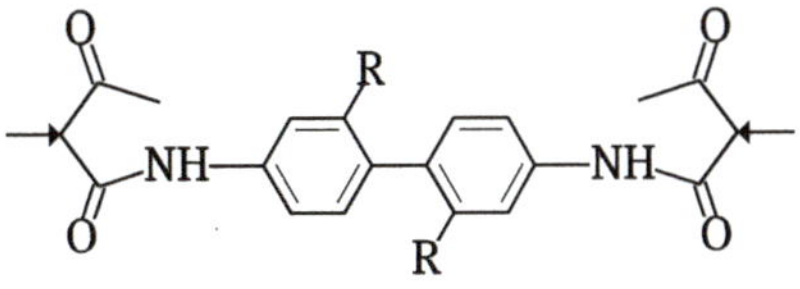

IV, R = CH_3, OCH_3, Cl
Winther: Z = E–B–E

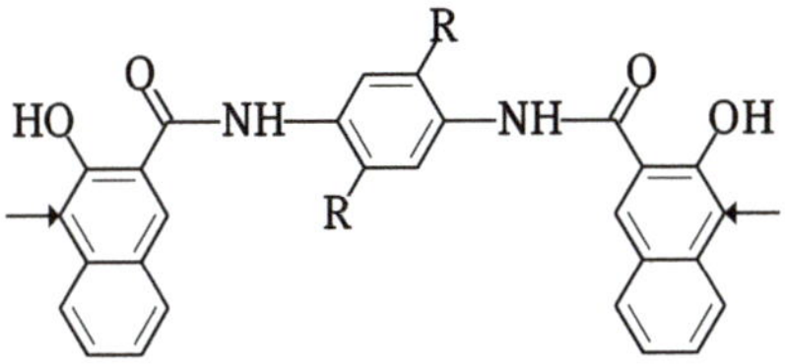

V, R = H, CH_3, OCH_3, Cl
Winther: Z = E–B–E

Combining monofunctional diazo and coupling components with their bifunctional counterparts results in different pigment types:

- *Diaryl yellow pigments* comprise a bifunctional diazo component I or II (Winther symbol “D”) and an acetoacetic arylide as a coupling component (Winther symbol “E”):

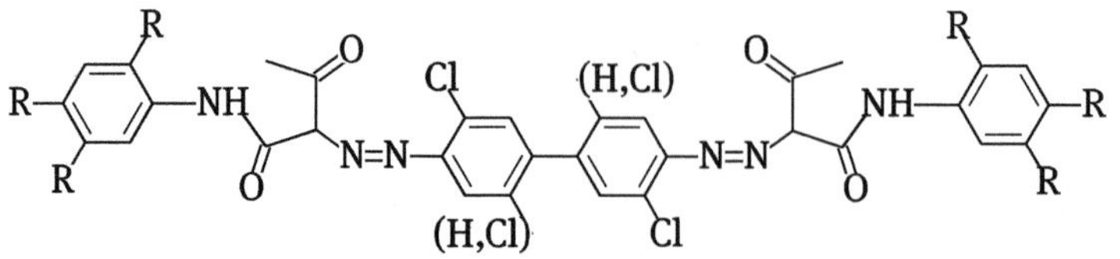

diaryl yellow pigment, R = H, Cl, CH_3, OCH_3, OC_2H_5
Winther: E←D→E

- *Disazopyrazolone pigments* (pyrazolone orange pigments, new: dihydrazone pyrazolone pigments) comprise a bifunctional diazo component I, II (Winther symbol “D”) and an arylpyrazolone as a coupling component (Winther symbol “E”):

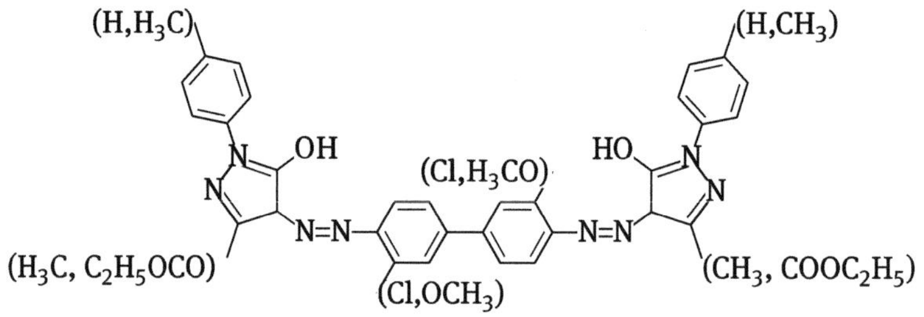

disazopyrazolone pigment, dihydrazone pyrazolone pigment
Winther: E←D→E

- *Bisacetoacetic arylide pigments* comprise a monofunctional diazo component (Winther symbol “A”) and a bifunctional coupling component III, IV (Winther symbol “Z”):

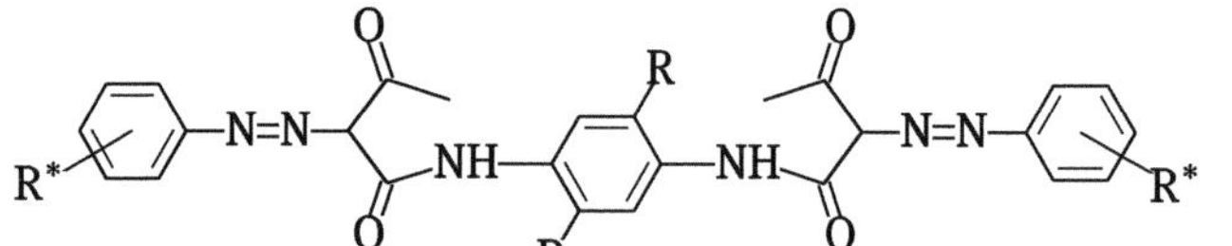

bisacetoacetic arylide pigment
R = H, Cl, CH_3, OCH_3; R* = CH_3, OCH_3, OC_2H_5, Cl, Br, NO_2, $COOC_2H_5$
Winther: A→Z←A = A→E–B–E←A

Diaryl yellow pigments have been known since 1911 and have been used commercially since 1938, having hues from very greenish-yellow to very reddish-yellow. They have high coloring power and are more transparent and stronger in color than the simple monoazo yellow pigments. Being their successors, they represent the dominant composition for yellow pigments in printing inks today, although their lightfastness is not very high either. Due to the molecular size, which is twice that of monoazo yellow pigments, their solubility in organic solvents is considerably lower. Examples of these modern yellow components for four-color printing are PY12 and PY13, and PY126 and PY127 as improved successors. PY83 is a successor of the classical Indian yellow:

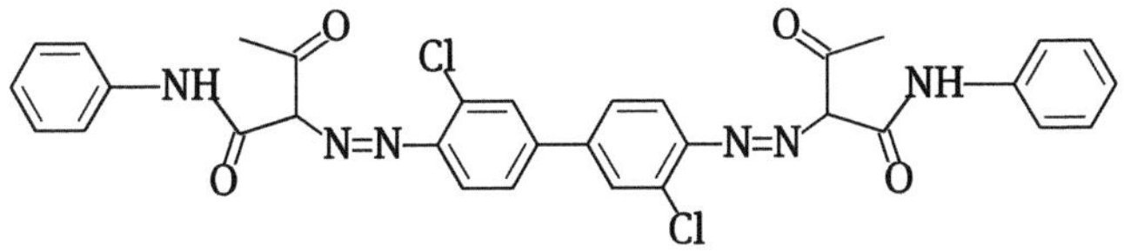

PY12, CI 21090
primary yellow for four-color printing

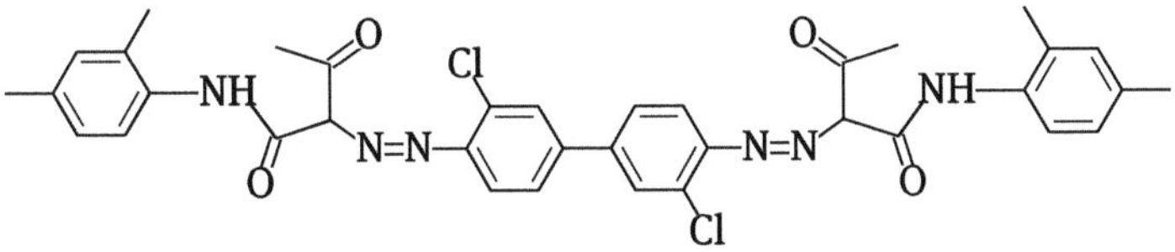

PY13, CI 21100
primary yellow for four-color printing

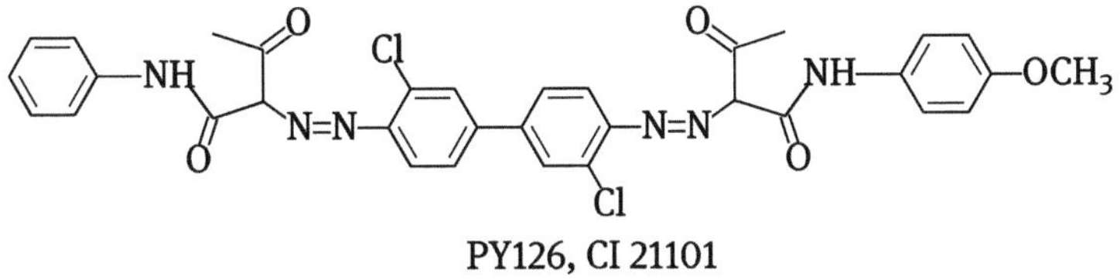

PY126, CI 21101
primary yellow for four-color printing

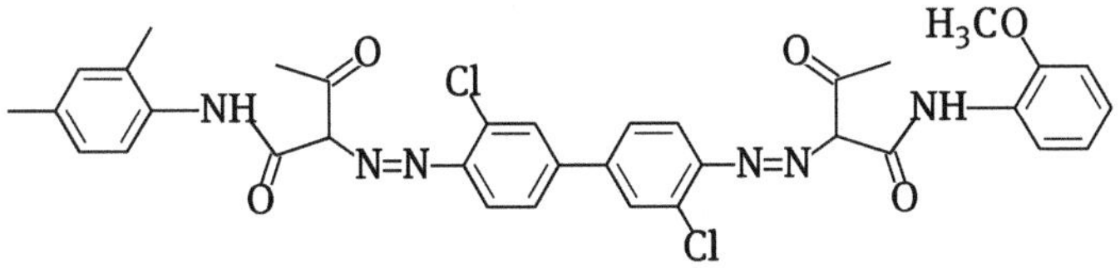

PY127, CI 21102
primary yellow for four-color printing

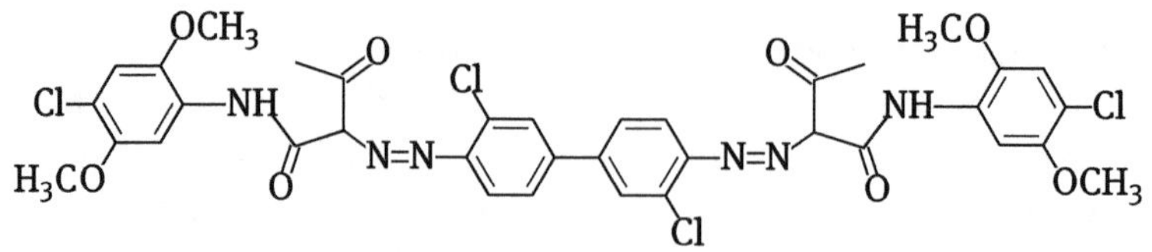

PY83
part of colorant mixtures
yellow ink, Indian yellow [DS]

The introduction of a pyrazolone ring into the molecule shifts the color range of the pigments from pure yellow to orange, red, and brown hues. Such pyrazolone pigments have been known since 1911, and some of them have been used commercially since 1950 because they combine high coloring power with good lightfastness and heat and solvent resistance. The examples show a pyrazolone orange pigment and a pyrazolone red pigment, in which the electron-rich ethyloxy carbonyl group replaces the methyl group on the pyrazolone ring:

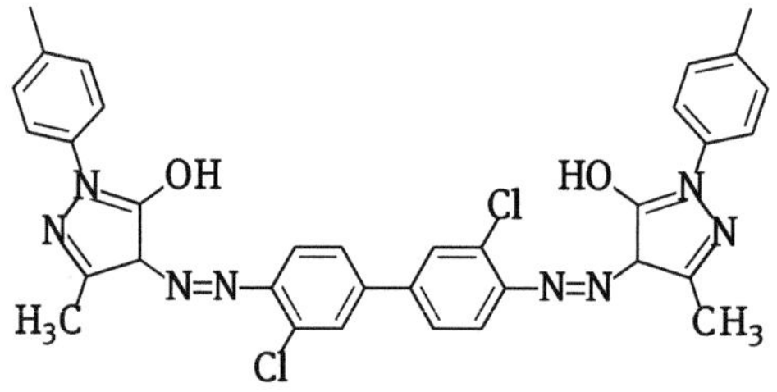

PO34, CI 21115
printing ink

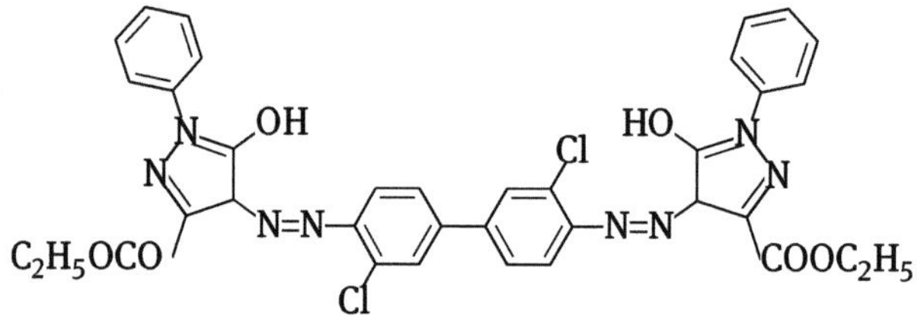

PR38, CI 21120

Yellow-colored examples of bisacetoacetic arylide pigments are PY128 and PY155, appearing in yellow inks, PY128 furthermore in artists' inks:

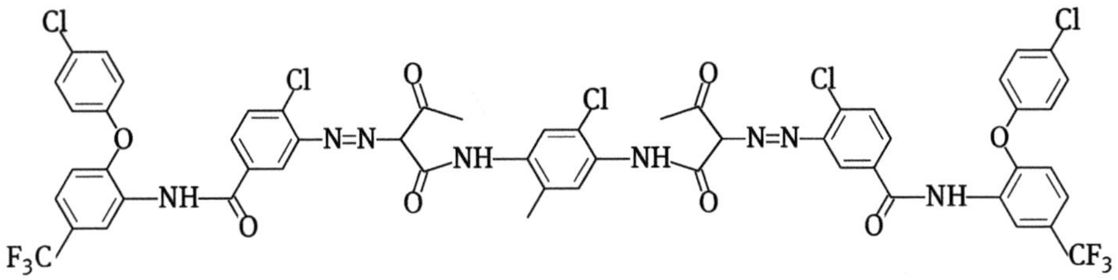

PY128, CI 20037
yellow inkjet inks, transparent brilliant yellow [SC], transparent yellow [WN]

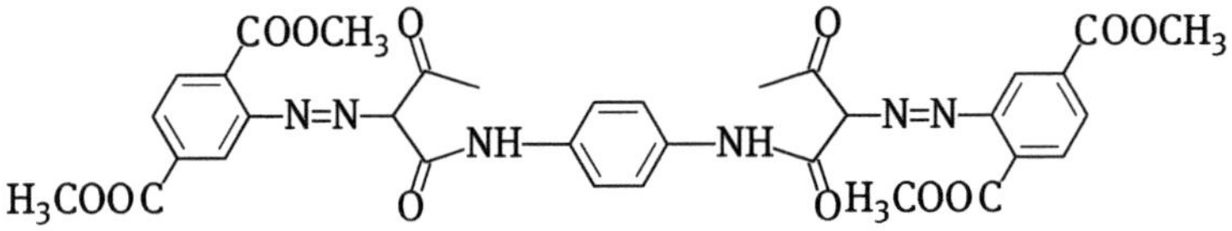

PY155
part of colorant mixtures
yellow ink, brilliant yellow [SC]

V and derivatives of V as coupling components extend the color range to orange, red, and brown, ►Figure 4.29:

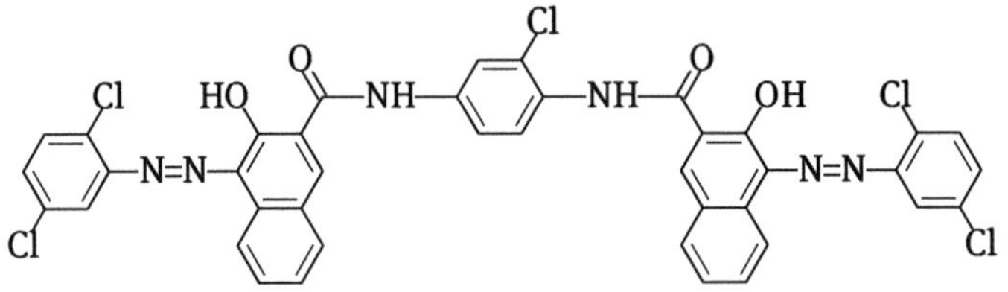

PR144, CI 20735
transparent red deep [SC]
CPT red [KR]

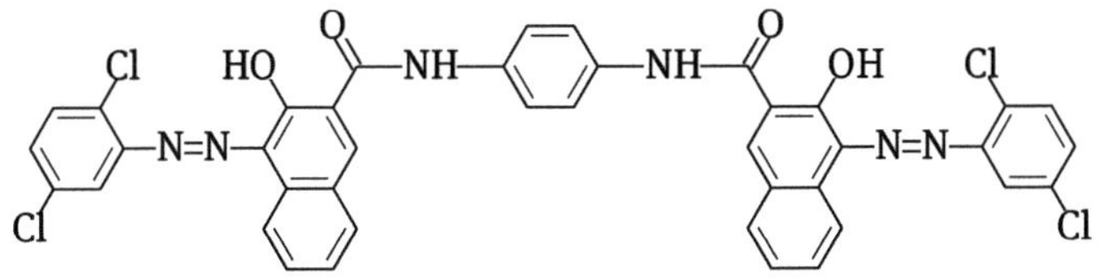

PR166, CI 20730
CPT scarlet [KR]

PR242, CI 20067
geranium red [SC], brilliant scarlet [SC]

PBr23, CI 20060
Gubbio red [KR]

PBr41
transparent brown [SC]

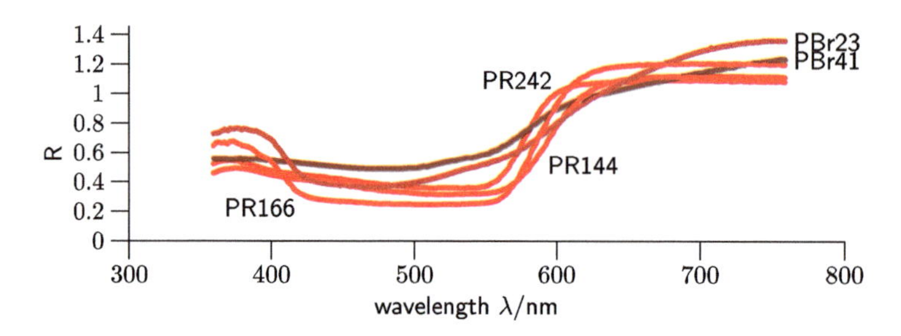

Figure 4.29: Reflectance spectrum of disazocondensation-based watercolors normalized to an arbitrary unit: PBr23 (Gubbio red, Kremer No. 234938), PBr41 (transparent brown, Schmincke Horadam No. 648), PR144 (transparent red deep, Schmincke Horadam No. 355), PR166 (CPT scarlet, Kremer No. 232028), and PR242 (geranium red, Schmincke Horadam No. 341).

4.11.4.3 β-Naphthol pigments

Orange **Red** β-Naphthol pigments [12, Chapter 2.5] are among the oldest synthetic dyes (1889, developmental dyes for cotton) and organic pigments (Para Red, 1885). In these, β-naphthol plays the role of the coupling component (Winther symbol "E"). Thus, their general structure corresponds to formula I:

I, β-naphthol pigments
$R = H, Cl, NO_2, CH_3, OCH_3$
Winther: A→E

PR1, CI 12070
Para Red

Their color ranges from yellowish-orange to brownish-red, and they were formerly used, e. g., for wallpaper printing. β-Naphthol pigments are weak in color and not particularly solvent-resistant, but waterproof. Nowadays, they are of little importance. However, water-soluble derivatives of β-naphthol are used as food colorants, AR18 also for red ink and E110 for yellow ink, ►Figure 4.30.

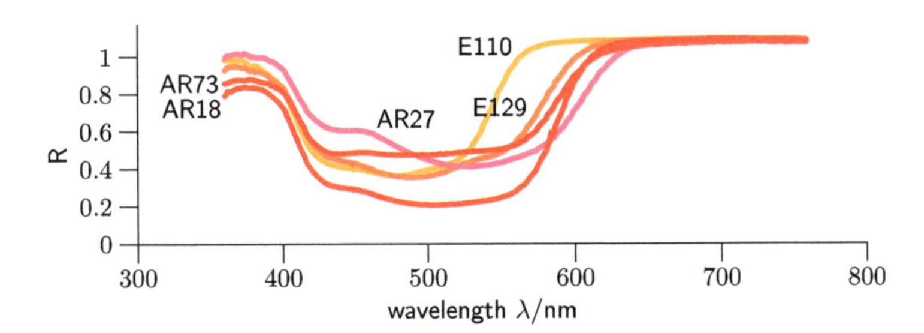

Figure 4.30: Reflectance spectrum of β-naphthol-based dyes normalized to an arbitrary unit: AR18 (cochineal red, E124), AR27 (amaranth, Food Red 9, E123), AR73 (brilliant red B), Food Red 17 (allura red AC, E129), and Food Yellow 3 (yellow orange S, E110).

Laked β-naphthol pigments carry a sulfonic acid group in the diazo component, forming insoluble salts with Ca^{II}, Ba^{II}, Mn^{II}, or Al^{III}. Compared to the unlaked β-naphthol pigments, they exhibit better solvent and migration fastness but are usually not very lightfast. Today, the red pigments are still of interest, as a red toner shows (toner being a synonym for insoluble metal salt pigments in general):

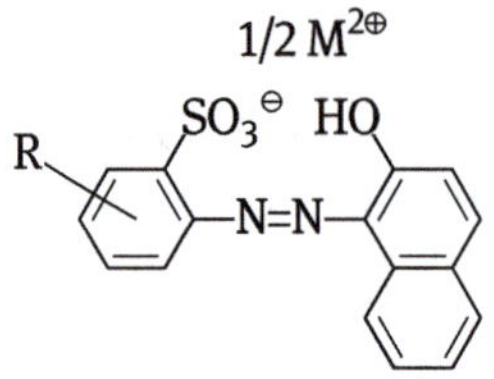

1/2 $M^{2\oplus}$ $SO_3^{\ominus}$ HO R N=N

β-naphthol lake

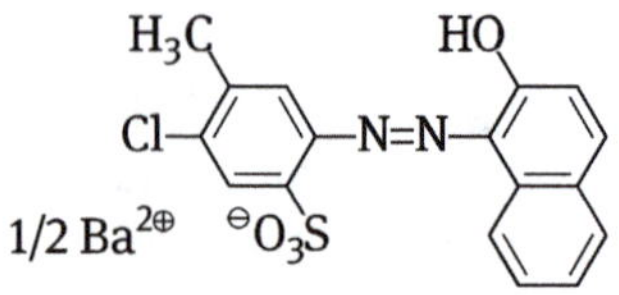

H_3C HO Cl N=N 1/2 $Ba^{2\oplus}$ $^{\ominus}O_3S$

PR53:1, CI 15585:1
primary magenta for four-color printing
(red toner)

Many red, laked *β*-naphthol pigments have *β*-oxy-naphthoic acid (BONA) II as a coupling component. The sulfonic acid group necessary for lake formation is located within the diazo component, a second substituent being H, Cl, or CH_3:

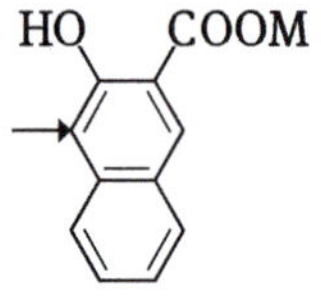

HO COOM

II, *β*-oxy-naphthoic acid BONA

$Ca^{2\oplus}$ $SO_3^{\ominus}$ HO $COO^{\ominus}$ H_3C N=N

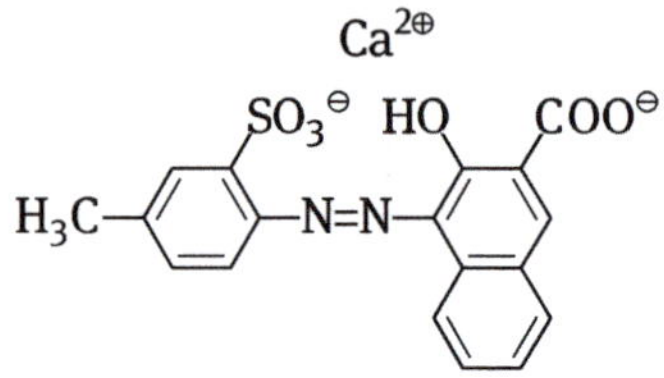

PR57:1, CI 15850
primary magenta for four-color printing

As ionic salts, laked pigments have a lower solubility in organic solvents, and they are therefore often used as red pigments in printing inks and coatings. For example, PR57:1 is a magenta component in four-color printing.

4.11.4.4 Naphthol AS pigments, naphthol red, arylamide red pigments

Red **Purple** These naphthol derivatives [12, Chapter 2.6] were discovered in 1892 and have been in use since 1909. They are also derived from a *β*-naphthol derivative as a coupling component (Winther symbol "E"), namely, an arylide of *β*-oxy-naphthoic acid (BONA), more precisely 2-hydroxy-3-naphthoic acid. The suffix "AS" indicates this fact: "Amid einer Säure" (German, for "amide of an acid"). The general structure of Naphthol AS pigments is as follows:

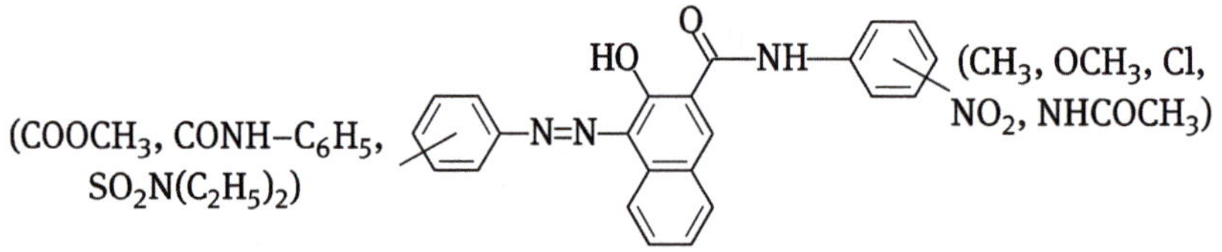

O HO NH (CH_3, OCH_3, Cl, NO_2, $NHCOCH_3$)
($COOCH_3$, CONH–C_6H_5, $SO_2N(C_2H_5)_2$) N=N

Naphthol AS pigments
Winther: A→E

The pigments exhibit yellowish to bluish shades of red, extending to purple and brown. They are intensive in color and transparent, but their fastness against light is not very high. However, due to introducing a carbonamide group and their molecule size, their fastness against solvents is high.

These superior properties, compared to those of simple naphthol pigments, have resulted in arylamide red pigments playing a significant role today in artists' paints and printing inks, ►Figure 4.31:

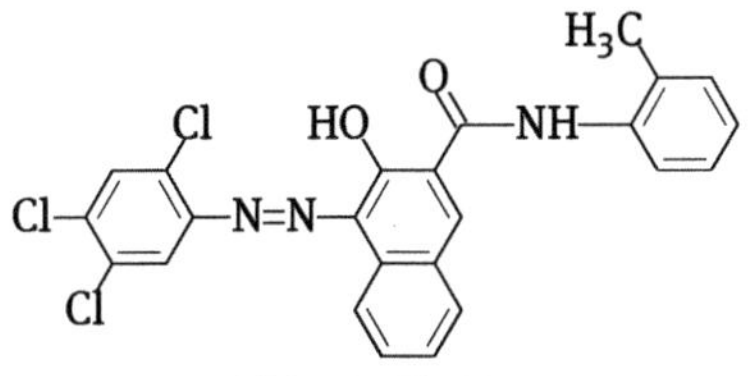

PR112, CI 12370
alizarin madder lake [KR]
Naphthol red light [WN]

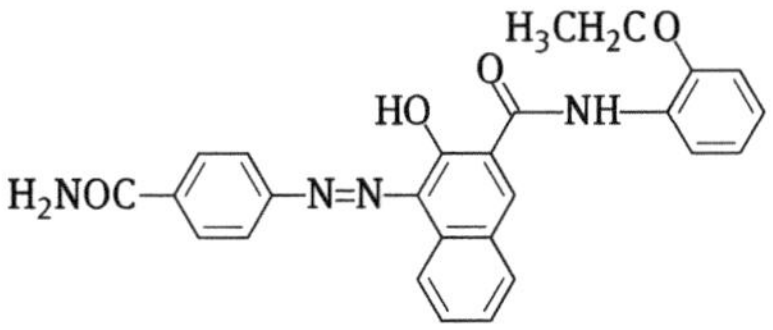

PR170, CI 12475
permanent red [KR][DS], permanent red deep [DS]
Naphthol red medium [WN]

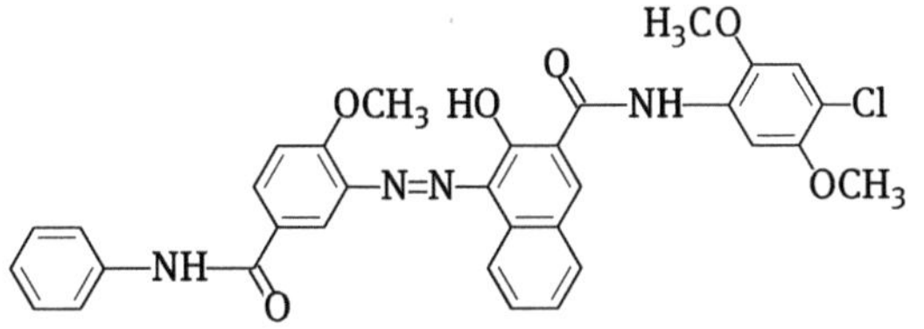

PR184, CI 12487
primary magenta for four-color printing

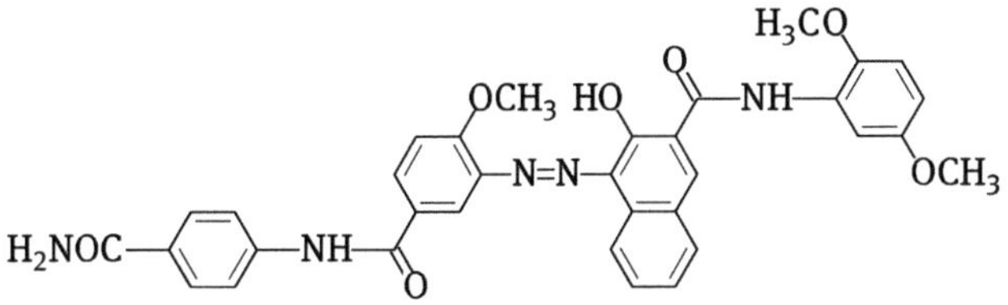

PR187, CI 12486
bordeaux [SC], ruby madder alizarin [WN]

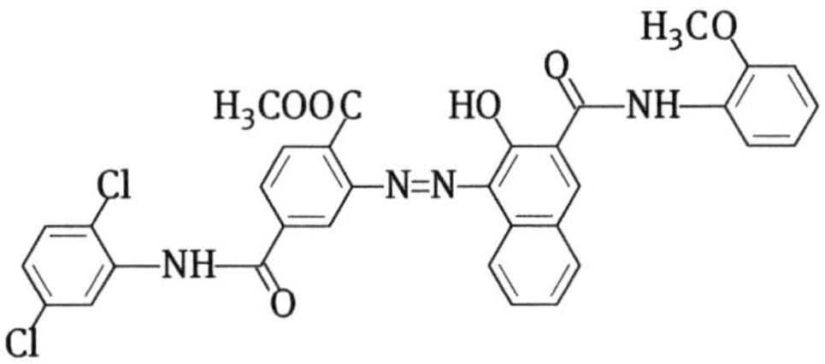

PR188, CI 12467
vermilion light [SC], scarlet lake [WN], organic vermilion [DS]

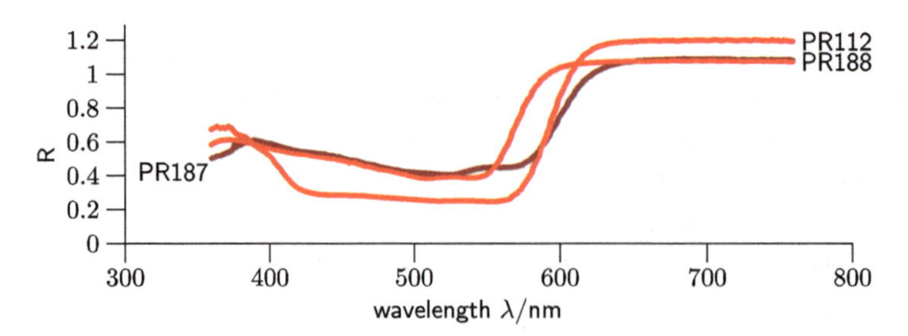

Figure 4.31: Reflectance spectrum of naphthol AS-based watercolors normalized to an arbitrary unit: PR112 (alizarin madder lake, Kremer No. 23600), PR187 (bordeaux, Schmincke Horadam no. 362), and PR188 (vermilion light, Schmincke Horadam no. 342).

Naphthol AS pigments can also be applied as lakes, but only some representatives are actually in use.

4.11.4.5 Benzimidazolone pigments (benzimidazolone hydrazone pigments)

Yellow Orange Red Purple Azo pigments' solvent and migration fastness can be improved by introducing a five- or six-membered heterocyclic group, similar to the carbonamide group in naphthol AS pigments [12, Chapter 2.8]. A suitable heterocycle is benzimidazolone I, used as 5-aminocarbonylbenzimidazolone II in the coupling component (Winther symbol "E"). Other heterocycles are tetrahydro-quinazoline-2,4-dione III and tetrahydro-quinoxaline-2,3-dione IV:

I, benzimidazolone

II
Winther: part of E

III, tetrahydroquinazoline-2,4-dione
Winther: part of E

IV, tetrahydroquinoxaline-2,3-dione
Winther: part of E

The heterocycles improve the pigments' fastnesses to migration, solvents, light, and weather so that they belong to the pigments with the highest fastnesses. By choosing 5-acetacetylamino-benzimidazolone V and 5-(2'-hydroxy-3'-naphthoyl)-amino-benzimidazolone VI as coupling components, we obtain yellow or orange VII and red or brown pigments VIII:

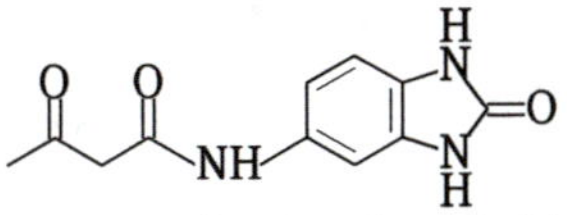

V, 5-acetacetylamino-benzimidazolone
Winther: E

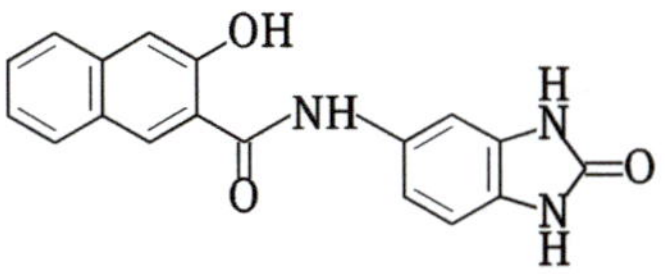

VI, 5-(2'-hydroxy-3'-naphthoyl)-aminobenzimidazolone
Winther: E

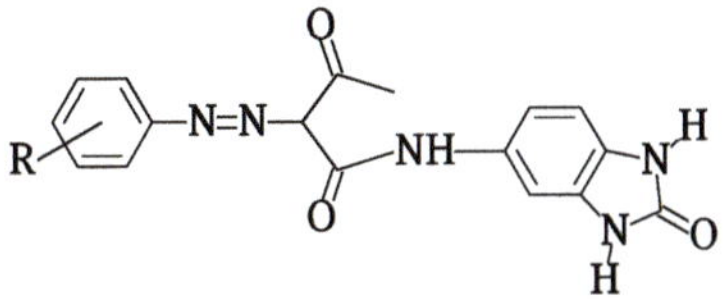

VII, yellow/orange, R = Cl, Br, CH_3, NO_2, OCH_3, COOH, COO–Alk, CONH(H, C_6H_5), SO_2NH–Alk
Winther: A→E

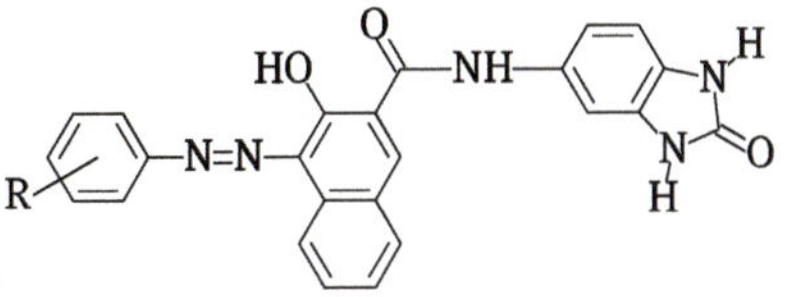

VIII, red/brown, R dto.
Winther: A→E

The coupling components (Winther symbol "E") of both pigment types correspond to the coupling components of acetoacetic arylides (monoazo yellow pigments) and naphthol AS pigments, as do the colors obtained.

In the field of artists' paints, we find a whole range of yellow to orange benzimidazolone pigments of type VII (►Figure 4.32), as well as some yellow-pigmented inks:

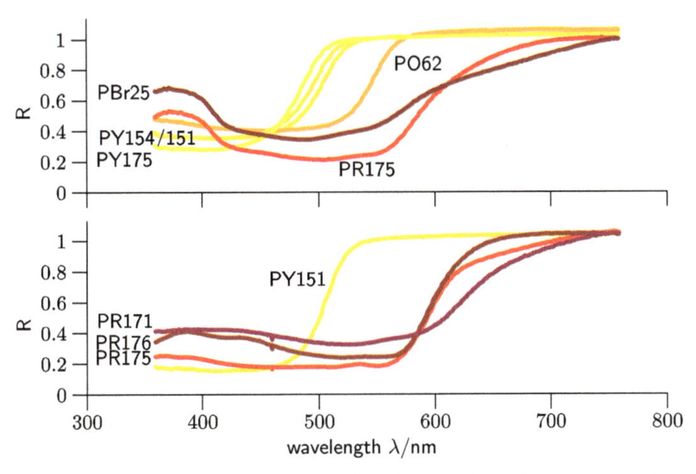

Figure 4.32: Reflectance spectrum of benzimidazolone-based watercolors normalized to an arbitrary unit: PBr25 (mahogany brown, Kremer No. 23495), PO62 (chromium orange hue, Schmincke Horadam no. 214), PR171 (naphthamide maroon, D. Smith Extra Fine Watercolor No. 059), PR175 (purple, Kremer No. 23490), PR175 (deep scarlet, D. Smith Extra Fine Watercolor No. 033), PR176 (carmine, D. Smith Extra Fine Watercolor No. 020), PY151 (aureolin hue, Schmincke Horadam No. 208), PY151 (azo yellow, D. Smith Extra Fine Watercolor No. 215), PY154 (pure yellow, Schmincke Horadam no. 216), and PY175 (chromium yellow hue lemon, Schmincke Horadam No. 211). PY175 has the coolest yellow hue, and PY151, PY154, and PY175 are similar in hue.

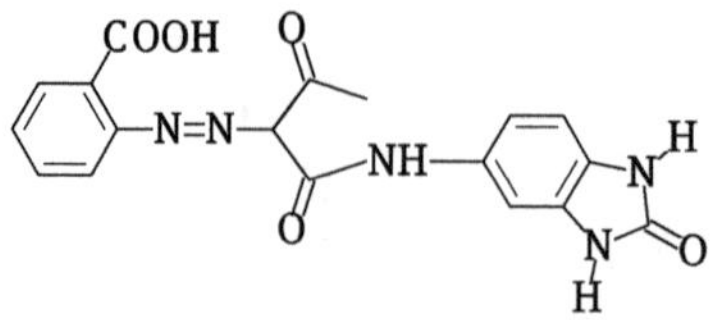

COOH O
N=N
NH
H
N
O
N
O
H

PY151, CI 13980
azo yellow [DS][WN], aureolin hue [SC], yellow ink

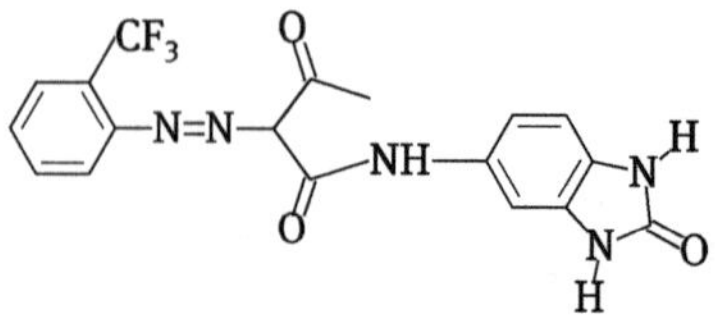

CF_3 O
N=N
NH
H
N
O
N
O
H

PY154, CI 11781
pure yellow [SC], Winsor yellow [WN]
brilliant yellow [SC], yellow ink

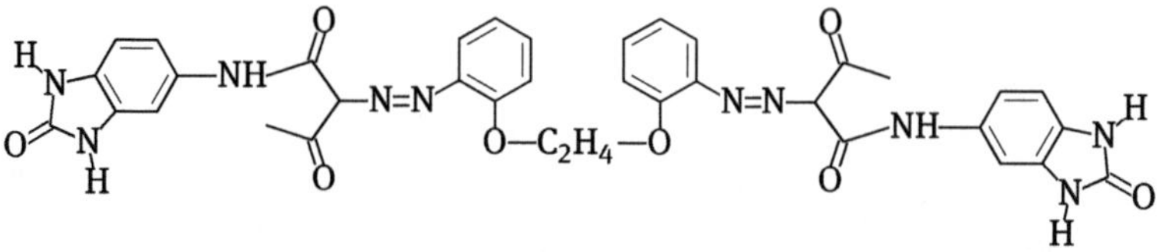

H
N
O
N
H
NH
O
N=N
O
O–C_2H_4–O
N=N
O
NH
O
H
N
N
O
H

PY180
yellow inkjet ink

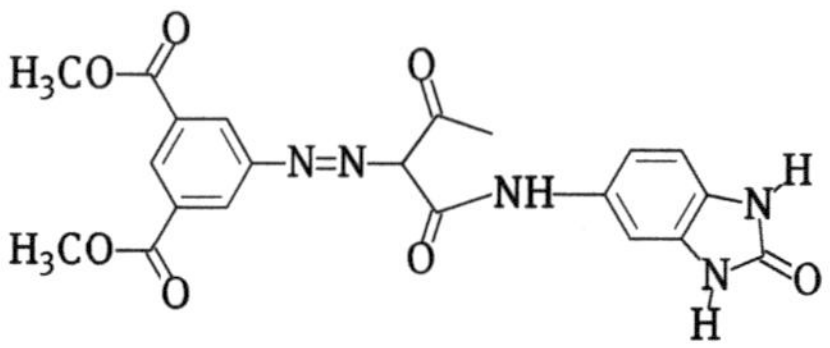

H_3CO O
H_3CO O
N=N
O
NH
O
H
N
N
O
H

PY120
yellow ink

$COOCH_3$ O
N=N
H_3COOC
NH
O
H
N
N
O
H

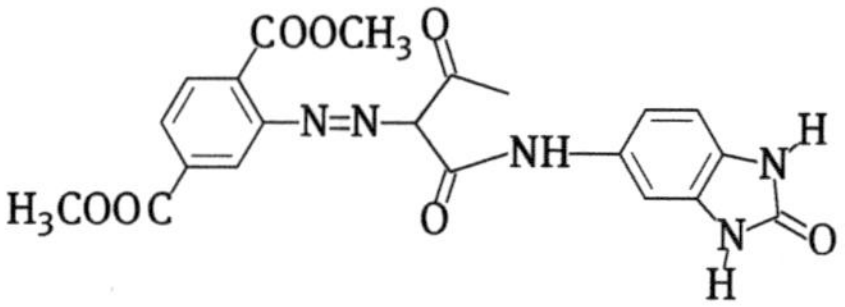

PY175, CI 11784
lemon yellow [DS], Winsor lemon [WN]
chromium yellow hue lemon [SC]

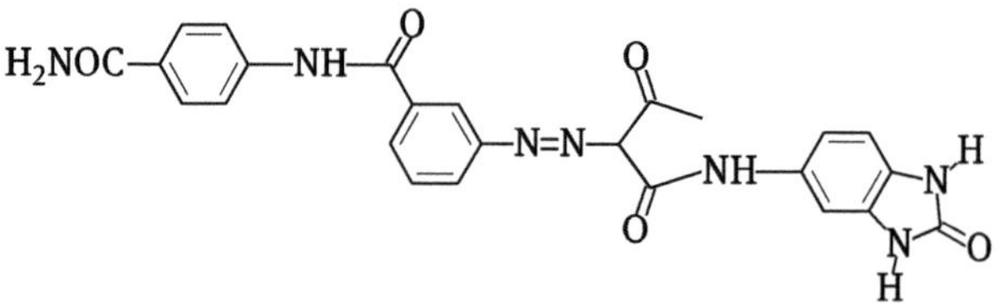

H_2NOC
NH
O
N=N
O
NH
O
H
N
N
O
H

PY181, CI 11777
yellow ink

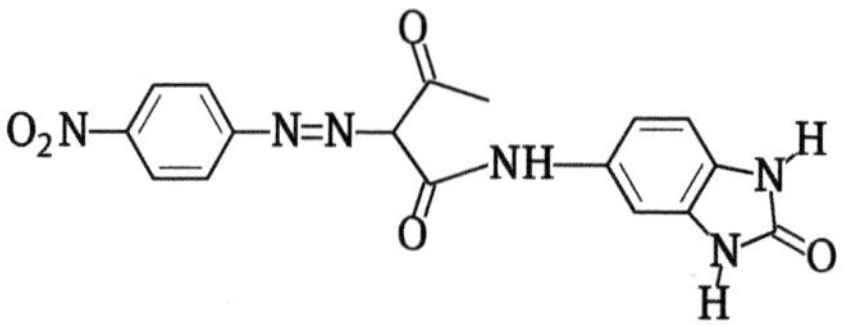

O_2N
N=N
O
NH
O
H
N
N
O
H

PO62, CI 11775
permanent orange [DS], Winsor orange [WN]
chromium orange hue [SC]

VIII-based red pigments also are frequently found in artists' paints and printing inks, ►Figure 4.32:

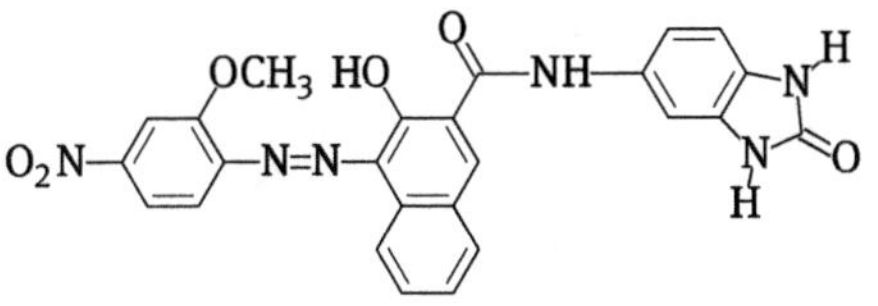

PR171, CI 12512
naphthamide maroon [DS]

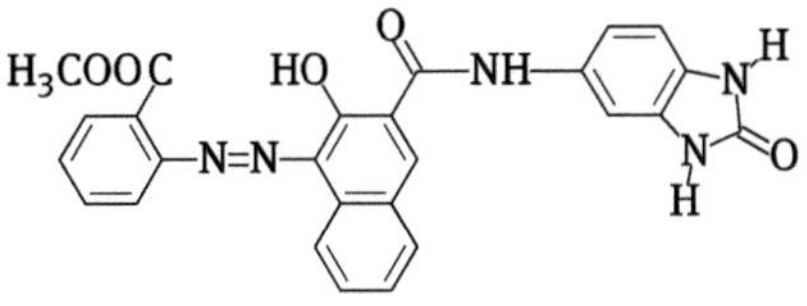

PR175, CI 12513
deep scarlet [DS], purple [KR]

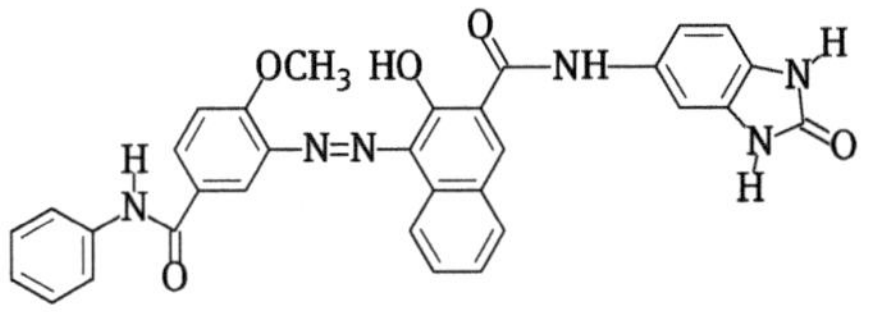

PR176, CI 12515
carmine [DS]

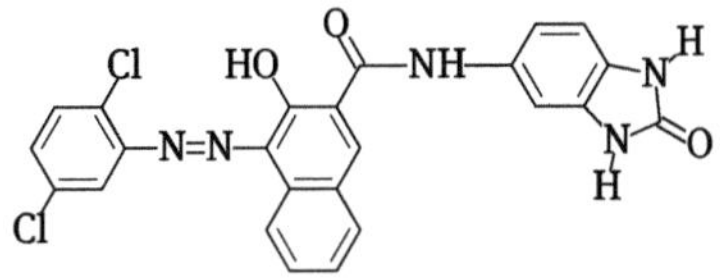

PBr25, CI 12510
Mahogany brown [KR], Indian red deep [WN]
permanent brown [DS]
transparent maroon [WN]

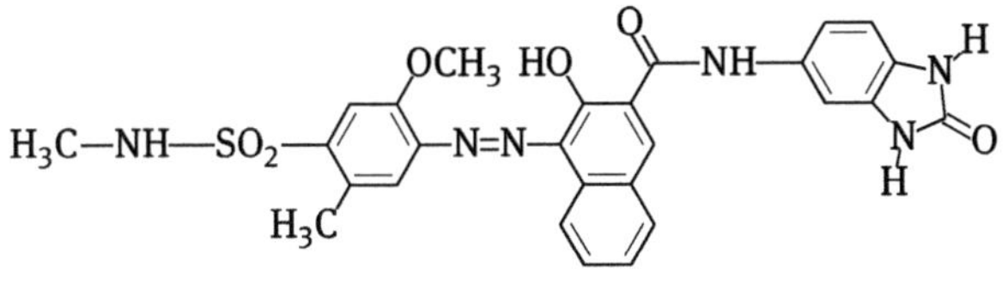

PR185, CI 12516
primary magenta for four-color printing

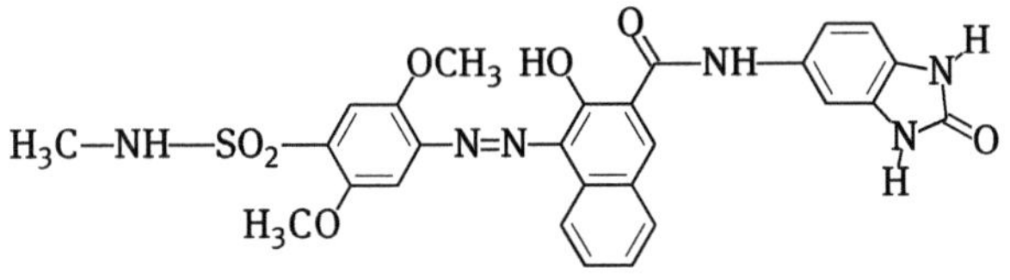

PV32, CI 12517
bordeaux [DS]

IV is rarely used as a base for pigments, and there is only one commercial product from these *quinoxalindione pigments*, ►Figure 4.33:

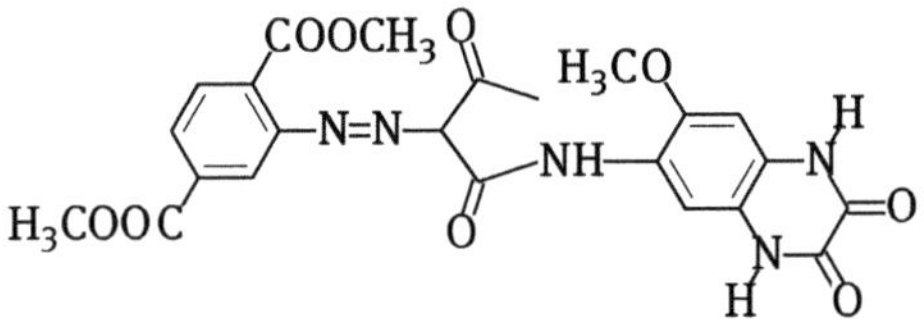

PY213, CI 117875
isoindolinone yellow [KR]

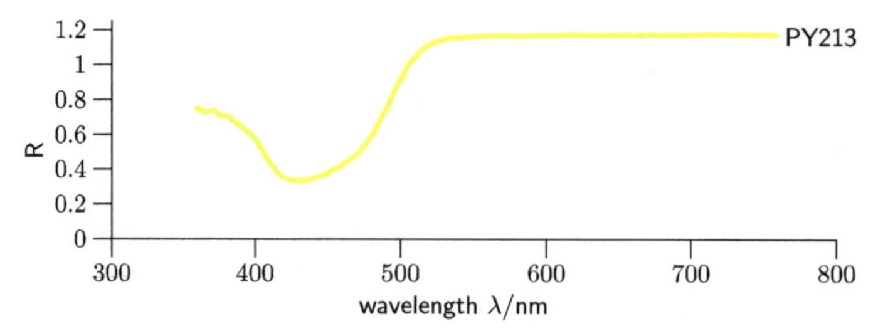

Figure 4.33: Reflectance spectrum of quinoxalindione-based watercolors normalized to an arbitrary unit: PY213 (isoindolinone yellow, Kremer no. 23660).

4.11.5 Cause of chromaticity, blue and green azo colorants

The chromophore of azo colorants is the diazo group. Due to its two nitrogen atoms, it is capable of $n \rightarrow \pi^*$ transitions, $\lambda \approx 350$ nm [211]. However, these transitions are of low intensity and do not contribute to the color of common azo colorants. The distinct color of commercial azo colorants is due to intense $\pi \rightarrow \pi^*$ transitions, which occur when we link the azo group to strong electron donors and acceptors and introduce an aromatic π electron system to achieve a bathochromic shift into the visible spectral range:

A–N=N–D

Acceptors and donors from the early era of azo chemistry could only shift the absorption band into the blue and green spectral range. Thus azo colorants are often associated with yellow, orange, and red colors. There is no blue or green azo pigment, but since about 1950, azo dyes have existed for this spectral range. For green and blue hues, donor and acceptor capabilities must be increased, and mesomeric systems must be enlarged, ►Table 4.15.

Table 4.15: Influence of substituents or donor and acceptor strength, respectively, on the absorption frequency of the diazo chromophore. Indicated is also the theoretical color of the compound (complementary color to the given absorbed wavelength) [4].

Colorant	$\lambda(\pi \rightarrow \pi^*)$ [nm]	$\lambda(n \rightarrow \pi^*)$ [nm]
–N=N–	< 200	350
C_6H_5–N=N–C_6H_5	318	444
C_6H_5–N=N–C_6H_4–OH	349	
C_6H_5–N=N–C_6H_4–NMe_2	408	
O_2N–C_6H_4–N=N–C_6H_4–NMe_2	478	
2,4-$(NC)_2$–C_6H_4–N=N–C_6H_4–NEt_2	515	
2,4,6-$(NC)_3$–C_6H_4–N=N–C_6H_4–NEt_2	562	

The progress achieved so far opens up the spectrum from greenish-yellow to a very greenish-blue or blue-green, [8, p. 1] giving an overview. Essentially, we can distinguish four basic patterns:

- strong electron acceptors
- heterocyclic diazo components
- extended conjugated π systems
- a subtractive mixture of a yellow and a blue dye within the same molecule

Strong electron acceptors

Strong electron acceptors in the diazo component (Winther symbol “A”) such as cyano, sulfonic acid methyl ester, or nitro groups cause large bathochromic shifts:

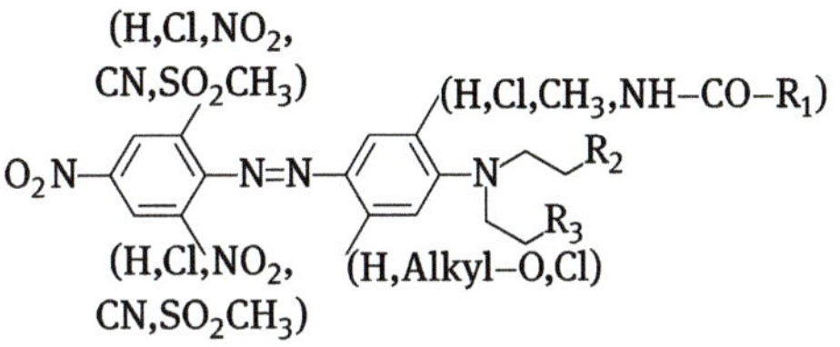

I
Winther: A→E

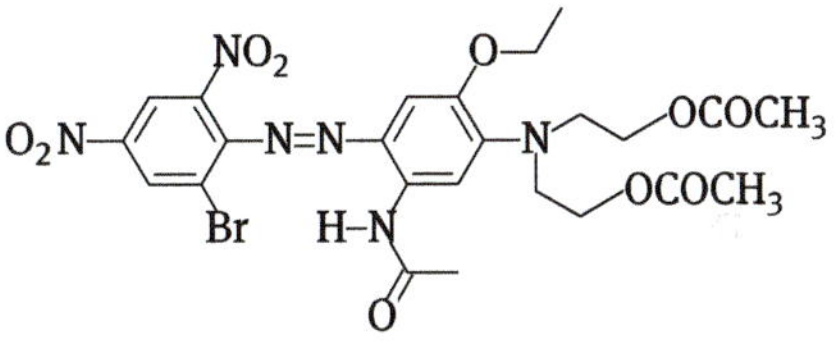

Disperse Blue 79, CI 11345
Winther: A→E

The example of Disperse Blue 79 shows a simple structure, resembling more historical than modern azo dyes. These blue azo colorants are not used as pigments but as dyes, predominantly as disperse and acid dyes, ►Section 5.7.

Heterocyclic diazo components

The second basic pattern introduces small heterocycles into the diazo components (Winther symbol “A”), extending the hues beyond purple and brown to blue and blue-green:

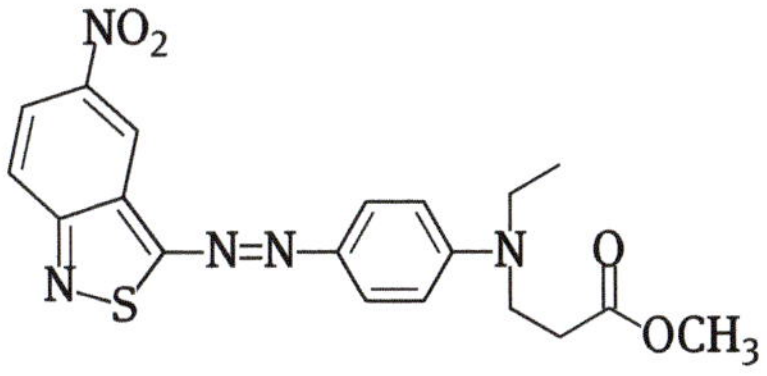

Disperse Blue 148, CI 11124
Winther: A→E

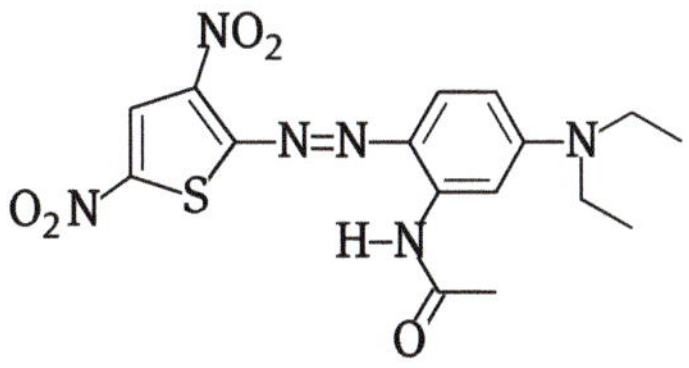

Disperse Green 9, CI 110795
Winther: A→E

Most of these heterocycles are five-membered, but six-membered rings also have been examined. Heteroatoms are sulfur and nitrogen in varying numbers and positions. This group of azo colorants includes only dyes, too.

Enlargement of π system

Classically, we achieve bathochromic shifts by enlarging the π system. Regarding azo colorants, this means:

- Lateral expansion of aromatic systems, typically by replacing benzene rings I with naphthalene II.

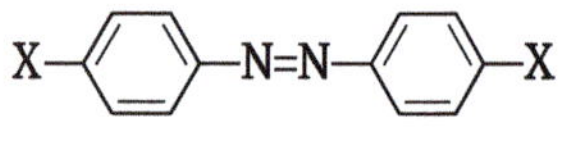

I

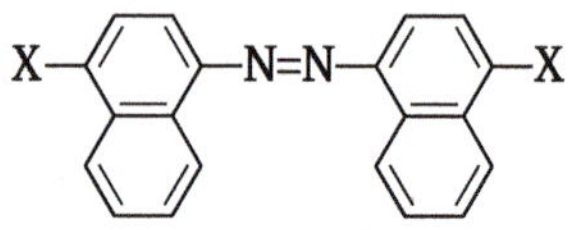

II

- Linear extension of aromatic systems, typically by linking aromatic rings across several azo groups, yielding disazo, trisazo, and polykisazo dyes.

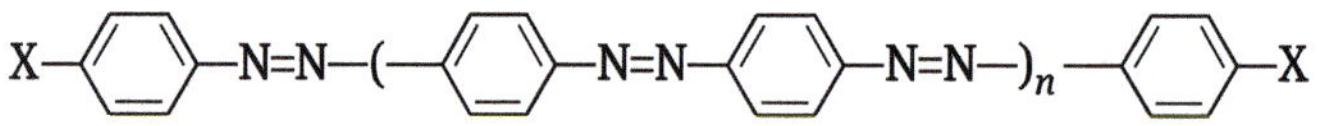

III

The bathochromic shifts achieved are considerable; ►Table 4.16 lists values for monoazo compounds I and II and polykis-azo compounds III.

Table 4.16: Bathochromic shifts achievable with polycyclic diazo (I, II) [5] and polykisazo (III) compounds and the theoretical color of the compound (complementary color to the given absorbed wavelength). A higher n implies a higher bathochromism.

n	λ [nm], X=H	λ [nm], X=NH_2	λ [nm], X=H	λ [nm], X=NH_2
0		318 (III)	390 (III)	
1	318 (I)	390 (I)	359 (III)	474 (III)
2	405 (II)	490 (II)	380 (III)	490 (III)
3		400 (III)	500 (III)	

The color series from red to purple to blue in some monoazo dyes illustrates the described effects. The introduction of an acetylamino group, a strong electron acceptor, shifts the hue of the red AR1 to purple in AV7. Replacing the benzene rings with naphthalene rings shifts the hue to blue in AB92:

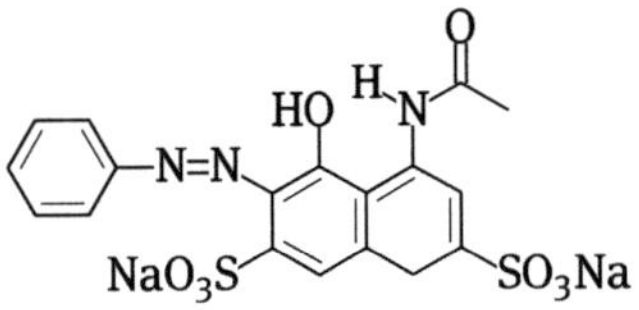

AR1
Winther: A→E

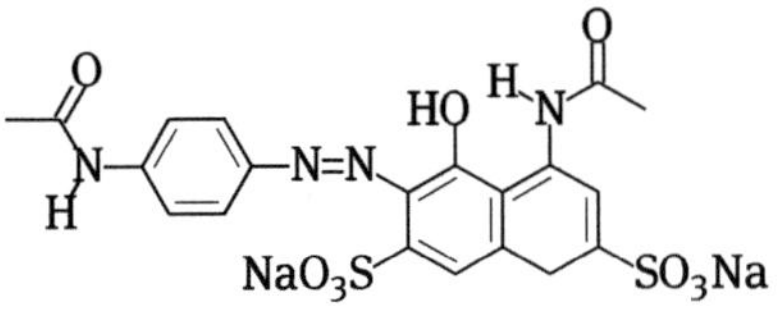

AV7
Winther: A→E

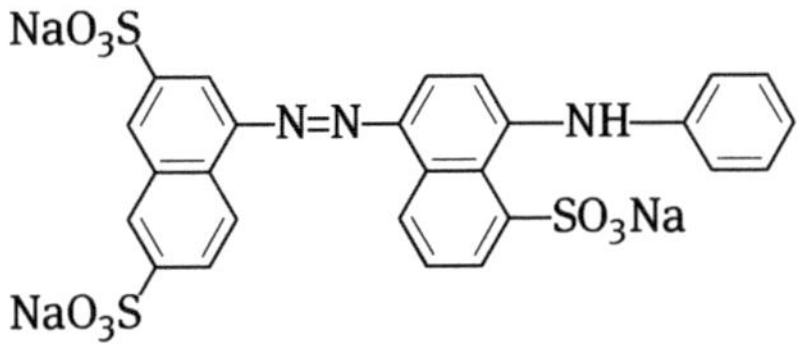

AB92
Winther: A→E

For synthesizing bisazo colorants and higher, building blocks come into play, which Winther denoted by the letter "M": a coupling component comprising an amino group, which can be diazotized after coupling to a monoazo colorant so that subsequent coupling reactions can follow. Bis and trisazo dyes are usually blue, green, and black, as DB67, DG13, and DBk19 demonstrate. By replacing the benzene rings with naphthalene rings, the red disazo dye is converted into a blue one. Enlarging the π system further leads to a green hue. The black ink DBk19 is an example of a tetrakisazo compound:

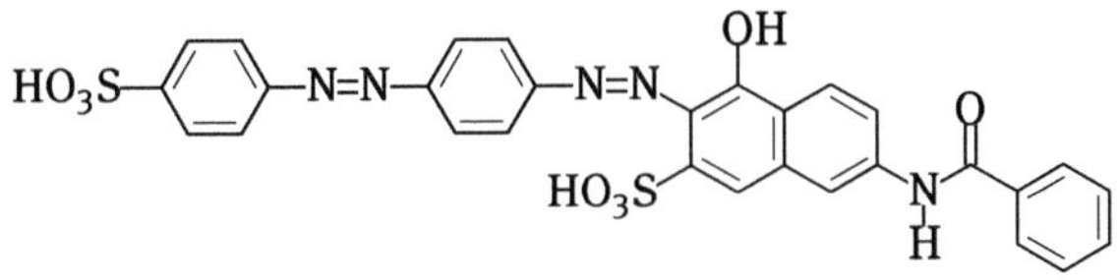

DR81, Sirius Red 4B
Winther: A→M→E

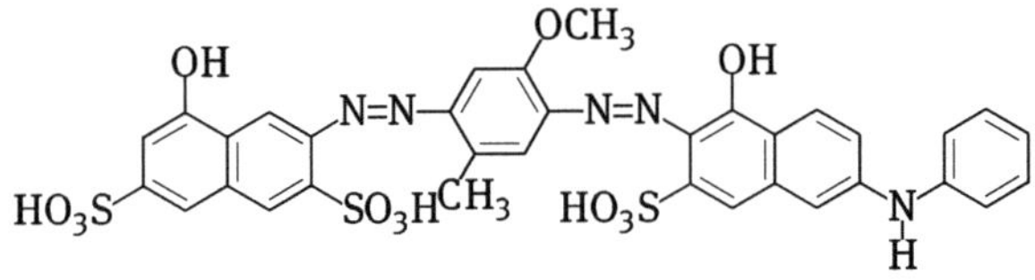

DB67, Sirius Supra Blue F3B
Winther: A→M→E

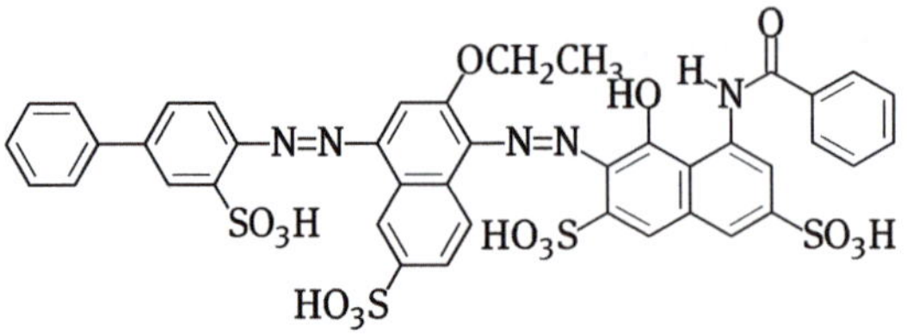

DG13, Brillant Benzo Fast Green GL
Winther: A→M→E

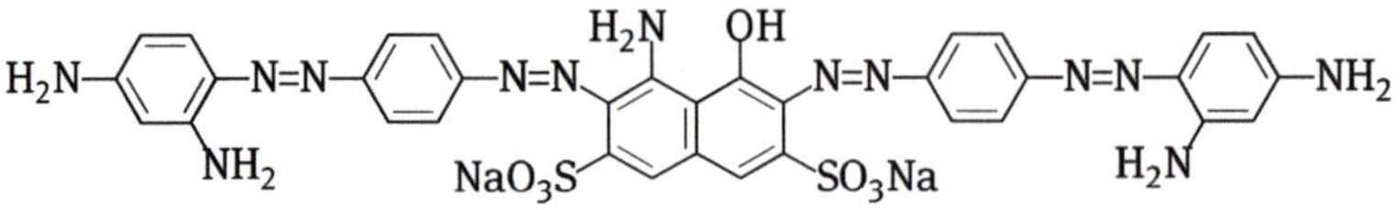

DBk19
black ink
Winther: E←M←D→M→E

Due to the size of their nonpolar region, polykisazo dyes are direct dyes for cellulose, e. g., as black writing and printing inks, as shown by DBk19 or DBk168, ►Section 5.4.

Mixed chromophores

A simple method of obtaining green colors is the subtractive mixing of yellow and blue pigments. Mixing can be performed by combining *two* chromophores within the same colorant. For example, in DG26, a blue disazo and a yellow monoazo dye are connected by a bridge, separating both chromophores' π systems to maintain their individual brilliance. A triazine ring known from reactive colorants is well suited for this task:

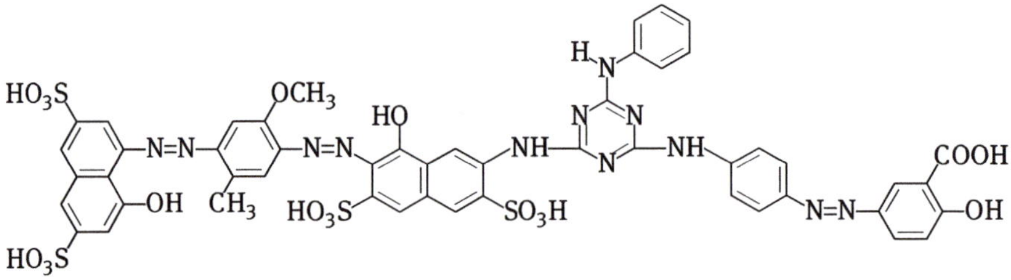

DG26, Chloranil Fast Green BLL
Winther: A→M→E and A→E

Other influencing factors

The formation of the π electron system in industrial azo colorants is promoted by planar molecular geometries so that the overlap of π orbitals is maximized. Stabilizing factors are oxo-substituents in the neighborhood of an azo group that can form hydrogen bonds with the nitrogen and, therefore, stable rings. A prominent structure for this is an acetoacetic acid amide, used as a coupling component in numerous types of azo colorants:

Equally stabilizing is the formation of rings due to azo-hydrazone tautomerism, ►Section 4.11.1.

4.11.6 Chronix toxicity, carcinogenicity

Some azo colorants are suspected of being chronically toxic, even carcinogenic; namely, those delivering known or supposed carcinogenic aromatic amines and diamines [12, Chapter 5.4.5], [781, Chapter 3.7]. Of particular interest are benzidine derivatives or simple aromatic amines (see [781, Table 2.1] for a full list). Due to their solubility in aqueous media, azo dyes are particularly under observation, but azo pigments were also examined.

Aromatic amines form from azo colorants by reductive cleavage of the diazo group, ►Figure 4.34. The reaction can be performed chemically (e. g., by using Zn/HCl or sodium dithionite $Na_2S_2O_4$), or biologically (enzymatic in mammals or by bacteria). Substrates can be any azo colorant, particularly diaryl yellow pigments, and disazopyrazolone pigments, delivering the benzidine, o-tolidine, or o-dianisidine moiety. Concerning artists' paints and printing inks, pigments such as PY12, PY13, PY126, PY127, PY83, PO34, and PR38 are of interest since they deliver the 3,3'-dichlorobenzidine unit. Other situations where aromatic amines are created are given in [781, Chapter 3.7.1], mainly thermal decomposition processes at high temperatures.

Many of the pigments mentioned above were thoroughly tested for the release of aromatic amines, and it was found that no biodegradation to 3,3'-dichlorobenzidine takes place [12, Chapter 5.4.5]. Insolubility and migration fastness decrease bioavailability and resorption of azo pigments to a level that is regarded as acceptable [781, Chapter 3.7.3], [783, 784]. Thermal decomposition, which might also deliver carcinogenic compounds, does not occur under application and usage conditions, and by carefully controlling manufacturing processes and observation, impurities of aromatic amines can be reduced below the detection limit [781, Chapter 3.7.2]. Furthermore, modern azo pigments are designed to deliver noncarcinogenic amino fragments, which suitable substituting all aromatic rings can achieve. As a result, azo pigments are regarded as safe pigments.

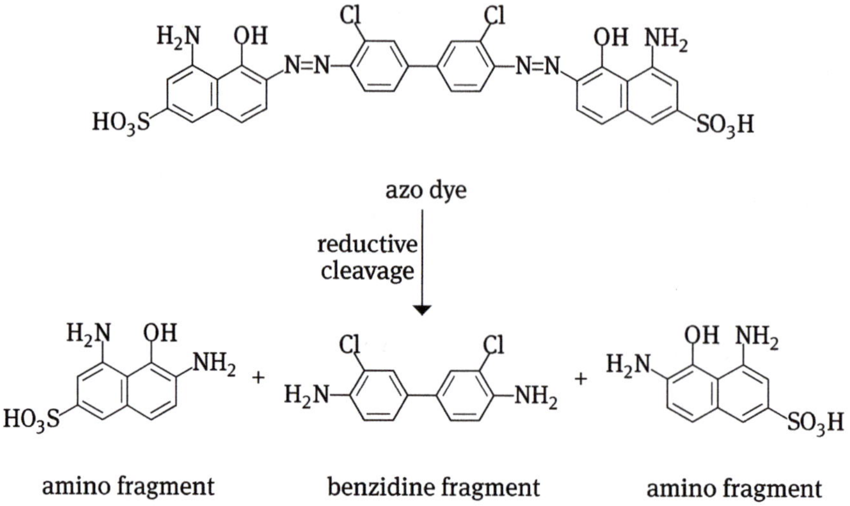

Figure 4.34: Reductive cleavage of the diazo group, delivering potential carcinogenic aromatic mono- and diamines, particularly 3,3'-disubstituted benzidines [781, Chapter 3.4].

4.12 Quinacridone pigments

Yellow Orange Red Magenta Purple Quinacridone pigments [12, Chapter 3.2] are derived from the linear trans-quinacridone I. Although several isomers exist (*cis-trans* isomers due to the carbonyl and amino groups, as well as angular and linear ring arrangements), only *trans*-quinacridone I is relevant for colorants. The unsubstituted structure I is intensely red with a blue tint and was first synthesized in 1935. However, its relevance for pigment chemistry was not recognized until 1955. Pigments derived from I are halogenated or methylated on the outer rings II. The color of the resulting pigments depends not only on their chemical structure but decisively on the pigment crystals' size and structure:

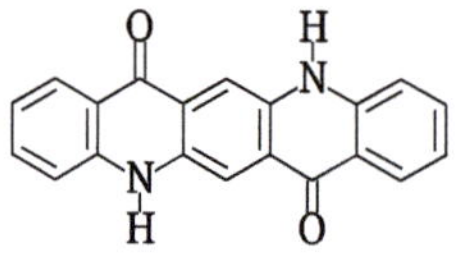

PV19, CI 73900
I, trans-quinacridone
ruby red [SC], permanent carmine [SC][WN]
permanent magenta [WN], permanent rose [WN]
quinacridone red [DS][SC], quinacridone rose [DS]
quinacridone violet [DS][SC], Caesar purple [SC]
quinacridone magenta [WN]

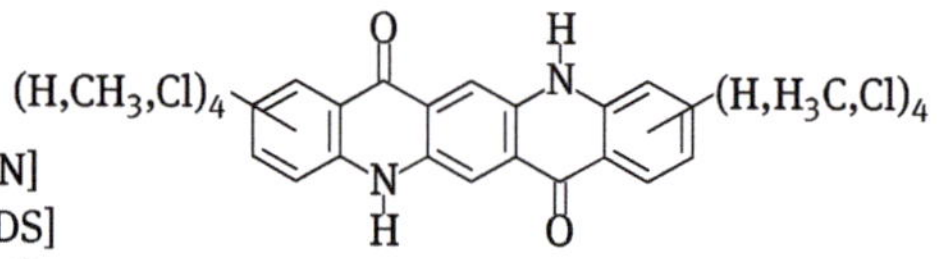

II, quinacridone pigments

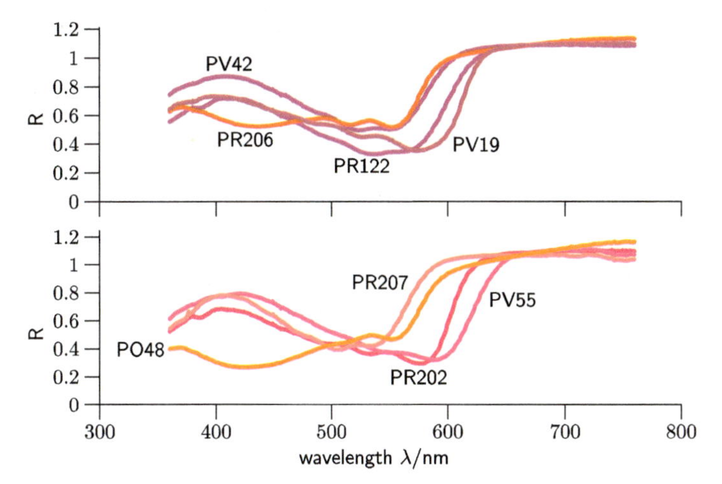

Figure 4.35: Reflectance spectrum of quinacridone-based watercolors normalized to an arbitrary unit: PO48 (quinacridone burnt orange, D. Smith Extra Fine Watercolor No. 086), PR122 (purple magenta, Schmincke Horadam No. 367), PR202 (quinacridone magenta, Schmincke Horadam No. 369), PR206 (madder brown, Schmincke Horadam No. 670), PR207 (quinacridone red light, Schmincke Horadam No. 343), PV19 (quinacridone violet, Schmincke Horadam No. 368), PV42 (magenta, Schmincke Horadam No. 352, primary magenta), and PV55 (quinacridone purple, Schmincke Horadam No. 472).

Quinacridone pigments are intensively colored and exhibit yellow to purple hues. They possess excellent fastnesses to migration, light, and weather and are among the high-quality pigments used in lacquers and printing inks for high demands. Numerous artists' paints contain PV19 and other quinacridone pigments (►Figure 4.35), as do red inkjet inks and primary magenta colors for four-color printing:

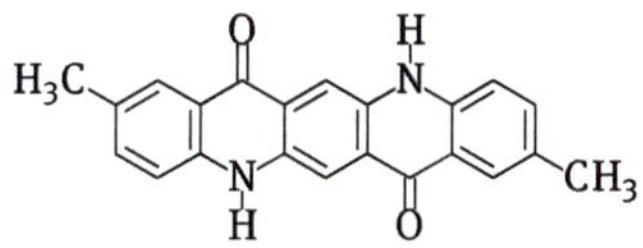

H_3C, CH_3

PR122, CI 73915
opera pink [DS], quinacridone magenta [WN][DS][SC]
purple magenta [SC], quinacridone lilac [DS]
opera rose [WN], transparent magenta [SC]
quinacridone violet [WN], primary magenta four-color

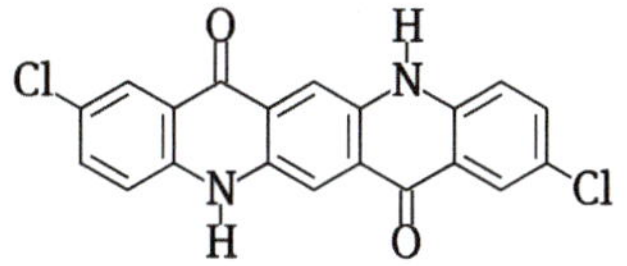

PR202, CI 73907
quinacridone magenta [DS][SC]
quinacridone fuchsia [DS], inkjet inks

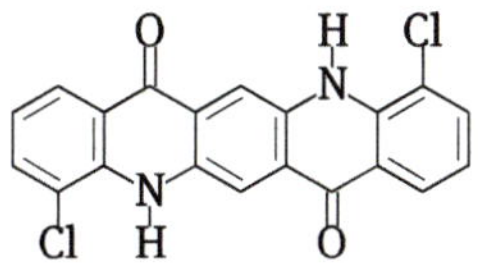

PR207, CI 73906/73900 (solid solution with PV19)
madder lake brilliant [SC], madder brilliant [SC]
quinacridone red light [SC]

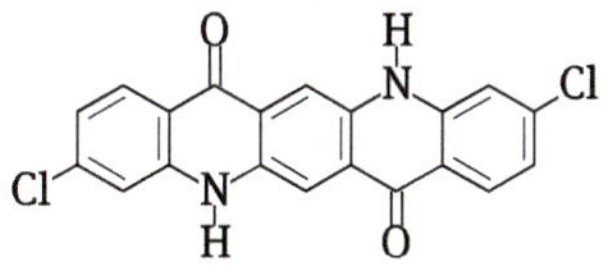

PR209, CI 73905
quinacridone red [WN]
quinacridone coral [DS]

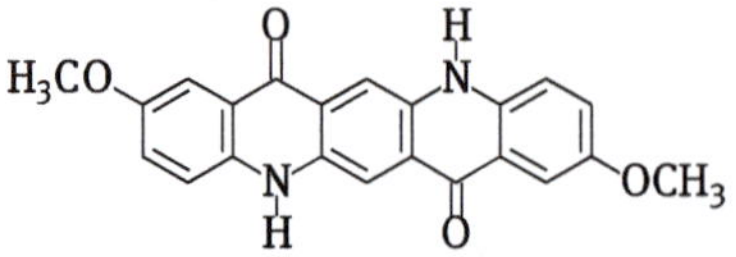

PV55 (solid solution with PR202 as host)
quinacridone purple [SC][DS]
quinacridone violet [WN]

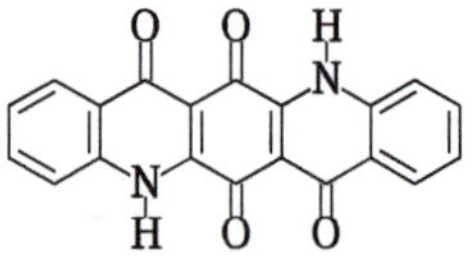

quinacridone quinone
PR206 (as mixed crystal with PV19)
quinacridone burnt scarlet [DS]
madder brown [SC][WN], madder root red [SC]
quinacridone burnt orange [WN]
PO48 (as solid solution with quinacridone)
quinacridone burnt orange [DS]

PV42 (structure not yet disclosed) is found in "magenta" [SC]. A derivative, quinacridone quinone, is only a weakly yellow and not entirely stable. However, its fastnesses are enhanced considerably by forming mixed crystals with quinacridone. Simultaneously, the hue shifts from purple to a dull orange of metallic character. Especially in combination with transparent iron oxide yellow, gold-colored and copper-colored pigments with metallic effects are obtained. Mixed crystals of PV19 and quinacridone quinone show as PR206, a shade of yellow-brown copper. As PO49, they have a shade of gold and were also included in artists' paints. The blend PO48 is used by violin makers due to its warm red-gold shade and also by artists.

Cause of color

The chromophore of quinacridones is structure I:

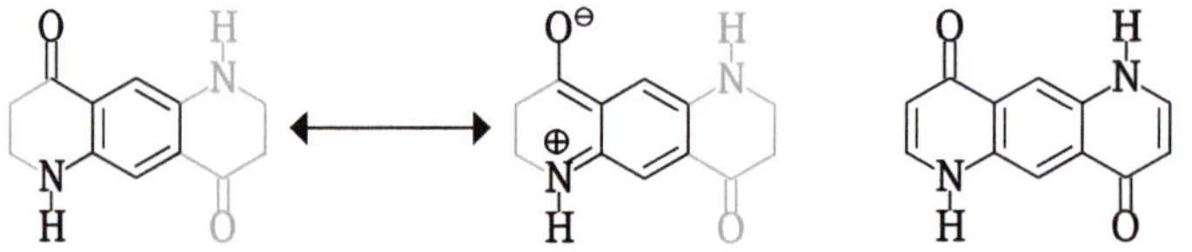

I

II

The carbonyl groups act as electron acceptors, and the amino groups as donors, similar to donor-substituted quinones [5]. Structure I with a saturated bridge (or an aromatic ring system) shows absorption at a longer wavelength than II, in which the unsaturated bridge stabilizes the ground state.

However, it has been discovered that molecules in solution are only weakly colored and that the crystal lattice is of great importance for the hue. It is concluded that the deep hue is induced by an interaction of individual π systems, carbonyl groups, and amino groups over more extensive parts of the crystal lattice [12]. Indigo is a similar case; the isolated chromophore imparts a red color to the indigo molecule, while the electron delocalization over the complete crystal lattice contributes the decisive part of its deep blue hue. Therefore, substituents in quinacridones do not directly influence the hue but change the preferred spatial structure of the crystal, which in turn determines the final hue.

4.13 Perylene pigments

Red Purple Black The class of perylene pigments [12, Chapter 3.4] is derived from perylene tetracarboxylic acid I. Numerous derivatives have been produced as vat dyes since 1912. It was not until 1950, however, that the di-imides II, especially the methyl- and aryl-substituted ones, were discovered to exhibit qualities interesting for paint-makers:

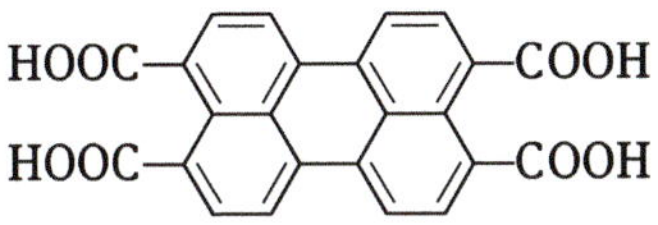

I
perylene tetracarboxylic acid

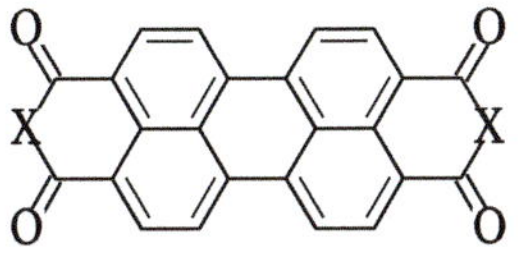

II
perylene pigments, X = N-R
PR224, CI 71127, X = O

Perylenes exhibit excellent properties and are employed in high-quality coatings. Interestingly, we find black pigments among them. Pigments used in artists' paints (►Figure 4.36) are:

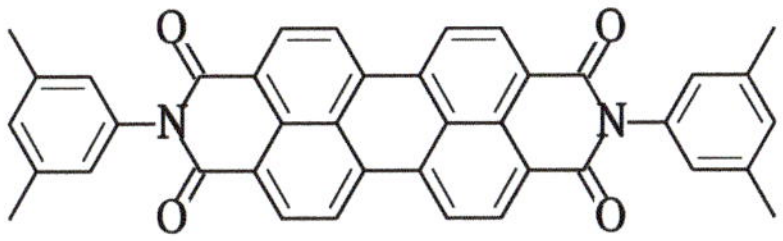

PR149, CI 71137
perylene scarlet [DS], Winsor red deep [WN]
perylene red [WN]

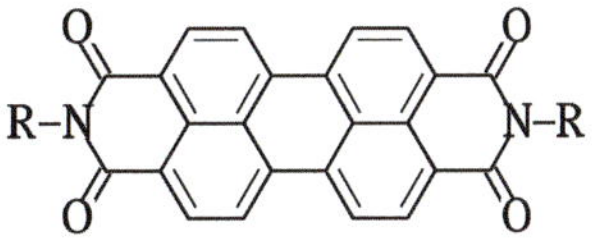

PR178, CI 71155, R = –Ph–N=N–Ph
perylene dark red [SC], perylene red [DS]

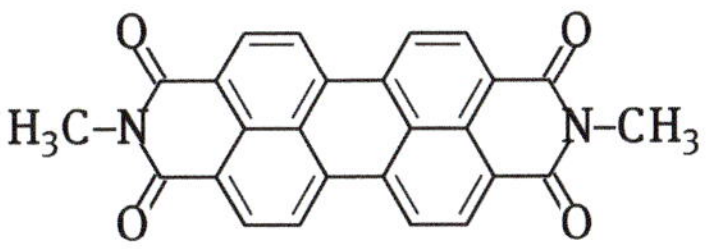

PR179, CI 71130
perylene maroon [SC][DS][WN], Florentine red [SC]
permanent alizarin crimson [WN], brown madder [WN]
alizarin crimson hue [SC]

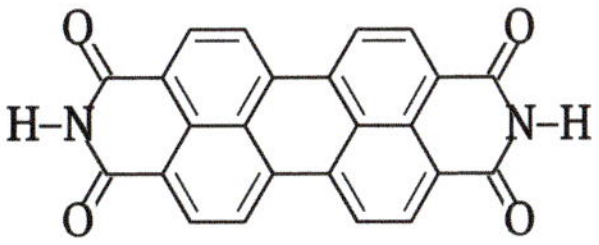

PV29, CI 71129
perylene violet [SC][DS][WN]

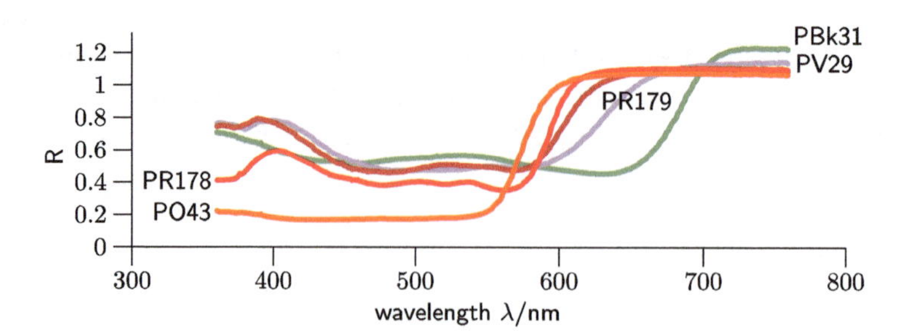

Figure 4.36: Reflectance spectrum of perylene-based watercolors normalized to an arbitrary unit: PBk31 (perylene green, Schmincke Horadam no. 784), PO43 (perinone orange, D. Smith Extra Fine Watercolor No. 066), PR178 (perylene dark red, Schmincke Horadam No. 344), PR179 (perylene maroon, Schmincke Horadam No. 366), and PV29 (perylene violet, Schmincke Horadam no. 371).

PBk31, CI 71132
perylene green [SC][DS][WN], atrament black [SC]
perylene black [WN]

PO43, CI 71105
perinone orange [DS]

Perinone pigments show a simpler structure [12, Chapter 3.4]. Their production started as early as 1926, but only two are in use today, one of them (PO43) in artists' paints.

Cause of color

The color-bearing system of perylene diimides is 1,8-naphthalimide I. Its carboximide group represents a weak electron acceptor [5], while the naphthalene system acts as a donor:

I, 1,8-naphthalimid, λ=334 nm (R=–H)
λ=372 nm (R=–NHCOCH$_3$)

R=–H   R=–NHCOCH$_3$

This chromophore does not sufficiently explain specific color changes of perylene pigments. If, e. g., the N-methyl substituent is replaced with an N-ethyl substituent, it should have little effect on the carboximide chromophore; nevertheless, the color changes from red to black [12, Chapter 3.4.1.3]. Therefore, it is assumed that the crystal structure also has a decisive influence on the pigment's color, similar to quinacridones and indigoid colorants. Depending on the steric requirements of the substituents, the molecules in the crystal adopt different arrangements relative to one another. In the example given, the ethyl substituent forces an arrangement that allows more extensive overlaps of the individual π systems, and thus leads to a black color, which common chromophores cannot achieve.

4.14 Diketopyrrolo-pyrrole (DPP) pigments

Orange Red Purple Diketo-pyrrolo-pyrrole pigments (DPP pigments) [12, Chapter 3.5] were first discovered around 1970. They are based on compound I and cover the color range from orange and red to purple. The typical substitution pattern is shown in II:

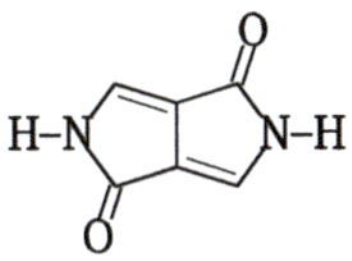

(alkyl, aryl, Cl, Br, CF_3)

(alkyl, aryl, Cl, Br, CF_3)

I

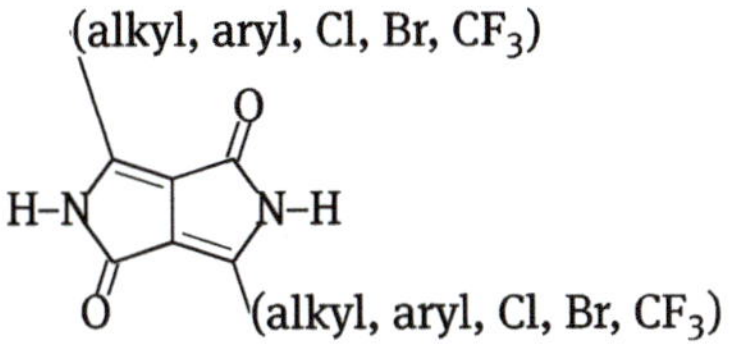

II, DPP pigments

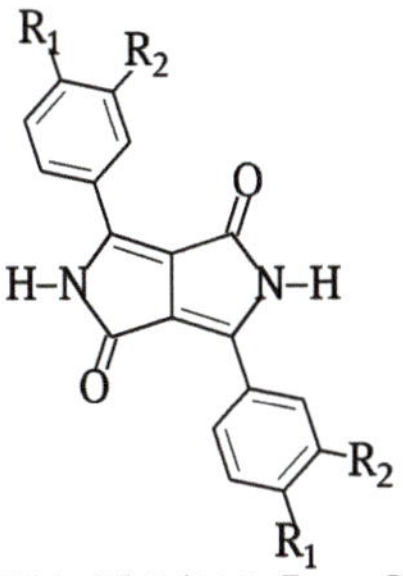

PR254, CI 56110, R_1 = Cl, R_2 = H
scarlet red [SC], pyrrol red [DS][WN], Winsor red [WN], bright red [WN]
PR255, CI 561050, R_1 = R_2 = H
vermillion (red) [SC], pyrrol scarlet [DS], scarlet lake [WN], pyrrole red light [WN]
PR264, CI 561300, R_1 = C_6H_5, R_2 = H
ruby red deep [SC], pyrrol crimson [DS], Winsor red deep [WN], ruby [SC]
PO71, CI 561200, R_1 = H, R_2 = CN
transparent orange [SC], transparent pyrrol orange [DS]
PO73, CI 561170, R_1 = p-C_6H_4-*tert*Bu, R_2 = H
pyrrol orange [DS][WN], Winsor orange red shade [WN], Winsor orange [WN]

Like perylenes, they exhibit good properties and are used in high-quality coatings and artists' paints, ►Figure 4.37.

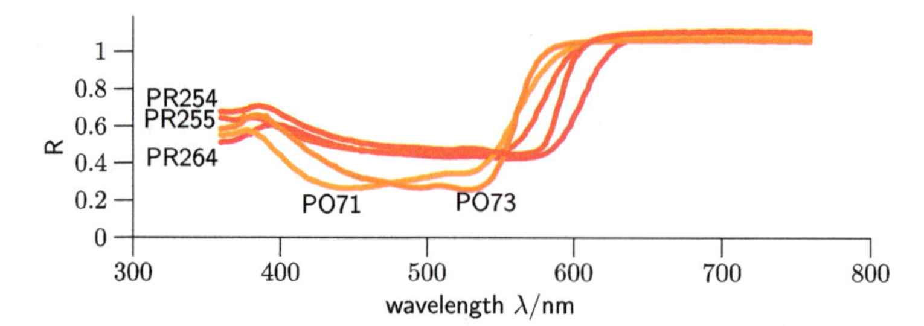

Figure 4.37: Reflectance spectrum of diketo-pyrrolo-pyrrole-based watercolors normalized to an arbitrary unit: PO71 (transparent orange, Schmincke Horadam No. 218), PO73 (Winsor orange red shade, Winsor and Newton Professional Watercolor No. 723), PR254 (scarlet red, Schmincke Horadam No. 363), PR255 (vermilion, Schmincke Horadam No. 365), and PR264 (ruby red deep, Schmincke Horadam No. 346).

Cause of color

Despite the two five-membered rings, DPP pigments are planar molecules so that their carbonyl and amino groups can form a merocyanine system, ►Section 2.5.5. Similar to indigo, in which the crossed arrangement of two merocyanines results in a strong bathochromic shift (►Section 2.5.3.2), the phenyl rings act as secondary electron donors, turning the very short-chained merocyanine into a strong chromophore:

I

In addition, intermolecular interactions occur in the crystal between carbonyl and amino groups and between the π systems of neighboring rings, influencing the physical and color properties considerably. We already had seen similar interactions at quinacridone pigments, indigo derivatives, and perylene pigments.

4.15 Azomethine, methine or isoindoline pigments

Yellow Orange Red Isoindoline pigments [12, Chapter 3.10] are derived from 1,3-disubstituted isoindoline I. According to the substituents, we distinguish azomethine pigments (II, III) and methine pigments (IV). Azomethines are substituted with an amino group and a second amino or an oxo group, whereas methines carry two carbon substituents, attached by double bonds:

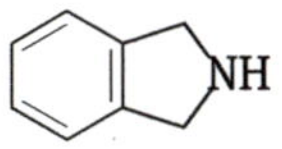

I, isoindoline 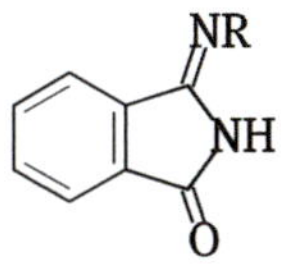 | II, azomethine 3-imino-isoindolinone 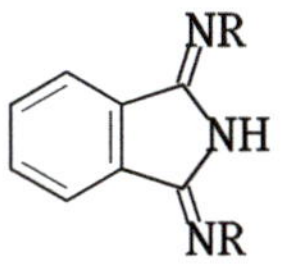 | III, azomethine 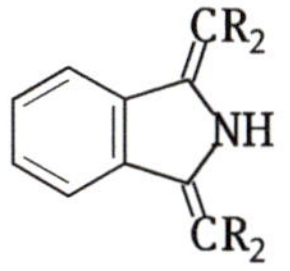 | IV, methine

Azomethine (isoindolinone) pigments

In practice, azomethine pigments are derivatives of structure II. Due to their oxo group, they are also called isoindolinone pigments. A phenyl bridge R = $-C_6H_4-$ connects two molecules. The pigments are yellow and orange, but their technical application is limited. A significant step forward was the introduction of tetrachloro isoindolinone pigments V, R= –H, –Cl, $-CH_3$, $-OCH_3$:

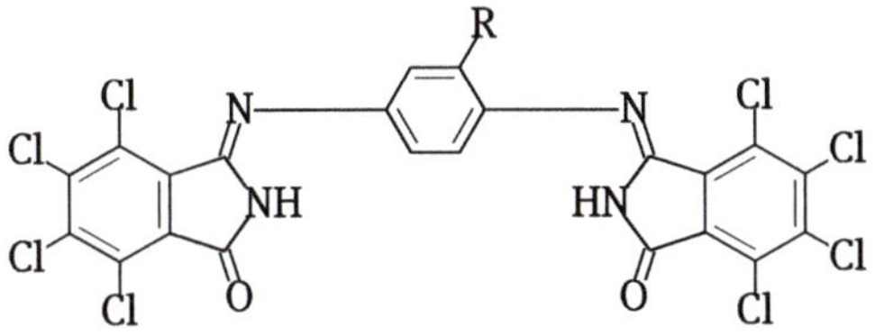

V, tetrachloro-isoindolinone pigments

These pigments display shades from yellow and red to brown, the most important being yellow. Besides having high tinting strength, they possess greatly improved migration fastness, are very stable against acids and alkalis, and show resistance to heat, light, and weather. Therefore, they satisfy high demands, which puts them into the group of high-quality pigments. A few also appear in artists' paints, ►Figure 4.38:

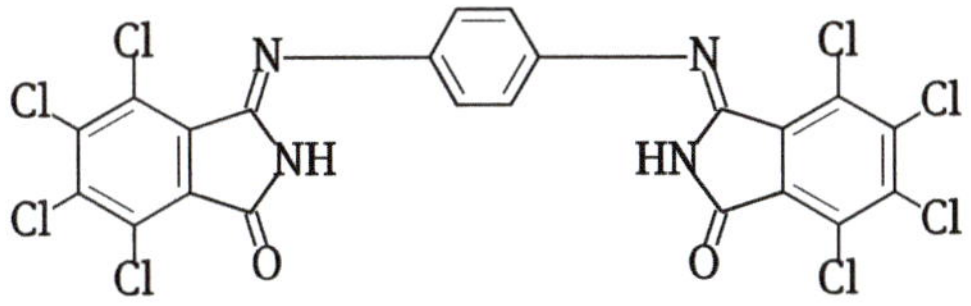

PY110, CI 56280
yellow orange [SC], permanent yellow deep [DS]

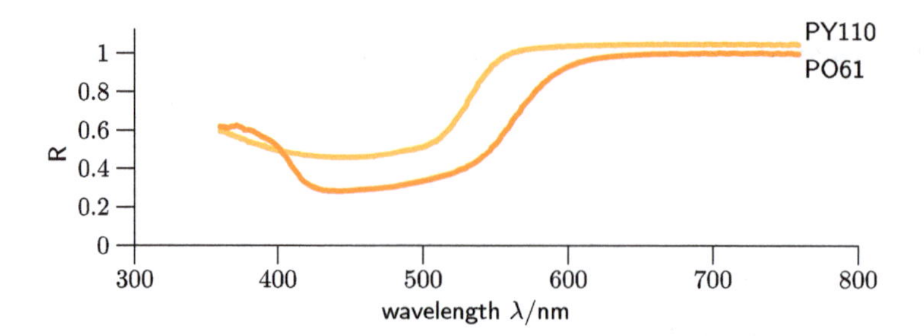

Figure 4.38: Reflectance spectrum of isoindolinone-based watercolors normalized to an arbitrary unit: PO61 (isoindole orange, Kremer No. 23800) and PY110 (yellow orange, Schmincke Horadam No. 222).

PO61, CI 11265
isoindole orange [KR]

Methine pigments

Methine pigments are derived from structure IV, R= –CN, –CONH–Alkyl, and –CONH–Aryl. Heterocycles such as trioxo-diazine can replace two groups R. Methine pigments are yellow, orange, red, or brown and are as resistant as azomethine pigments to migration, light, weathering, and heat. Therefore, they are of the same high quality. Below we see a yellow, an orange, and a red example of the methine type. Artists' paints include the yellow methine pigment:

PY139, CI 56298
isoindoline yellow [DS]

PO66, CI 48210

PR260, CI 56295

Cause of color

In azomethines I, there is a donor–acceptor system in which halogens and the aromatic ring are electron donors, and the oxo group an electron acceptor. In methines II, cyano groups act as an acceptor. Upon excitation by light, charges can be transferred from the aromatic ring and other donors to the acceptors. This process is facilitated by a high number of acceptors and by halogenation of the aromatic indole ring:

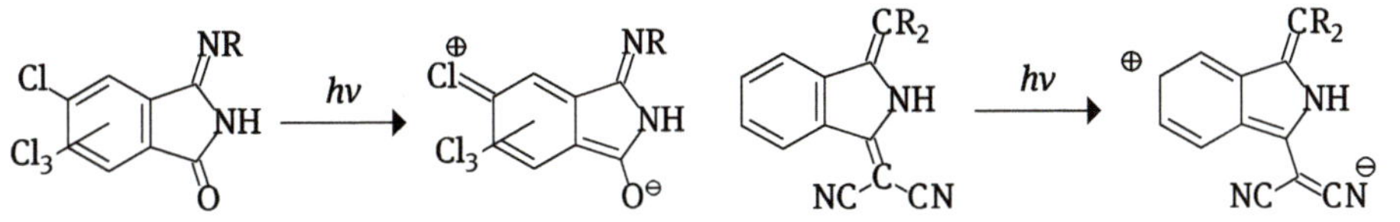

Cl, Cl_3, NR, NH, O, $h\nu$, ⊕ Cl, NR, NH, $O^{\ominus}$

I

CR_2, NH, NC, C, CN, $h\nu$, ⊕, CR_2, NH, NC, $CN^{\ominus}$

II

Bibliography

[1] *Colour Index*, several volumes, Society of Dyers and Colourists, Bradford, 1971.

[2] H. Zollinger, *Color Chemistry*, 3rd ed., Wiley-VCH, Weinheim, 2003. ISBN 978-3906390239.

[3] P. F. Gordon, P. Gregory, *Organic Chemistry in Colour*, Springer, 1983. ISBN 3-540-11748-2.

[4] J. Griffiths, *Colour and Constitution of Organic Molecules*, Academic Press Inc., London, 1976. ISBN 0-12-303550-3.

[5] J. Fabian, H. Hartmann, *Light Absorption of Organic Colorants*, Springer, 1980. ISBN 3-540-09914-X.

[6] P. Bamfield, *Chromic Phenomena*, The Royal Society of Chemistry, 2001. ISBN 0-85404-474-4.

[7] A. Bartecki, J. Burgess, *The Colour of Metal Compounds*, Gordon and Breach Science Publishers, Amsterdam, 2000. ISBN 90-5699-250-3.

[8] A. T. Peters, H. S. Freeman (eds.), *Colour Chemistry*, Elsevier Applied Science, London and, New York, 1991. ISBN 1-85166-577-3.

[9] R. Johnston-Feller, *Color Science in the Examination of Museum Objects*, The Getty Conservation Institute, Los Angeles, 2001. ISBN 0-89236-586-2. https://www.getty.edu/conservation/publications_resources/pdf_publications/pdf/color_science.pdf, accessed on 2022-06-20.

[10] S. Bucklow, *The Alchemy of Paint: Art, Science and Secrets from the Middle Ages*, Marion Boyars Publishers Ltd, 2009. ISBN 978-0714531724.

[11] G. Buxbaum, G. Pfaff, *Industrial Inorganic Pigments*, WILEY-VCH, Weinheim, 2005. ISBN 3-527-30363-4.

[12] K. Hunger, M. U. Schmidt, *Industrial Organic Pigments*, 4th ed., WILEY-VCH, Weinheim, 2018. ISBN 978-3-527-32608-2.

[13] K. Hunger (ed.), *Industrial Dyes*, WILEY-VCH, Weinheim, 2003. ISBN 3-527-30426-6.

[14] H. M. Smith, *High Performance Pigments*, WILEY-VCH, Weinheim, 2002. ISBN 3-527-30204-2.

[15] H. Endriß, *Aktuelle anorganische Bunt-Pigmente*, Curt R. Vincentz-Verlag, Hannover, 1997. ISBN 3-87870-440-2.

[16] G. Pfaff, *Inorganic Pigments*, Walter de Gruyter GmbH, Berlin, 2017. ISBN 978-3-11-048450-2.

[17] H. G. Völz, *Industrielle Farbprüfung*, 2nd ed., Wiley VCH, 2001. ISBN 3-527-30418-5.

[18] D. R. Waring, G. Hallas (eds.), *The Chemistry and Application of Dyes*, Plenum Press, New York, 1990. ISBN 0-306-43278-1.

[19] A. D. Broadbent, *Basic Principles of Textile Coloration, Society of Dyers and Colourists Thanet*, Press Ltd, London, 2001. ISBN 0-901956-76-7.

[20] W. Ingamells, *Colours for Textiles*, Society of Dyers and Colourists, Bradford, 1993. ISBN 0-901956-56-2.

[21] A. K. R. Choudhury, *Textile Preparation and Dyeing*, Science Publishers, 2006. ISBN 978-1578084043.

[22] M. Clark (ed.), *Handbook of Textile and Industrial Dyeing: Principles, Processes and Types of Dyes*, 1st ed., Woodhead Publishing Series in Textiles, vol. 116, Woodhead Publishing, 2011. ISBN 978-1845696955.

[23] H. Schweppe, *Handbuch der Naturfarbstoffe, Nikol-Verlag*, 1993. ISBN 978-3933203465.

[24] T. W. Goodwin (ed.), *Plant Pigments*, Academic Press, 1988. ISBN 0-12-289847-8.

[25] T. W. Goodwin (ed.), *Chemistry and Biochemistry of Plant Pigments*, Academic Press, 1976. ISBN 0-12-289901-6.

[26] K. M. Davies (ed.), *Plant Pigments and their Manipulation*, Annual Plant Reviews, vol. 14, Blackwell Publishing Ltd., 2004. ISBN 1-4051-1737-0.

[27] P. Kurzmann, *Mittelalterliche Glastechnologie*, Peter Lang GmbH, Europäischer Verlag der Wissenschaften, Frankfurt/Main, 2004. ISBN 3-631-52994-5.

https://doi.org/10.1515/9783110777116-005

[28] K. H. Wedepohl, *Glas in Antike und Mittelalter*, Schweizerbart'sche Verlagsbuchhandlung, Stuttgart, 2003. ISBN 3-510-65207-X.
[29] M. T. Wypyski, *Compositional Study of Medieval Islamic Enameled Glass from The Metropolitan Museum of Art*, in *Metropolitan Museum Studies in Art, Science, and Technology*, vol. 1/2010, pp. 109–132, The Metropolitan Museum of Art, New York, 2010. ISBN 978-1-58839-365-4.
[30] W. Vogel, *Glass Chemistry*, 2nd ed., Springer, 1992. ISBN 3-540-57572-3.
[31] H. Scholze, *Glas. Natur, Struktur und Eigenschaften*, 3rd ed., Springer, 1988. ISBN 3-540-18977-7.
[32] M. B. Volf, *Chemical Approach to Glass*, Elsevier, Amsterdam, 1984. ISBN 0-444-99635-4.
[33] C. R. Bamford, *Colour Generation and Control in Glas*, Elsevier Scientific Publishing Company, 1977. ISBN 0-444-41614-5.
[34] W. A. Weyl, *Coloured Glasses*, The Society of Glass Technology, Sheffield, 1951.
[35] W. Spiegl, *Die Geschichte vom Glasmachen 1550 bis 1700*, Version Oktober 2002, http://www.glas-forschung.info/pageone/pdf/cristallo.pdf, accessed on 2022-06-10.
[36] Webseite des Centre International Du Vitrail, Chartres, https://www.centre-vitrail.org/de/die-technik-der-glasmalerei-/, accessed on 2022-06-10.
[37] A. Y. Spengler, *Zur Technik der Glasmalerei*, Webseite des Bezirksmuseums Wien, Mariahilf, https://www.bezirksmuseum.at/de/bezirksmuseum_6/bezirksmuseum/glasmuseum_mariahilf/technik/, accessed on 2022-06-10.
[38] P. Sciau, *Nanoparticles in Ancient Materials: The Metallic Lustre Decorations of Medieval Ceramics*, in A. A. Hashim (ed.) *The Delivery of Nanoparticles*, InTech, 2012. ISBN 978-953-51-0615-9, https://cdn.intechopen.com/pdfs/36899.pdf, accessed on 2022-06-20.
[39] M. A. Pollard, C. Heron, *Archaeological Chemistry*, 2nd ed., Royal Society of Chemistry, 2008. ISBN 978-0854042623.
[40] *CPMA Classification of Chemical Descriptions and Usage of the Complex Inorganic Color Pigments*, 4th ed., Color Pigments Manufacturers Association, Inc., Alexandria, 2013.
[41] W. Noll, *Alte Keramiken und ihre Pigmente*, E. Schweizerbartsche Verlagsbuchhandlung, Stuttgart, 1991. ISBN 3-510-65145-6.
[42] *Colour Pigments and Colouring in Ceramics*, Italian Ceramic Society SALA, Modena, 2003.
[43] S. Stefanov, S. Batschwarov, *Keramik-Glasuren*, Bauverlag GmbH, Wiesbaden, 1988. ISBN 3-7652-2600-1.
[44] R. Hopper, *The Ceramic Spectrum. A Simplified Approach to Glaze and Color Development*, Krause publications, 2001. ISBN 0-87341-821-2.
[45] K. und, W. Lehnhäuser, *Keramische Glasuren und ihre Farben*, 4th ed., Ritterbach Verlag GmbH, Frechen, 2000. ISBN 3-89314-659-8.
[46] P. Gregory, *Metal Complexes as Speciality Dyes and Pigments*, in J. A. McCleverty (ed.) *Comprehensive Coordination Chemistry II*, vol. 9, pp. 549–579, Elsevier, 2004. ISBN 0-08-0437486.
[47] M. Doerner, T. Hoppe, *Malmaterial und seine Verwendung im Bilde*, 18th ed., Verlag Enke, Stuttgart, 1994. ISBN 3-363-00800-7.
[48] V. Herausgeber, *Artist's Pigments Vol. 1–4*, Oxford University Press, New York, 1997. ISBN 0-89468-256-3.
[49] N. Eastaugh, V. Walsh, T. Chaplin, R. Siddall, *Pigment Compendium. A Dictionary of Historical Pigments*, Elsevier Butterworth-Heinemann, Oxford, 2004. ISBN 0-7506-57499.
[50] K. Lutzenberger, *Künstlerfarben im Wandel: Synthetische organische Pigmente des 20. Jahrhunders und Möglichkeiten ihrer zerstörungsarmen, analytischen Identifizierung*, Herbert Utz-Verlag, 2009. ISBN 9-783-83160903-1.
[51] T. C. Patton (ed.), *Pigments Handbook*, vol. 1, Wiley-InterScience, 1973. ISBN 0-471-67123-1.

[52] R. D. Harley, *Artistś Pigments c. 1600–1835, Butterworth Scientific*, 2nd ed., 1982. ISBN 0-408-70945-6.

[53] G. R. Rapp, *Archaeomineralogy (Natural Science in Archaeology)*, Springer, Berlin, 2009. ISBN 978-3540785934.

[54] H. Howard, *Pigments of English Medieval wall Painting*, Archetypes Publications Ltd., London, 2003. ISBN 1-873132-48-4.

[55] H. Berke, *Chemie im Altertum – Die Erfindung von blauen und purpurnen Farbpigmenten*, Universitätsverlag Konstanz, Konstanz, 2006. ISBN 978-3-87940-802-3.

[56] G. Kremer, Web-Informationen zum Pigment 10400 Vivianit, https://www.kremer-pigmente.com/de/shop/pigmente/104000-vivianit.html, accessed on 2022-06-20.

[57] L. Resenberg, *Zinnober – zurück zu den Quellen*, Anton Sigl, München, 2005. ISBN 3-935643-30-6.

[58] G. Kremer, Web-Informationen zum Pigment 10620 Zinnober, https://www.kremer-pigmente.com/de/shop/pigmente/10620-zinnober.html, accessed on 2022-01-10.

[59] forum member “lemmi”, *Synthese von Zinnober – ein alchemistisches Experiment*, https://illumina-chemie.de/viewtopic.php?t=3336, accessed on 2022-06-20.

[60] D. L. Douglass, C. Shing, G. Wang, *The light-induced alteration of realgar to pararealgar*, Amer. Mineral., 77 (1992), 1266–1274.

[61] G. Kremer, website informationen for pigment 10200 Azurit, https://www.kremer-pigmente.com/de/shop/pigmente/10200-azurit-natur-standard.html and https://www.kremer-pigmente.com/elements/resources/products/files/10200-10280.pdf, accessed on 2022-06-20 and personal communication.

[62] G. Kremer, website informationen for pigment 46400 Gofun Shirayuki, https://www.kremer-pigmente.com/de/shop/pigmente/eigene-und-historische-pigmente/46400-gofun-shirayuki.html, accessed on 2014-06-20.

[63] J. Klaas, *Studien zu Ägyptisch Grün. Bildungsbedingungen und Identifizierung der chemischen Zusammensetzung, S. Gerzer, Kaadener Grün. Lagerstätte, Gewinnung und Verwendung der böhmischen Grünen Erde*, Anton Sigl, München, 2006. ISBN 3-935643-30-6.

[64] I. Stösser, *Rote Farblacke in der Malerei. Herstellung und Verwendung im deutschsprachigen Raum zwischen ca. 1400 und 1850*, Diplomarbeit, Institut für Technologie der Malerei der Staatlichen Akademie der Bildenden Künste, Stuttgart, 1985.

[65] R. M. Cornell, U. Schwertmann, *The Iron Oxides*, Wiley-VCH, 2003. ISBN 3-527-30274-3.

[66] U. Schwertmann, R. M. Cornell, *Iron Oxides in the Laboratory*, Wiley-VCH, 2000. ISBN 3-527-29669-7.

[67] C. Blänsdorf, *Blaue Farblacke in Staffeleimalerei und Skulpturenfassung*, Schriftenreihe Konservierung und Restaurierung, Hochschule der Künste, Bern, 2004. ISBN 3-9522804-3-7.

[68] R. Chenciner, *Madder Red: a History of Luxury and Trade, Curzon*, 2001. ISBN 0-7007-1259-3.

[69] H. Kühn, *Die Pigmente in den Gemälden der Schack-Galerie*, in *Gemäldekataloge. Band II: Schack-Galerie*, Bayerische Staatsgemäldesammlungen, München, 1969.

[70] J. H. Hofenk de Graaff, *The Colourful Past: Origins, Chemistry and Identification of Natural Dyestuffs*, Riggisberg, Abegg-Stiftung, 2004. ISBN 3-905014-25-4.

[71] H. Roosen-Runge, *Farbe, Farbmittel der abendländischen mittelalterlichen Buchmalerei*, in *Reallexikon zur Deutschen Kunstgeschichte*, vol. 6, pp. 1463–1492, Verlag C. H. Beck, 1986. ISBN 978-3406140068. Webzugriff über https://www.rdklabor.de/wiki/Farbe,_Farbmittel_der_abendl%C3%A4ndischen_ma._Buchmalerei, accessed on 2022-06-20.

[72] I. Sandner, B. Bünsche, H.-P. Schramm, G. Meier, J. Voss, *Konservierung von Gemälden und Holzskulpturen*, Deutscher Verlag der Wissenschaften, 1990. ISBN 3-326-00524-5.

[73] J. Hill Stoner, R. Rushfield (eds.), *Conservation of Easel Paintings*, Routledge, 2012. ISBN 978-0-7506-8199-5.

[74] M. D. Gottsegen, *The Painter's Handbook, Watson-Guptill Publications*, 2006. ISBN 0-8230-3496-3.

[75] R. Mayer, *The Artist's Handbook of Materials and Techniques*, Faber & Faber, 1991. ISBN 978-0571143313.

[76] P. Taggart, *The Essential Painting Guide*, Thunder Bay Press, 2002. ISBN 978-1571455635.

[77] H. Speed, *The Practice and Science of Drawing*, Dover Publications Inc., 1988. ISBN 978-0486228709.

[78] J. Trevelyan, *Tiefdruckgraphik heute*, O. Maier Verlag, Ravensburg, 1966.

[79] W. Authenrieth, *Neue und alte Techniken der Radierung und Edeldruckverfahren, Authenriet*, 6th ed., 2010. ISBN 978-3000356193.

[80] F. Hollenberg, W. Rabe (Bearbeiter), *Radierung. Ätzkunst und Kupfertiefdruck, O*, Maier Verlag, Ravensburg, 1970. ISBN 3-473-61107-7.

[81] V. Preissig, *Zur Technik der farbigen Radierung und des farbigen Kupferstichs. I. Teil.*, Karl W. Hiersemann, Leipzip, 1909.

[82] R. Simmons, K. Clemson, *DuMont's Handbuch Holz- und Linolschnitt*, Dt. Erstausgabe, DuMont Buchverlag, Köln, 1990. ISBN 3-7701-2468-5.

[83] J. Stobart, *Einfach drucken*, Haupt Verlag, 2003. ISBN 3-258-06608-6.

[84] B. Fick, B. Grabowski, *Printmaking: A Complete Guide to Materials & Process*, Laurence King Publishing, 2015. ISBN 978-1780671949.

[85] V. Mooncie, *Print Making Book: Projects and Techniques in the Art of Hand-printing, Guild of Master Craftsman*, Publications Ltd, 2014. ISBN 978-1861089212.

[86] B. Butts, L. Hendrix, *Painting on Light: Drawings and Stained Glass in the Age of Durer and Holbein*, J. Paul Getty Museum, 2000. ISBN 978-0892365784. https://www.getty.edu/publications/resources/virtuallibrary/089236579X.pdf, accessed on 2022-06-20.

[87] F. Wirtl, *Hinterglasmalerei*, Otto Maier Verlag, Ravensburg, 1980. ISBN 3-473-43052-8.

[88] A. K. Leonhard, *Hinterglasmalerei*, 5th ed., Frech-Verlag Stuttgart, 1977. ISBN 3-7724-0272-0.

[89] S. Bretz, *Hinterglasmalerei*, Klinkhardt & Biermann Verlag, München, 2013. ISBN 978-3-943616-12-5.

[90] K.-P. Schäffel, *Book Wasserfarben aus Naturmaterialien selber machen*, Basler Papiermühle, 2007.

[91] G. Ziesemann, M. Krampfer, H. Knieriemen, *Natürliche Farben. Anstriche und Verputze selber herstellen*, 4th ed., AT Verlag, 2000. ISBN 3-85502-523-1.

[92] L. Edwards, J. Lawless, *Naturfarben Handbuch, edition antis im Ökobuch Verlag*, 1st ed., 2003. ISBN 3-922964-92-3.

[93] W. Blanke, in *Malen mit Pigmenten und variablen Bindemitteln*, Knaur, 2003. ISBN 978-3-426-66956-3.

[94] H. Arendt, *Werkstatt Pflanzenfarben. Natürliche Malfarben selbst herstellen und anwenden*, 2nd ed., AT Verlag, 2010. ISBN 978-3-03800-407-3.

[95] P. Seymour, *Making Soft Pastels Using Dry Pigments*, Lee Press, London, 1999. ISBN 0-9524727-2-4.

[96] G. P. A. Turner, *Introduction to paint chemistry and principles of paint technology*, St Edmundsbury Press Ltd, 1988. ISBN 0-412-29440-0.

[97] T. J. S. Learner, *Analysis of Modern Paints, Getty Publications*, 2005. ISBN 978-0-89236-779-5.

[98] T. J. S. Learner, P. Smithen, J. W. Krueger M. R, *Schilling (eds.), Modern Paints Uncovered*, in *Proceedings from the Modern Paints Uncovered Symposium, Tate Modern London, 2006*, Getty Publications, 2007. ISBN 978-0-89236-906-5.

[99] H.-P. Schramm, B. Hering, *Historische Malmaterialien und ihre Identifizierung*, Bücher des Restaurators, vol. 1, Hrg. U. Schießl, Verlag Enke, Stuttgart, 1995. ISBN 3-432-27041-0.

[100] M. F. Striegel, J. Hill, *Thin-Layer Chromatography for Binding Media Analysis*, The Getty Conservation Institute, Los Angeles, 1996. ISBN 0-89236-390-8. https://www.getty.edu/conservation/publications_resources/pdf_publications/pdf/thin_layer.pdf, accessed on 2022-06-20.

[101] M. P. Colombini, F. Modugno (eds.), *Organic Mass Spectrometry in Art and Archaeology*, John Wiley & Sons, Ltd., 2009. ISBN 978-0-470-51703-1.

[102] J. S. Mills R, *White The Organic Chemistry of Museum Objects, Butterworth-Heinemann Ltd.*, 2nd ed., 1994. ISBN 0-7506-4693-4.

[103] J. Riederer, *Archäologie und Chemie*, in *Ausstellung des Rathgen-Forschungslabors SMPK 1987/1988*, Staatliche Museen Preußischer Kulturbesitz, Berlin, 1987. ISBN 3-88609-212-7.

[104] C. Ehrenfort, *Aquarell- und Gouachefarben. Beiträge zu Materialzusammensetzung, Veränderungen, Schäden*, Diplomarbeit 1993, Institut für Technologie der Malerei, Staatliche Akademie der Bildenden Künste, Stuttgart.

[105] A. Roy, P. Smith (eds.), *Painting Techniques. History, Materials and Studio Practice*, The International Institute for Conservation of Historic and Artistic Works, 1998. ISBN 0-9500525-8-2.

[106] A. Wallert, M. Rey, *Historical painting techniques, materials, and studio practice*, preprints of a symposium, University of Leiden, the Netherlands, 26–29 June 1995, The Getty Conservation Institute, Los Angeles, 1995. ISBN 0-89236-322-3. https://www.getty.edu/conservation/publications_resources/pdf_publications/pdf/historical_paintings.pdf, accessed on 2022-06-20.

[107] J. Riederer, *Kunstwerke chemisch betrachtet*, Springer, 1981. ISBN 3-540-10552-2.

[108] A. Bartl, C. Krekel, M. Lautenschlager, D. Oltrogge, *Der Liber illuministarum aus dem Kloster Tegernsee*, Veröffentlichung des Instituts für Kunsttechnik und Konservierung im Germanischen Nationalmuseum, vol. 8, Franz Steiner Verlag, 2005. ISBN 3-515-08472-X.

[109] D. A. Scott, *Copper and Bronze in Art. Corrosion, Colorants, Conservation., Getty Conservation Institute*, 2002. ISBN 0-89236-638-9.

[110] P. Villa, L. Pollarolo, I. Degano, L. Birolo, M. Pasero, C. Biagioni, K. Douka, R. Vinciguerra, J. J. Lucejko, L. Wadley, *A Milk and Ochre Paint Mixture Used 49,000 Years Ago at Sibudu, South Africa*, PLoS ONE, 10(6), e0131273 (2015). https://doi.org/10.1371/journal.pone.0131273, accessed on 2022-06-20.

[111] Spektrum der Wissenschaft (online editors), *Frühe Kunstkenner*, news from 2000-05-04, Spektrum der Wissenschaft Verlag, https://www.spektrum.de/news/fruehe-kunstkenner/344026, accessed on 2022-06-20.

[112] Spektrum der Wissenschaft (online editors), *Die Urväter der Kunst*, news from 2000-10-23, Spektrum der Wissenschaft Verlag, https://www.spektrum.de/news/die-urvaeter-der-kunst/344691, accessed on 2022-06-20.

[113] G. Bataille, *Die vorgeschichtliche Malerei. Lascaux oder Die Geburt der Kunst, Editions dÁrt Albert Skira Genf/Klett-Cotta Stuttgart*, 1983. ISBN 3-88447-069-8.

[114] G. Rietschel, et al., *Lascaux. Höhle der Eiszeit*, in *Ausstellungskatalog des Roemer- und Pelizaeus-Museum Hildesheim*, Verlag Philipp von Zabern, Mainz, 1982. ISBN 3-8053-0593-1.

[115] M. Lorblanchet, *Höhlenmalerei, Jan Thorbecke Verlag*, 2nd ed., 2000. ISBN 3-7995-9025-0.

[116] R. H. der künstlerichen Techniken. Band 1, *Farbmittel, Buchmalerei, Tafel- und Leinwandmalerei., Philipp Reclam jun*, GmbH & Co., Stuttgart, 1988. ISBN 3-15-010322-3.

[117] H. Roosen-Runge, *Tüchleinmalerei, Tafelmalerei*, in [116].

[118] H. Roosen-Runge, *Buchmalerei*, in [116], pp. 25–124.

[119] *Reclams Handbuch der künstlerischen Techniken. Band 2: Wandmalerei, Mosaik*, Philipp Reclam jun. GmbH & Co. Stuttgart, 1990. ISBN 3-15-010345-2.

[120] A. Knoepfli, O. Emmenegger, *Wandmalerei bis zum Ende des Mittelalters*, in [119].

[121] R. H. der künstlerischen Techniken. Band 3, *Glas, Keramik, Porzellan, Möbel, Intarsie und Rahmen, Lackkunst, Leder., Philipp Reclam jun*, GmbH & Co., Stuttgart, 1986. ISBN 3-15-030015-0.

[122] A. Nunn, *Handbuch der Orientalistik, 7. Abteilung, Die Wandmalerei und der glasierte Wandschmuck im alten Orient*, J. Brill, Leiden, 1988. ISBN 90-04-08428-2.

[123] M. A. Corzo, M. Afshar, *Art and Eternity. The Nefertari Wall Paintings Conservation Project 1986--1992*, J. Paul Getty Trust, 1993. ISBN 0-89236-130-1.

[124] D. Stulik, E. Porta, A. Palet, *Analysis of Pigments, Binding Media and Varnishes*, in [123], pp. 55–65.

[125] S. Hood, *The Arts in Prehistoric Greece*, Penguin Books, 1990. ISBN 0-14-0561-42-0.

[126] R. Higgins, *Minoan and Mycenaen Art (World of Art)*, Verlag Thames & Hudson Ltd, 1981. ISBN 978-0500201848.

[127] I. Scheibler, *Griechische Malerei der Antike*, Beck, München, 1994. ISBN 3-406-38491-9.

[128] T. Spiteris, *Griechische und Etruskische Malerei, Weltgeschichte der Malerei*, vol. 3, Editions Rencontre, Lausanne, 1966.

[129] T. Spiteris, *Bunte Götter. Die Farbigkeit antiker Skulptur*, 4th ed., Ausstellungskatalog des Museums für Kunst und Gewerbe Hamburg, 2007.

[130] I. Baldassarre, A. Pontrandolfo, A. Rouveret, M. Salvadori, *Römische Malerei*, DuMont, 2003. ISBN 3-8321-7210-6.

[131] H. Mielsch, *Römische Wandmalerei*, Wissenschaftliche Buchgesellschaft, Darmstadt, 2001.

[132] S. Augusti, *I colori pompeiani*, De Luca Editore, Istituto Grafico Tiberino, Roma, 1967.

[133] M. Exner (ed.), *Wandmalerei des frühen Mittelalters. Bestand, Maltechnik, Konservierung*, ICOMOS Hefte des Deutschen Nationalkomitees XXIII, 1998. ISBN 3-87490-663-9.

[134] J. und, K. Hecht, *Die frühmittelalterliche Wandmalerei des Bodenseegebietes, 2 Bände*, J. Thorbecke Verlag, Sigmaringen, 1979. ISBN 3-7995-7008-X.

[135] H. Stampfer, T. Steppan, *Die romanische Wandmalerei in Tirol*, 1st ed., Schnell & Steiner, 2007. ISBN 978-3795415747.

[136] D. Jakobs, *Sankt Georg in Reichenau-Oberzell, 2*, Band, Konrad-Theiss-Verlag, Stuttgart, 1999. ISBN 978-3806214628.

[137] A. Burmester, C. Heilmann, M. F. Zimmermann, *Barbizon. Malerei der Natur – Natur der Malerei*, Klinghardt & Biermann Verlagsbuchhandlung GmbH, München, 1999. ISBN 3-7814-ß424-2.

[138] C. Syre, *Tintoretto. Der Gonzaga-Zyklus*, Ausstellungskatalog, Bayerische Staatsgemäldesammlungen, Hatje Cantz Verlag, München, 2000. ISBN 3-7757-0887-1.

[139] D. Bomford, J. Kirby, A. Roy, A. Rüger, R. White, *Art in the Making. Rembrandt*, National Gallery Company, London, 2006. ISBN 978-1-85709-356-8.

[140] C. Brown, *Making & Meaning. Rubens's Landscapes*, National Gallery Publications, London, 1996. ISBN 1-85709-155-8.

[141] R. Spronk, *More than Meets the Eye: An Introduction to Technical Examination of Early Netherlandish Paintings at the Fogg Art Museum*, Harv. Univ. Art Mus. Bull., 5(1) (1996), 1–3, 6–64.

[142] I. Gaskell, M. Jonker, *Vermeer Studies*, in *Studies in the History of Art 55*, National Gallery of Art, Washington, 1998. ISBN 0-300-07521-9.

[143] J. Townsend, *Turner's Painting Techniques*, 2nd ed., Tate Gallery London, 1996. ISBN 1-85437-202-5.

[144] C. Peres, M. Hoyle, L. van Tilborgh (eds.), *A Closer Look: Technical and art-historical Studies on Works by Van Gogh and Gauguin, Rijksmuseum Vincent van Gogh, Cahier Vincent 3*, Waanders Publishers, Zwolle, 1991.

[145] D. Bomford, J. Dunkerton, D. Gordon, A. Roy, J. Kirby, *Art in the Making: Italian Painting before 1400*, National Gallery, London, 1989, 1992. ISBN 0-947645-67-5.

[146] D. Bomford, J. Kirby, J. Leighton, A. Roy, *Art in the Making. Impressionism*, National Gallery Company, London, 1990. ISBN 0-300-05035-6.
[147] G. Goldberg, B. Heimberg, M. Schawe, *Albrecht Dürer. Die Gemälde der Alten Pinakothek, Bayerische Staatsgemäldesammlungen*, Edition Braus, München, 1998. ISBN 3-89466-216-6.
[148] J. D. Sanders, *Pigments for Inkmakers*, SITA Technology, London, 1989. ISBN 0-947798-07-02.
[149] S. Magdassi, *The Chemistry of Inkjet Inks*, World Scientific Publishing Co. Pte. Ltd., Singapur, 2010. ISBN 978-981-281-921-8.
[150] S. F. Pond, *Inkjet Technology and Product Development Strategies*, Torrey Pines Research, 2000. ISBN 0-9700860-0-8.
[151] J. T. Kunjappu, *Essays in Ink Chemistry*, Nova Science Publishers Inc., 2001. ISBN 1-59033-111-7.
[152] J. A. G. Drake (ed.), *Chemical Technology in Printing and Imaging Systems*, The Royal Chemical Society of Chemistry, 1993. ISBN 0-85186-655-7.
[153] N. Underwood, T. V. Sullivan, *The Chemistry and Technology of Printing Inks*, D. Van Nostrand Company, New York, 1915. ISBN 978-1103807611, Reprint von BiblioBazaar 2009.
[154] C. Maywald, *Schreibtinten. Einführung und Übersicht*, Robert Wiegner Verlag, 2010. ISBN 978-3-931775-15-5.
[155] A. Schopen, *Tinten und Tuschen des arabisch-islamischen Mittelalters*, Vandenhoeck & Ruprecht GmbH & Co. KG, Göttingen, 2006. ISBN 978-3-525-82541-9.
[156] M. Zerdoun Bat-Yehouda, *Les Encres Noires au Moyen Âge*, Centre National de la recherche Scientifique, Paris, 1983. ISBN 2-222-02972-4.
[157] J. Needham, *Science and Civilization in China: Volume 5, Chemistry and Chemical Technology, Part 1, Paper and Printing*, 6th ed., Cambridge University Press, Cambridge, 2020. ISBN 978-0-521-08690-5.
[158] K. S. Clair, *The secret Lives of Colour*, John Murray (Publishers), 2016. ISBN 978-1-473-63082-6.
[159] M. Lyons, *Books: A Living History*, Thames & Hudson Ltd, 2013. ISBN 978-0500291153.
[160] D. N. Carvalho, *Forty Centuries of Ink*, Project Gutenberg Release #1483, https://onlinebooks.library.upenn.edu/webbin/gutbook/lookup?num=1483, accessed on 2022-06-13.
[161] M. B. Woods, M. Woods, *Ancient Communication: From Grunts to Graffiti*, Lerner Publishing Group, 2001. ISBN 978-0822529965.
[162] J. A. Smith, *Pen & Ink Book: Materials and Techniques for Today's Artist*, Watson-Guptill Publications Inc., U. S., 1999. ISBN 978-0823039869.
[163] R. H. Leach (ed.), *The Printing Ink Manual*, 4th ed. (Softcover reprint of 1st ed. 1988), Springer, 2012. ISBN 978-9401170994.
[164] W. Baumann, T. Rothardt, *Druckereichemikalien: Daten und Fakten zum Umweltschutz*, 2nd ed., Springer, 1999. ISBN 3-540-66046-1.
[165] V. C. Malshe, M. A. Sikchi, *Basics of Paint Technology Part I*, 2nd ed., Antar Prakash Centre for Yoga, 2004. ISBN 978-8190329859.
[166] Oil and Colour Chemists' Association, *Surface Coatings: Volume 1 Raw Materials and their Usage*, 3rd ed., Springer Science & Business Media, 1993. ISBN 0-412-55210-8.
[167] N. Equiamicus, *Kleines Rezeptbuch der historischen Tinten*, Bohmeier Verlag, 2011. ISBN 978-3-89094-593-4.
[168] G. Weibel, *Miniaturisiertes Dosiersystem zur geregelten Dosierung von Tinte in einem Schreibgerät*, Dissertation, Universität Stuttgart, 2008, http://dx.doi.org/10.18419/opus-4142, accessed on 2015-02-09.
[169] M. Ek, G. Gellerstedt, G. Henriksson (eds.), *Pulp and Paper Chemistry and Technologie. 1. Wood Chemistry and Wood Biotechnology*, Walter de Gruyter GmbH, 2009. ISBN 978-3-11-048195-2.

[170] M. Ek, G. Gellerstedt, G. Henriksson (eds.), *Pulp and Paper Chemistry and Technologie. 2. Pulping Chemistry and Technology*, Walter de Gruyter GmbH, 2009. ISBN 978-3-11-048342-0.

[171] M. Ek, G. Gellerstedt, G. Henriksson (eds.), *Pulp and Paper Chemistry and Technologie. 3. Paper Chemistry and Technology*, Walter de Gruyter GmbH, 2009. ISBN 978-3-11-048343-7.

[172] M. Ek, G. Gellerstedt, G. Henriksson (eds.), *Pulp and Paper Chemistry and Technologie. 4. Papier Products Physics and Technology*, Walter de Gruyter GmbH, 2009. ISBN 978-3-11-048346-8.

[173] D. Zerler, H.-U. Kästner, *Werkstoffkunde für den Papiermacher*, 1st ed., Schriftenreihe der Papiermacherschule, vol. 19, Papiermacherschule Gernsbach, 1997.

[174] J. C. Roberts (ed.), *Paper Chemistry*, Blackie & Son Ltd., Glasgow, 1991. ISBN 0-216-92909-1.

[175] H. Holik (ed.), *Handbook of Paper and Board*, WILEY-VCH Verlag Chemie, 2006. ISBN 978-3-527-30997-9.

[176] W. Baumann, B. Herberg-Liedtke, *Papierchemikalien: Daten und Fakten zum Umweltschutz*, Springer, 1994. ISBN 3-540-57593-6.

[177] J. P. Casey (ed.), *Pulp and Paper, Chemistry and Chemical Technology, Vol. III*, 3rd ed., John Wiley & Sons, 1981. ISBN 0-471-03177-1.

[178] J. Blechschmidt, *Papierverarbeitungstechnik*, Carl Hanser Fachbuchverlag, 2013. ISBN 978-3-446-43071-6.

[179] J. Blechschmidt (ed.), *Taschenbuch der Papiertechnik*, 2nd ed., Carl Hanser Fachbuchverlag, 2013. ISBN 978-3-446-43802-6.

[180] M. Lewin (ed.), *Handbook of Fiber Chemistry*, 3.rd ed., CRC Press, 2006. ISBN 0-8247-2565-4.

[181] H. Kittel, *Lehrbuch der Lacke und Beschichtungen, mehrere Bände S*, Hirzel Verlag, Stuttgart, 2007. ISBN 978-3-7776-1014-6.

[182] J. L. Keddie, A. F. Routh, *Fundamentals of Latex Film Formation*, Springer, 2010. ISBN 978-90-481-2844-0.

[183] D. Stoye, W. Freitag (eds.), *Paints, Coatings and Solvents*, 2nd ed., WILEY-VCH, 1998. ISBN 3-527-28863-5.

[184] D. Urban, K. Takamura, *Polymer Dispersions and Their Industrial Applications*, WILEY-VCH, 2002. ISBN 3-527-30286-7.

[185] Z. W. Wicks, F. N. Jones, S. P. Pappas, *Organic Coatings – Science and Technology, vol. 1: Film Formation, Components and Appearance*, John Wiley & Sons, Inc., New York, 1992. ISBN 0-471-61406-8.

[186] J. Bieleman, *Lackadditive*, Wiley-VCH, 1998. ISBN 3-527-28819-9.

[187] B. Müller, *Lackadditive kompakt erklärt*, 2nd ed., Vincentz Network Hannover, Farbe und Lack Bibliothek, 2018. ISBN 978-3-74830-011-4.

[188] C. V. Horie, *Materials for Conservation*, Butterworth & Co. Ltd., 1987. ISBN 0-408-01531-4.

[189] H. Briehl, *Chemie der Werkstoffe*, 2nd ed., Vieweg und Teubner Verlag, 2007. ISBN 978-3835102231.

[190] A. Goldschmidt, H.-J. Streitberger, *BASF Handbook Basics of Coating Technology*, 2nd revised ed., BASF Coatings AG, Münster/Germany, Vincentz Network, 2007. ISBN 978-3-86630-903-6.

[191] T. Brock, *Lehrbuch der Lacktechnologie*, 2nd ed., Vincentz-Verlag Hannover, 1998. ISBN 3-87870-569-7.

[192] D. Stoye, W. Freitag, *Lackharze: Chemie, Eigenschaften und Anwendungen*, Hanser-Verlag, 1996. ISBN 3-446-17475-3.

[193] M. El-Sayed Abdel-Raouf, A.-R. Mahmoud Abdul-Raheim, *Rosin: Chemistry, Derivatives, and Applications: a review*, BAOJ Chem., 4, 039 (2018).

[194] *PERIODIC TABLE OF ELEMENTS*, official website of IUPAC regarding the periodic table of elements, https://iupac.org/what-we-do/periodic-table-of-elements/, accessed on 2022-01-20.

[195] A. F. Hollemann, N. Wiberg, *Lehrbuch der Anorganischen Chemie*, 103rd ed., Walter de Gruyter, Berlin, 2017, vol. 1: ISBN 978-3-11-026932-1, vol. 2: ISBN 978-3-11-049573-7.
[196] N. N. Greenwood, A. Earnshaw, *Chemie der Elemente*, 1. Nachdr. d. 1. ed., VCH, 1990. ISBN 3-527-26169-9.
[197] R. Steudel, *Chemie der Nichtmetalle*, 3rd ed., Walter de Gruyter, Berlin, 2008. ISBN 978-3-11-019448-7.
[198] M. Winter, *WebElements: the periodic table on the WWW. Binary compounds and complexes*, University of Sheffield and WebElements Ltd, UK, http://www.webelements.com/compounds.html, accessed on 2014-06-06.
[199] F. Ullmann, *Ullmanns Encyclopedia of industrial chemistry*, Wiley-VCH Verlag GmbH, Weinheim, 2005, electronic ed.
[200] F. Ullmann, *Ullmanns Encyklopädie der technischen Chemie*, 4th ed., Verlag Chemie, Weinheim, 1972–1984.
[201] R. Dittmeyer, W. Keim, G. Kreysa, A. Oberholz, *Chemische Technik, Band 7: Industrieprodukte*, 5th ed., Wiley-VCH, 2004. ISBN 9783527307722.
[202] J. I. Kroschwitz (ed.), *Encyclopedia of Chemical Technology*, 5th ed., Wiley & Sons, New York, 2004. ISBN 0-471-48494-6.
[203] A. Wollrab, *Organische Chemie*, 2nd ed., Springer, 2002. ISBN 3-540-43998-6.
[204] H. P. Latscha, U. Kazmaier, A. Klein, *Organische Chemie: Chemie-Basiswissen II*, 6th ed., Springer, 2008. ISBN 978-3-540-77106-7.
[205] E. Breitmaier, G. Jung, *Organische Chemie*, 5th ed., Georg Thieme Verlag, 2005. ISBN 3-13-541505-8.
[206] W. Kutzelnigg, *Einführung in die Theoretische Chemie*, WILEY-VCH, 2002. ISBN 3-527-30609-9.
[207] M. J. S. Dewar, R. C. Dougherty, *The PMO Theory of Organic Chemistry*, Plenum Publishing Company, New York, 1975. ISBN 0-306-20010-4.
[208] E. Heilbronner, H. Bock, *Das HMO-Modell und seine Anwendung. Grundlagen und Handhabung*, Verlag Chemie, Weinheim, 1968, Verlagsnummer 68 109.
[209] A. Rauk, *Orbital Interaction Theory of Organic Chemistry*, 2nd ed., Wiley-Interscience, 2001. ISBN 0-471-35833-9.
[210] T. A. Albright, J. K. Burdett, M.-H. Whangbo, *Orbital Interactions in Chemistry*, 2nd ed., John Wiley & Sons, Inc., 2013. ISBN 978-0-471-08039-8.
[211] M. Hesse, H. Meier, B. Zeeh, *Spektroskopische Methoden in der organischen Chemie*, 7th ed., Thieme Verlag Stuttgart, 2005. ISBN 978-3135761077.
[212] G. Lagaly, O. Schulz, R. Zimehl, *Dispersionen und Emulsionen, eine Einführung in die Kolloidik feinverteilter Stoffe*, Dr. Dietrich Steinkopff Verlag, Darmstadt, 1997. ISBN 3-7985-1087-3.
[213] R. M. Pashley, M. E. Karaman, *Applied Colloid and Surface Chemistry*, John Wiley & Sons Ltd., 2004. ISBN 0-470-86883-X.
[214] R. J. Hunter, *Introduction to Modern Colloid Science*, Oxford University Press Inc., 1993. ISBN 0-19-855387-0.
[215] E. J. W. Verwey, J. Th. G. Overbeek, *Theory of the Stability of Lyophobic Colloids*, Elsevier Publishing Company Inc., 1948.
[216] K. S. Birdi, *Surface Tension and Interfacial Tension of Liquids*, in *Handbook of Colloid and Surface Chemistry*, CRC Press, 2009. ISBN 978-1-4200-0720-6.
[217] T. F. Tadros, *Colloids in Paints*, Wiley-VCH, 2010. ISBN 978-3-527-31466-9.
[218] T. F. Tadros, *An Introduction to Surfactants*, Walter de Gruyter, 2014. ISBN 978-3-11-031212-6.
[219] D. Distler (ed.), *Wäßrige Polymerdispersionen*, Wiley-VCH, 1999. ISBN 3-527-29587-9.
[220] E. Dickinson, D. J. McClements, *Advances in Food Colloids*, 1st ed., Blackie Academic & Professsional, 1995. ISBN 0-7514-0203-6.

[221] M. Ladd, *Symmetry and Group Theory in Chemistry*, Horwood Chemical Science Series, Horwood Publishing Ltd., 1998. ISBN 1-898563-39-X.

[222] M. S. Dresselhaus, G. Dresselhaus, A. Jorio, *Group theory: application to the physics of condensed matter*, Springer, 2008. ISBN 978-3-540-32897-1.

[223] G. Steffen, *Farbe und Lumineszenz von Mineralien*, Enke Verlag, Stuttgart, 2000. ISBN 3-13-118341-1.

[224] R. G. Burns, *Mineralogical applications of crystal field theory*, Cambridge University Press, 1993. ISBN 0-521-430771.

[225] D. Reinen, *Kationenverteilung zweiwertiger $3d^n$-Ionen in oxidischen Spinell-, Granat- und anderen Strukturen*, in P. Hemmerich (ed.) *Structure and Bonding, Band 7/1970*, pp. 114–154, Springer, 1970. ISBN 978-3-540-05022-3.

[226] A. B. P. Lever, *Inorganic Elektronic Spectroscopy*, Elsevier, 1984. ISBN 0-444-42389-3.

[227] B. N. Figgis, M. A. Hitchman, *Ligand field theory and its applications*, Wiley-VCH, 2000. ISBN 0-471-31776-4.

[228] F. Kober, *Grundlagen der Komplexchemie*, Otto Salle Verlag, 1979. ISBN 3-7935-5482-1.

[229] J. R. Gispert, *Coordination Chemistry*, 1st ed., Wiley-VCH, 2008. ISBN 978-3-527-31802-5.

[230] L. H. Gade, *Koordinationschemie*, 1st ed., Wiley-VCH, 1998. ISBN 3-527-29503-8.

[231] E. König, S. Kremer, *Ligand field energy diagrams*, Plenum Press, New York, 1977. ISBN 0-306-30946-7.

[232] B. Di Bartolo, *Magnetic Ions in Solids*, in B. Di (ed.) *Bartolo, Optical Properties of Ions in Solids*, pp. 15–61, Plenum Press, New York, 1975. ISBN 0-306-35708-9.

[233] A. S. Marfunin, *Physics of Minerals and Inorganic Materials*, Springer, 1979. ISBN 3-540-08982-9.

[234] I. B. Bersuker, *Electronic Structure and Properties of Transition Metal Compounds*, 2nd ed., John Wiley & Sons, 2010. ISBN 978-0-470-18023-5.

[235] R. S. Mulliken, W. B. Person, *Molecular complexes: a Lecture and Reprint Volume*, Wiley, New York, 1969.

[236] C. Janiak, T. M. Klapötke, E. Riedel, *Moderne anorganische Chemie*, Walter de Gruyter, 2003. ISBN 3-11-015672-5.

[237] U. Müller, *Anorganische Strukturchemie*, 6th ed., Vieweg+Teubner Verlag, 2008. ISBN 978-3834806260.

[238] A. Krüger, *Neue Kohlenstoffmaterialien: Eine Einführung*, Vieweg+Teubner Verlag, 2007. ISBN 978-3519005100.

[239] I. Glassman, R. A. Yetter, *Combustion*, 4th ed., Academic Press, 2008. ISBN 978-0-12-088573-2.

[240] H. Bockhorn (ed.), *Soot Formation in Combustion*, Springer, 1994. ISBN 3-540-58398-X.

[241] H. Richter, J. B. Howard, *Formation of polycyclic aromatic hydrocarbons and their growth to soot—a review of chemical reaction pathways*, Prog. Energy Combust. Sci., 26 (2000), 565–608.

[242] W. Pejpichestakula, A. Frassoldatia, A. Parenteb, T. Faravelli, *Kinetic modeling of soot formation in premixed burner-stabilized stagnation ethylene flames at heavily sooting condition*, Fuel, 234 (2018), 199–206.

[243] J. D. Herdman, J. Houston Miller, *Intermolecular Potential Calculations for Polynuclear Aromatic Hydrocarbon Clusters*, J. Phys. Chem. A, 112 (2008), 6249–6256.

[244] H. Wang, *Formation of nascent soot and other condensed-phase materials in flames*, Proceedings of the Combustion Institute, 33 (2011), 41–67.

[245] R. A. Dobbins, R. A. Fletcher, W. Lu, *Laser Microprobe Analysis of Soot Precursor Particles and Carbonaceous Soot*, Combustion and Flame, 100 (1995), 301–309.

[246] T. Totton, *Investigations into the Formation of Soot Particles*, PhD Thesis, Department of Chemical Engineering and Biotechnology, University of Cambridge, 2009

[247] T. Totton, A. J. Misquitta, M. Kraft, *A quantitative study of the clustering of polycyclic aromatic hydrocarbons at high temperatures*, Phys. Chem. Chem. Phys., 14 (2012), 4081–4094.

[248] M. Sirignano, A. Ciajolo, A. D'Anna, C. Russo, *Particle formation in premixed ethylene-benzene flames: An experimental and modeling study*, Combustion and Flame, 200 (2019), 23–31.

[249] R. A. Dobbins, R. A. Fletcher, H.-C. Chang, *The Evolution of Soot Precursor Particles in a Diffusion Flame*, Combustion and Flame, 115 (1998), 285–298.

[250] S. Iavarone, L. Pascazio, M. Sirignano, A. De Candia, A. Fierro, L. de Arcangelis, A. D'Anna, *Molecular dynamics simulations of incipient carbonaceous nanoparticle formation at flame conditions*, Combustion Theory and Modelling, https://10.1080/13647830.2016.1242156

[251] F. A. Atiku, A. R. Lea-Langton, K. D. Bartle, J. M. Jones, A. Williams, I. Burns, G. Humphries, *Some Aspects of the Mechanism of Formation of Smoke from the Combustion of Wood*, Energy Fuels, 31 (2017), 1935–1944. https://doi.org/10.1021/acs.energyfuels.6b02639.

[252] R. T. Toledo, *Wood Smoke Components and Functional Properties*, in D. E. Kramer, L. Brown (eds.) *International Smoked Seafood Conference Proceedings*, pp. 55–61, Alaska Sea Grant College Program, Fairbanks, 20008. https://doi.org/10.4027/isscp.2008.12.

[253] A. C. Eriksson, E. Z. Nordin, R. Nyström, E. Pettersson, E. Swietlicki, C. Bergvall, R. Westerholm, C. Boman, J. H. Pagels, *Particulate PAH Emissions from Residential Biomass Combustion: Time-Resolved Analysis with Aerosol Mass Spectrometry*, Environ. Sci. Technol., 48 (2014), 7143–7150. https://doi.org/10.1021/es500486j.

[254] S. Sinja, A. Jhalani, M. R. Ravi, A. Ray, *Modelling of Pyrolysis in Wood: A Review*, Indian Institute of Technology Delhi, 2004, https://web.iitd.ac.in/~ravimr/Publications/IndianJournals/sesi-sanjiv.pdf, accessed on 2022-06-20.

[255] D. Chen, K. Cen, X. Zhuang, Z. Gan, J. Zhou, Y. Zhang, H. Zhang, *Insight into biomass pyrolysis mechanism based on cellulose, hemicellulose, and lignin: Evolution of volatiles and kinetics, elucidation of reaction pathways, and characterization of gas, biochar and bio-oil*, Combust. Flame, 242, 112142 (2022).

[256] S. D. Stefanidis, K. G. Kalogiannis, E. F. Iliopoulou, C. M. Michailof, P. A. Pilavachi, A. A. Lappas, *A study of lignocellulosic biomass pyrolysis via the pyrolysis of cellulose, hemicellulose and lignin*, J. Anal. Appl. Pyrolysis, 105 (2014), 143–150. https://doi.org/10.1016/j.jaap.2013.10.013. ISSN 0165-2370.

[257] C. A. Zaror, D. L. Pyle, *The pyrolysis of biomass: A general review*, Proc. Indian Acad. Sci., Eng. Sci., 5(4) (1982), 269–285.

[258] P. Kazimierski, D. Kardas, *Influence of Temperature on Composition of Wood Pyrolysis Products*, Drvna Industrija, 2017. https://doi.org/10.5552/drind.2017.1714.

[259] E. Khalimov, et al., *Some peculiarities of burnt birch wood pyrolysis*, IOP Conf. Ser. Earth Environ. Sci., 316 (2019), 012019.

[260] H. W. Emmons, A. Atreya, *The science of wood combustion*, Proc. Indian Acad. Sci., Eng. Sci., 5(4) (1982), 259–268.

[261] F. L. Browne, *Theories of the Combustion of Wood and Its Control. A Survey of the Literature*, Forest Products Laboratory, Forest Service, U. S. Department of Agriculture, 1958, https://www.fpl.fs.fed.us/documnts/fplmisc/rpt2136.pdf, accessed on 2022-06-20.

[262] F. Shafizadeh, *Introduction to pyrolysis of biomass*, J. Anal. Appl. Pyrolysis, 3 (1982), 283–305.

[263] N. S. Hassan, A. A. Jalil, C. N. C. Hitam, D. V. N. Vo, W. Nabgan, *Biofuels and renewable chemicals production by catalytic pyrolysis of cellulose: a review*, Environ. Chem. Lett., 18 (2020), 1625–1648. https://doi.org/10.1007/s10311-020-01040-7.

[264] H. Kawamoto, *Lignin pyrolysis reactions*, J. Wood Sci., 63 (2017), 117–132. https://doi.org/10.1007/s10086-016-1606-z.

[265] V. Gargiulo, A. I. Ferreiro, P. Giudicianni, S. Tomaselli, M. Costa, R. Ragucci, M. Alfe, *Insights about the effect of composition, branching and molecular weight on the slow pyrolysis of xylose-based polysaccharides*, J. Anal. Appl. Pyrolysis, 161, 105369 (2022).
[266] W. Zhou, W. Li, R. Mabon, L. J. Broadbelt, *A Critical Review on Hemicellulose Pyrolysis*, Energy Technol. 5 (2017), 52–79. https://doi.org/10.1002/ente.201600327.
[267] J. Reinhold, *Quantentheorie der Moleküle*, 3rd ed., Teubner Verlag, 2006. ISBN 978-3-8351-0037-4.
[268] O. Krätz, *7000 Jahre Chemie*, Nikol Verlag, 2000. ISBN 978-3933203205.
[269] H. Hager, W. Blaschek, K. Keller, *Hagers Handbuch der Pharmazeutischen Praxis*, vol. 3, Springer, 1998. ISBN 978-3540616191.
[270] E. Breitmaier, *Terpene: Aromen, Düfte, Pharmaka, Pheromone*, Teubner Studienbücher Chemie, 1999. ISBN 3-519-03548-0.
[271] G. Michal (ed.), *Biochemical Pathways*, Spektrum, Akademischer Verlag, 1999. ISBN 3-86025-239-9.
[272] G. Richter, *Stoffwechselphysiologie der Pflanzen*, 4th ed., Georg Thieme Verlag, Stuttgart, 1981. ISBN 3-13-442004-X.
[273] H.-W. Heldt, B. Piechulla, *Plant Biochemistry*, 4th ed., Academic Press, 2011. ISBN 978-0-12-384986-1.
[274] G. Tegge (ed.), *Stärke und Stärkederivate*, 3rd ed., Behr's Verlag, 2004. ISBN 3-89947-075-3.
[275] J. N. BeMiller, *Carbohydrate Chemistry for Food Scientists. AACC International Inc.*, St. Paul, 2007, ISBN 978-1-891127-53-3.
[276] A. M. Stephen, G. O. Philips, P. A. Williams, *Food Polysaccharides and Their Applications*, Marcel Dekker Inc., 2006. ISBN 978-0824759223.
[277] H.-D. Belitz, W. Grosch, *Lehrbuch der Lebensmittelchemie*, vol. 4, Springer, 1992. ISBN 3-540-55449-1.
[278] W. Ternes, *Lebensmittel-Lexikon*, Behrs Verlag, 2005. ISBN 978-3899471656.
[279] N. P. Wong, R. Jenness, *Fundamentals of Dairy Chemistry*, Aspen Publishers, 1999. ISBN 0-8342-1360-5.
[280] P. Walstra, R. Jenness, *Dairy Chemistry and Physics*, J. Wiley & Sons, 1984. ISBN 0-471-09779-9.
[281] P. Nuhn, *Naturstoffchemie*, 3rd ed., S. Hirzel Verlag Stuttgart, 1997. ISBN 3-7776-0613-8.
[282] Y. Margalit, *Concepts in Wine Chemistry*, The Wine Appreciation Guild, San Francisco, 2004. ISBN 1-891267-74-4.
[283] J. F. Zayas, *Functionality of Proteins in Food*, Springer, 1997. ISBN 3-540-60252-6.
[284] J. C. Cheftel, J. L. Cuq, D. Lorient, *Lebensmittelproteine*, Behr's Verlag, 1992. ISBN 3-86022-071-3.
[285] S. Damodaran, A. Paraf (eds.), *Food Proteins and their Applications*, Marcel Dekker Inc., New York, 1997. ISBN 0-8247-9820-1.
[286] A. Gennadios (ed.), *Protein-based Films and Coatings*, CRC Press, 2002. ISBN 978-1587161070.
[287] T. Yamamoto, L. R. Juneja, H. Hatta, M. Kim (eds.), *Hen Eggs. Their Basic and Applied Science*, CRC Press, 1997. ISBN 0-8493-4005-5.
[288] W. Ternes, L. Acker, S. Scholtyssek, *Ei und Eiprodukte*, Verlag Paul Parey, 1994. ISBN 3-489-63114-5.
[289] P. Fratzl (ed.), *Collagen. Structure and Mechanics*, Springer, 2008. ISBN 978-0-387-73905-2.
[290] E. N. Frankel, *Lipid Oxidation*, The Oily Press, Dundee, 1998. ISBN 0-9514171-9-3.
[291] H. W.-S. Chan (ed.), *Autoxidation of Unsatured Lipids*, Academic Press Inc., London, 1987. ISBN 0-12-167630-7.
[292] A. Brossi (Hrg.), *The Alkaloids: Chemistry and Pharmacology: 39*, Academic Press Inc., 1990. ISBN 978-0124695399.

[293] E. Haslam, *Practical Polyphenolics. From Structure to Molecular Recognition and Physiological Action*, Cambridge University Press, 1998. ISBN 0-521-46513-3.
[294] M. Luckner, *Secondary Metabolism in Microorganisms, Plants, and Animals*, 3rd ed., Springer, 1990. ISBN 3-540-50287-4.
[295] B. A. Bohm, *Introduction to Flavonoids*, Harwood Academic Publishers, 1998. ISBN 90-5702-353-9.
[296] W. I. Taylor, A. R. Battersby, *Oxidative Coupling of Phenols*, Marcel Dekker Inc., New York, 1967.
[297] H. A. Stafford, *Flavonoid Metabolism*, CRC Press Inc, 1990. ISBN 978-0849360855.
[298] F. Constabel, I. K. Vasil (eds.), *Cell Culture and Somatic Cell Genetics of Plants, vol. 5: Phytochemicals in Plant Cell Cultures*, Academic Press Inc., 1988. ISBN 0-12-715005-6.
[299] D. S. Seigler, *Plant Secondary Metabolism*, Kluwer Academic Publishers, 1998. ISBN 0-412-01981-7.
[300] F. A. Tomás-Barberán, R. J. Robins, *Phytochemistry of Fruit and Vegetables*, Proceedings of the Phytochemical Society of Europe, Clarendon Press, Oxford, 1997. ISBN 0-19-857790-7.
[301] K. R. Markham, *Techniques of Flavonoid Identification*, Academic Press, 1982. ISBN 0-12-472680-1.
[302] G. Prota, *Melanins and Melanogenesis*, Academic Press Inc., 1992. ISBN 0-12-565970-9.
[303] D. W. Lee, *Nature's Palette: the Science of Plant Color*, The University of Chicago Press, 2007. ISBN 0-226-47052-0.
[304] A. S. Diamond, D. S. Weiss, *Handbook of Imaging Materials*, 2nd. ed., Marcel Dekker Inc., 2002. ISBN 0-8247-8903-2.
[305] M. Zander, *Polycyclische Aromaten*, Teubner Studienbücher, Stuttgart, 1995. ISBN 3-519-03537-5.
[306] I. Gutman, S. J. Cyvin, *Introduction to the Theory of Benzenoid Hydrocarbons*, Springer, 1989. ISBN 3-540-51139-3.
[307] E. Clar, *Polycyclic Hydrocarbons, 2 volumes*, Academic Press Inc., London, 1964.
[308] D. Dolphin, *The Porphyrins. Vol. III. Physical Chemistry, Part A*, Academic Press Inc., 1978. ISBN 0-12-220103-5.
[309] C. C. Leznoff, A. B. P. Lever (eds.), *Phthalocyanines. Properties and Applications, volume I*, VCH, 1989. ISBN 3-527-26955-X. Volume II: VCH, 1993. ISBN 3-527-89544-2. volume III: VCH, 1993. ISBN 3-527-89638-4.
[310] J. Jiang (ed.), *Functional phthalocyanine molecular materials*, Springer, 2010. ISBN 978-3-642-04751-0.
[311] K. Nassau, *The physics and chemistry of color*, 2nd ed., Wiley serie in pure and applied optics, 2001. ISBN 0-471-39106-9.
[312] M. Fox, *Optical Properties of Solids*, Oxford University Press, New York, 2008. ISBN 978-0-19-850612-6.
[313] E. Hecht, *Optik*, 7th ed., Walter de Gruyter GmbH, 2018. ISBN 978-3-11-052664-6.
[314] J. G. Solé, L. E. Bausá, D. Jaque, *An Introduction to the Optical Spectroscopy of Inorganic Solids*, John Wiley & Sons, Ltd, 2005. ISBN 0-470-86886-4.
[315] D. L. Greenaway, G. Harbeke, *Optical Properties and Band Structure of Semiconductors*, 1st ed., Pergamon Press, Oxford, 1968.
[316] M. Aven, J. S. Prener (eds.), *Physics and Chemistry of II-VI Compounds*, North-Holland Publishing Company, Amsterdam, 1967.
[317] S. Adachi, *Properties of Semiconductor Alloys*, Wiley, 2009. ISBN 978-0-470-74369-0.
[318] R. Sauer, *Halbleiterphysik*, Oldenbourg Wissenschaftsverlag, München, 2009. ISBN 978-3-486-58863-7.
[319] M. Grundmann, *The Physics of Semiconductors*, 2nd ed., Springer, 2010. ISBN 978-3-642-13883-6.

[320] B. Sapoval, C. Hermann, *Physics of Semiconductors*, Springer, New York, 2003. ISBN 0-387-40630-1.
[321] S. Hunklinger, *Festkörperphysik*, Oldenbourg, 2007. ISBN 978-3-486-57562-0.
[322] C. Kittel, *Einführung in die Festkörperphysik*, Oldenbourg, München, 1999. ISBN 3-486-23843-4.
[323] A.-B. Chen, A. Sher, *Semiconductor Alloys*, Plenum Press, New York, 1995. ISBN 0-306-45052-6.
[324] M. L. Cohen, J. R. Chelikowsky, *Electronic Structure and Optical Properties of Semiconductors*, 2nd ed., Springer, 1989. ISBN 3-540-51391-4.
[325] L. I. Berger, *Semiconductor materials*, CRC Press Inc., 2000. ISBN 0-8493-8912-7.
[326] P. K. Basu, *Theory of Optical Processes in Semiconductors*, Oxford University Press Inc., 1997. ISBN 0-19-851788-2.
[327] M. Kerker, *The Scattering of Light and Other Electromagnetic Radiation*, Academic Press, 1969.
[328] U. Kreibig, M. Vollmer, *Optical Properties of Metal Clusters*, Springer, 1995. ISBN 3-540-57836-6.
[329] M. Quinten, *Optical Properties of Nanoparticle Systems*, Wiley-VCH, 2011. ISBN 978-3-527-41043-9.
[330] H. Raether, *Surface Plasmons on Smooth and Rough Surfaces and on Gratings*, Springer Tracts in Modern Physics, vol. 111, Springer, 1988. ISBN 3-540-17363-3.
[331] H. Raether, *Excitation of Plasmons and Interband Transitions by Electrons*, Springer Tracts in Modern Physics, vol. 88, Springer, 1980. ISBN 3-540-09677-9.
[332] W. Kleber, *Einführung in die Kristallographie*, 16th ed., VEB Verlag Technik, Berlin, 1985.
[333] P. Laven, computer programm Mieplot, http://www.philiplaven.com/mieplot.htm, accessed on 2022-06-20.
[334] M. Thompson, computer programm ArgusLab, http://www.arguslab.com/arguslab.com/Welcome.html, accessed on 2022-09-10.
[335] D. Greig, *Elektronen in Metallen und Halbleitern*, R. Oldenbourg Verlag, 1972. ISBN 3-486-33551-0.
[336] G. Mie, *Beiträge zur Optik trüber Medien, speziell kolloidaler Metallösungen*, Ann. Phys., 25(3) (1908), 377–445.
[337] P. Walstra, *Approximation formulae for the light scattering coefficient of dielectric spheres*, Br. J. Appl. Phys., 15 (1964), 1545–1552.
[338] M. Halik, *2,2-Difluor-1,3,2-(2H)-dioxaborine als Bausteine zur Darstellung von langwellig absorbierenden Methinfarbstoffen*, Dissertation der Universität Halle-Wittenberg, http://sundoc.bibliothek.uni-halle.de/diss-online/98/99H017/prom.pdf, accessed on 2022-06-20.
[339] J. S. Dewar, *Modern Theories of Colour*, Spec. Publ., Geochem. Soc., 4 (1956), 64–82.
[340] Y. Tanabe, S. Sugano, *On the absorption spectra of complex ions I–III*, J. Phys. Soc. Jpn., 9(5) (1954), 753.
[341] K.-A. Kovar, W. Mayer, *Elektronen-Donator-Acceptor-Komplexe*, Pharmazie Unserer Zeit, 8(2) (1979), 46–53.
[342] G. Briegleb, *"Charge-Transfer"-Spektren von Komplexen zwischen Neutralmolekülen*, Pure Appl. Chem., 4(1) (1962), 105–120.
[343] W. Kaim, S. Ernst, S. Kohlmann, *Farbige Komplexe: das Charge-Transfer-Phänomen*, Chem. Unserer Zeit, 21(2) (1987), 50–58.
[344] G. Briegleb, J. Czekalla, *Elektronenüberführung durch Lichtabsorption und -emission in Elektronen-Donator-Acceptor-Komplexen*, Angew. Chem., 72 (1960), 401–413.
[345] R. G. Burns, *Intervalence Transitions in Mixed-Valence Minerals of Iron and Titanium*, Annu. Rev. Earth Planet. Sci., 9 (1981), 345–383.

[346] A. G. Marinopoulos, L. Reining, A. Rubio, V. Olevano, *Ab initio study of the optical absorption and wave-vector-dependent dielectric response of graphite*, Phys. Rev. B, 69, 245419 (2004). https://doi.org/10.1103/PhysRevB.69.245419.

[347] A. K. Solanki, A. Kashyap, T. Nautiyal, S. Auluck, M. A. Khan, *Band structure and optical properties of graphite*, Solid State Commun., 100(9) (1996), 645–649.

[348] R. J. Papoular, R. Papoular, *Some optical properties of graphite from IR to millimetric wavelengths*, Mon. Not. R. Astron. Soc., 443(4) (2014), 2974–2982. https://doi.org/10.1093/mnras/stu1348.

[349] L. P. Bouckaert, R. Smoluchowski, E. Wigner, *Theory of Brillouin Zones and Symmetry Properties of Wave Functions in Crystals*, Phys. Rev., 50 (1936), 58–67.

[350] A. P. Alivisatos, *Semiconductor Clusters, Nanocrystals, and Quantum Dots*, Science, 271(5251) (1996), 933–937.

[351] A. P. Alivisatos, *Perspectives on the Physical Chemistry of Semiconductor Nanocrystals*, J. Phys. Chem., 100 (1996), 13226–13239.

[352] C. B. Murray, D. J. Norris, M. G. Bawendi, *Synthesis and Characterization of Nearly Monodisperse CdE (E = S, Se, Te) Semiconductor Nanocrystallites*, J. Am. Chem. Soc., 115 (1993), 8706–8715.

[353] H. Weller, *Kolloidale Halbleiter-Q-Teilchen: Chemie im Übergangsbereich zwischen Festkörper und Molekül*, Angew. Chem., 105 (1993), 43–55.

[354] A. Riss, O. Diwald, *Gestalten in der Nanowelt. Partikel-Morphologien und Festkörper-Eigenschaften*, Chem. Unserer Zeit, 43 (2009), 84–92.

[355] R. A. Abram, G. J. Rees, B. L. H. Wilson, *Heavily doped semiconductors and devices*, Adv. Phys., 27(6) (1978), 799–892.

[356] A. L. Efros, *Density of states and interband absorption of light in strongly doped semiconductors*, Sov. Phys., Usp., 16(6) (1974), 789–805.

[357] H. C. Casey, F. Stern, *Concentration-dependent absorption and spontaneous emission of heavily doped GaAs*, J. Appl. Phys., 47 (1976), 631–643.

[358] H. C. Casey, D. D. Sell, K. W. Wecht, *Concentration dependence of the absorption coefficient for n- and p-type GaAs between 1.3 and 1.6 eV*, J. Appl. Phys., 46 (1975), 250–257.

[359] F. Bassani, D. Brust, *Effect of Alloying and Pressure on the Band Structure of Germanium and Silicon*, Phys. Rev., 131(4) (1963), 1524–1529.

[360] R. Braunstein, A. R. Moore, F. Herman, *Intrinsic Optical Absorption in Germanium-Silicon Alloys*, Phys. Rev., 109(3) (1958), 695–710.

[361] N. R. J. Poolton, K. B. Ozanyan, J. Wallinga, A. S. Murray, L. Botter-Jensen, *Electrons in feldspar II: a consideration of the influence of conduction band-tail states on luminescence processes*, Phys. Chem. Miner., 29 (2002), 217–225.

[362] C. X. Wu, Y. M. Lu, D. Z. Shen, X. W. Fan, *Effect of Mg content on the structural and optical properties of $Mg_xZn_{1-x}O$ alloys*, Chin. Sci. Bull., 55(1) (2010), 90–93.

[363] J. Dong, D. A. Drabold, *Band-tail states and the localized-to-extended transition in amorphous diamond*, Phys. Rev. B, 54 (1996), 10284–10287.

[364] N. A. Noor, N. Ikram, S. Ali, S. Nazir, S. M. Alay-e-Abbas, A. Shaukat, *First-principles calculations of structural, electronic and optical properties of $Cd_xZn_{1-x}S$ alloys*, J. Alloys Compd., 507 (2010), 356–363.

[365] S. Mecabih, N. Amrane, B. Belgoumène, H. Aourag, *Opto-electronic properties of the ternary alloy $Hg_{1-x}Cd_xTe$*, Phys. A, 276 (2000), 495–507.

[366] S.-H. Weia, S. B. Zhang, A. Zunger, *First-principles calculation of band offsets, optical bowings, and defects in CdS, CdSe, CdTe, and their alloys*, J. Appl. Phys., 87(3) (2000), 1304–1311.

[367] L. Visscher, K. G. Dyall, *Dirac-Fock atomic electronic structure calculations using different nuclear charge distributions*, At. Data Nucl. Data Tables, 67 (1997), 207–224, see also link given in text or http://www.chem.vu.nl/~visscher/FiniteNuclei/FiniteNuclei.htm, particularly orbital energies: http://www.chem.vu.nl/~visscher/FiniteNuclei/Table3.html (all no longer accessible).

[368] M. Klessinger, *Konstitution und Lichtabsorption organischer Farbstoffe*, Chem. Unserer Zeit, 12(1) (1978), 1–11.

[369] J. W. Armit, R. Robinson, *Polynuclear Heterocyclic Aromatic Types. Part II. Some Anhydronium Bases*, J. Chem. Soc. (1925), 1604.

[370] E. Clar, M. Zander, *1:12-2:3-10:11-Tribenzoperylen*, J. Chem. Soc. (1958), 1861.

[371] H. B. Klevens, J. R. Platt, *Spectral Resemblances of Cata-Condensed Hydrocarbons*, J. Chem. Phys., 17 (1949), 470.

[372] J. R. Platt, *Classification of Spectra of Cata-Condensed Hydrocarbons*, J. Chem. Phys., 17 (1949), 484.

[373] E. Clar, *Ein einfaches Prinzip des Aufbaues der aromatischen Kohlenwassenstoffe und ihrer Absorptionsspektren (Aromatische Kohlenwasserstoffe, 20. Mitteil.)*, Ber. Dtsch. Chem. Ges., 69 (1936), 607.

[374] K. Toyota, J-y. Hasegawa, H. Nakatsuji, *SAC-CI study of the excited states of free base tetrazaporphin*, Chem. Phys. Lett., 250 (1996), 437–442.

[375] K. Toyota, J-y. Hasegawa, H. Nakatsuji, *Excited States of Free Base Phthalocyanine Studied by the SAC-CI Method*, J. Phys. Chem. A, 101 (1997), 446–451.

[376] M. Gouterman, *Optical Spectra and Electronic Structure of Porphyrins and Related Rings*, in [308], pp. 1–166.

[377] M. J. Stillman, T. Nyokong, *Absorption and Magnetic Circular Dichroism Spectral Properties of Phthalocyanines Part 1: Complexes of the Dianion PC(-2)*, in [309], vol. I, pp. 133–290.

[378] N. Kobayashi, *Synthesis and Spectroscopic Properties of Phthalocyanine Analogues*, in [309], vol. II, pp. 97–162.

[379] M. J. Stillman, *Absorption and Magnetic Circular Dichroism Spectral Properties of Phthalocyanines Part 2: Ring-Oxidized and Ring-Reduced Complexes*, in [309], vol. III, pp. 227–296.

[380] T. Nyokong, *Electronic Spectral and Electrochemical Behavior of Near Infrared Absorbing Metallophthalocyanines*, in [310], pp. 45–88.

[381] M. Gouterman, *Study of the Effects of Substitution on the Absorption Spectra of Porphin*, J. Chem. Phys., 30 (1959), 1139.

[382] M. Gouterman, *Spectra of Porphyrins*, J. Mol. Spectrosc., 6 (1961), 138–163.

[383] M. Gouterman, G. H. Wagnière, L. C. Snyder, *Spectra of Porphyrins. Part II. For Orbital Model*, J. Mol. Spectrosc., 11 (1963), 108–127.

[384] M. Kasha, H. R. Rawls, M. A. El-Bayoumi, *The Exciton Model in Molecular Spectroscopy*, Pure Appl. Chem., 11(3–4) (1965), 371–392.

[385] M. B. Robin, P. Day, *Mixed Valence Chemistry – a Survey and Classification*, Adv. Inorg. Chem. Radiochem., 10 (1967), 247–433.

[386] R. J. Gettens, H. Kühn, W. T. Chase, *Lead White*, Stud. Conserv., 12(4) (1967), 125–139.

[387] R. zum Hagen, *Darstellung des Saftgrüns*, Pharmazeutisches Zentralblatt für 1831, 2. Band (Nr. 26 bis Nr. 55), Nr. 32, pp. 513–514.

[388] F. Hund, *Abhängigkeit der Farbe roter Eisen(III)-oxide von Teilchengrösse und Teilchengrössenverteilung*, Chem. Ing. Tech., 38(4) (1966), 423–428.

[389] L. Resenberg, *Das Pigment Zinnober*, Restauro, 5 (2005), 362–372.

[390] M. Spring, R. Grout, *The Blackening of Vermilion: An Analytical Study of the Process in Paintings*, Natl. Gallery Tech. Bull., 23 (2002), 50–61. ISBN 978-1-85709-941-9.

[391] J. Kirby, M. Spring, C. Higgitt, *The Technology of Eighteenth- and Nineteenth-Century Red Lake Pigments*, Natl. Gallery Tech. Bull., 28 (2007), 69–95. ISBN 978-1-85709-357-5.

[392] J. Kirby, M. Spring, C. Higgitt, *The Technology of Red Lake Pigment Manufacture: Study of the Dyestuff Substrate*, Natl. Gallery Tech. Bull., 26 (2005), 71–87. ISBN 1-85709-341-0.

[393] J. Kirby, R. White, *The Identification of Red Lake Pigment Dyestuffs and a Discussion of their Use*, Natl. Gallery Tech. Bull., 17 (1996), 56–79. ISBN 1-85709-113-2.

[394] J. Kirby, *A Spectrophotometric Method for the Identification of Lake Pigment Dyestuffs*, Natl. Gallery Tech. Bull., 1 (1977), 35–48.

[395] D. Saunders, J. Kirby, *Light-induced Colour Changes in Red and Yellow Lake Pigments*, Natl. Gallery Tech. Bull., 15 (1994), 79–97. ISBN 1-85709-049-7.

[396] J. Kirby, D. Saunders, *Sixteenth- to Eighteenth-Century Green Colours in Landscape And Flower Paintings: Composition and Deterioration*, in [105], pp. 155–159.

[397] A. Burnstock, *The Fading of the Virgin's Robe in Lorenzo Monaco's 'Coronation of the Virgin'*, Natl. Gallery Tech. Bull., 12 (1988), 58–65. ISBN 0-947645-63-2.

[398] R. White, *Brown and Black Organic Glazes, Pigments and Paints*, Natl. Gallery Tech. Bull., 10 (1986), 58–71. ISBN 0-947645-10-1.

[399] J. WinterReviewed, *The Characterization of Pigments Based on Carbon*, Stud. Conserv., 28(2) (1983), 49–66.

[400] J. Lehmann, *Seit wann wird mit Zinkweiß gemalt? Restauro*, 5 (2000), 356–361.

[401] C. Rötter, *Auripigment*, Restauro, 6 (2003), 408–413.

[402] A. Wallert, *Orpiment and Realgar*, Restauro, 4 (1984), 45–57.

[403] H. Kühn, *Lead-Tin-Yellow*, Stud. Conserv., 13(1) (1968), 7–33.

[404] E. Martin, A. Duval, *Le deux varietes de jaune de plomb et détain. Etude chronologique*, Stud. Conserv., 35(3) (1990), 117–136.

[405] M.-C. Corbeil, J.-P. Charland, E. A. Moffatt, *The Characterization of Cobalt Violet Pigments*, Stud. Conserv., 47(4) (2002), 237–249.

[406] H. Pfleiderer, *Cadmiumzinnober, eine neue Klasse von Pigmenten*, Fette Seifen Anstrichm., 58 (1956), 12.

[407] R. J. Gettens, E. W. Fitzhugh, *Malachite and Green Verditer*, Stud. Conserv., 19(1) (1974), 2–23.

[408] A. Burmester, L. Resenberg, *Von Berggrün, Schiefergrün und Steingrün aus Ungarn*, Restauro, 3 (2003), 180–187.

[409] H. Kühn, *Verdigris and Copper Resinat*, Stud. Conserv., 15(1) (1970), 12–36.

[410] R. Woudhuysen-Keller, *Aspects of Painting Technique in the Use of Verdigris and Copper Resinate*, in [106], pp. 65–69.

[411] H. Andreas, *Schweinfurter Grün – das brillante Gift*, Chem. Unserer Zeit, 1 (1996), 23–31.

[412] C. Marrder, *Schweinfurter Grün. Teil 1: Geschichte eines künstlichen Farbpigments*, Restauro, 5 (2004), 326–331.

[413] C. Marrder, *Schweinfurter Grün. Teil 2: Eigenschaften, naturwissenschaftliche Untersuchungen*, Restauro, 8 (2004), 543–547.

[414] W. Feitknecht, K. Maget, *Zur Chemie und Morphologie der basischen Salze zweiwertiger Metalle. XIV. Die Hydroxychloride des Kupfers*, Helv. Chim. Acta, 32 (1949), 1639–1653.

[415] G. Banik, *Untersuchungen über den Abbau von Papier durch Kupferpigmente. Ein Beitrag zur Aufklärung des Kupferfraßes in graphischen Kunstwerken*, Habilitationsschrift, Institut für Farbenlehre und Farbenchemie an der Akademie der Bildenden Künste, Wien, 1982.

[416] F. Schweizer, B. Mühlethaler, *Einige grüne und blaue Kupferpigmente*, Farbe und Lack, 74(12) (1968), 1159–1173.

[417] F. Ellwanger-Eckel, *Herstellung und Verwendung künstlicher grüner und blauer Kupferpigmente in der Malerei*, Diplomarbeit (1979), Institut für Technologie der Malerei, Staatliche Akademie der Bildenden Künste, Stuttgart.

[418] B. Mühlethaler, J. Thissen, *Smalt*, Stud. Conserv., 14(2) (1969), 47–61.

[419] J. A. Darrah, *Connections and Coincidences: Three Pigments*, in [106], pp. 70–77.

[420] M. Spring, C. Higgitt, D. Saunders, *Investigation of Pigment-Medium Interaction Processes in Oil Paint containing Degraded Smalt*, Natl. Gallery Tech. Bull., 26 (2005), 50–61. ISBN 1-85709-341-0.

[421] J. J. Boon, K. Keune, J. van der Weerd, M. Geldof, J. R. J. van Asperen de Boer, *Imaging Microspectroscopic, Secondary Ion Mass Spectrometric and Electron Microscopic Studies on Discoloured and Partially Discoloured Smalt in Cross-Sections of 16th Century Paintints*, Chimia, 55(11) (2001), 952–960.

[422] L. Robinet, M. Spring, S. Pages, D. Vantelon, N. Trcera, *Investigation of the Discoloration of Smalt Pigment in Historic Paintings by Micro-X-ray Absorption Spectroscopy at the Co K-Edge*, Anal. Chem., 83(13) (2011), 5145–5152.

[423] A. Ludi, *Berliner Blau*, Chem. Unserer Zeit, 22(4) (1988), 123–127.

[424] H. J. Buser, D. Schwarzenbach, W. Petter, A. Ludi, *The Crystal Structure of Prussian Blue:* $Fe_4[Fe(CN)_6]_3 \cdot xH_2O$, Inorg. Chem., 16(11) (1977), 2704–2710.

[425] A. Börner, *Berlinerblau für die Retusche? Restauro*, 5 (2002), 327–331.

[426] J. Kirby, D. Saunders, *Fading and Colour Change of Prussian Blue: Methods of Manufacture and the Influence of Extenders*, Natl. Gallery Tech. Bull., 25 (2004), 73–99. ISBN 1-85709-320-8.

[427] J. Kirby, *Fading and Colour Change of Prussian Blue: Occurrences and Early Reports*, Natl. Gallery Tech. Bull., 14 (1993), 62–71. ISBN 1-85709-033-0.

[428] M. Prinzmeier, *Dem Bronzieren auf der Spur*, Farbe und Lack, 118(2) (2012), 24–27.

[429] H. Berke, *Chemie im Altertum: die Erfindung von blauen und purpurnen Farbpigmenten*, Angew. Chem., 114(14) (2002), 2595–2600.

[430] L. A. Polette-Niewold, et al., *Organic/inorganic complex pigments: Ancient colors Maya Blue*, J. Inorg. Biochem., 101 (2007), 1958–1973.

[431] G. Chiari, et al., *Pre-columbian nanotechnology: reconciling the mysteries of the maya blue pigment*, Appl. Phys. A, 90 (2008), 3–7.

[432] P. Kempter, G. Wilker, *Blaues bleibt Blau. Dioxazin-Blau: ein neues lösemittelechtes Pigment*, Farbe und Lack, 107(11) (2001), 29–31.

[433] N. Tozzi, *Glaze and Body Pigments and Stains in the Ceramic Tile Industry*, http://digitalfire.com/4sight/education/glaze_and_body_pigments_and_stains_in_the_ceramic_tile_industry_342.html, accessed on 2011-07-05.

[434] J. Riederer, *Keramik*, in [103], pp. 175–201.

[435] P. Kleinschmit, *Ein Kapitel angewandte Festkörperchemie: Zirkonsilicat-Farbkörper*, Chem. Unserer Zeit, 20(6) (1986), 182–190.

[436] R. W. Batchelor, *Modern Inorganic Pigments*, Trans. Br. Ceram. Soc., 73 (1974), 297–301.

[437] M. Trojan, *Synthesis of a Blue Zircon Pigment*, Dyes Pigments, 9 (1988), 221–232.

[438] M. Dondi, M. Ardit, G. Cruciani, *Turquoise zirkon: new diffraction and optical data — an overview of* $V^{4\oplus}$ *crystal chemistry and optical spectroscopy in minerals*, Period. Mineral. (2015), 65–66.

[439] M. Ardit, G. Cruciani, F. Di Benedetto, L. Sorace, M. Dondi, *New spectroscopic and diffraction data to solve the vanadium-doped zircon pigment conundrum*, J. Eur. Ceram. Soc., 38 (2018), 5234–5245. https://doi.org/10.1016/j.jeurceramsoc.2018.07.033.

[440] A. Niesert, M. Hanrath, S. Siggel, M. Jansen, K. Langer, *Theoretical study of the polarized electronic absorption spectra of vanadium-doped zircon*, J. Solid State Chem., 169(1) (2002). https://doi.org/10.1016/S0022-4596(02)00010-5.

[441] A. Siggel, M. Jansen, *Röntgenographische Untersuchungen zur Bestimmung der Einbauposition von Seltenen Erden (Pr, Tb) und Vanadium in Zirkonpigmenten*, ZAAC, 583(1) (1990), 67–77. https://doi.org/10.1002/zaac.19905830109.

[442] M. Trojan, *Synthesis of a Yellow Zircon Pigment*, Dyes Pigments, 9 (1988), 261–273.

[443] M. Trojan, *Synthesis of a Pink Zircon Pigment*, Dyes Pigments, 9 (1988), 329–342.
[444] P. Šulcova, M. Trojan, *Synthesis of a Pink Zircon Pigment*, Dyes Pigments, 40 (1999), 83–86.
[445] R. Aguiar, D. Logvinovich, A. Weidenkaff, A. Rachel, A. Reller, S. G. Ebbinghaus, *The vast colour spectrum of ternary metal oxynitride pigments*, Dyes Pigments, 76 (2008), 70–75.
[446] M. Janson, H. P. Letschert, *Inorganic yellow-red pigments without toxic metals*, Nature, 404 (2000), 980–982.
[447] E. Günther, R. Hagenmayer, M. Jansen, *Strukturuntersuchungen an den Oxidnitriden $SrTaO_2N$, $CaTaO_2N$ und $LaTaON_2$ mittels Neutronen- und Röntgenbeugung*, Z. Anorg. Allg. Chem., 626 (2000), 1519–1525.
[448] E. Günther, R. Hagenmayer, M. Jansen, *Nitrides and Oxynitrides: Chemistry and Properties Preparation, Crystal*, J. Eur. Ceram. Soc., 8 (1991), 197–213.
[449] R. A. Eppler, *Inverse Spinel Pigments*, J. Amer. Ceram. Soc., 66(11) (1983), 794–801.
[450] W. Lüttke, M. Klessinger, *Infrarot- und Lichtabsorptionsspektren einfacher Indigofarbstoffe*, Chem. Ber., 97 (1964), 2342–2357.
[451] M. Klessinger, W. Lüttke, *Das chromophore System der Indigo-Farbstoffe*, Tetrahedron, 19(2) (1963), 315–335.
[452] M. Klessinger, *PPP-Rechnungen am Indigo-Chromophor*, Tetrahedron, 22 (1966), 3355–3365.
[453] M. Klessinger, W. Lüttke, *Der Einfluss zwischenmolekularer Wasserstoffbrücken auf die Spektren von Indigo im festen Zustand*, Chem. Ber., 99 (1966), 2136–2145.
[454] Z. C. Koren, *A New HPLC-PDA Method for the Analysis of Tyrian Purple Components*, Dyes Hist. Archaeol., 21 (2008), 26–35.
[455] C. J. Cooksey, *Tyrian Purple: 6, 6'-Dibromoindigo and Related Compounds*, Molecules, 6 (2001), 736–769.
[456] H. Fouquet, H.-J. Bielig, *Biological precursors and genesis of Tyrian purple*, Angew. Chem. Int. Ed., 19(11) (1971), 816–817.
[457] R. R. Melzer, P. Brandhuber, T. Zimmermann, U. Smola, *Der Purpur. Farben aus dem Meer*, Biol. Unserer Zeit, 31(1) (2001), 30–39.
[458] D. Reinen, G.-G. Lindner, *The nature of the chalcogen colour centres in ultramarine-type solids*, Chem. Soc. Rev., 28 (1999), 75–84.
[459] F. Seel, G. Schäfer, H.-J. Güttler, G. Simon, *Des Geheimnis des Lapis lazuli*, Chem. Unserer Zeit, 8 (1974), 65–71.
[460] F. A. Cotton, J. B. Harmon, R. M. Hedges, *Calculation of the Ground State Electronic Structures and Electronic Spectra of Di- and Trisulfide Radical Anions by the Scattered Wave-SCF-Xα Method*, J. Am. Chem. Soc., 98 (1976), 1417–1424.
[461] W. Koch, J. Natterer, C. Heinemann, *Quantum chemical study on the equilibrium geometries of S_3 and S_3^{2-}, The electron affinity of S_3 and the low lying electronic states of S_3^{2-}*, J. Chem. Phys., 102(15) (1995), 6159–6167.
[462] G.-G. Lindner, D. Reinen, *Synthese, Farbe und Struktur eines roten Selen-Ultramarins*, Z. Anorg. Allg. Chem., 620 (1994), 1321–1328.
[463] H. Schlaich, G.-G. Lindner, J. Feldmann, E. O. Göbel, D. Reinen, *Optical Properties of $Se_2^{\ominus}$ and Se_2 Color Centers in the Red Selenium Ultramarine with the Sodalite Structure*, Inorg. Chem., 39 (2000), 2740–2746.
[464] C. Heinemann, W. Koch, G.-G. Lindner, D. Reinen, P.-O. Widmark, *Ground- and excited-state properties of neutral and anionic selenium dimers and trimers*, Phys. Rev. A, 54(3) (1996), 1979–1993.
[465] G.-G. Lindner, K. Witke, H. Schlaich, D. Reinen, *Blue-green ultramarine-type zeolites with dimeric tellurium colour centres*, Inorg. Chim. Acta, 252 (1996), 39–45.
[466] S. E. Tarling, P. Barnes, J. Klinowski, *The Structure and Si, Al Distribution of the Ultramarines*, Acta Crystallogr., Sect. B, 44 (1988), 128–135.

[467] F. Hund, P. Köhl, J. Kemper, D. Reinen, *Der Chromat-Nosean $Na_8[Al_6Si_6O_{24}](CrO_4)$ – ein zeolithisches Gelbpigment? Z. Anorg. Allg. Chem.*, 628 (2002), 1457–1458.
[468] S. Brühlmann, *Kobaltblau. Werkstoffgeschichte und Werkstofftechnologie*, Schriftenreihe Konservierung und Restaurierung, Hochschule der Künste Bern, pp. 11–34.
[469] D. Reinen, T. C. Brunold, H. U. Güdel, N. D. Yordanov, *Die Natur des Farbzentrums im Manganblau*, Z. Anorg. Allg. Chem., 624 (1998), 438–442.
[470] F. Hund, *Anorganische Pigmente durch iso-, homöo- und heterotype Mischphasenbildung*, Farbe und Lack, 73(2) (1967), 111–120.
[471] E. Korinth, *Neuere anorganische Pigmente*, Angew. Chem., 64 (1952), 265–269.
[472] F. Hund, *Mischphasenpigmente mit Rutil-Struktur*, Angew. Chem., 74 (1962), 23–27.
[473] G. Pfaff, *Perlglanzpigmente*, Chem. Unserer Zeit, 31(1) (1997), 6–16.
[474] P. Sayer, *General Aspects of Organic Pigment Technology*, in [152], pp. 92–106.
[475] Brandenburgisches Landesamt für Denkmalpflege und Archäologisches Landesmuseum, *Umweltbedingte Pigmentveränderungen an mittelalterlichen Wandmalereien*, Arbeitshefte Nr. 24, Wernersche Verlagsgesellschaft mbH 2009. ISBN 978-3-88462-290-2.
[476] Website of the caves of Lascaux, https://www.lascaux.fr/en, accessed on 2022-06-20.
[477] Heilbrunn Timeline of Art History, vom Metropolitan Museum of Art, https://www.metmuseum.org/toah, accessed on 2022-06-20.
[478] J. Clottes, M. Menu, Ph. Walter, *La préparation des peintures magdaléniennes des cavernes ariégeoises*, Bull. Soc. Préhist. Fr., 87(6) (1990), 170–192.
[479] D. Garate, É. Laval, M. Menu, *Étude de la matière colorante de la grotte d'Arenaza (Galdames, Pays Basques, Espagne)*, L'anthropologie, 108 (2004), 251–289.
[480] A. El Goresy, *Ancient Pigments in Wall Paintings of Egyptian Tombs and Temples. An Archaeometric Project*, Max-Planck-Institut für Kernphysik, Heidelberg, 1986, MPI H-1986-V 12.
[481] H. Jaksch, *Farbpigmente aus Wandmalereien altägyptischer Gräber und Tempel: Technologien der Herstellung und mögliche Herkunftsbeziehungen*, Dissertation Universität Heidelberg, 1985.
[482] A. Brysbaert, *The Power of Technology in the Bronze Age Eastern Mediterranean*, Equinox Publishing Ltd., 2008. ISBN 978-1-84553-433-2.
[483] P. Westlake, P. Siozos, A. Philippidis, et al., *Studying pigments on painted plaster in Minoan, Roman and Early Byzantine Crete. A multi-analytical technique approach*, Anal. Bioanal. Chem., 402 (2012), 1413. https://doi.org/10.1007/s00216-011-5281-z.
[484] R. E. Jones, E. Photos-Jones, *Technical studies of Aegean Bronze Age wall painting: methods, results and future prospects*, in *British School at Athens Studies Vol. 13, Aegean Wall Painting: A Tribute to Mark Cameron*, pp. 199–228, 2005.
[485] F. Blakolmer, *Minoan Wall-Painting: The Transformation of a Craft into an Art Form*, in R. Laffineur, P. P. Betancourt (eds.) *TEXNH Craftsmen, Craftswomen and Craftmanship in the Aegean Bronze Age, Proceedings of the 6th International Aegean Conference, Philadelphia, Temple University, 18–21 April 1996*, pp. 95–113, 1996.
[486] A. Vlachopoulos, S. Sotiropoulou, *The Blue Colour on the Akrotiri Wall-Paintings: From the Palette of the Theran Painter to the Laboratory Analysis*, in A. Papadopoulos (ed.) *TALANTA. Proceedings of the Dutch Archaeological and Historical Society*, vol. XLIV, pp. 245–272, 2012.
[487] J. Riederer, *Pigmente in der Antike*, Prax. Nat.wiss., Chem., 7(37) (1988), 3–10.
[488] W. Noll, *Chemie vor unserer Zeit: Antike Pigmente*, Chem. Unserer Zeit, 14(2) (1980), 37–43.
[489] W. Noll, *Anorganische Pigmente in Vorgeschichte und Antike*, Fortschr. Mineral., 57 (1979), 203–263.
[490] R. Fuchs, *Farbmittel in der mittelalterlichen Buchmalerei – Untersuchungen zur Konservierung geschädigter Handschriften*, Prax. Nat.wiss., Chem., 37 (1988), 20–29.
[491] J. Riederer, *Die Pigmente der antiken Malerei*, Naturwiss., 69 (1982), 82–86.

[492] W. Noll, R. Holm, L. Born, *Chemie und Technik altkretischer Vasenmalerei vom Kamares-Typ*, Naturwiss., 58 (1971), 615–618.
[493] W. Noll, L. Born, R. Holm, *Chemie und Technik altkretischer Vasenmalerei vom Kamares-Typ, II*, Naturwiss., 61 (1974), 361–362.
[494] W. Noll, R. Holm, L. Born, *Bemalung antiker Keramik*, Angew. Chem., 87(18) (1975), 639–682.
[495] J. Riederer, *Die Pigmente der antiken Malerei*, in [103], pp. 202–213.
[496] R. Germer, *Die Textilfärberei und die Verwendung gefärbter Textilien im alten Ägypten*, Otto Harrassowitz, Wiesbaden, 1992. ISBN 3-447-03183-2.
[497] N. Welter, Untersuchungen von Pigmenten in römischer Wandmalerei und antiken Gläsern, Dissertation Universität Würzburg, 2008.
[498] N. Riedel, *Provinzialrömische Wandmalerei in Deutschland*, Dissertation Universität Bamberg, 2007.
[499] B. Mazzei, *Preservation and Use of the Religious Sites Case Study of the Roman Catacombs*, Eur. J. Sci. Theol., 11(2) (2015), 33–43, http://www.ejst.tuiasi.ro/Files/51/5_Mazzei.pdf, accessed on 2022-06-17.
[500] D. Tapete, F. Fratini, B. Mazzei, E. Cantisani, E. Pecchioni, *Petrographic study of lime-based mortars and carbonate incrustation processes of mural paintings in Roman catacombs*, Period. Mineral., 82(3) (2013), 503–527.
[501] A. Zucchiatti, P. Prati, A. Bouquillon, L. Giuntini, M. Massi, A. Migliori, A. Cagnana, S. Roascio, *Characterisation of early medieval frescoes by p-PIXE, SEM and Raman spectroscopy*, Nucl. Instrum. Methods Phys. Res., Sect. B, 219–220 (2004), 20–25.
[502] A. Cagnana, S. Roascio, A. Zucchiatti, A. d'Allesandro, P. Prati, *Gli affreschi altomedievali del Tempietto di Cividale: Nuovi dati da recenti analisi di laboratorio*, in *Forum Iulii. Annuario Del Museo Archeologico Nazionale Di Cividale Del Friuli, Archivi e Biblioteca, XXVII*, pp. 143–153, 2003.
[503] S. Demailly, P. Hugon, M. Stefanaggi, W. Nowik, *The Technique of the Mural Paintingsin the Choir of Angers Cathedral*, in [105], pp. 10–15.
[504] M. R. Katz, *The Mediaeval Polychromy of the Majestic West Portal of Toro, Spain*, in [105], pp. 27–34.
[505] H. Howard, T. Manning, S. Stewart, *Late Mediaeval Wall Painting Techniques at Farleigh Hungerford Castle and Their Context*, in [105], pp. 59–64.
[506] G. Horta, I. Ribeiro, L. Afonso, *The 'Calvary' of S. Francisco's Church in Leiria: Workshop Practice in a Portuguese Late Gothic Wall Painting*, in [105], pp. 65–69.
[507] H. C. Howard, Techniques of the Romanesque and Gothic Wall Paintings in the Holy Sepulche Chapel, Winchester Cathedral, in [106], pp. 91–104.
[508] S. Cather, D. Park, P. Williamson (eds.), *Early Medieval Wall Painting and Painted Sculpture in England*, B. A. R. Oxford, BAR, vol. 216, 1990. ISBN 0-86054-719-1.
[509] H. Howard, *'Blue' in the Lewes Group*, in [508], pp. 195–203.
[510] R. R. Milner-Gulland, *Clayton Church and the Anglo-Saxon heritage*, in [508], pp. 205–219.
[511] R. J. Cramp, J. Cronyn, *Anglo-Saxon polychrome plaster and other materials from the excavations of Monkwearmouth and Jarrow: an interim report*, in [508], pp. 17–30.
[512] S. Rickerby, *Kempley: a technical examination of the Romanesque wall paintints*, in [508], pp. 249–262.
[513] H. Howard, *Pigments of English Medieval Wall Painting*, Archetype Publications Ltd., 2003. ISBN 978-1873132487.
[514] J. Rollier-Hanselmann, *D'Auxerre à Cluny: technique de la peinture murale entre le VIIIe et le XIIe siècle en Bourgogne*, in *Cahiers de civilisation médiévale, 40e année (n°157), Janvier–mars 1997*, pp. 57–90, 1997, https://www.persee.fr/doc/ccmed_0007-9731_1997_num_40_157_2674, accessed on 2016-04-10.

[515] O. Emmenegger, *Karolingische und romanische Wandmalereien in der Klosterkirche. Technik, Restaurierungsprobleme, Massnahmen*, in A. Wyss, H. Rutishauser, M. A. Nay (eds.) *Die mittelalterlichen Wandmalereien im Kloster Müstair*, pp. 77–140, vdf, Hochschul-Verlag an der ETH, Zürich, 2002. ISBN 3-7281-2803-1.
[516] N. Turner, *The Recipe Collection of Johannes Alcherius and the Painting Materials Used in Manuscript Illumination in France and Northern Italy, c. 1380–1420*, in [105], pp. 45–50.
[517] B. Guineau, I. Villela-Petit, R. Akrich, J. Vezin, *Painting Technique in the Boucicaut Hours and in Jacques Coene's Colour Recipes as Found in Jean Lebegue's Libri Colorum*, in [105], pp. 51–54.
[518] B. New, H. Howard, R. Billinge, H. Tomlinson, D. Peggie, D. Gordon, *Niccolò di Pietro Gerini's Baptism Altarpiece: Technique, Conservation and Original Design*, vol. 33, National Gallery, London, 2012. ISBN 978-1-85709-549-4.
[519] J. Dunkerton, A. Roy, *The materials of a Group of Late Fifteenth-century Florentine Panel Paintings*, Natl. Gallery Tech. Bull., 17 (1996), 20–31. ISBN 1-85709-113-2.
[520] J. Dunkerton, A. Roy, *Uccello's Saint George and the Dragon: Technical Evidence Re-evaluated*, Natl. Gallery Tech. Bull., 19 (1998), 26–30. ISBN 1-85709-220-1.
[521] A. Roy, D. Gordon, *Uccello's Battle of San Romano*, Natl. Gallery Tech. Bull., 22 (2001), 4–17. ISBN 978-1-85709-926-5.
[522] J. Dunkerton, C. Plazzotta, *Vincenzo Foppa's Adoration of the Kings*, Natl. Gallery Tech. Bull., 22 (2001), 18–28. ISBN 978-1-85709-926-5.
[523] J. Dunkerton, S. Foister, M. Spring, *The Virgin and Child Enthroned with Angels and Saints attributed to Michael Pacher*, Natl. Gallery Tech. Bull., 21 (2000), 4–19. ISBN 978-1-85709-251-1.
[524] J. Dunkerton, L. Syson, *In Search of Verrocchio the Painter: The Cleaning and Examination of The Virgin and Child with Two Angels*, Natl. Gallery Tech. Bull., 31 (2010), 4–41. ISBN 978-1-85709-495-4.
[525] J. Dunkerton, N. Penny, A. Roy, *Two Paintings by Lorenzo Lotto in the National Gallery*, Natl. Gallery Tech. Bull., 19 (1998), 52–63. ISBN 1-85709-220-1.
[526] *Methods and materials of Northern European painting in the National Gallery, 1400--1550*, Natl. Gallery Tech. Bull., 18 (1997), 6–55. ISBN 1-85709-178-7.
[527] *The materials and technique of five paintings by Rogier van der Weyden and his workshop*, Natl. Gallery Tech. Bull., 18 (1997), 68–86. ISBN 1-85709-178-7.
[528] *A double-sided panel by Stephan Lochner*, Natl. Gallery Tech. Bull., 18 (1997), 56–67. ISBN 1-85709-178-7.
[529] A. Smith, M. Wyld, *Altdorfer's 'Christ taking Leave of His Mother'*, Natl. Gallery Tech. Bull., 7 (1983), 51–64. ISBN 0-901791-88-1.
[530] W. Huber's, *Christ taking Leave of His Mother*, Natl. Gallery Tech. Bull., 18 (1997), 98–112. ISBN 1-85709-178-7.
[531] J. Dunkerton, N. Penny, M. Spring, *The Technique of Garofalo's Paintings at the National Gallery*, Natl. Gallery Tech. Bull., 23 (2002), 20–41. ISBN 978-1-85709-941-9.
[532] A. Roy, M. Spring, C. Plazzotta, *Raphael's Early Work in the National Gallery: Paintings before Rome*, Natl. Gallery Tech. Bull., 25 (2004), 4–35. ISBN 1-85709-320-8.
[533] K. H. Weber, *Die Sixtinische Madonna*, Restauro, 90(4) (1984), 9–28.
[534] L. Keith, A. Roy, R. Morrison, P. Schade, *Leonardo da Vinci's "Virgin of the Rocks": Treatment, Technique and Display*, vol. 32, National Gallery, London, 2011. ISBN 978-1-85709-530-2.
[535] M. Spring, A. Mazzotta, A. Roy, R. Billinge, D. Peggie, *Painting Practice in Milan in the 1490s: The Influence of Leonardo*, vol. 32, National Gallery, London, 2011. ISBN 978-1-85709-530-2.
[536] J. Dunkerton, S. Foister, D. Gordon, N. Penny, *Giotto to Dürer. Early Renaissance Painting in The National Gallery*, National Gallery Publications, Limited, 1991. ISBN 978-0-300-05082-0.

[537] P. Joannides, J. Dunkerton, *A Boy with a Bird in the National Gallery: Two Responses to a Titian Question*, Natl. Gallery Tech. Bull., 28 (2007), 36–57. ISBN 978-1-85709-357-5.
[538] A. Lucas, J. Plesters, *Titian's "Bacchus and Ariadne"*, Natl. Gallery Tech. Bull., 2 (1978), 25–47.
[539] J. Dunkerton, M. Spring, *The Technique and Materials of Titian's Early Paintings in the National Gallery, London*, in *Titian, Jacopo Pesaro being presented by Pope Alexander VI to Saint Peter*, Restoration Vol. 3 No. 1, Journal of Koninklijk Museum, 2003, pp. 9–21.
[540] H. Dubois, A. Wallert, *Titian's Painting Technique in Jacopo Pesaro being presented by Pope Alexander VI to Saint Peter*, in *Titian, Jacopo Pesaro being presented by Pope Alexander VI to Saint Peter*, Restoration Vol. 3 No. 1, Journal of Koninklijk Museum, 2003, pp. 22–37.
[541] U. Birkmaier, A. Wallert, A. Rothe, *Technical Examinations of Titian's Venus and Adonis: A Note on Early Italian Oil Painting Technique*, in [106], pp. 117–126.
[542] J. Grenberg, W. Kireewa, S. Pisarewa, *Das Gemälde "Die Heilige Familie" von Bronzino*, Restauro, 3 (1992), 180–184.
[543] A. Burmester, C. Krekel, *"Azurri oltramarini, lacche et altri colori fini": Auf der Suche nach der verlorenen Farbe*, in [138], pp. 193–211.
[544] J. Plesters, *Tintoretto's Paintings in the National Gallery*, Natl. Gallery Tech. Bull., 3 (1979), 3–24.
[545] J. Plesters, *Tintoretto's Paintings in the National Gallery*, Natl. Gallery Tech. Bull., 4 (1980), 32–48. ISBN 0-901791-72-5.
[546] N. Penny, M. Spring, *Veronese's Paintings in the National Gallery Technique and Materials, Part I*, Natl. Gallery Tech. Bull., 16 (1995), 4–29. ISBN 1-85709-071-3.
[547] N. Penny, A. Roy, M. Spring, *Veronese's Paintings in the National Gallery Technique and Materials, Part II*, Natl. Gallery Tech. Bull., 17 (1996), 32–55. ISBN 1-85709-113-2.
[548] J. Dunkerton, *The Technique and Restoration of Bramantino's Adoration of the Kings*, Natl. Gallery Tech. Bull., 14 (1993), 42–61. ISBN 1-85709-033-0.
[549] B. Heimberg, *Zur Maltechnik Albrecht Dürers*, in [147], pp. 32–54.
[550] A. Burmester, C. Krekel, *Von Dürers Farben*, in [147], pp. 55–101.
[551] P. Ackroyd, S. Foister, M. Spring, R. White, R. Billinge, *A Virgin and Child from the Workshop of Albrecht Dürer? Natl. Gallery Tech. Bull.*, 21 (2000), 29–42. ISBN 978-1-85709-251-1.
[552] S. Foister, M. Wyld, A. Roy, *Hans Holbein's A Lady with a Squirrel and a Starling*, Natl. Gallery Tech. Bull., 15 (1994), 6–19. ISBN 1-85709-049-7.
[553] M. Spring, N. Penny, R. White, M. Wyld, *Colour Change in The Conversion of the Magdalen attributed to Pedro Campana*, Natl. Gallery Tech. Bull., 22 (2001), 54–63. ISBN 978-1-85709-926-5.
[554] L. Keith, *Three Paintings by Caravaggio*, Natl. Gallery Tech. Bull., 19 (1998), 37–51. ISBN 1-85709-220-1.
[555] J. Plesters, *'Samson and Delilah': Rubens and the Art and Craft of Painting on Panel*, Natl. Gallery Tech. Bull., 7 (1983), 30–50. ISBN 0-901791-88-1.
[556] A. Roy, *Rubens's 'Peace and War'*, Natl. Gallery Tech. Bull., 20 (1999), 89–95. ISBN 1-85709-251-1.
[557] C. Christensen, M. Palmer, M. Swicklik, *Van Dyck's Painting Technique, His Writings, and Three Paintings in the National Gallery of Art*, in A. K. Wheelock, S. J. Barnes, J. S. Held (eds.) *Anthony van Dyck, National Gallery of Art Washington*, pp. 45–52, Harry N. Abrams, Inc., Publishers, New York, 1990. ISBN 0-89468-155-9.
[558] A. Roy, *The National Gallery Van Dycks: Technique and Development*, Natl. Gallery Tech. Bull., 20 (1999), 50–83. ISBN 1-85709-251-1.
[559] M. Spring, *Pigments and Colour Change in the Paintings of Aelbert Cuyp*, in A. Wheelock (ed.) *Aelbert Cuyp*, pp. 65–73, National Gallery of Art Washington, 2001. ISBN 0-89468-286-5.

[560] N. Costaras, *A Study of the Materials and Techniques of Johannes Vermeer*, in [142], pp. 145–165.
[561] K. N. Groen, I. D. Van der Werf, K. J. Van den Berg, J. J. Boon, *Scientific Examination of Vermeer's Girl with a Pearl Earring*, in [142], pp. 169–183.
[562] A. Burmester, *Die Palette Giovanni Battista und Giovanni Domenico Tiepolo*, in P. O. Krückmann (ed.) *Der Himmel auf Erden. Tiepolo in der Residenz Würzburg. II: Aufsätze, Ausstellungskatalog 1996*, pp. 160–162, Prestel-Verlag, München, 1996. ISBN 3-7913-1660-5.
[563] I. Stümmer, *Die Maltechnik von Giovanni Battista Tiepolo*, Restauro, 3 (1996), 194–199.
[564] W. Helmberger, A. Staschull, *Tiepolos Welt. Das Deckenfresko im Treppenhaus des Residenz Würzburg*, Bayerische Verwaltung der staatlichen Schlösser, Gärten und Seen, 2006. ISBN 978-3-932982-73-6.
[565] D. Bomford, A. Roy, *Canaletto's 'Stonemason's Yard' and 'San Simeone Piccolo'*, Natl. Gallery Tech. Bull., 14 (1993), 35–41. ISBN 1-85709-033-0.
[566] D. Bomford, A. Roy, *Canaletto's Venice: The Feastday of S. Roch*, Natl. Gallery Tech. Bull., 6 (1982), 40–44. ISBN 0-901791-84-9.
[567] S. L. Fischer, *The Examination and Treatment of Watteau' Italian Comedians*, in M. Morgan Grasselli, P. Rosenberg (eds.) *Watteau, Ausstellungskatalog*, pp. 465–467, National Gallery of Art Washington, 1984. ISBN 0-89468-074-9.
[568] P. Ackroyd, A. Roy, H. Wine, *Nicolas Lancret's The Four Times of Day*, Natl. Gallery Tech. Bull., 25 (2004), 48–61. ISBN 1-85709-320-8.
[569] M. Kirby Talley, *"Tous les bons tableaux se craquellent". La technique picturale et látelier de Sir Joshua Reynolds*, in N. Penny, P. Rosenberg (eds.) *Sir Joshua Reynolds, Ausstellungskatalog*, pp. 83–114, Éditions de la Réunion des musées nationaux, Paris, 1985. ISBN 2-7118-2 017-3.
[570] G. 'Dr, R. Schomberg', D. Bomford, A. Roy, D. Saunders, Natl. Gallery Tech. Bull., 12 (1988), 44–57. ISBN 0-947645-63-2.
[571] M. Leonard, A. Roy, S. Schaefer, *Two Versions of The Fountain of Love by Jean-Honoré Fragonard: A Comparative Study*, Natl. Gallery Tech. Bull., 29 (2008), 31–45. ISBN 978-1-85709-419-0.
[572] D. Bomford, A. Roy, *Hogarth's 'Marriaga á la Mode'*, Natl. Gallery Tech. Bull., 6 (1982), 45–67. ISBN 0-901791-84-9.
[573] J. Leighton, A. Reeve, A. Burnstock, *A 'Winter Landscape' by Caspar David Friedrich*, Natl. Gallery Tech. Bull., 13 (1989), 44–60. ISBN 0-947645-70-5.
[574] J. H. Townsend, *The Materials of J. M. W. Turner: Pigments*, Stud. Conserv., 38(4) (1993), 231–254.
[575] N. W. Hanson, *Some Painting Materials of J. M. W. Turner*, Stud. Conserv., 1(4) (1954), 162–173.
[576] J. H. Townsend, *A Re-Assessment of Hanson's Paper on the Painting Materials of J. M. W. Turner*, Stud. Conserv., 50(4) (2005), 316–318.
[577] J. H. Townsend, *Painting Technique and Materials of Turner and Other British Artists 1775–1875*, in [106], pp. 176–185.
[578] S. Cove, *Mixing and Mingling: John Constable's Oil Paint Mediums c. 1802–-1837, Including the Analysis of the 'Manton' Paint Box*, in [105], pp. 211–216.
[579] J. Kirby, A. Roy, *Paul Delaroche: A Case Study of Academic Painting*, in [106], pp. 166–175.
[580] L. Sheldon, *Methods and Materials of the Pre-Raphaelite Circle in The 1850s*, in [105], pp. 229–234.
[581] S. Herring, *Six Paintings by Corot: Methods, Materials and Sources*, Natl. Gallery Tech. Bull., 30 (2009), 86–111. ISBN 978-1-85709-420-6.
[582] R. und, P. Woudhuysen-Keller, J. Cuttle, C. Hurst, *Zwei Gemälde von Camille Corot*, Restauro, 5 (1993), 306–315.

[583] E. Klein, *An Investigation into the Painting Technique of the Two Versions of Corot's St. Sebastian Succoured by Holy Women*, in [137], pp. 178–191.

[584] R. Woudhuysen-Keller, *Observations Concerning Corot's Late Painting Technique*, in [137], pp. 192–200.

[585] A. Burmester, C. Denk, *Comment ils inventaient ces verts chatoyants? Blau, Gelb, Grün und die Landschaftsmalerei von Barbizon*, in [137], pp. 295–329.

[586] A. Burmester, C. Denk, *Blue, Yellow and Green on the Barbizon Palette, Kunsttechnologie und Konservierung*, pp. 79–87, Wernersche Verlagsgesellschaft, Worms, 1999.

[587] S. Hackney, *Art for Art's Sake: The Materials and Technique of James McNeill Whistler*, in [106], pp. 186–190.

[588] J. Leighton, A. Reeve, A. Roy, R. White, *Vincent Van Gogh's 'A Cornfield with Cypresses'*, Natl. Gallery Tech. Bull., 11 (1987), 42–59. ISBN 0-947645-35-7.

[589] D. Bomford, A. Roy, *Manet's 'The Waitress': An Investigation into its Origin and Development*, Natl. Gallery Tech. Bull., 7 (1983), 3–20. ISBN 0-901791-88-1.

[590] A. Roy, *The Palettes of Three Impressionist Paintings*, Natl. Gallery Tech. Bull., 9 (1985), 12–20. ISBN 0-901791-97-0.

[591] E. Reissner, *Ways of Making: Practive and Innovation in Cézanne's Paintings in the National Gallery*, Natl. Gallery Tech. Bull., 29 (2008), 4–30. ISBN 978-1-85709-419-0.

[592] A. Roy, *Monet's Palette in the Twentieth Century: Water-Lilies and Irises*, Natl. Gallery Tech. Bull., 28 (2007), 58–68. ISBN 978-1-85709-357-5.

[593] A. Roy, *Monet and the Nineteenth Century Palette*, Natl. Gallery Tech. Bull., 5 (1981), 22–26. ISBN 0-901791-76-8.

[594] J. Kirby, K. Stonor, A. Roy, A. Burnstock, R. Grout, R. White, *Seurat's Painting Practice: Theory, Development and Technology*, Natl. Gallery Tech. Bull., 24 (2003), 4–37. ISBN 1-85709-997-4.

[595] K. Stonor, R. Morrison, *Adolphe Monticelli: The Materials and Techniques of an Unfashionable Artist*, vol. 33, National Gallery, London, 2012. ISBN 978-1-85709-549-4.

[596] T. Froysaker, M. Liu, *Four (or eleven) unvarnished oil paintings on canvas by Edvard Much in the Aula of Oslo University*, Restauro, 1 (2009), 44–61.

[597] A. Robbins, K. Stonor, *Past, Present, Memories: Analysing Edouard Vuillard's La Terrasse at Vasouy*, vol. 33, National Gallery, London, 2012. ISBN 978-1-85709-549-4.

[598] R. D. Harley, *Oil colour containers: Development work by artists and colourmen in the nineteenth century*, Ann. of Sci., 27(1) (1971), 1–12. https://doi.org/10.1080/00033797100203607, accessed on 2022-05-06.

[599] Fa. Winsor&Newton, *From the Archives: The History of the Metal Paint Tube*, online article, published 06-JAN-2015, https://www.winsornewton.com/row/articles/art-history/history-metal-paint-tube/, accessed on 2022-05-09.

[600] A. Roy, *Barbizon Painters: Tradition and Innovation in Artist's Materials*, in [137], pp. 330–342.

[601] A. Callen, *The Work of Art: Plein Air Painting and Artistic Identity in Nineteenth-century France*, Reaktion Books, 2015. ISBN 978-1780233550.

[602] J. Zeidler, *Lithographie und Steindruck*, Christophorus Verlag, 2008. ISBN 978-3-419-53486-1.

[603] G. Baars, *Das Färben von Naturfasern mit Naturfarbstoffen*, Prax. Nat.wiss., Chem., 47 (1998), 24–32.

[604] H. M. Koch, P. Pfeifer, *Naturfarbstoffe im Unterricht, Anthrachinonfarbstoffe der Krappwurzel*, Nat.wiss. Unterr., Chem., 10(52) (1999), 25–29.

[605] G. H. Frischat, *Glas – Struktur und Eigenschaften*, Chem. Unserer Zeit, 11(3) (1977), 65–74.

[606] C. Rüssel, D. Ehrt, *Neue Entwicklungen in der Glaschemie*, Chem. Unserer Zeit, 32(3) (1998), 126–135.

[607] R. J. Schwankner, M. Eigenstetter, R. Laubinger, M. Schmidt, *Strahlende Kostbarkeiten. Uran als Farbkörper in Gläsern und Glasuren*, Phys. Unserer Zeit, 36(4) (2005), 160–167.
[608] E. W. J. L. Oomen, A. M. A. van Dongen, *EUROPIUM (III) IN OXIDE GLASSES. Dependence of the emission spectrum upon glass composition*, J. Non-Cryst. Solids, 111 (1989), 205–213.
[609] R. Reisfeld, N. Lieblich, *OPTICAL SPECTRA AND RELAXATION OF Eu +3 IN GERMANATE GLASSES*, J. Phys. Chem. Solids, 34 (1973), 1467–1476.
[610] R. W. Douglas, M. S. Zaman, *The chromophore in iron-sulphur amber glasses*, Phys. Chem. Glasses, 10 (1969), 125.
[611] H. D. Schreiber, S. J. Kozak, C. W. Schreiber, D. G. Wetmore, M. W. Riethmiller, *Sulfur chemistry in a borosilicate melt. Part 3. Iron-sulfur interactions and the amber chromophore*, Glastech. Ber., 63(3) (1990), 49–60.
[612] D. C. W. Sanderson, J. B. Hutchings, *The origins and measurement of colour in archaeological glasses*, Glass Technol., 28(2) (1987), 99–105.
[613] L. R. Green, F. A. Hart, *Colour and Chemical Composition in Ancient Glass: An Examination of some Roman and Wealden Glass by means of Ultraviolet-Visible-Infra-red Spectrometry and Electron Microprobe Analysis*, J. Archaeol. Sci., 14 (1987), 271–282.
[614] P. Colomban, *The Use of Metal Nanoparticles to Produce Yellow, Red and Iridescent Colour, from Bronze Age to Present Times in Lustre Pottery and Glass: Solid State Chemistry, Spectroscopy and Nanostructure*, J. Nanopart. Res., 8 (2009), 109–132.
[615] M. Garcia-Vallès, M. Vendrell-Saz, *The glasses of the transept's rosette of the cathedral of Tarragona: characterisation, classification and decay*, Bol. Soc. Esp. Cerám. Vidrio, 41(2) (2002), 217–224.
[616] E. Thilo, J. Jander, H. Seemann, *Die Farbe des Rubins und der (Al, Cr)$_2$O$_3$-Mischkristalle*, Z. Anorg. Allg. Chem., 279(1–2) (1955), 2–17.
[617] C. J. Ballhausen, *Intensities in Inorganic Complexes Part II. Tetrahedral Complexes*, J. Mol. Spectrosc., 2 (1958), 342–360.
[618] L. H. Hillier, V. R. Saunders, *Ab Initio Calculations of the Excited States of the Permanganate and Chromate Ions*, Chem. Phys. Lett., 9 (1971), 219–221.
[619] K. H. Johnson, F. C. Smith Jr., *Scattered-wave Model for the Electronic Structure and Optical Properties of the Permanganate Ion*, Chem. Phys. Lett., 10 (1971), 219–223.
[620] H. Johansen, *SCF LCAO MO calculation for $MnO_4^{\ominus}$*, Chem. Phys. Lett., 17 (1972), 569–573.
[621] A. P. Mortola, H. Basch, J. W. Moskowitz, *An Ab Initio Study of the Permanganate Ion*, Int. J. Quantum Chem., 7 (1973), 725–737.
[622] H.-L. Hsu, C. Peterson, R. M. Pitzer, *Calculations on the permanganate ion in the ground and excited states*, J. Chem. Phys., 64(2) (1976), 791–795.
[623] J. A. Connor, I. H. Hillier, V. R. Saunders, M. H. Wood, M. Barber, *Ab initio molecular orbital calculations of transition metal complexes*, Mol. Phys., 24 (1972), 497–509.
[624] H. Johansen, S. Rettrup, *Limited Configuration Interaction Calculation of the Optical Spectrum for the Permanganate Ion*, Chem. Phys., 74 (1983), 77–81.
[625] S. Jitsuhiro, H. Nakai, M. Hada, H. Nakatsuji, *Theoretical study on the ground and excited states of the chromate anion $CrO_4^{2\ominus}$*, J. Chem. Phys., 101(2) (1994), 1029–1036.
[626] M. Wolfsberg, L. Helmholz, *The Spectra and Electronic Structure of the Tetrahedral Ions $MnO_4^{\ominus}$, $CrO_4^{2\ominus}$, and $ClO_4^{\ominus}$*, J. Chem. Phys., 20(5) (1952), 837–843.
[627] R. M. Millier, D. S. Tinti, D. A. Case, *Comparison of the Electronic Structures of Chromate, Halochromates, and Chromyl Halides by the Xα Method*, Inorg. Chem., 28 (1989), 2738–2743.
[628] O. Schmitz-DuMont, H. Brokopf, K. Burkhardt, *Farbe und Konstitution bei anorganischen Feststoffen. I. Die Lichtabsorptions des zweiwertigen Kobalts in oxydischen Koordinationsgittern*, Z. Anorg. Allg. Chem., 295 (1958), 7–35.
[629] D. Reinen, O. Schmitz-DuMont, *Farbe und Konstitution bei anorganischen Feststoffen. IV. Die Lichtabsorptions als Mittel zur Auffindung struktureller Feinheiten von Kristallgittern*, Z. Anorg. Allg. Chem., 312 (1961), 121–134.

[630] D. M. Sherman, *The Electronic Structures of $Fe^{3\oplus}$ Coordination Sites in Iron Oxides, Applications to Spectra, Bonding, and Magnetism*, Phys. Chem. Miner., 12 (1985), 161–175.
[631] D. M. Sherman, T. D. Waite, *Electronic spectra of $Fe^{3\oplus}$ oxides and oxide hydroxides in the near IR to near UV*, Amer. Mineral., 70 (1985), 1262–1269.
[632] D. M. Sherman, *Crystal Chemistry, Electronic Structures, and Spectra of Fe Sites in Clay Minerals*, in L. Coyne et al. (eds.) *Spectroscopic Characterization of Minerals and Their Surfaces*, ACS Symposium Series, American Chemical Society, Washington, DC, 1990.
[633] *Correlation between structural features and vis-NIR spectra of α-Fe_2O_3 hematite and AFe_2O_4 spinel oxides (A=Mg, Zn)*, J. Solid State Chem., 181 (2008), 1040–1047.
[634] G. Kämpf, *Korngrössenverteilung und optische Eigenschaften von Pigmenten*, Farbe und Lack, 71(5) (1965), 353–365.
[635] M. Dondi, G. Cruciani, G. Guarini, F. Matteucci, M. Raimondo, *The role of counterions (Mo, Nb, Sb, W) in Cr-, Mn-, Ni- and V-doped rutile ceramic pigments. Part 2. Colour and technological properties*, Ceram. Int., 32 (2006), 393–405.
[636] K. M. Glassford, J. R. Chelikowsky, *Structural and electronic properties of titanium dioxide*, Phys. Rev. B, 46 (1992), 1284–1298.
[637] F. Thomazi, L. Stolz Roman, A. Ferreira da Silva, C. Persson, *Optical absorption of rutile SnO_2 and TiO_2*, Phys. Status Solidi C, 6(12) (2009), 2740–2742.
[638] M. C. Comstock, *Complex Inorganic Colored Pigments: Comparison of options and relative properties when faced with elemental restrictions*, J. Surf. Coat. Aust. (2016), 10–30.
[639] C. L. Dong, C. Persson, L. Vayssieres, A. Augustsson, T. Schmitt, M. Mattesini, R. Ahuja, C. L. Chang, J.-H. Guo, *Electronic structure of nanostructured ZnO from x-ray absorption and emission spectroscopy and the local density approximation*, Phys. Rev. B, 70, 195325 (2004).
[640] A. Kobayashi, O. F. Sankey, J. D. Dow, *Deep energy levels of defects in the wurtzite semiconductors AlN, CdS, CdSe, ZnS and ZnO*, Phys. Rev. B, 28 (1983), 946–956.
[641] M. Rohlfing, P. Krüger, J. Pollmann, *Quasiparticle Band Structure of CdS*, Phys. Rev. Lett., 75(19) (1995), 3489–3492.
[642] V. E. Henrich, *The surfaces of metal oxides*, Rep. Progr. Phys., 48 (1985), 1481–1541.
[643] S.-R. Sun, Y.-H. Dong, *The optical properties of HgS under high pressures*, Solid State Commun., 138 (2006), 476–479.
[644] E. Doni, L. Resca, S. Rodriguez, W. M. Becker, *Electronic energy levels of cinnabar (α-HgS)*, Phys. Rev. B, 20 (1979), 1663–1668.
[645] X. Zhong, Y. Feng, W. Knoll, M. Han, *Alloyed $Zn_xCd_{1-x}S$ Nanocrystals with Highly Narrow Luminescence Spectral Width*, J. Am. Chem. Soc., 125 (2003), 13559–13563.
[646] X. Zhong, M. Han, Z. Dong, T. J. White, W. Knoll, *Composition-Tunable $Zn_xCd_{1-x}Se$ Nanocrystals with High Luminescence and Stability*, J. Am. Chem. Soc., 125 (2003), 8589–8594.
[647] X. Zhong, Y. Feng, *New strategy for band-gap tuning in semiconductor nanocrystals*, Res. Chem. Intermed., 34(2–3) (2008), 287–298.
[648] K. Glassford, J. Chelikowsky, *Electronic Structure of TiO_2:Ru*, Phys. Rev. B, 47(19) (1993), 12550–12553.
[649] J. C. Phillips, *The fundamental optical spectra of solids*, Solid State Phys., 18 (1966), 55–164.
[650] R. A. Evarestov, V. A. Veryazov, *The Electronic Structure of Crystalline Lead Oxides. I. Crystal Structure and LUC-CNDO Calculations*, Phys. Status Solidi (b), 165 (1991), 401–410.
[651] D. L. Perry, T. J. Wilkinson, *Synthesis of High-Purity alpha-and beta-PbO and Possible Applications to Synthesis and Processing of Other Lead Oxide Materials*, Lawrence Berkeley National Laboratory, 2011, https://escholarship.org/uc/item/5xb2g68h, accessed on 2022-06-20.
[652] G. Trinquier, R. Hoffmann, *Lead Monoxide. Electronic Structure and Bonding*, J. Phys. Chem., 88 (1984), 6696–6711.

[653] G. A. Bordovskii, M. L. Gordeev, A. N. Ermoshkin, V. A. Izvozchikov, R. A. Evarestov, *Energy Band Structure of Rhombic Lead Monoxide*, Phys. Status Solidi (b), 111 (1982), K123–K127.

[654] G. A. Bordovskii, M. L. Gordeev, A. N. Ermoshkin, V. A. Izvozchikov, R. A. Evarestov, *Energy Band Structure of Rhombic Lead Monoxide*, Phys. Status Solidi (b), 115 (1983), K15–K19.

[655] J. Robertson, *Electronic structure of SnO_2, GeO_2, PbO_2, TeO_2 and MgF*, J. Phys. C, Solid State Phys., 12 (1979), 4767.

[656] G. Pfaff und andere, *Rare earth metal sulfide pigments useful in paint, lacquer, printing ink, plastics and cosmetics*, Deutsches Patent DE19810317A1, https://patents.google.com/patent/DE19810317A1/en?oq=DE19810317A1, accessed on 2022-04-25.

[657] G. Pfaff und andere, *SULPHIDE AND OXYSULPHIDE PIGMENTS*, WIPO Patent Application WO/1999/046336, https://www.sumobrain.com/patents/wipo/Sulphide-oxysulphide-pigments/WO1999046336.html, accessed on 2022-04-25.

[658] T. Corbiere, Preparation and Methods of Characterisation of the Barium Copper Silicates $BaCuSi_4O_{10}$ (Chinese Blue), $BaCuSi_2O_6$ (Chinese Purple), $Ba_2CuSi_2O_7$, $BaCu_2Si_2O_7$ Used as Blue and Purple Pigments, Dissertation, Universität Zürich, 2009.

[659] C. Burda, X. Chen, R. Narayanan, M. A. El-Sayed, *Chemistry and Properties of Nanocrystals of Different Shapes*, Chem. Rev., 105 (2005), 1025–1102.

[660] E. N. Economou, K. L. Ngai, *Surface plasma oscillations and related surface effects in solids*, Adv. Chem. Phys., 27 (1974), 265–354.

[661] S. Link, M. A. El-Sayeda, *Shape and size dependence of radiative, non-radiative and photothermal properties of gold nanocrystals*, Int. Rev. Phys. Chem., 19(3) (2000), 409–453.

[662] S. Link, M. A. El-Sayeda, *Optical Properties and Ultrafast Dynamics of Metallic Nanocrystals*, Annu. Rev. Phys. Chem., 54 (2003), 331–366.

[663] M. A. Garcia, *Surface plasmons in metallic nanoparticles: fundamentals and applications*, J. Phys. D, Appl. Phys., 44(28) 283001 (2011), ISSN 0022-3727.

[664] E. F. Wesp, W. R. Brode, *The Absorption Spectra of Ferric Compounds. I. The Ferric Chloride-Phenol Reaction*, J. Am. Chem. Soc., 56 (1934), 1037–1042.

[665] W. Mayer, E. H. Hoffmann, N. Losch, H. Wow, B. Wolter, G. Schilling, *Dehydrierungsreaktionen mit Gallussäureestern*, Justus Liebigs Ann. Chem. (1984), 929–938.

[666] L. Sommer, *Über die genaue spektrophotometrische Eisen(III)- und Titan(IV)-Bestimmung mit Polyphenolen und verwandten Verbindungen*, Acta Chim. Acad. Sci. Hung., 33 (1962), 23–30.

[667] L. Sommer, *Über einige analytische Reaktionen der Polyphenole, Fresenius'*, Z. Anal. Chem., 187(1) (1962), 7–16.

[668] P. H. Gore, P. J. Newman, *Quantitative aspects of the colour reaction between iron(III) and phenols*, Anal. Chim. Acta, 31 (1964), 111–120.

[669] G. Ackermann, D. Hesse, *Über Eisen(III)-Komplexe mit Phenolen. I. Reaktion zwischen Eisen(III)-Ionen und Monophenolderivaten*, Z. Anorg. Allg. Chem., 367 (1969), 243–248.

[670] G. Ackermann, D. Hesse, *Über Eisen(III)-Komplexe mit Phenolen. II. Chelate mit Salicylsäure- und Brenzkatechinderivaten*, Z. Anorg. Allg. Chem., 368 (1969), 25–30.

[671] H. Kipton, J. Powell, M. C. Taylor, *Interactions of Iron(II) and Iron(III) with Gallic Acid and its Homologues: a Potentiormetric and Spectrophotometric Study*, Aust. J. Chem., 35 (1982), 739–756.

[672] R. C. Hider, A. Rahim Mohd-Nor, J. Silver, I. E. G. Morrison, L. V. C. Rees, *Model Compounds for Microbial Iron-transport Compounds. Part I. Solution Chemistry and Mössbauer Study of Iron(II) and Iron(III) Complexes from Phenolic and Catecholic Systems*, J. Chem. Soc., Dalton (1981), 609–622.

[673] R. C. Hider, B. Howlin, J. R. Miller, A. Rahim Mohd-Nor, J. Silver, *Model Compounds for Microbial Iron-transport Compounds. Part IV. Further Solution Chemistry and Mössbauer Studies on Iron(II) and Iron(III) Catechol Complexes*, Inorg. Chim. Acta, 80 (1983), 51–56.

[674] B. Howlin, A. Rahim Mohd-Nor, J. Silver, *Model Compounds for Microbial Iron-transport Compounds. Part V. Substituent Effects in the Catechol/FeCl$_3$ System*, Inorg. Chim. Acta, 91 (1984), 153–160.

[675] M. Brenes, C. Romero, P. Garcia, A. Garrido, *Effect of pH on the Colour Formed by Fe-Phenolic Complexes in Ripe Olives*, J. Sci. Food Agric., 67 (1995), 35–41.

[676] A. Krilov, R. Gref, *Mechanism of sawblade corrosion by polyphenolic compounds*, Wood Sci. Technol., 20 (1986), 369–375.

[677] L. J. Porter, K. R. Markham, *The Aluminium(III) Complexes of Hydroxy-flavones in Absolute Methanol. Part I. Ligands containing Only One Chelating Site*, J. Chem. Soc. (C) (1970), 344–349.

[678] L. J. Porter, K. R. Markham, *The Aluminium(III) Complexes of Hydroxy-flavones in Absolute Methanol. Part II. higands containing More than One Chelating Site*, J. Chem. Soc. (C) (1970), 1309–1313.

[679] M. Katyal, *Flavones as Analytical Reagents – a Review*, Talanta, 15 (1968), 95–106.

[680] M. Katyal, S. Prakash, *Analytical Reactions of Hydroxyflavones*, Talanta, 24 (1977), 367–375.

[681] L. Jurd, T. A. Geissman, *Absorption Spectra of Metal Complexes of Flavonoid Compounds*, J. Org. Chem., 21(12) (1956), 1395–1401.

[682] H. Deng, G. J. van Berkel, *Electrospray Mass Spectrometry and UV/Visible Spectrophotometry Studies of Complexes*, J. Mass Spectrom., 33 (1998), 1080–1087.

[683] S. H. Etaiw, M. M. Abou Sekkina, G. B. El-Hefnawey, S. S. Assar, *Spectral behaviour and solvent effects of some aminoanthraquinone dyes*, Can. J. Chem., 60 (1982), 304–307.

[684] R. H. Peters, H. H. Sumner, *Spectra of Anthraquinone Derivatives*, J. Chem. Soc. (1953), 2101–2110.

[685] H. Musso, *Die Trennung des Orceins in seine Komponenten (II. Mitteilung über Orceinfarbstoffe)*, Chem. Ber., 89 (1956), 1659–1673.

[686] H. Musso, C. Rajtjen, *Über Orceinfarbstoffe, X. Lichtabsorption und Chromophor des Lackmus*, Chem. Ber., 92 (1959), 751–753.

[687] H. Musso, *Zur Konstitution der Orceinfarbstoffe*, Angew. Chem., 69(5) (1957), 178.

[688] H. Musso, H. Beecken, *Über Orceinfarbstoffe, VI. Die Konstitution von α-, β- und γ-Amino-Orcein*, Chem. Ber., 90 (1957), 2190–2196.

[689] H. Musso, *Orcein- und Lackmusfarbstoffe. Konstitutionsermittlung und Konstitutionsbeweis durch die Synthese*, Planta Med., 8 (1960), 432–446.

[690] H. Beecken, E.-M. Gottschalk, U. v. Gizycki, H. Krämer, D. Maassen, H.-G. Matthies, H. Musso, C. Rathjen, U. I. Zahorszky, *Orcein und Lackmus*, Angew. Chem., 73(20) (1961), 665–673.

[691] H. Auterhoff, H. Frauendorf, W. Liesenklas, C. Schwandt, *Der Hauptbestandteil des Guttiharzes. 1. Mitt.: Chemie des Gummigutts*, Arch. Pharm., 295(11) (1962), 833–846.

[692] H. Auterhoff, W. Liesenklas, *Der Hauptbestandteil des Guttiharzes. 3. Mitt.: Chemie des Gummigutts*, Arch. Pharm., 299(1) (1966), 91–96.

[693] H. Auterhoff, W. Liesenklas, *Die Konstitution der Gambogasäure und ihre Isomerisierung. 4. Mitt.: Chemie des Gummigutts*, Arch. Pharm., 299(9) (1966), 797–798.

[694] W. D. Ollis, M. V. J. Ramsay, I. O. Sutherland, *The Constitution of Gambogic Acid*, Tetrahedron, 21 (1965), 1453–1470.

[695] L.-J. Lin, L.-Z. Lin, J. M. Pezzuto, G. A. Cordell, N. Ruangrungsi, *Isogambogic acid and Isomorellinol from Garcinia Hanbuyi*, Magn. Reson. Chem., 31(4) (1993), 340–347.

[696] J.-Z. Song, Y.-K. Yip, Q.-B. Han, C.-F. Qiao, H.-X. Xu, *Rapid determination of polyprenylated xanthones in gamboges resin of Garcinia hanburyi by HPLC*, J. Sep. Sci., 30 (2007), 304–309.

[697] A. López-Montes, R. Blanc García, T. Espejo, J. F. Huertas-Perez, A. Navalón, J. L. Vílchez, *Simultaneous identification of natural dyes in the collection of drawings and maps from The Royal Chancellery Archives in Granada (Spain) by CE*, Electrophoresis, 28 (2007), 1243–1251.

[698] K. Roth, *Fingerfarben, Ideal für kleine Künstler*, Chem. Unserer Zeit, 40 (2006), 260–267.

[699] H. Baumann, H. R. Hensel, *Neue Metallkomplexfarbstoffe, Struktur und färberische Eigenschaften*, Fortschr. Chem. Forsch., 7(4) (1967), 643–783.
[700] F. Beffa, G. Back, *Metal-complex Dyes for Wool and Nylon – 1930 to date*, Rev. Prog. Color., 14 (1984), 33–42.
[701] H. S. Freeman, L. C. Edwards, *Iron-Complexed Dyes: Colorants in Green Chemistry*, in P. Anastas et al. (eds.) *Green Chemical Syntheses and Processes*, ACS Symposium Series, pp. 18–32, American Chemical Society, Washington, DC, 2000.
[702] J. Sokolowska-Gajda, H. S. Freeman, A. Reife, *Synthetic Dyes Based on Environmental Considerations: Part I: Iron Complexes for Protein and Polyamide Fibers*, Tex. Res. J., 64(7) (1994), 388–396.
[703] Y. Yagi, *Studies of the Absorption Spectra of Azo Dyes and their Metal-Complexes. I. The Absorption Spectra of Phenylazo-acetoacetanilide and its Related Compounds*, Bull. Chem. Soc. Jpn., 36 (1963), 487–492, *II. The Absorption Spectra of Phenolazoacetoacetamide and its Derivatives*, Bull. Chem. Soc. Jpn., 36 (1963), 492–500, *III. The Electronic Absorption Spectra of Metal-Complexes Derived from Phenolazoacetoacetanilides*, Bull. Chem. Soc. Jpn., 36 (1963), 500–506, *IV. The Effects of Substituents on the Absorption Spectra of Chromium(III)- and Cobalt(III)-Complexes of Phenolazoacetoacetanilides*, Bull. Chem. Soc. Jpn., 36 (1963), 506–512, *V. The Electronic Absorption Spectra of Metal-complexed Phenolazoacetoacetbenzylamides*, Bull. Chem. Soc. Jpn., 36 (1963), 512–517, *VII. The Absorption Spectra of Metal-complexes Derived from Phenolazo-α- and Phenolazo-β-naphthols*, Bull. Chem. Soc. Jpn., 37 (1964), 1878–1880, *VIII. The Absorption Spectra of Azo Compounds Derived from 5, 8-Dichloro-1-naphthol and Their Metal-complexes*, Bull. Chem. Soc. Jpn., 37 (1964), 1881–1883.
[704] J. Griffiths, D. Rhodes, *The Prediction of Colour Change in Dye Equilibria. II – Metal Complexes of Ortho-Hydroxyazobenzenes*, J. Soc. Dyers Colour., 88 (1972), 400.
[705] P. S. Vankar, *Chemistry of Natural Dyes*, Resonance (2000), 73–80.
[706] E. G. Kiel, P. M. Heertjes, *Metal Complexes of Alizarin. I. The Structure of the Calcium-Aluminium-Lake of Alizarin*, J. Soc. Dyers Colour., 79 (1963), 21–27.
[707] E. G. Kiel, P. M. Heertjes, *Metal Complexes of Alizarin. II. The Structure of some Metal Complexes of Alizarin other than Turkey Red*, J. Soc. Dyers Colour., 79 (1963), 61–64.
[708] E. G. Kiel, P. M. Heertjes, *Metal Complexes of Alizarin. IV. The Structure of the Potassium and Calcium Salts of Alizarin and of 3-Nitroalizarin*, J. Soc. Dyers Colour., 79 (1963), 363–367.
[709] E. G. Kiel, P. M. Heertjes, *Metal Complexes of Alizarin. V. Investigations of Alizarin-dyed Cotton Fabrics*, J. Soc. Dyers Colour., 81 (1965), 98–102.
[710] C.-H. Wunderlich, G. Bergerhoff, *Konstitution und Farbe von Alizarin- und Purpurin-Farblacken*, Chem. Ber., 127 (1994), 1185–1190.
[711] C.-H. Wunderlich, *Krapplack und Türkischrot*, Dissertation, Universität Bonn, 1993.
[712] P. Soubayrol, G. Dana, *Aluminium-27 Solid-state NMR Study of Aluminium Coordination Complexes of Alizarin*, Magn. Reson. Chem., 34 (1996), 638–645.
[713] H. Harms, *Zur Chemie des Karmins und seiner in der Mikroskopie verwendeten Lösungen*, Naturwiss., 44(11) (1957), 327.
[714] S. N. Meloan, L. S. Valentine, H. Puchtler, *On the Structure of Carminic Acid and Carmine*, Histochem., 27 (1971), 87–95.
[715] C. Clementi, et al., *Vibrational and electronic properties of painting lakes*, Appl. Phys. A, 92 (2008), 25–33.
[716] R. A. Dixon, D.-Y. Xie, S. B. Sharma, *Proanthocyanidins – a final frontier in flavonoid research? New Phytol.*, 165 (2005), 9–28.
[717] L. Pourcel, J.-M. Routaboul, V. Cheynier, L. Lepiniec, I. Debeaujon, *Flavonoid oxidation in plants: from biochemical properties to physiological functions*, Trends Plant Sci., 12(1) (2007), 29–36.

[718] F. He, Q.-H. Pan, Y. Shi, C.-Q. Duan, *Biosynthesis and Genetic Regulation of Proanthocyanidins in Plants*, Molecules, 13 (2008), 2674–2703.
[719] C. Santos-Buelga, A. Scalbert, *Proanthocyanidins and tannin-like compounds – nature, occurrence, dietary intake and effects on nutrition and health*, J. Sci. Food Agric., 80 (2000), 1094–1117.
[720] H. Musso, *Phenol Oxidation Reactions*, Angew. Chem., Int. Ed. Engl., 2(12) (1963), 723–735.
[721] R. M. Desentis-Mendoza, H. Hernández-Sánchez, A. Moreno, E. R. del C, L. Chel-Guerrero, J. Tamariz, M. E. Jaramillo-Flores, *Enzymatic Polymerization of Phenolic Compounds Using Laccase and Tyrosinase from Ustilago maydis*, Biomacromolecules, 7 (2006), 1845–1854.
[722] S. Guyot, J. Vercauteren, V. Cheynier, *Structural Determination of Colourless and Yellow Dimers resulting from (+)-Catechin Coupling Catalysed by Grape Polyphenoxidase*, Phytochemistry, 42(5) (1996), 1279–1288.
[723] R. Niemetz, G. G. Gross, *Enzymology of gallotannin and ellagitannin biosynthesis*, Phytochemistry, 66 (2005), 2001–2011.
[724] P. Grundhöfer, R. Niemetz, G. Schilling, G. G. Gross, *Biosynthesis and subcellular distribution of hydrolyzable tannins*, Phytochemistry, 57 (2001), 915–927.
[725] V. de Freitas, N. Mateus, *Chemical transformations of anthocyanins yielding a variety of colours (Review)*, Environ. Chem. Lett., 4 (2006), 175–183.
[726] T. Shoji, *Polyphenols as Natural Food Pigments: Changes during Food Processing*, Amer. J. Food Technol., 2(7) (2007), 570–581.
[727] T. C. Somers, *The Polymeric Nature of Wine Pigments*, Phytochemistry, 10 (1971), 2175–2186.
[728] G.-I. Nonaka, O. Kawahara, I. Nishioka, *Tannins and Related Compounds. XV. A New Class of Dimeric Flavan-3-ol Gallates, Theasinensins A and B, and Proanthocyanidin Gallates from Green Tea Leaf. (1)*, Chem. Pharm. Bull., 31(11) (1983), 3906–3914.
[729] E. Haslam, *Thoughts on thearubigins*, Phytochemistry, 64 (2003), 61–73.
[730] Y. Wang, C.-T. Ho, *Functional Contribution of Polyphenols in Black Tea*, in N. Da Costa et al. (eds.) *Flavors in Noncarbonated Beverages*, ACS Symposium Series, American Chemical Society, Washington, DC, 2010.
[731] H. Li, A. Guo, H. Wang, *Mechanisms of oxidative browning of wine*, Food Chem., 108 (2008), 1–13.
[732] B. Winkel-Shirley, *Flavonoid biosynthesis. A colorful model for genetics, biochemistry, cell biology, and biotechnology*, Plant Physiol., 126 (2001), 485–493.
[733] C.-T. Ho, S. Sang, J. W. Jhoo, *Chemistry of Theaflavins: The Astringent Taste Compounds of Black Tea*, in T. Hofmann (ed.) *Challenges in Taste Chemistry and Biology*, ACS Symposium Series, American Chemical Society, Washington, DC, 2003.
[734] C. F. Timberlake, *Anthocyanins – Occurence, Extraction and Chemistry*, Food Chem., 5 (1980), 69–80.
[735] K. Robards, P. D. Prenzler, G. Tucker, P. Swatsitang, W. Glover, *Phenolic compounds and their role in oxidative processes in fruits*, Food Chem., 66 (1999), 401–436.
[736] N. G. Irani, J. M. Hernandez, E. Grotewold, *Regulation of Anthocyanin Pigmentation*, Recent Adv. Phytochem., 38 (2003), 59–78.
[737] G. Forkmann, *Flavonoids as Flower Pigments: The Formation of the Natural Spectrum and its Extension by Genetic Engineering*, Plant Breed., 106 (1991), 1–26.
[738] R. W. Hemingway, P. E. Laks, *Plant Polyphenols*, Plenum Press, New York, 1992.
[739] G. G. Gross, *Enzymes in the Biosynthesis of Hydrolyzable Tannins*, in [738], pp. 43–60.
[740] E. Haslam, *Gallic Acid and its Metabolites*, in [738], pp. 169–194.
[741] J. Nicolas, V. Cheynier, A. Fleuriet, M.-A. Rouet-Mayer, *Polyphenols and Enzymatic Browning*, in A. Scalbert (ed.) *Polyphenolic Phenomena*, pp. 165–175, Institut Nationale de la Recherche Agronomique, Paris, 1993.
[742] H. Musso, *Phenol Coupling*, in [296], chapter 1, pp. 1–94.

[743] A. I. Scott, *Some Natural Products Derived by Phenol Oxidation*, in [296], chapter 2, pp. 95–117.
[744] B. R. Brown, *Biochemical Aspects of Oxidative Couplings of Phenols*, in [296], chapter 4, pp. 167–201.
[745] A. J. Taylor, F. M. Clydesdale, *Potential of Oxidised Phenolics as Food Colourants*, Food Chem., 24 (1987), 301–313.
[746] G. Jungbluth, *Wechselwirkung von Polyphenolen mit* $Cu^{2\oplus}$, $Fe^{3\oplus}$, $Fe^{2\oplus}$ *und* $A1^{3\oplus}$. *Analyse der Reaktionsprodukte mittels HPLC*, Dissertation, Universität Hannover, 2000.
[747] P. S. Vankar, D. Shukla, *Spectrum of colors from reseda luteola and other natural yellow dyes*, J. Text. Eng. Fash. Technol., 4(2) (2018), 106–119.
[748] N. Kuhnert, *Unraveling the structure of the black tea thearubigins*, Arch. Biochem. Biophys., 501 (2010), 37–51.
[749] N. Kuhnert, J. W. Drynan, J. Obuchowicz, M. N. Clifford, M. Witt, *Mass spectrometric characterization of black tea thearubigins leading to an oxidative cascade hypothesis for thearubigin formation*, Rapid Commun. Mass Spectrom., 24 (2010), 3387–3404.
[750] H. Obara, J. Onodera, *Structure of Carthamin*, Chem. Lett. (1979), 201–204.
[751] Y. Takahashi, et al., *Constitution of two coloring matters in the flower petals of C. tinctorius*, Tetrahedron, 23 (1982), 5163–5166.
[752] K. Kazuma, et al., *Quinochalcones and Flavonoids from Fresh Florets in Different Cultivars of C. tinctorius L*, Biosci. Biotechnol. Biochem., 64(8) (2000), 1588–1599.
[753] G. Britton, D. Hornero-Méndez, *Carotenoids and colour in fruit and vegetables*, in [300], chapter 2, pp. 11–27.
[754] R. Brouillard, P. Figueiredo, M. Elhabiri, O. Dangles, *Molecular interactions of phenolic compounds in relation to the colour of fruit and vegetables*, in [300], chapter 3, pp. 29–49.
[755] M. J. Amiot, A. Fleuriet, V. Cheynier, J. Nicolas, *Phenolic compounds and oxidative mechanism in fruit and vegetables*, in [300], chapter 4, pp. 51–85.
[756] J. M. Harkin, *Lignin – A Natural Polymeric Product of Phenol Oxidation*, in [296], chapter 6, pp. 243–321.
[757] K. Weinges, R. Spänig, *Lignans and Cyclolignans*, in [296], chapter 7, pp. 323–355.
[758] B. B. Adhyaru, N. G. Akhmedov, A. R. Katritzky, C. R. Bowers, *Solid-state cross-polarization magic angle spinning 13C and 15N NMR characterization of Sepia melanin, Sepia melanin free acid and Human hair melanin in comparison with several model compounds*, Magn. Reson. Chem., 41 (2003), 466–474.
[759] A. Lopez-Montes, R. Blanc, T. Espejo, A. Navalon, J. L. Vilchez, *Characterization of sepia ink in ancient graphic documents by capillary electrophoresis*, Microchem. J., 93 (2009), 121–126.
[760] M. G. Peter, *Chemische Modifikation von Biopolymeren durch Chinone und Chinonmethide*, Angew. Chem., 101 (1989), 572–587.
[761] R. H. Thomson, *Pigmente aus rötlichen Haaren und Federn*, Angew. Chem., 86 (1974), 355–386.
[762] M. Piattelli, R. A. Nicolaus, *The structure of melanins and melanogenesis – I. The structure of melanin in Sepia*, Tetrahedron, 15 (1961), 66–75.
[763] J. J. Nordlund (ed.), *The Pigmentary System: Physiology and Pathphysiology*, 2nd ed., Oxford University Press, 2006. ISBN 978-1-4051-2034-0.
[764] S. Ito, K. Wakamatsu, *Chemistry of Melanins*, in [763], pp. 282–310.
[765] M. L. Wolbarsht, A. W. Walsh, G. George, *Melanin, a unique biological absorber*, Appl. Opt., 20(13/1) (1981), 2184–2186.
[766] M. G. Peter, *Chemische Modifikation von Biopolymeren durch Chinone und Chinonmethide*, Angew. Chem., 101 (1989), 572–587.
[767] A. Vetter, *Bereitstellung von Naturfarbstoffen aus Färberpflanzen*, 2004, Abschlussbericht des Projektes Projekt/Förderkennzeichen (FKZ): 00NR185 der Thüringer Landesanstalt für Landwirtschaft.

[768] U. Kynast, Vorlesungsscript *Optische Eigenschaften der Materialien II: Farbe und Pigmente*, FH Münster, https://www.fh-muenster.de/fb1/downloads/personal/VLGMWFarbPig1.pdf.
[769] H. Michaelsen, A. Unger, C.-H. Fischer, *Blaugrüne Färbung an Intarsienhölzern des 16. bis 18. Jahrhunderts*, Restauro, 1 (1992), 17–25.
[770] P. Kopp, H. Piening, *Wiederentdeckte Farbigkeit der Renaissance*, Restauro, 2 (2009), 107–111.
[771] G. Buschle-Diller, A. Unger, *Schreib- und Malmittel auf Papyrus*, Restauro, 2 (1996), 108–114.
[772] C. Jenkins, *Analyzing Pigments in the Book of the Dead Using XRF Spectroscopy*, Blog-Beitrag vom 26.01.2011, Brooklyn Museum https://www.brooklynmuseum.org/community/blogosphere/2011/01/26/analyzing-pigments-in-the-book-of-the-dead-using-xrf-spectroscopy/, accessed on 2022-06-20.
[773] O. Wächter, *Die Restaurierung und Konservierung von Miniaturen*, Beitrag zum IADA-Kongress 1975 Kopenhagen, https://cool.culturalheritage.org/iada/ta75_047.pdf.
[774] A. von Bohlen, P. Vandenabeerle, M. De Reu, L. Moens, R. Klockenkämper, B. Dekeyzer, B. Cardon, *Pigmente und Tinten in mittelalterlichen Handschriften*, Restauro, 2 (2003), 118–122.
[775] P. Gregory, *Colorants for Electronic Printers*, in [152], pp. 150–167.
[776] R. B. McKay, *Control of the Physical Character and Performance of Organic Pigments for Inks*, in [152], pp. 107–126.
[777] R. Fraas, *Aqueous ink with prolonged cap-off time and method of making it*, Europäisches Patent EP0897961, https://patents.google.com/patent/EP0897961B1/en?oq=EP0897961, accessed on 2022-06-10.
[778] H. Asami, A. Owariasahi-shi, *Wässerige Kugelschreiber-Tintezusammensetzung*, Deutsches Patent DE69910670T2, https://patents.google.com/patent/DE69910670T2/de, accessed on 2022-06-10.
[779] EuPIA, European Printing Ink Association, https://www.eupia.org/, Onlineinformationen zu Druckfarben, *EXCLUSION LIST FOR PRINTING INKS AND RELATED PRODUCTS*, 4th ed., 2021, https://www.eupia.org/fileadmin/Documents/Our_commitment/20210310_-Exclusion_Policy_for_Printing_Inks_and_Related_Products_final_March_2021.pdf, accessed on 2022-06-20.
[780] T. Brüning, H. U. Käfferlein, A. Slowicki, *Azofarbmittel und deren Hautgängigkeit beim Menschen*, BGFA-Report 2, Februar 2009, BGFA – Forschungsinstitut für Arbeitsmedizin der Deutschen Gesetzlichen Unfallversicherung, Institut der Ruhr-Universität Bochum, ISSN 1867-9358, https://dguv.de/medien/ipa/publikationen/ipa-reporte/09-02-27_bgfa-report2_azofarbstoffe.pdf, accessed on 2022-06-20.
[781] M. Beth-Hübner, B. Brandt, R. Rupp, V. Neumann, U. Eickmann, G. Lindner, et al., *Aromatische Amine – Eine Arbeitshilfe in Berufskrankheiten-Ermittlungsverfahren (BK-Report 1/2019)*, Deutsche Gesetzliche Unfallversicherung e. V., Berlin, 2019. ISBN 978-3-948657-03-1, https://publikationen.dguv.de/widgets/pdf/download/article/3520, accessed on 2022-05-29.
[782] H. S. Freeman, *Aromatic amines: use in azo dye chemistry*, Front. Biosci., 18 (2013), 145–164.
[783] P. Sagelsdorff, R. Haenggi, B. Heuberger, R. Joppich-Kuhn, R. Jung, H. J. Weideli, M. Joppich, *Lack of bioavability of dichlorobenzidine from diarylide azo pigments: molecular dosimetry for hemoglobin and DNA adducts*, Carcinogenesis, 17(3) (1996), 507–514.
[784] K. Golka, S. Kopps, Z. W. Myslak, *Carcinogenicity of azo colorants: influence of solubility and bioavailability*, Toxicol. Lett., 15(1) (2004), 203–210.
[785] F. Nakaya, Y. Fukase, *Laser Printer*, in N. Ohta, M. Rosen (eds.) *Color Desktop Printer Technology*, pp. 157–192, CRC Taylor and Francis, 2006. ISBN 0-8247-5364-X.
[786] H.-T. Macholdt, *Organische Pigmente für Photokopierer und Laserdrucker*, Chem. Unserer Zeit, 24(4) (1990), 176–181.
[787] P. Gregory, *Colorants for High Technology*, in A. T. Peters, H. S. Freeman (eds.) *Colour Chemistry: The Design and Synthesis of Organic Dyes and Pigments*, pp. 193–223, Elsevier, New York, 1991. ISBN 978-1851665778.

[788] B. E. Springett, *A Brief Introduction to Electrophotography*, in [304], chapter 4, pp. 145–171.
[789] P. C. Julien, R. J. Gruber, *Dry Toner Technology*, in [304], chapter 5, pp. 173–208.
[790] L. O. Jones, *Carrier Materials for Imaging*, in [304], chapter 6, pp. 209–238.
[791] J. R. Larson, G. A. Gibson, S. P. Schmidt, *Liquid Toner Materials*, in [304], chapter 7, pp. 239–264.
[792] G. Banik, H. Weber, *Tintenfrassschäden und ihre Behandlung, Werkhefte der staatlichen Archivverwaltung Baden-Württemberg, Heft 10*, Verlag W. Kohlhammer, Stuttgart, 1999. ISBN 3-17-015377-3.
[793] C. Krekel, *Chemische Struktur historischer Eisengallustinten*, in [792], pp. 25–36.
[794] C. Krekel, *Chemische Untersuchungen an Eisengallustinten und Anwendung der Ergebnisse bei der Begutachtung einer mittelalterlichen Handschrift*, Diplomarbeit Universität Göttingen, 1990.
[795] C.-H. Wunderlich, *Geschichte und Chemie der Eisengallustinte*, Restauro, 100 (1994), 414–421.
[796] R. Fuchs, *Der Tintenfrass historischer Tinten und Tuschen – ein komplexes, nie enden wollendes Problem*, in [792], pp. 37–75.
[797] C.-H. Wunderlich, R. Weber, G. Bergerhoff, *Über Eisengallustinte*, Z. Anorg. Allg. Chem., 598/599 (1991), 371–376.
[798] F. Zetsche, G. Vieli, G. Lilljeqvist, A. Loosli, *Bildung und Altern der Schriftzüge, die primären Tintensalze der Eisentinten*, Justus Liebigs Ann. Chem., 435 (1924), 233–264.
[799] F. Zetsche, A. Loosli, *Bildung und Altern der Schriftzüge, 2. Mitteilung*, Liebigs Ann. Chem., 445 (1926), 283–296.
[800] A. S. Lee, P. J. Mahon, D. C. Creagh, *Raman analysis of iron gall inks on parchment*, Vib. Spectrosc., 41 (2006), 170–175.
[801] J. G. Neevel, *Phytate: a Potential Conservation Agent for the Treatment of Ink Corrosion Caused by Irongall Inks*, Restaurator, 16 (1995), 143–160.
[802] G. Farusi, *Monastic Inks: Linking Chemistry and History*, Sci. School, 6 (2007), 36–40.
[803] G. Pfingstag, *Colorants in Inks for Writing, Drawing and Marking*, JSDC, 109 (1993), 188–192.
[804] W. J. Wnek, M. A. Andreottola, P. F. Doll, S. M. Kelly, *Ink Jet Ink Technology*, in [304], chapter 14, pp. 531–602.
[805] D. E. Bugner, *Papers and Films for Ink Jet Printing*, in [304], chapter 15, pp. 603–627.
[806] E. Kuckert, *Tinten für Ink-Jet Verfahren*, presentation Bayer AG 1999, http://www.f07.fh-koeln.de/einrichtungen/imp/forschung_und_gestaltung/showroom/00161/index.html, accessed on 2011-06-10, no longer available.
[807] M. C. Jürgens, *Preservation of Ink Jet Hardcopies. An Investigation by Martin C. Jürgens for the Capstone Project, Cross-Disciplinary Studies, at Rochester Institute of Technology*, Rochester, NY, August 27, 1999, https://www.ica.org/sites/default/files/WG_1999_PAAG-preservation-of-ink-jet-hardcopies_EN.pdf, accessed on 2022-06-20.
[808] W. Li, J. Danio, *Fundamentals in the Development of High Performance Inkjet Receptive Coatings*, TAPPI Conference Paper, 2008 Advanced Coating Fundamentals Symposium, Product Code 08ADV16.
[809] E. Svanholm, *Printability and Ink-Coating Interactions in Inkjet Printing*, Dissertation, Karlstad University Studies 2007:2, http://www.diva-portal.org/smash/get/diva2:6256/FULLTEXT01.pdf, accessed on 2022-06-20.
[810] T. Lamminmäki, J. Kettle, P. Puukko, J. Ketoja, P. Gane, *The role of binder type in determining inkjet print quality*, Nord. Pulp Pap. Res. J., 25(3) (2010), 380–390.
[811] D. L. Briley, *Polyester Media Development for Inkjet Printers*, Hewlett-Packard J. (1994), 28–34.
[812] R. W. Kenyon, *Beyond the Black – Novel High Waterfast Dyes for Colour Ink Jet Printing*, in I. Rezanka, R. Eschbach (eds.) *Recent Progress in Ink Jet Technologies*, pp. 278–280, Society for Imaging Science and Technology, 1996. ISBN 0-89208-192-9.

[813] W. Bauer, J. Ritter, *Novel Dyes for Ink Jet Applications*, in I. Rezanka, R. Eschbach (eds.) *Recent Progress in Ink Jet Technologies*, pp. 295–296, Society for Imaging Science and Technology, 1996. ISBN 0-89208-192-9.
[814] K. Lussky, P. Engel, S. Pentzien, J. Krüger, *Analyse häufig verwendeter europäischer Papiere*, Restauro, 7 (2007), 472–478.
[815] Application information of the Competence Center Specialities (Lanxess), *Anorganische Pigmente für die Papierindustrie*, Januar 2006, and *Buntpigmente zur Einfärbung von Dekorpapier*, edition 09/2008.
[816] A. Pingel Keuth, *Von Zellstoff zu Filtertüte, Schreibpapier,...Papierproduktion*, Chem. Unserer Zeit, 39 (2005), 402–409.
[817] K. Roth, *Chemie kontra Papierzerfall*, Chem. Unserer Zeit, 40 (2006), 54–62.
[818] W. Griebenow, *Alterserscheinungen bei Papier – vorwiegend aus chemischer Sicht, Teil 1*, Restauro, 5 (1991), 329–335.
[819] W. Griebenow, *Alterserscheinungen bei Papier – vorwiegend aus chemischer Sicht, Teil 2*, Restauro, 6 (1991), 409–415.
[820] E. Gruber, *Papier-Chemie*, Scriptum der Dualen Hochschule Karlsruhe für den Lehrgang Papiertechnologie, http://www.gruberscript.net/papierchemie.html, accessed on 2022-06-10.
[821] E. Bojaxhiu, *Untersuchung der Abbauprodukte der chemischen und elektrochemischen Reduktion ausgewählter Dispersionsfarbstoffe mittels LC- und GCxGC-(TOF)MS*, Dissertation, Wuppertal 2008, http://elpub.bib.uni-wuppertal.de/servlets/DerivateServlet/Derivate-896/dc0811.pdf, accessed on 2022-06-10.
[822] *Water Soluble Polymers*, Brochure by SNF, https://www.snf.com/wp-content/uploads/2019/12/Water-Soluble-Polymers-EN.pdf, accessed on 2022-06-10.
[823] B. Kießler, *Verbesserung der Papierfestigkeiten und der Prozesssicherheit bei der Oberflächenbehandlung von Wellpappenrohpapieren durch spezifischen Stärkeaufschluss*, research report to project InnoWatt-Projekt IW 050275 by Papiertechnischen Stiftung PTS, http://www.ptspaper.de/fileadmin/PTS/Dokumente/Forschung/Forschungsprojekte/IW_050275.pdf, no longer accessible.
[824] M. Suhr, G. Klein, I. Kourti, M. R. Gonzalo, G. G. Santonja, S. Roudier, L. D. Sancho, *Best Available Techniques (BAT) Reference Document for the Production of Pulp, Paper and Board*, European Union, 2015, https://op.europa.eu/en/publication-detail/-/publication/fcaab1a5-d287-40af-b21c-115e529685fc/language-en/format-PDF, accessed on 2022-06-02.
[825] *MedienStandard Druck 2018. Technische Richtlinien für Daten, Prüfdruck und Auflagendruck*, Bundesverband Druck und Medien e. V. (bvdm), 2016–2018, Art.-Nr. 86035, https://www.bvdm-online.de/fileadmin/Themen/T_F/R_H/MedienStandard_Druck_2018.pdf, accessed on 2022-06-18.
[826] *Paper Categorization Meeting ICC/ISO TC130*, meeting minutes, ICC DevCon '06, Leeds, 15.06.2006, https://www.color.org/papermeetingminutes.pdf, accessed on 2022-06-18.
[827] D. Simmert, *Acrylharzkünstlerfarben, Kunsttechnologie und Konservierung*, pp. 78–105, Wernersche Verlagsgesellschaft, Worms, 1995.
[828] R. E. Harren, *History and Development of Acrylic Latex Coatings*, in R. B. Seymour, H. F. Mark (eds.) *Organic Coatings: Their Origin and Development*, Proceedings of the International Symposium on the History of Organic Coatings, pp. 297–314, Elsevier, 1990.
[829] R. N. Hildred, *Acrylate als Additive für wäßrige Beschichtungsstoffe*, farbe+lack, 96 (1990), 857–859.
[830] E. S. Daniels, A. Klein, *Development of cohesive strength in polymer films from latices: effect of polymer chain interdiffusion and crosslinking*, Prog. Org. Coat., 19 (1991), 359–378.
[831] P.-B. Eipper, *Problemfall Acrylfarbenoberflächen. Teil 1: Hintergründe zur Beständigkeit von Acrylfarben*, Restauro, 1 (2009), 30–35.

[832] P.-B. Eipper, *Problemfall Acrylfarbenoberflächen. Teil 2: Schäden und Restaurierungsmöglichkeiten*, Restauro, 2 (2009), 112–122.

[833] P.-B. Eipper, *Problemfall Acrylfarbenoberflächen. Teil 3: Untersuchungen von Veränderungen nach einer Oberflächenreinigung und Entwicklung eines Reinigungsablaufs*, Restauro, 4 (2009), 244–251.

[834] E. Jablonski, T. Learner, J. Hayes, M. Golden, *Conservation Concerns for Acrylic Emulsion Paints: A Literature Review*, Tate Papers, TATE'S ONLINE RESEARCH JOURNAL, ISSN 1753-9854, Autumn 2004, https://www.tate.org.uk/research/tate-papers/02/conservation-concerns-for-acrylic-emulsion-paints-literature-review, accessed on 2022-06-10.

[835] M. Golden, *Mural Paints: Current and Future Formulations*, The Getty Conservation Institute, Los Angeles, 2004, https://www.getty.edu/conservation/publications_resources/pdf_publications/pdf/golden.pdf, accessed on 2022-06-20.

[836] A. van Tent, K. te Nijenhuis, *The Film Formation of Polymer Particles in Drying Thin Films of Aqueous Acrylic Latices. II. Coalescence, Studied with Transmission Spectrophotometry*, J. Colloid Interface Sci., 232 (2000), 350–363.

[837] F. Dobler, Y. Holl, *Mechanisms of Particle Deformation During Latex Film Formation*, in *T. Provder et al., Film Formation in Waterborne Coatings*, ACS Symposium Series, American Chemical Society, Washington, DC, 1996.

[838] J. L. Keddie, *Film formation of latex*, Mater. Sci. Eng., 21 (1997), 101–170.

[839] M. A. Winnik, *Latex film formation*, Curr. Opin. Colloid Interface Sci., 2(2) (1997), 192–199.

[840] P. A. Steward, J. Hearn, M. C. Wilkinson, *An overview of polymer latex film formation and properties*, Adv. Colloid Interface Sci., 86(3) (2000), 195–267.

[841] R. E. Dillon, L. A. Matheson, E. B. Bradford, *Sintering of Synthetic Latex Particles*, Colloid Sci., 6 (1951), 108.

[842] G. L. Brown, *Formation of Films from Polymer Dispersions*, J. Polym. Sci., 22 (1956), 423–434.

[843] D. P. Sheetz, E. C. Britton, *Formation of Films by Drying of Latex*, J. Appl. Polym. Sci., 9(11) (1965), 3759–3773.

[844] F. Dobler, T. Pith, M. Lambla, Y. Holl, *Coalescence Mechanisms of Polymer Colloids. I. Coalescence under the Influence of Particle-Water Interfacial Tension*, J. Colloid Interface Sci., 152(1) (1992).

[845] P. A. Lovell, M. S. El-Aasser, *Emulsion Polymerization and Emulsion Polymers*, John Wiley & Sons, 1997. ISBN 0-471-96746-7.

[846] M. A. Winnik, *The Formation and Properties of Latex Films*, in [845], pp. 467–518.

[847] P. M. Lesko, P. R. Sperry, *Acrylic and Styrene-Acrylic Polymers*, in [845], pp. 619–655.

[848] C. Heyn, *Mikrobieller Angriff auf synthetische Polymere*, Dissertation 2002, Universität Oldenburg, https://oops.uni-oldenburg.de/181/, accessed on 2022-06-20.

[849] M. Laschet, *Rheo-mechanische und rheo-optische Charakterisierung wäßriger Lösungen von Hydroxyethylcellulosen und deren hydrophob modifizierter Derivate im Hinblick auf supramolekulare Strukturen*, Dissertation 2002, Universität Hamburg.

[850] J. E. Glass (ed.), *Associative Polymers in Aqueous Media*, ACS Symposium Series, vol. 765, Washington D. C., 2000. ISBN 0-8412-3659-3.

[851] V. Tirtaatmadja, K. C. Tam, R. D. Jenkins, *Rheological Properties of Model Alkali-Soluble Associative (HASE) Polymers: Effect of Varying Hydrophobe Chain Length*, Macromolecules, 30 (1997), 3271–3282.

[852] A. A. Abdala, K. Olesen, S. A. Khan, *Solution rheology of hydrophobically modified associative polymers: Solvent quality and hydrophobic interactions*, J. Rheol., 47(2) (2003), 497–511.

[853] R. J. English, S. R. Raghavan, R. D. Jenkins, S. A. Khan, *Associative polymers bearing n-alkyl hydrophobes: Rheological evidence for microgel-like behavior*, J. Rheol., 43(5) (September/October 1999), 1175–1194.

[854] R. J. English, H. S. Gulati, R. D. Jenkins, S. A. Khan, *Solution rheology of a hydrophobically modified soluble associative polymer*, J. Rheol., 41(2) (March/April 1997), 427–444.
[855] H. J. Spinelli, *Polymeric Dispersants in Ink Jet Technology*, Adv. Mater., 10(15) (1998), 1215–1218.
[856] C. M. Miller, K. R. Olesen, G. D. Shay, *Determination of the Thickening Mechanism of a Hydrophobically Modified Alkali Soluble Emulsion Using Dynamic Viscosity Measurements*, in [850], pp. 338–350.
[857] W. P. Seng, K. C. Tam, R. D. Jenkins, D. R. Bassett, *The Network Strength and Junction Density of a Model HASE Polymer in Non-Ionic Surfactant Solutions*, in [850], pp. 351–368.
[858] R. J. English, R. D. Jenkins, D. R. Bassett, S. A. Khan, *Rheology of a HASE Associative Polymer and Its Interaction with Non-Ionic Surfactants*, in [850], pp. 369–380.
[859] P. T. Elliott, L. Xing, W. H. Wetzel, J. E. Glass, *Behavior of Branched-Terminal, Hydrophobe-Modified, Ethoxylated Urethanes*, in [850], pp. 163–178.
[860] H. N. Naé, W. W. Reichert, *Rheological properties of lightly crosslinked carboxy copolymers in aqueous solutions*, Rheol. Acta, 31 (1992), 351–360.
[861] L. A. De Graaf, *Denaturation of proteins from a non-food perspective*, J. Biotechnol., 79 (2000), 299–306.
[862] D. E. Graham, M. C. Phillips, *Proteins at Liquid Interfaces III. Molecular Structures of Adsorbed Films*, J. Colloid Interface Sci., 70(3) (1979), 427–439.
[863] P. Walstra, A. L. de Roos, *Proteins at air-water and oil-water interfaces: static and dynamic aspects*, Food Rev. Int., 9(4) (1993), 503–525.
[864] S. Damodaran, *Interfaces, Protein Films, and Foams*, Adv. Food Nutr. Res., 34 (1990), 1–79.
[865] E. Dickinson, Y. Matsumura, *Time-dependent polymerization of β-lactoglobulin through disulphide bonds at the oil-water interface in emulsions*, Int. J. Biol. Macromol., 13 (1991), 26–30.
[866] E. A. Hauser, L. E. Swearingen, *The Aging of Surfaces of Aqueous Solutions of Egg Albumin*, J. Phys. Chem., 45(4) (1941), 644–659.
[867] A. E. Mirsky, *Sulfhydryl Groups in Films of Egg Albumin*, J. Gen. Physiol., 24(6) (1941), 725–733.
[868] H. B. Bull, *Studies on Surface Denaturation of Egg Albumin*, J. Biol. Chem., 123 (1938), 17–30.
[869] A. Gennadios, C. L. Weller, M. A. Hanna, G. W. Froning, *Mechanical and Barrier Properties of Egg Albumen Films*, J. Food Sci., 61(3) (1996), 585–589.
[870] K. Broersen, A. M. M. van Teffelen, A. Vries, A. G. J. Voragen, R. J. Hamer, H. H. J. de Jongh, *Do Sulfhydryl Groups Affect Aggregation and Gelation Properties of Ovalbumin? J. Agric. Food Chem.*, 54 (2006), 5166–5174.
[871] D. Fukushima, J. van Buren, *Mechanism od Protein Insolubilization during the Drying of Soy Milk. Role of Disulfide and Hydrophobic Bonds*, Am. Assoc. Cereal Chem., 47 (1970), 687–696.
[872] N. Kitabatake, E. Doi, *Conformational Change of Hen Egg Ovalbumin during Foam Formation Detected by 5, 5'-Dithiobis(2-nitrobenzoic acid)*, J. Agric. Food Chem., 35 (1987), 953–957.
[873] A. C. C. Alleoni, *Albumen Protein and Functional Properties of Gelation and Foaming*, Sci. Agric., 63(3) (2006), 291–298.
[874] H. J. T. Beveridge, *A Study of some Physical and Chemical Properties of Egg White*, Thesis, University of British Columbia, 1973, https://open.library.ubc.ca/soa/cIRcle/collections/ubctheses/831/items/1.0100825, accessed on 2022-06-20.
[875] A. Deuerling, *Die Physik und Chemie der "Mousse au Chocolat"*, Schriftliche Hausarbeit für die erste Staatsexamensprüfung, Julius-Maximilian-Universität Würzburg, 2010, http://www.thomas-wilhelm.net/arbeiten/Mousse_au_Chocolat.pdf, accessed on 2022-06-20.
[876] F. E. Cunningham, *Properties of Egg White Foam Drainage*, Poultry Sci., 55 (1976), 738–743.

[877] S. Poole, S. I. West, C. L. Walters, *Protein-Protein Interactions: Their Importance in the Foaming of Heterogeneous Protein Systems*, J. Sci. Food Agric., 35 (1984), 701–711.
[878] A. Hoppe, *Examination of egg white proteins and effects of high pressure on select physical and functional properties*, PhD Thesis, University of Nebraska, Lincoln, 2010, https://digitalcommons.unl.edu/foodscidiss/7/, accessed on 2022-06-20.
[879] B. Brodsky, J. A. Werkmeister, J. A. M. Ramshaw, *Collagens and Gelatins, Band 8*, in S. R. Fahnestock, A. Steinbüchel (eds.) *Biopolymers*, WILEY-VCH, 2003. ISBN 3-527-30223-9.
[880] M. Djabourov, J. Leblond, *Thermally Reversible Gelation of the Gelatin-Water System*, in *Reversible Polymeric Gels and Related Systems*, ACS Symposium Series, vol. 350, pp. 211–223, American Chemical Society, 1987. ISBN 9-780841214156.
[881] A. G. Ward, *The physical properties of gelatin solutions and gels*, Br. J. Appl. Phys., 5 (1954), 85–90.
[882] M. Bansal, C. Ramakrishnan, G. N. Ramachandran, *Stabilization of the collagen structure by hydroxyproline residues*, Proc. Indian Acad. Sci. A, 82(4) (1975), 152–164.
[883] S. Zumbühl, *Proteinische Leime – Ein vertrauter Werkstoff?* Z. Kunsttechnol. Konserv. (2003), 95–104. Wernersche Verlagsgesellschaft Worms.
[884] E. Reinkowski-Häfner, *Tempera*, Z. Kunsttechnol. Konserv., 2 (1994), 297–317. Wernersche Verlagsgesellschaft Worms.
[885] M. Lazzari, O. Chiantore, *Drying and oxidative degradation of linseed oil*, Polym. Degrad. Stab., 65 (1999), 303–313.
[886] J. Mallégol, J.-L. Gardette, J. Lemaire, *Long-Term Behavior of Oil-Based Varnishes and Paints I. Spectroscopic Analysis of Curing Drying Oils*, J. Am. Oil Chem. Soc., 76(8) (1999), 967–976.
[887] J. Mallégol, J.-L. Gardette, J. Lemaire, *Long-Term Behavior of Oil-Based Varnishes and Paints. Fate of Hydroperoxides in Drying Oils*, J. Am. Oil Chem. Soc., 77(3) (2000), 249–255.
[888] H. P. Kaufmann, *Oxydation und Verfilmung trocknender Öle*, Fette Seifen Anstrichm., 59(3) (1957), 153–162.
[889] J. Mallégol, J. Lemaire, J.-L. Gardette, *Drier influence on the curing of linseed oil*, Prog. Org. Coat., 39 (2000), 107–113.
[890] E. Ioakimoglou, S. Boyatzis, P. Argitis, A. Fostiridou, K. Papapanagiotou, N. Yannovits, *Thin-Film Study on the Oxidation of Linseed Oil in the Presence of Selected Copper Pigments*, Chem. Mater., 11 (1999), 2013–2022.
[891] E. N. Frankel, *Lipid Oxidation: Mechanisms, Products and Biological Significance*, J. Am. Oil Chem. Soc., 61(12) (1984), 1908–1917.
[892] E. N. Frankel, *Lipid Oxidation*, Prog. Lipid Res., 19 (1980), 1–22.
[893] E. N. Frankel, *Recent Advances in Lipid Oxidation*, J. Sci. Food Agric., 54 (1991), 495–511.
[894] E. N. Frankel, *Secondary Products of Lipid Oxidation*, Chem. Phys. Lipids, 44 (1987), 73–85.
[895] N. A. Porter, *Mechanisms for the Autoxidation of Polyunsaturated Lipids*, Ace. Chem. Res., 19 (1986), 262–268.
[896] W. J. Muizebelt, M. W. F. Nielen, *Oxidative Crosslinking of Unsaturated Fatty Acids Studied with Mass Spectrometry*, J. Mass Spectrom., 31 (1996), 545–554.
[897] W. J. Muizebelt, J. C. Hubert, R. A. M. Venderbosch, *Mechanistic study of drying of alkyd resins using ethyl linoleate as a model substance*, Prog. Org. Coat., 24 (1994), 263–279.
[898] D. Erhardt, C. S. Tumosa, M. F. Mecklenburg, *Long-Term Chemical and Physical Processes in Oil Paint Films*, Stud. Conserv., 50(2) (2005), 143–150.
[899] J. Mallégol, J.-L. Gardette, J. Lemaire, *Long-Term Behavior of Oil-Based Varnishes and Paints. Photo- and Thermooxidation of Cured Linseed-Oil*, J. Am. Oil Chem. Soc., 77(3) (2000), 257–263.
[900] C. Boelhouwer, Th. Knegtel, M. Tels, *On the Mechanism of the Thermal Polymerization of Linseed Oil, Fette*, Seifen, Anstrichmittel, 69(6) (1967), 432–436.
[901] H. Wexler, *Polymerization of Drying Oil*, Chem. Rev., 64(6) (1964), 591–611.

[902] G. Dobson, W. W. Christie, J. L. Sebedio, *Saturated Bicyclic Fatty Acids formed in Heated Sunflower Oils*, Chem. Phys. Lipids, 87 (1997), 137–147.

[903] G. Dobson, J. L. Sebedio, *Monocyclic dienoic fatty acids formed from γ-linolenic acid in heated evening primrose oil*, Chem. Phys. Lipids, 97 (1999), 105–118.

[904] H. P. Kaufmann, *Diels-Alder-Reaktionen auf dem Fettgebiet, Fette*, Seifen, Anstrichmittel, 64(12) (1962), 1115–1126.

[905] G. Billek, *Die Veränderungen von Nahrungsfetten bei höheren Temperaturen*, Fat. Sci. Technol., 94(5) (1992), 161–172.

[906] W. C. Ault, J. C. Cowan, J. P. Kass, J. E. Jackson, *Polymerization of Drying Oils*, Ind. Eng. Chem., 34 (1942), 1120–1123.

[907] J. Koller, U. Baumer, *Die Benzinlösemittel (des Restaurators)*, Restauro, 7 (2001), 528–536.

[908] F. Casadio, K. Keune, P. Noble, A. van Loon, E. Hendriks, S. A. Centeno, G. Osmond (eds.), *Metal Soaps in Art: Conservation and Research*, Cultural Heritage Science, Crown, 2019. ISBN 978-3-319-90616-4.

[909] P. Noble, A. van Loon, J. J. Boon, *Selective darkening of ground layers associated with the wood grain in 17th-century panel paintings*, in J. Townsend, T. Doherty, G. Heydenreich (eds.) *Preparation for painting: the Artist's choice and its consequences*, pp. 68–78, Archetype, London, 2008, https://www.researchgate.net/publication/290997156, accessed on 2022-06-12.

[910] P. Noble, A. van Loon, J. J. Boon, *Chemical changes in old master paintings II: darkening due to increased transparency as a result of metal soap formation*, in I. Verger James, James (eds.) *Preprints of the ICOM committee for conservation 14th triennial meeting, The Hague, 12–16 Sep 2005*, pp. 496–503, 2005, https://www.researchgate.net/publication/305687557, accessed on 2022-06-12.

[911] J. J. Boon, K. Keune, J. van der Weerd, P. Noble, *Mechanical and chemical changes in Old Master Paintings: dissolution, metal soap formation and remineralization processes in lead pigmented ground/intermediate paint layers of 17th century paintings*, in R. Vontobel (ed.) *Icom-cc 13th Triennial Meeting Preprints, Rio de Janeiro, 22–27 Sep 2002*, pp. 401–406, 2002. https://www.researchgate.net/publication/283363916, accessed on 2022-06-12.

[912] A. van Loon, *Color changes and chemical reactivity in seventeenth-century oil paintings*, PhD thesis, University of Amsterdam, 2008, https://hdl.handle.net/11245/1.292118, accessed on 2022-06-12.

[913] J. J. Hermans, *Metal soaps in oil paint: Structure, mechanisms and dynamics*, PhD thesis, University of Amsterdam, 2017, https://hdl.handle.net/11245.1/53663926-183c-40aa-b7b3-e6027979cb7d, accessed on 2022-06-12.

[914] J. Koller, U. Baumer, *Die Bindemittel der Schule von Barbizon*, in [137], pp. 343–369.

[915] J. Koller, I. Fiedler, U. Baumer, *Die Bindemittel auf Dürers Tafelgemälden*, in [147], pp. 102–119.

[916] J. Koller, U. Baumer, *Jacopo Tintoretto: Die Bindemittel seiner Schnellmalerei*, in [138], pp. 213–225.

[917] R. White, J. Kirby, *Rembrandt and his Circle: Seventeenth Century Dutch Paint Media Re-examined*, Natl. Gallery Tech. Bull., 15 (1994), 64–78. ISBN 1-85709-049-7.

[918] R. White, *The Characterization of Proteinaceous Binders in Art Objects*, Natl. Gallery Tech. Bull., 8 (1984), 5–14. ISBN 0-901791-94-6.

[919] A. Spyros, D. Anglos, *Studies of organic paint binders by NMR spectroscopy*, Appl. Phys. A, 83 (2006), 705–708.

[920] A. Lluveras-Tenorio, J. Mazurek, A. Restivo, M. P. Colombini, I. Bonaduce, *The Development of a New Analytical Model for the Identification of Saccharide Binders in Paint Samples*, PLoS ONE, 7(11) (2012), e49383. https://journals.plos.org/plosone/article?id=10.1371/journal.pone.0049383, accessed on 2022-06-20.

[921] I. Bonaduce, L. Carlyle, M. P. Colombini, C. Duce, C. Ferrari, et al., *New Insights into the Ageing of Linseed Oil Paint Binder: A Qualitative and Quantitative Analytical Study*, PLoS ONE, 7(11), e49333 (2012). https://doi.org/10.1371/journal.pone.0049333, accessed on 2022-06-20.
[922] R. White, J. Pilc, *Analyses of paint media*, Natl. Gallery Tech. Bull., 16 (1995), 85–95. ISBN 1-85709-071-3.
[923] R. White, J. Pilc, *Analyses of paint media*, Natl. Gallery Tech. Bull., 17 (1996), 91–103. ISBN 1-85709-113-2.
[924] C. Higgitt, R. White, *Analyses of Paint Media: New Studies of Italian Paintings of the Fifteenth and Sixteenth Centuries*, Natl. Gallery Tech. Bull., 26 (2005), 88–97. ISBN 1-85709-341-0.
[925] J. Mills, R. White, *Paint Media Analyses*, Natl. Gallery Tech. Bull., 13 (1989), 69–71. ISBN 0-947645-705.
[926] R. White, J. Pilc, J. Kirby, *Analyses of Paint Media*, Natl. Gallery Tech. Bull., 19 (1998), 74–95. ISBN 1-85709-220-1.
[927] J. H. Townsend, L. Carlyle, A. Burnstock, M. Odlyha, J. J. Boon, Nineteenth-Century Paint Media: the Formulation and Properties of Megilps, in [105], pp. 205–210.
[928] J. Mills, R. White, *Analyses of Paint Media*, Natl. Gallery Tech. Bull., 5 (1981), 66–67. ISBN 0-901791-76-8.
[929] J. Mills, R. White, *Analyses of Paint Media*, Natl. Gallery Tech. Bull., 1 (1977), 57–59.
[930] J. Mills, R. White, *Analyses of Paint Media*, Natl. Gallery Tech. Bull., 3 (1979), 66–67.
[931] J. Mills, R. White, *Analyses of Paint Media*, Natl. Gallery Tech. Bull., 4 (1980), 65–67. ISBN 0-901791-72-5.
[932] J. Mills, R. White, *Analyses of Paint Media*, Natl. Gallery Tech. Bull., 7 (1983), 65–67. ISBN 0-901791-88-1.
[933] J. Mills, R. White, *Analyses of Paint Media*, Natl. Gallery Tech. Bull., 9 (1985), 70–71. ISBN 0-901791-97-0.
[934] J. Mills, R. White, *Analyses of Paint Media*, Natl. Gallery Tech. Bull., 11 (1987), 92–95. ISBN 0-947645-35-7.
[935] J. Mills, R. White, *Analyses of Paint Media*, Natl. Gallery Tech. Bull., 12 (1988), 78–79. ISBN 0-947645-63-2.
[936] R. White, *Van Dyck's Paint Medium*, Natl. Gallery Tech. Bull., 20 (1999), 84–88. ISBN 1-85709-251-1.
[937] R. G. Weissenhorn, *Chemie in der Freskomalerei*, Chem. Unserer Zeit, 36(5) (2002), 310–316.
[938] J. Koller, U. Baumer, *Kunstharzfirnisse. Teil I: Geschichte und Verwendung von Naturharzen und Kunstharzen in Firnissen*, Restauro, 7 (2000), 534–537.
[939] J. Koller, U. Baumer, *Kunstharzfirnisse. Teil II: Mechanische Eigenschaften, Alterungsverhalten und Löslichkeit von polymeren Kunstharzfirnissen*, Restauro, 8 (2000), 616–625.
[940] J. Koller, U. Baumer, *Kunstharzfirnisse. Teil III: Die niedermolekularen (nichtpolymeren) Kunstharzfirnisse*, Restauro, 1 (2001), 26–38.
[941] S. Zumbühl, R. D. Knochenmuss, S. Wülfert, *Rissig und blind werden in relativ kurzer Zeit alle Harzessenzfirnisse*, Z. Kunsttechnol. Konserv., 12 (1998), 205–219.
[942] E. Wenders, *Dammar als Gemäldefirnis*, Z. Kunsttechnol. Konserv., 15 (2001), 133–162.
[943] P. Dietemann, C. Higgitt, M. Kälin, M. J. Edelmann, R. Knochenmuss, R. Zenobi, *Aging and yellowing of triterpenoid resin varnishes – Influence of aging conditions and resin composition*, J. Cult. Herit., 10 (2009), 30–40.
[944] G. A. van der Doelen, *Molecular studies of fresh and aged triterpenoid varnishes*, Dissertation Universiteit van Amsterdam, 1999.
[945] G. A. van der Doelen, K. J. van den Berg, J. J. Boon, *Comparative Chromatographic and Mass-Spectrometric Studies of Triterpenoid Varnishes: Fresh Material and Aged Samples from Paintings*, Stud. Conserv., 43 (1998), 249–264.

[946] S. Zumbühl, R. Knochenmuss, S. Wülfert, F. Dubois, M. J. Dale, R. Zenobi, *A Graphite-Assisted Laser Desorption/Ionization Study of Light-Induced Aging in Triterpene Dammar and Mastic Varnishes*, Anal. Chem., 70 (1998), 707–715.
[947] E. R. de la Rie, A. M. Shedrinsky, *The Chemistry of Ketone Resins and the Synthesis of a Derivative with Increased Stability and Flexibility*, Stud. Conserv., 34 (1989), 9–19.
[948] K. J. van den Berg, J. van der Horst, J. J. Boon, O. O. Sudmeijer, *cis-l, 4-poly-β-myrcene, the Structure of the Polymeric Fraction of Mastic Resin (Pistacia lentiscus L.) Elucidated*, Tetrahedron Lett., 39 (1998), 2645–2648.
[949] E. R. de la Rie, *Photochemical and Thermal Degradation of Films of Dammar Resin*, Stud. Conserv., 33 (1988), 53–70.
[950] P. Dietemann, M. Kälin, S. Zumbühl, R. Knochenmuss, S. Wülfert, R. Zenobi, *A Mass Spectrometry and Electron Paramagnetic Resonance Study of Photochemical and Thermal Aging of Triterpenoid Varnishes*, Anal. Chem., 73 (2001), 2087–2096.
[951] E. R. de la Rie, *Old Master Paintings: A Study of the Varnish Problem*, Anal. Chem., 61(21) (1989), 1228A–1240A.
[952] R. White, J. Kirby, *A Survey of Nineteenth and early Twentieth Century Varnish Compositions found on a Selection of Paintings in the National Gallery Collection*, Natl. Gallery Tech. Bull., 22 (2001), 64–84. ISBN 978-1-85709-926-5.
[953] M. Stappel, *Schellack*, Restauro, 8 (2001), 596–603.
[954] S. K. Sharma, S. K. Shukla, D. N. Vaid, *Shellac – Structure, Characteristics & Modifications*, Def. Sci. J., 33 (1983), 261–271, https://publications.drdo.gov.in/ojs/index.php/dsj/article/view/6181/0, accessed on 2022-06-20.
[955] K. S. Brown, *The Chemistry of Aphids and Scale Insects*, Chem. Soc. Rev., 4 (1975), 263–288.
[956] J. S. Mills, R. White, *Natural Resins of Art and Archaeology. Their Sources, Chemistry, and Identification*, Stud. Conserv., 22(1) (1977), 12–31.
[957] R. M. Carman, D. E. Cowley, R. A. Marty, *Diterpenoids XXV. Dundathic acid and Polycommunic Acid*, Aust. J. Chem., 23 (1970), 1656–1665.
[958] G.-F. Chen, *Developments in the field of rosin chemistry and its implications in coatings*, Prog. Org. Coat., 20 (1992), 139–167.
[959] Brochure *Handbuch Cognis Kit für Haushalt & Kosmetik*, Cognis Deutschland GmbH & Co. KG, CFM Technical Training, 2002.
[960] Pigments Finder by Clariant, https://www.clariant.com/de/Business-Units/Pigments/Pigments-Finder, accessed on 2022-06-20.
[961] *Sales range for the printing, paints, plastics industries and special applications*, Clariant, Dokument DP5525, https://www.clariant.com/-/media/Files/Business-Units/Pigments/Clariant-Brochure-Sales-Range-202003-EN.pdf, accessed on 2022-06-26.
[962] *Clariant Creative Colors – Solutions for Coloring your Work Environment*, https://www.clariant.com/de/Business-Units/Pigments/Special-Applications/Stationery, accessed on 2022-06-26.
[963] Organic pigments (general purpose pigments) by DIC Corporation, https://www.dic-global.com/en/products/general_pigments/ (pigments for printing inks), accessed on 2022-06-24.
[964] Pigments for digital printing by DIC Corporation, https://www.dic-global.com/en/products/nip_pigments/, accessed on 2022-06-24.
[965] *Neozapon Farbstoffe*, BASF, Dokument EVP 001905d, no longer available.
[966] *Pigmente für Flexo-, Tiefdruck- und Offset-Verpackungsdruckfarben*, BASF, Dokument EVP 000204d, no longer available.
[967] *Basacid Farbstoffe*, no longer available.
[968] *Lanxess Farbstoffe für Schreib- und Inkjettinten: Bayscript, Pyranin, Spezial*, https://lanxess.com/en/Products-and-Solutions/Industries/Colorants/Spezialfarbstoffe (index page), https://lanxess.com/-/media/Project/Lanxess/Corporate-Internet/Products-and-Solutions/Industries/Colorants/10249_BRO_PLA_ColorantsInk_EN_web.pdf (inkjet and stationary inks), accessed on 2022-06-20.

[969] *Pigments for digital printing*, https://www.basf.com/us/documents/en/general-business-topics/pigments/industries/printing/BASF-Colors-and-Effects_Brochure_Pigments-for-digital-printing.pdf, accessed on 2022-06-24.

[970] *Pulp & Paper Dyes and Pigments*, ORCO Organic Dyes and Pigments, https://www.organicdye.com/industries/paper-dyes/, no longer accessible.

[971] *Ink and Toner Dyes & Pigments*, ORCO Organic Dyes and Pigments, https://www.organicdye.com/industries/inks-toner-dyes/, accessed on 2022-06-21.

[972] *Paper And Pulp*, Dayglo Color Corporation, http://www.dayglo.in/paper_industry.html, accessed on 2022-06-21.

[973] *DYCROPULP Dyes for Paper*, DYCRO Jagson Colorchem Limited, https://www.jagson.com/dycropulp-dyes-for-paper.php, accessed on 2022-06-21.

[974] *Dyes for Ink*, DYCRO Jagson Colorchem Limited, https://www.jagson.com/ink-dyes.php, accessed on 2022-06-21.

[975] Information about Carbopol™, Lubrizol, https://www.lubrizol.com/Home-Care/Products/Carbopol-Polymers, accessed on 2022-06-20.

[976] Datasheets for ACRYSOL™ASE-60, Dow, https://www.dow.com/en-us/pdp.acrysol-ase-60er-rheology-modifier.136370z.html?productCatalogFlag=1#overview, accessed on 2022-06-20.

[977] Information about Klucel™products, Ashland, https://www.ashland.com/industries/paints-and-coatings/specialty-and-industrial-coatings/klucel-hydroxypropylcellulose (index page for HPC for coatings), https://www.ashland.com/file_source/Ashland/Product/Documents/Pharmaceutical/PC_11229_Klucel_HPC.pdf (HPC structure and properties for pharmaceutics), accessed on 2022-06-20.

[978] Information about Tafigel™and similar products, Münzing, https://www.munzing.com/de/downloads/product-brochures/, https://www.munzing.com/static/bed874f982e762433cc2e88af82651db/Dispersing_Technologies_MUNZING_2016_fffffe5d54.pdf (information about dispersing technologies), all accessed on 2022-06-24.

[979] Information about Tylose™products, SE Tylose GmbH & Co. KG, https://www.setylose.com/en/knowledge-base/industrial/technical-data-sheets (index page), https://www.setylose.com/fileadmin/user_upload/Tylose%20Paints_EN.pdf (brochure about cellulose ethers), accessed on 2022-06-20.

[980] *SIEGEL UND SIEGELLACK*, website information by *Scriptorium*, http://www.kalligraphie.com/443-0-Siegel.html, accessed on 2022-06-21.

[981] ASTM International Standards Worldwide https://www.astm.org/.

[982] Verband der deutschen Lack- und Druckfarbenindustrie, Fachgruppe Druckfarben, https://www.wirsindfarbe.de/service-publikationen/informationsmaterial-druckfarben/allgemeine-informationen-ueber-druckfarben/, https://www.wirsindfarbe.de/fileadmin/user_upload/Dokumente/Druckfarben/37_2110_Fluessige_Druckfarben.pdf (liquid printing inks), https://www.wirsindfarbe.de/fileadmin/user_upload/Dokumente/Druckfarben/38_2110_Pastoese_bis_dickfluessige_Druckfarben.pdf (viscous printing inks), https://www.wirsindfarbe.de/fileadmin/user_upload/Dokumente/Druckfarben/39_2110_Siebdruckfarben.pdf (screen printing inks), https://www.wirsindfarbe.de/service-publikationen/informationsmaterial-druckfarben/allgemeine-informationen-ueber-druckfarben/die-auswirkungen-von-druckfarben-auf-die-umwelt (environmental consequences of printing inks), all accessed on 2022-06-21.

[983] Various information by D. Smith: https://danielsmith.com/product/original-oils/ (composition artists' paint "Original Oils"), https://danielsmith.com/product/daniel-smith-extra-fine-watercolors/ (composition artists' paint "Extra Fine™ Watercolors"), accessed on 2022-06-10.

[984] Various information by Winsor&Newton: https://www.winsornewton.com/row/paint/oil/artists-oil/#product-info-colours and https://www.winsornewton.com/na/education/composition-permanence/artists-oil-colour/ (composition artists' oil color), https://www.winsornewton.com/row/paint/watercolour/professional-watercolour/#product-info-colours and https://www.winsornewton.com/na/education/composition-permanence/professional-water-colour/ (composition professional water colors), https://www.winsornewton.com/row/paint/acrylic/professional-acrylic/#product-info-colours and https://www.winsornewton.com/na/education/composition-permanence/professional-acrylic-colour/ (composition professional acrylics), all accessed on 2022-06-21, article "The Science behind Artists' Acrylic and its benefits" (resource center, http://www.winsornewton.com/resource-centre/product-articles/) accessed on 2019-05-15, http://www.winsornewton.com/assets/HealthandSafetyDataSheets/OIL%20COLOUR/Griffin%20Alkyd/04912258.pdf (MSDS artists' alkyd colors), accessed on 2013-06-10.

[985] Various information by Schmincke: https://www.schmincke.de/fileadmin/downloads/pdf/Broschueren_2016/MUSSINI_D_EN.pdf (composition MUSSINI oil colors), https://www.schmincke.de/fileadmin/downloads/pdf/Broschueren_2016/HORADAM_AQUARELL_D_EN.pdf (composition HORADAM water colors), https://www.schmincke.de/fileadmin/downloads/pdf/Broschueren_2016/PRIMAcryl_DE_EN.pdf (composition PRIMAcryl acrylics), https://www.schmincke.de/fileadmin/downloads/pdf/Broschueren_2016/aqua_Linoldruck_DE_EN.pdf (composition aqua LINOPRINT inks), all accessed on 2022-06-21.

[986] Various information by Charbonnel, https://www.charbonnelshop.fr/row/product-category/metal-plate-printing-intaglio/intaglio-ink/ (intaglio/engraving- and gravure printing inks), https://www.charbonnelshop.fr/row/product-category/metal-plate-printing-intaglio/aquawash-intaglio/ (engraving- and gravure printing inks solvent-free (Aqua Wash)), accessed on 2022-06-21.

[987] Various information by Lascaux: https://lascaux.ch/dbFile/4392/u-1749/Lascaux_Polysaccharides_Cellulose_Starch.pdf, https://lascaux.ch/dbFile/4606/u-20a2/Lascaux%20Synthetic%20Resins%20and%20Dispersions.pdf, https://lascaux.ch/en/products/painting-mediums, accessed on 2022-06-21.

[988] Various information by Dow: http://www.dow.com/, https://www.dow.com/documents/en-us/mark-prod-info/884/884-02316-01-tamol-dispersants-product-solutions-guide.pdf (dispersants), https://www.dow.com/en-us/pdp.orotan-731a-er-dispersant.242161z.html?productCatalogFlag=1#overview (Orotan 731A), https://www.dow.com/en-us/pdp.tamol-731a-dispersant.242189z#overview (Tamol 731A), all accessed on 2022-06-21.

[989] J. B. Clarke, C. R. Walker, *Dispersants for emulsion paints*, Europäisches Patent EP 0802951 A1, 1997, http://www.google.com/patents/EP0802951A1?cl=en, accessed on 2022-06-24.

[990] J. B. Blackburn, A. W. Field, *Pigment preparation*, US-Patent 4089699, 1978, https://patents.justia.com/patent/4089699, accessed on 2022-06-24.

[991] Fa. Kremer Pigmente, *Orotan 731 K Material Safety Data Sheet*, https://www.kremer-pigmente.com/elements/resources/products/files/78032_SHD.pdf, accessed on 2022-06-21.

[992] Various information by Celanese: http://www.celanese.com/msds/pdf/993-64529418.pdf (MSDS für Mowilith LDM 7412), accessed on 2011-07-20.

[993] Hahnemühle FineArt GmbH, *Künstlerpapiere*, website to product portfolio, https://www.hahnemuehle.com/de/kuenstlerpapiere.html, in particular https://www.hahnemuehle.com/de/kuenstlerpapiere/haeufig-gestellte-fragen.html (FAQ), https://www.hahnemuehle.com/de/kuenstlerpapiere/aquarell.html (watercolor paper), https://www.hahnemuehle.com/de/kuenstlerpapiere/klassische-druckverfahren.html (printing paper), https://www.hahnemuehle.com/de/kuenstlerpapiere/skizze-zeichnen/skizzenpapiere.html (drawing paper), https://www.hahnemuehle.com/de/kuenstlerpapiere/grafik-illustration.html (layout- and design paper), https://www.hahnemuehle.com/de/kuenstlerpapiere/pastell.html (pastell drawing paper), all accessed on 2022-06-21.

[994] Kemira pigments and coatings for paper: https://www.kemira.com/products/?industry=none&oil_gas=none&pulp_paper=436&water=none and https://www.kemira.com/products/?industry=none&oil_gas=none&pulp_paper=432&water=none, acccessed on 2022-06-21, http://www.kemira.com/regions/germany/SiteCollectionDocuments/Broschüren_PulpPaper/Farbmittel.pdf (colorants, no longer accessible).
[995] Various websites of federal and state archives to the topic, a selection: https://afz.lvr.de/de/bestandserhaltung_2/bestandserhaltung_1/vergleich_din_6738_und_din_en_iso_9706/vergleich_der_papiernormen_din_6738_und_din_en_iso_9706.html, https://www.bundesarchiv.de/DE/Content/Downloads/KLA/positionspapier-alterungsbestaendiges-papier.pdf?__blob=publicationFile, https://www.klug-conservation.com/medien/Wissen/Wissens_Folder/wissen8_blauer_engel_de.pdf, accessed on 2022-06-21.
[996] Various information by Ferro, https://www.ferro.com/-/media/files/news-and-events/events/ceramitec-2022/structural/fk02_e_overview_ceramic_pigments.pdf (Technical Information FK02 Ceramic Stains), accessed on 2022-06-21.
[997] Sekisui Specialty Chemicals, *Informationen zu Selvol™Polyvinylalkohol*, http://www.sekisui-sc.com/products/selvol/productline.html, accessed on 2022-06-21.

Index

Page numbers in blue refer to vol. 1.

https://doi.org/10.1515/9783110777116-006